Genomics in Drug Discovery and Development

Genomics in Drug Discovery and Development

Dimitri Semizarov, Ph.D.
Eric Blomme, D.V.M., Ph.D.

Abbott Laboratores
Abbott Park, Illinois

WILEY

A JOHN WILEY & SONS, INC., PUBLICATION

Published by John Wiley & Sons, Inc., Hoboken, New Jersey
Published simultaneously in Canada.

Library of Congress Cataloging-in-Publication Data:

Semizarov, Dimitri.
Genomics in drug discovery and development / Dimitri Semizarov, Eric Blomme.
p. ; cm.
Includes bibliographical references and index.
ISBN 978-0-470-09604-8 (cloth)
1. Pharmacogenomics. 2. Drug development. 3. Genetic toxicology. 4. DNA microarrays. I. Blomme, Eric. II. Title.
[DNLM: 1. Pharmacogenetics–methods. 2. Biomarkers, Pharmacological. 3. Drug Design. QV 38 S471g 2008]
RM301.3.G45S45 2008
615'.19–dc22

2008021434

Printed in the United States of America

10 9 8 7 6 5 4 3 2 1

Contents

Preface

Most human diseases are manifested through extremely complex phenotypes that reflect contributions from germ line alterations in the patient's genome, somatic genetic aberrations in the diseased tissue, and environmental factors. One of the best studied examples is cancer, a disease of the genome characterized by tremendous heterogeneity in clinical manifestation and prognosis, which is a consequence of the multitude of genetic alterations in the tumor and the patient's germ line. The heterogeneity of human disease is an extremely important subject in drug discovery research, as it determines, among other factors, the widely observed variability in response to pharmaceutical intervention. In the past several decades, the genetic alterations driving many diseases have been identified and the genetic basis for variability in drug efficacy and toxicity has been extensively studied.

This increased awareness has given rise to the widely publicized concept of **personalized medicine**, which implies the use of information on the patient's genetic makeup in making individualized treatment decisions. Intuitively, personalized medicine may only become reality if drug discovery and development are reorganized to incorporate early identification of genomic markers predictive of drug efficacy and safety. This new paradigm has been particularly well embraced in oncology, largely because of the significant progress made in the area of cancer genomics. The success of the new targeted drug discovery paradigm in oncology is illustrated by such remarkable advances as the development of imatinib (Gleevec®) for the treatment of chronic myeloid leukemia and trastuzumab (Herceptin®) for breast cancer.

In this book, we cover several critical and rapidly developing areas of drug discovery and development that enable personalized medicine, namely, biomarker research, toxicogenomics, and pharmacogenomics. These three fields have been widely recognized as tranformational in drug discovery and development, but despite a significant synergy between their applications they have not yet been considered together in a single text. This monograph is an attempt to review the current state of the three areas of research, emphasizing the synergies between them. Indeed, as the development of genome-wide screening technologies enables routine profiling of clinical samples for gene copy number abnormalities, mutations, gene expression, and germ line polymorphisms, concurrent application of these technologies in clinical trials will certainly facilitate the discovery of genomic patterns associated with better drug response and lower toxicity. These integrated genomic markers would then be used to rationally select subjects for treatment and individually tailor pharmacological intervention to appropriate populations, thus advancing the concept of personalized medicine for the benefit of the patients.

In today's environment in the pharmaceutical industry, which is characterized by exponentially rising R&D costs and a steadily decreasing number of new approved drugs, the economic impact of biomarkers, toxicogenomics, and pharmacogenomics may become a critical factor that would allow a firm to establish a competitive advantage. Indeed, stratification of the patient population to identify potential responders who would not manifest toxicity can reduce expected development time and costs, expedite the drug's approval, and improve its life cycle. The development costs will be lower because patient stratification allows one to focus on a subpopulation in which the response rates are expected to be higher, thus reducing the size and the number of clinical trials. Higher response rates will facilitate regulatory approval, thus shortening the review times and improving the life cycle of the drug. Throughout the book, we emphasize the potential of the genomics technologies to impact the drug discovery and development process.

We hope that this book will be of interest to a varied audience, from biologists in academia and the pharmaceutical industry, who wish to broaden their knowledge of genomics, to representatives of adjacent fields, namely, pharmacologists, toxicologists, chemists, and biochemists, as well as regulatory professionals in the industry, who would like to better understand the scientific advances driving the transformational processes that occur in today's drug discovery and development. We also anticipate that this manuscript will be useful to R&D managers responsible for strategically incorporating biomarker, toxicogenomics, and pharmacogenomics programs into drug discovery and development organizations, thus eventually adapting them to the demands of the era of personalized medicine. Finally, investment research professionals who analyze pharmaceutical and biotechnology sectors will find in this book an instructive summary of the key concepts and scientific definitions for several of the most financially impactful areas of drug discovery and development.

ACKNOWLEDGMENTS

The authors would like to acknowledge the intellectual and moral support of many of our colleagues at Abbott. We would like to recognize the special contribution of Dr. Anahita Bhathena, who has contributed a chapter on pharmacogenetics of drug disposition (Chapter 8). We are particularly grateful to Drs. Rick Lesniewski and Steve Fesik for creating the intellectually stimulating environment that has enabled us to complete this book. We are also indebted to several colleagues for critically reviewing parts of this book. Dr. Rick Lesniewski reviewed Chapters 1–3 and 7, Dr. Brian Spear reviewed Chapters 8 and 9, and Drs. David Katz and Jeffrey Baker reviewed Chapter 8. We thank Michael Liguori and Rita Ciurlionis for help in creating several figures. Outside of our professional environment, we are extremely thankful to all our family members, friends, and colleagues, whose encouragement, patience, and moral support allowed us to concentrate on this work for over a year and a half.

Chapter 1

Introduction: Genomics and Personalized Medicine

1.1. FUNDAMENTALS OF GENOMICS

The **genotype** is the genetic constitution of an organism that determines its phenotype by directing protein synthesis in the cell. The term **phenotype** is used to refer to the observable characteristics of a biological entity, regardless of its complexity, and may encompass the morphology of a single cell or a set of complex behaviors of an individual. Because it is the phenotypes that define our environment, our quality of life, and our susceptibility to diseases, and because it is the genotype that holds the key to the phenotypic variability observed on our planet, it is not at all surprising that a very significant share of the biology research in the past decades was devoted to the elucidation of the genotype–phenotype relationship. Understanding this association became the central task of a novel discipline born in the twentieth century, molecular biology. The exploration of the mechanisms of expression of genetic information was initiated by the discovery of DNA as a molecular entity by Avery and coworkers in 1944, followed by the determination of its structure by Watson and Crick in 1953.

All phenotypic characteristics of a multicellular organism are determined by the collection of proteins contained in its cells and the associated intracellular space. Owing to a series of breakthrough discoveries that took place in the second half of the twentieth century, the basic mechanism whereby the genetic information contained in DNA is translated into proteins is now well known. The DNA sequence is copied by specialized enzymes termed RNA polymerases into RNA molecules during a process called **transcription**. The basic unit of genetic information is a gene. According to recent estimates, the human genome appears to contain 20,000 to 25,000 protein-coding genes (1). As one gene is transcribed,

Genomics in Drug Discovery and Development, by Dimitri Semizarov and Eric Blomme

an RNA molecule is formed that is similar in length to the gene. It is then processed through **splicing** to produce a mature **transcript**, which is exported into the cytoplasm. The transcript, or messenger RNA (mRNA), serves as a template for protein synthesis by ribosomes in a process termed **translation**. When the gene is transcribed to produce RNA, it is said to be **expressed**, and when a gene is not transcribed, it is said to be **repressed**. While all normal cells in an organism have the same set of genes, the spectrum of expressed genes (often referred to as the transcriptome) varies among different cell types and changes with the phases of the cell cycle and the stage of cell differentiation. It is thus gene expression that controls the fate of the cell and determines the phenotypic diversity of cells.

While molecular biology was able to elucidate the processes responsible for expression of individual genes, the question of how the structure and function of the *entire* genome determines the phenotype remained unanswered. However, in the past two decades the development of powerful high-throughput technologies for determining the DNA sequence and measuring gene expression has enabled genome-wide studies relating genotypes to specific phenotypes, such as genetic diseases. This has given rise to a new scientific discipline termed **genomics**. A particularly notable milestone in genomics was the complete sequencing of the human genome (2, 3), a remarkable achievement that has received publicity unprecedented for a biological discovery. The determination of the genome sequence has made possible the design of tools to interrogate genomic variation and gene expression on the whole-genome scale, so-called **DNA microarrays**, which are introduced in Chapter 2 of this book. This technological development in turn led to the emergence of **functional genomics**, a genome-wide study of gene function, and opened a new era in the study of genetic diversity.

In the context of drug discovery and development, these groundbreaking scientific advances have opened new opportunities for study of human diseases and design of targeted therapeutics. Complex phenotypes associated with diseased human tissue, just like normal phenotypes, can be explained by gene expression patterns of the cells in the tissue. It is particularly instructive to consider the example of cancer, which is widely recognized to be a disease of the genome. Cancer cells are known to have numerous **structural aberrations of the genome**, such as changes in the chromosome number and structure, changes in gene copy number, and mutations. Structural changes often result in **functional genomic abnormalities**, namely, changes in the gene expression patterns of individual cells. These gene expression changes ultimately lead to the complex cancer phenotypes, such as uncontrolled cell proliferation, evasion of apoptosis, and invasion.

Figure 1.1 illustrates some genomic alterations that are associated with human disease and therefore are commonly measured in a drug discovery setting. Common structural changes include mutations and larger structural chromosomal changes, such as gene copy number abnormalities. **Mutations** represent permanent and transmissible alterations of the genome sequence, which can be somatic or heritable in nature. Occasionally, the term “mutation” is used to refer to any changes in the genome structure, including copy number changes, but most

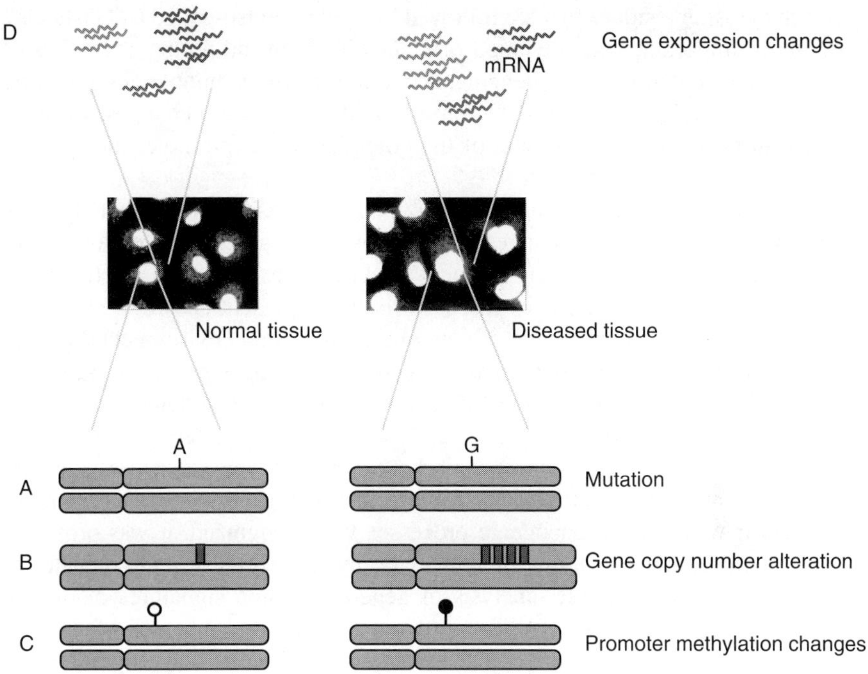

Figure 1.1 Genomic alterations found in diseased tissue. Common alterations at the DNA level include single-point mutations (A), gene copy number alterations (B), and epigenetic changes, such as abnormal promoter methylation (C). Single-point mutations represent insertions, substitutions, or deletions of individual base pairs in DNA. Copy number changes (gains or losses) may affect individual genes but may also involve large regions, such as entire chromosomal arms or whole chromosomes. One or both copies of a locus may be lost, resulting in a heterozygous or homozygous deletion, respectively. Copy number gains may vary in amplitude from one extra copy to dozens of additional copies. The amplified DNA sequences may either be incorporated into the mother chromosome or organized as extrachromosomal material. DNA methylation normally occurs at cytosine residues that are followed by a guanine (CpG islands). Methylation of CpG islands in the promoter regions of genes causes gene silencing. All these alterations at the DNA level may result to altered gene expression (D), thus affecting the phenotype of the cell. See color insert.

frequently it is used to designate point mutations or single base pair changes (substitutions, insertions, or deletions), as shown in Fig. 1.1A. A broad range of larger chromosomal aberrations has been detected in solid tumors, including changes in the number of entire chromosomes, balanced and unbalanced chromosomal translocations, and gains and losses of chromosomal fragments. **Copy number alterations** are gains or losses of DNA fragments, ranging in size from kilobases to entire chromosomes (Fig. 1.1B). Both single-point mutations and gene copy number alterations are comprehensively analyzed as genomic biomarkers in Chapter 3 of this book. Another DNA modification that is associated with disease is a **change in DNA methylation status** (Fig. 1.1C). DNA methylation normally occurs at cytosine residues that are followed by a guanine (so-called CpG islands). Particularly important is methylation of CpG islands in the promoter regions of genes, because it causes gene silencing. As cutting-edge methodologies are being developed for high-throughput detection of DNA methylation changes, we have included in Chapter 3 a discussion of the potential use of promoter methylation profiles of tissues as biomarkers.

The aforementioned structural genome modifications affect the phenotype by causing functional genome changes, namely, by altering gene expression. Other factors including the cellular environment affect gene expression as well. Therefore, gene expression patterns of diseased tissues represent sensitive molecular indicators reflecting the multitude of genomic changes and environmental factors affecting the cells in the tissue. The concept of a **gene expression signature** has been developed through pioneering studies in cancer genomics that were conducted in the late 1990s to early 2000s (for examples, see (4–15)). Gene expression signatures are composite markers comprised by the expression patterns of relevant genes that describe biological states in a quantitative manner. As the complexity of the oncogenic processes was recognized, it was proposed that gene expression signatures of tumors be used to classify and characterize human cancers. For example, analysis of gene expression signatures of diffuse large B-cell lymphoma has identified previously unknown subtypes of the disease (15–20). Figure 1.2 illustrates the utility of gene expression signatures in describing the genomic subtypes of diffuse large B-cell lymphoma (21). As relevant classifier genes (57 genes in Fig. 1.2) are selected from the entire list of genes measured, application of various clustering methods often results in formation of tight clusters denoting distinct subgroups of the disease. More broadly, gene expression signatures are now also used as a universal language to describe cellular processes and reflect perturbations associated with drug treatments, gene manipulations, etc. We comprehensively review these multiple applications of gene expression signatures in the subsequent chapters of this book.

The concept of using high-throughput genomic data to extract relevant signatures that may serve as "molecular phenotypes" has thus been pioneered for gene expression profiles. One may predict that in the future, **genomic signatures composed of copy number aberrations, mutations, promoter methylation profiles, and microRNA expression patterns** will become just as useful as gene expression profiles in characterizing disease subgroups and guiding drug discovery. Moreover, we believe that in oncology alterations at the DNA level will likely prove to be more reliable molecular descriptors, as they represent stable, fundamental events that are not affected by the extracellular environment. Currently, the limiting factor in developing these genomic signatures is the availability of mature technologies for genome-wide profiling for copy number alterations, DNA methylation, or microRNA expression. However, a number of microarray platforms have recently been commercialized for gene copy number detection, and technologies are rapidly being developed for high-throughput DNA methylation and microRNA expression profiling. Based on the current developments in the field, one may predict that different types of genomic signatures will be used

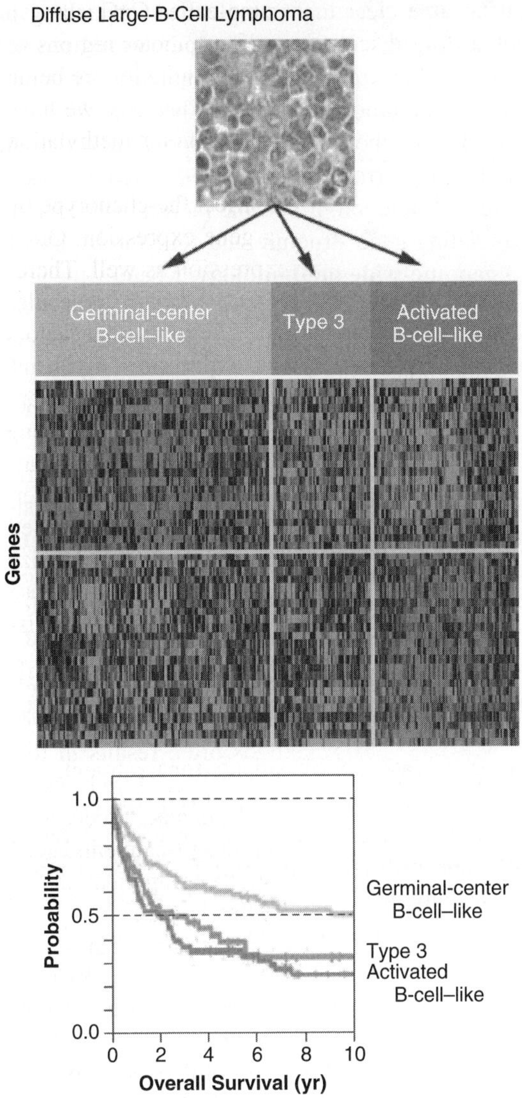

Figure 1.2 Utility of gene expression profiling in the identification of clinically relevant disease subtypes. Microarray-based profiling followed by selection of relevant genes and hierarchical clustering revealed three molecularly and clinically distinct subgroups of diffuse large B-cell lymphoma, a clinically heterogeneous disease. The heat map shows the expression levels of 57 genes that distinguish three subgroups of the disease, namely germinal-center B-cell–like (orange), type 3 (purple), and activated B-cell–like (blue). The Kaplan–Meier curve clearly demonstrates that overall survival after chemotherapy significantly differs among the subgroups, implying the clinical relevance of this genomic classification. Adopted with permission from L. Staudt 2003, *N Engl J Med* 348: 1777–1785. See color insert.

jointly as parts of integrated genomic data sets to characterize human diseases and guide pharmacological intervention.

1.2. THE CONCEPT OF PERSONALIZED MEDICINE

In the past decades, a substantial body of knowledge has been accumulated on the mechanisms of gene regulation in the cell and on the relationship between gene function and disease. For example, as evidence was gathered for multiple levels of gene deregulation in cancer, it became clear that complete understanding of the disease mechanism and targeted drug discovery in oncology would require extensive examination of gene copy number, transcriptional regulation, promoter methylation, and microRNA expression in tumors, as well as a better understanding of the germ line genetic factors affecting the disease and response to drugs. At the same time, as rapidly developing microarray technologies enabled a broader look at the human genome structure and function, it became increasingly evident that the most fruitful approach to relating gene structure and function to disease mechanism and drug response is a **genome-wide methodology**, whereby the gene copy number and expression, promoter methylation, and microRNA expression, as well as germ line polymorphisms are interrogated across the entire genome, as opposed to focusing on selected candidate genes.

As different types of microarray technologies were invented and improved, their value was demonstrated in numerous studies that used genomic data to classify and understand diseases, identify new drug targets, and predict drug sensitivity. This development coincided with a major paradigm shift in the pharmaceutical industry, which resulted in a new process of **targeted drug discovery**, guided by increased used of biomarkers to predict and monitor drug response. A term "personalized medicine" was introduced, which implies the use of information on the patient's genetic makeup in making treatment decisions. In this context, the term "genetic makeup" encompasses the entire complexity of the genome structure and function in both the diseased tissue and the germ line. It is noteworthy that the implementation of this concept requires appropriate genomic diagnostic tests to select the appropriate category of patients for treatment.

Intuitively, personalized medicine may only become reality if the processes of drug discovery and development are reorganized to involve **early determination of correlates of drug efficacy and safety** in patients and appropriate monitoring of drug effects. This is only possible through the discovery and implementation of biomarkers predictive of efficacy and toxicity for each new therapeutic developed. This new paradigm has been particularly well embraced in oncology, largely owing to the significant advances made in the area of cancer genomics. The success of the new targeted drug discovery paradigm in oncology is illustrated by such remarkable advances as the development of imatinib (Gleevec®) for the treatment of chronic myeloid leukemia (CML) (22, 23) and trastuzumab (Herceptin®) for breast cancer. Imatinib targets cells that carry a so-called Philadelphia chromosome, formed by fusion of chromosomes 9 and 22 (24). Trastuzumab specifically inhibits the proliferation of cells carrying an amplification of the HER2/neu oncogene, a copy number abnormality that leads to a significant overexpression of the HER2 protein, the target of the drug (25–27). The high response rates seen in patients receiving imatinib (95%) (28) and trastuzumab (~35%) (25) testify to the immense progress in oncology drug development initiated by the new paradigm of targeted drug discovery.

In the case of imatinib, the successful development story can be explained by three main factors (29). First, CML is the least complex cancer from the perspective of targeted drug development, because it is caused by a single oncogene (bcr-abl), as opposed to most other cancers that represent complex phenotypes initiated and supported by multiple oncogenic lesions in the genome. Second, the oncogenic lesion results in gain of function, so disease can be suppressed by inhibiting the protein produced by the oncogene. This is much easier than restoring a lost function, which is necessary when the disease is caused by a deletion or loss-of-function mutation. Finally, the chromosomal translocation (9; 22), which leads to the formation of the bcr-abl oncogene, can be readily detected by fluorescent in situ hybridization (FISH), thus enabling the development of a diagnostic test that can assist in the selection of patients for therapy. In the case of trastuzumab, the oncogenic event addressed is also a gain-of-function genetic lesion, but it is not the only one driving tumorigenesis in breast cancer cells. The complex breast cancer phenotypes involve multiple gene copy abnormalities and signaling changes, thus complicating pharmacological intervention. Accordingly, not all breast cancer patients benefit from trastusumab, as only 25–30% of them carry the HER2 amplification. Additionally, in the HER2-amplified category, the response rate is approximately 35% (25). As in the case of imatinib, molecular diagnostic tests have been developed to detect HER2 amplification, thus facilitating patient selection for treatment with trastuzumab.

As these drugs were discovered, the concept of genomics-based stratification was employed early in the discovery process, when the model systems used to test the compounds were selected based on the presence of the genetic lesions that later in development proved to be predictive of response. As trastuzumab was tested in vitro, its potency was much higher in breast cancer cell lines that carry a HER2 amplification (30). The established correlation between HER2 amplification and sensitivity to trastuzumab was later used in the clinical development of the drug (27). Had the HER2 amplification marker not been used to stratify patients in the clinical trial, the response rate to the drug would have been much lower, and the drug would have not reached the market. This and other examples emphasized the importance of early implementation of patient stratification markers in drug development and led to the formulation of the **therapeutic/diagnostic codevelopment** concept. As the optimal use of targeted therapeutics necessitates application of companion diagnostic tests, the drug development process would benefit from synchronization of the development efforts for the therapeutic and the diagnostic. The codevelopment efforts should begin as early as the drug discovery stage, as the drug should be tested in model systems that are sensitive to the drug. If candidate genomic biomarkers are discovered at the preclinical stage, they can then be tested and validated early in clinical development, so that they would direct the later stages of clinical trials by assisting in patient selection. This would significantly reduce the duration and cost of clinical trials by ensuring that only potential responders are enrolled. Thus, as we emphasize the importance of early incorporation of genomic biomarkers in the discovery process, we believe that it is appropriate to build upon the existing concept of therapeutic/diagnostic codevelopment and introduce a new paradigm of **therapeutic/diagnostic codiscovery**.

1.3. GENOMICS TECHNOLOGIES IN DRUG DISCOVERY

As these new concepts are being formulated and implemented by the pharmaceutical industry, what is the role of the genomic technologies in today's drug discovery? In this section, we attempt to systematically review the established and emerging applications of the microarray technologies covered in this book, emphasizing their critical role in various functional areas of pharmaceutical research and development. As can be seen in Figure 1.3, the first step of targeted drug discovery, **identification of therapeutic targets**, widely uses several microarray technologies. This is, in fact, one of the initial applications of gene expression microarrays that dates back to the early days of the microarray technology. Indeed, the concept is very simple: Genes overexpressed in the diseased tissue relative to the normal tissue are likely to be involved in the disease process. To date, dozens of therapeutic targets have been discovered in several major cancer types [for examples see refs. (5, 31–33)]. However, the involvement of the overexpressed genes in the disease process is not necessarily causal, as their deregulation may just be a consequence of disturbed intracellular signaling. This poses a limitation on the direct application of gene expression microarrays in target discovery, but also stimulates further development of bioinformatics approaches to microarray data analysis. Can the information on the entire body of deregulated genes be used to identify causal events in the disease? This type of analysis requires an algorithm that would map the up- and downregulated genes to intracellular pathways and thus enable the identification of signaling events that trigger the disease process. Multiple software packages were therefore developed that generate pathway information from gene expression patterns. They were used to perform pathway analysis in diseased cells and thus indirectly identify therapeutic targets. Many bioinformatics issues surrounding microarray data analysis are covered comprehensively in Chapter 2 of this book.

More recently, the development and improvement of comparative genomic hybridization (CGH) microarrays has permitted the application of this powerful technology in target identification. Array-based CGH involves hybridization of processed genomic DNA from the test and normal control sample onto microarrays carrying a representation of the genome. The methodology enables identification of changes in gene copy number on a genome-wide scale, so that amplifications and deletions of chromosomal regions are readily detected. Development of high-density oligonucleotide-based CGH microarrays has facilitated genome scanning at an increasingly high resolution, which in turn permitted identification of individual genes targeted by chromosomal aberrations. Gene copy number abnormalities play a causal role in a number of diseases and therefore represent attractive drug targets. In particular, cancer is a disease of the genome, whereby somatic gene amplifications and deletions represent fundamental events that drive tumorigenesis. In neuroscience, germ line gene copy number changes have also been shown to play a causal role in such disorders as Alzheimer's and

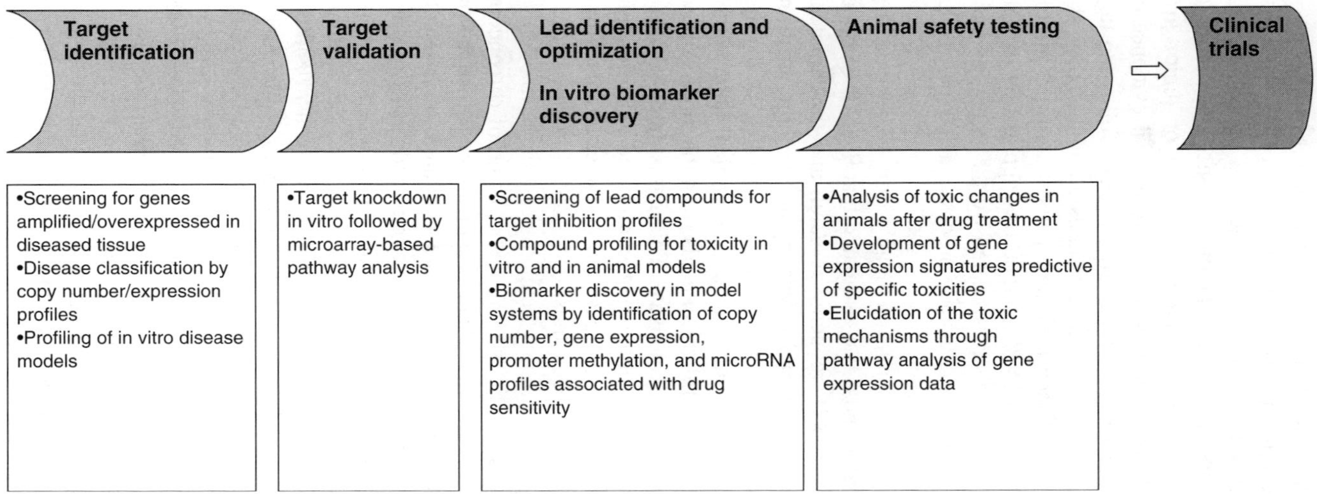

Figure 1.3 Applications of genomics technologies at different stages of drug discovery.

Parkinson's diseases (34). Genome-wide profiling of diseased tissues for copy number abnormalities has already been demonstrated to be a fruitful strategy in therapeutic target identification (for examples, see refs. (35, 36))

Validation of therapeutic targets (Fig. 1.3) typically requires a demonstration of a link between inhibition of the target and phenotypic changes associated with disease suppression. For example, in oncology inhibition of a target is expected to suppress cell proliferation in vitro or tumor growth in vivo, induce apoptosis, or decrease cell invasion. Additional evidence can be derived from microarray analysis of gene expression in cells following target inhibition, whether the target is suppressed with a candidate compound or ablated by short interfering RNA (siRNA). Since it is anticipated that target inhibition will suppress the pathways controlled by the target, this application of microarrays may elucidate the signaling mechanisms initiated by the target, and if these mechanisms are known mediators of the disease process, such experiments may provide additional validation of the target.

The most significant challenge of today's drug development process is the high failure rate of compounds. It is estimated that 99% of compounds are eliminated from the pipeline (37), reducing research and development productivity and increasing its costs. Particularly alarming are the high attrition rates in later stages of development (Phases IIb and III) (38), because of the high R&D costs incurred by the time a compound reaches late clinical development. Therefore, early elimination of unsuccessful compounds from the pipeline has become a top priority for the pharmaceutical industry. This has stimulated the investment in technologies that improve the process of **compound selection and characterization** (Fig. 1.3). Whereas in the past the major cause of compound attrition was poor pharmacokinetics, today most drugs are eliminated because of lack of efficacy or safety (38). As genomics technologies had proven their utility in early assessment of efficacy and toxicity in a number of proof-of-concept studies, they were widely adopted by drug discovery organizations across the industry.

When a target has been identified and validated, lead selection and optimization series usually involve testing of compound series in preclinical model systems, such as cell lines and animal models. Identification of gene expression changes associated with compound treatment in a model system may provide extremely useful information on the compound mechanism and the intracellular signaling changes associated with target inhibition (39–43). Since similarity of transcriptional responses to drugs usually indicates relatedness of the compounds' mechanisms, gene expression data are often used to classify compounds according to their mechanisms of action. Additionally, analysis of gene expression patterns associated with compound treatment may identify pharmacodynamic biomarkers that can be used to monitor drug efficacy. Taken together, these data may provide an early indication of target inhibition and potential compound efficacy. Biomarkers of efficacy identified in a model system may then be validated in the target tissue in patients, as the drug is administered in clinical trials.

Genomics tools play an increasingly important role in the assessment of drug toxicity, as they present an opportunity to evaluate compounds earlier and at a lower cost. Traditional toxicological evaluation through in vivo studies is lengthy and expensive and therefore creates a bottleneck in the R&D process. It also requires significant amounts of the compound. If therapeutic candidates are preselected at the discovery stage following a genomics-based evaluation, only those with adequate toxicological profiles will be subjected to traditional toxicology studies. The application of gene expression microarrays for toxicological evaluation of therapeutic candidates is the subject of an emerging discipline commonly referred to as **toxicogenomics**. Some of the recognized advantages of using toxicogenomics are: (i) low compound requirements (typically a quantity that would not require scale-up chemistry); (ii) high throughput; (iii) high sensitivity and improved mechanistic clarity; and (iv) relatively low cost.

A distinct application of gene expression microarrays is the **identification of stratification biomarkers** by analysis of baseline *pretreatment* expression profiles of cell lines that are used to test a therapeutic candidate. If differential sensitivity is observed when a panel of cell lines is used to screen a compound, the cell lines in a panel can be profiled, and their baseline gene expression patterns can be subjected to statistical analysis to identify a composite gene signature that is associated with drug sensitivity. This expression of the genes in the signature can then be tested in pretreatment clinical samples as the drug enters clinical trials. If certain genes in the signature prove to correlate with drug sensitivity in vivo, they will have utility in predicting response to the therapeutic and hence will represent useful stratification markers.

As CGH microarrays are adopted by the pharmaceutical industry, genome-wide scanning for copy number abnormalities is becoming an increasingly important tool for biomarker discovery. The copy number profiles of cell lines used to screen a candidate oncology compound may reveal gene amplifications or deletions associated with sensitivity to the drug. As changes at the chromosomal level represent stable events, they have a great potential as stratification markers, if their association with drug response is validated in clinical samples. Emerging microarray technologies, such as methylation and microRNA arrays, may also be considered for profiling of model systems. Initial studies on correlation of DNA methylation profiles in cancer cell lines and tumor samples with their response to drugs have yielded promising data [reviewed in ref. (44)], but the results remain to be validated in larger cell line panels and in clinical studies. It should be mentioned that analysis of clinical samples for promoter methylation is particularly difficult, because samples of normal tissue from the same organ need to be used as controls (DNA methylation patterns are tissue-specific). As of the day when this chapter is being written, no compelling data exists for the utility of microRNA profiles as predictors of drug sensitivity, but they have already been used to classify cancers (45, 46), and thus have demonstrated their potential as biomarkers.

As compounds undergo **safety evaluation in animal studies** (Fig. 1.3), genomic technologies may play a very important role in early detection of potential toxic liabilities and elucidation of the toxic mechanisms. Specifically, microarray-based gene expression profiling represents an extremely sensitive approach to detecting deregulation (either activation or inhibition) of specific intracellular signaling pathways in tissues following exposure to compounds. Importantly, it has been demonstrated that specific, toxicologically relevant transcriptional effects develop before the manifestation of the morphological and functional changes that are typically used to detect toxicity with clinical or pathological observations or histopathological examination (47, 48). This is consistent with our experience with the vast majority of toxic changes in well-studied tissues such as liver, kidney, spleen, or heart, which is comprehensively reviewed in Chapter 5. Largely owing to this phenomenon, toxicogenomics represents an extremely promising novel approach to toxicological assessment of compounds, as it enables early identification of toxic liabilities of compounds in the drug discovery process, thus potentially improving the productivity of drug discovery (49–51).

Early detection of toxicities through toxicogenomics is enabled through **development of predictive models of toxicity**, based on gene expression signatures associated with a specific toxic effect. Development of such models typically involves several key steps:

- Treatment of appropriately sized groups of animals with carefully selected doses of the test compound
- Gene expression profiling of carefully dissected organ of interest after several days of compound exposure
- Detection in the organ of interest of traditional toxicology end points, such as histopathology and clinical observations, after a sufficiently long exposure to the compound
- Identification of gene expression patterns in the organ of interest that are associated with future development of toxicity
- Validation of the resulting gene expression signature in an independent study and asssessment of its predictive power

Such predictive models assist compound assessment by providing early signals on potential toxic liabilities. Studies of this type are typically conducted with as little as 1–2 grams of test article, an amount that can be generated by medicinal chemists at the bench. Because of the lower compound requirement, such tests can usually be conducted 2–6 months earlier than traditional rat exploratory studies.

The second important benefit of toxicogenomics is the ability to **ascertain the molecular mechanism of a toxicity**. While traditional toxicology is primarily observational in nature and uses few end points with mechanistic value, toxicogenomics enables the analysis of deregulation of biological pathways associated with toxic changes through global assessment of gene expression. Gene expression signatures associated with a toxic effect may be interrogated in the context of biological pathways by using the multiple pathway analysis software programs reviewed in Chapters 2 and 4. This generates hypotheses that can be tested by functional experiments, such as gene silencing, forced expression, or creation of knockout animals. On the contrary, one may use gene expression signatures as molecular markers statistically associated with certain types of toxicity, without considering the association of the genes in the signature with specific toxic effects. This approach involves creation of a large database of gene expression signatures for known toxicants that generate well-studied toxic effects. Once gene expression profiles are generated for the test compounds, they can be analyzed together with the known toxicant signatures by using one of the available unsupervised or supervised statistical algorithms. In the simplest case, unsupervised hierarchical clustering can be applied to cluster the test compounds together with known toxicants and then hypothesize on the toxic mechanism of the test compounds based on their association with particular clusters. More sophisticated supervised algorithms can also be used when a sufficient number of reference profiles is available. Statistical procedures used to assess compounds are comprehensively reviewed in the bioinformatics-centered subsections of Chapters 2 and 4, while their applications in toxicogenomics are covered in Chapters 5 and 6.

In summary, the role of genomics technologies in drug discovery ranges from fundamental, as in target identification where they often enable the initial screen that generates list of potential targets, to auxiliary, as in compound characterization where they reduce the time and costs associated with efficacy and toxicity assessment and generate hypotheses on the compound mechanism. The utility of these technologies increases with our ability to extract information from genomic data, which is primarily driven by our bioinformatics and statistics capabilities as well as our knowledge of the genome. Therefore, it is very likely that the applications of genomics in drug discovery are going to expand in the coming decade, powered by further advances in the fields of bioinformatics and computational biology.

1.4. SCOPE OF THIS BOOK

In the subsequent chapters, we attempt to cover all of the applications reviewed above, emphasizing their transformational role in drug discovery. The entire Chapter 2 is devoted to the review of the state-of-the-art microarray methodologies, including gene expression microarrays, comparative genomic hybridization (CGH), and emerging DNA methylation and microRNA profiling arrays. Gene expression microarray technology is the oldest, most mature of the genomics technologies, with clearly defined advantages and disadvantages of the major platforms. Therefore, rather than cover all existing platforms, we briefly review the most frequently used short oligonucleotide microarrays and describe the relevant sample processing protocol. Today, sample processing and microarray hybridization have become routine procedures, with the main focus shifting to increasing the throughput and automation of the protocol. At the same time, as the density of gene expression information derived from microarrays continues to increase, data analysis has become the most challenging and labor-intensive step. As the statistical procedures used for expression microarray data analysis are continually refined, it is becoming evident that the validity of microarray data may be compromised by numerous factors (52). These observations prompted us to cover in detail various aspects of data analysis for gene expression microarrays. Therefore, a separate section in Chapter 2 deals with **expression microarray bioinformatics** and includes a discussion of the common issues that compromise the validity of microarray data, such as bias, overfitting, and generalizability.

Today, the analytical procedures used in microarray studies involve much more than just creating gene lists: Genes affected by a disease or a pharmacological treatment are now mapped to intracellular pathways, grouped according to their chromosomal location, or associated with a particular disease subtype or characteristic, such as drug sensitivity. Association of gene expression signatures with intracellular pathways expands the application of microarrays to analysis of the mechanism of action of therapeutic candidates and thus makes them into a useful tool for compound characterization. This requires highly specialized software programs that contain pathway reference information for all genes interrogated by the microarray. In the microarray bioinformatics subchapter, we comprehensively review the existing approaches to **pathway analysis** and analyze a number of relevant studies.

Gene expression microarrays have recently emerged as promising tools for **disease classification and outcome prediction**. In the late 1990s, it was demonstrated that gene expression profiles of clinical samples, or so-called gene expression *signatures*, can discover therapeutically relevant subclasses of several cancer types and predict disease outcome (5, 6, 8, 9, 53). They provided a fine classification, where the phenotype (disease outcome) was linked to a molecular signature that is thought to reflect the genetic underpinnings of the disease. As the ultimate goal of disease classification is to improve the treatment, these successes were followed by studies that correlated gene expression signatures with response to therapeutic agents (18, 54). For the pharmaceutical discovery organizations, these findings open up the possibility of using microarrays as tools to stratify patients for treatment during clinical trials of new therapeutics. Because of the potential of this approach, we have included in the gene expression bioinformatics subsection a discussion of approaches used in disease classification. We cover several most common supervised and unsupervised approaches and explain their applications to microarray data. A detailed analysis of the application follows in the RNA Biomarker subchapter of Chapter 3.

Comparative genomic hybridization is quickly gaining acceptance in the pharmaceutical industry as a tool to discover therapeutic targets and identify biomarkers predictive of drug response. This technology is not as mature as expression microarrays: Although it was first described in 1992, the first commercial CGH arrays did not enter the marketplace until 2004. Therefore, we devote a significant amount of space in Chapter 2 to the description of the most common CGH protocols and array designs. We cover both two-color and one-color protocols and describe arrays based on both oligonucleotides and bacterial artificial chromosomes. As genotyping arrays are widely used to measure gene copy number, their application in CGH is discussed separately. Although we realize that the technology will have evolved by the time this book is published, the improvements are likely to be driven by increases in array density or detection sensitivity, but they are not expected to change the principles of gene copy number analysis. Therefore, we believe that the reader will benefit from this analysis of the technology's fundamentals and the comparison of the major platforms.

As the main technology platforms for CGH are being adopted, the analysis of genome-wide copy number profiles is becoming an increasingly important area of bioinformatics. While the identification of copy number changes in individual samples has been facilitated by development of various smoothing and segmentation approaches, multisample analysis aimed at detection of patterns in data sets or discovery of class-specific aberrations still represents a largely unmet need. Therefore we devote a substantial amount of space in Chapter 2 to the **bioinformatics of gene copy number analysis**. The CGH Bioinformatics subchapter is organized so that the sample-level analysis is discussed first, followed by the experiment-level analysis. We comprehensively review the existing statistical approaches that enable these analyses and list the commonly used software programs along with their main functionalities.

MicroRNA and methylation microarrays are the most recently developed and therefore the least mature of the microarray technologies covered in Chapter 2. Therefore, we felt that it was necessary to first introduce the phenomena of promoter methylation and microRNA control of gene expression and then cover in detail the emerging experimental protocols. The biology of these phenomena and their significance in drug discovery are discussed. The promoter methylation status of several individual genes has been shown to correlate with the response of oncology patients to specific anticancer drugs. There is currently a great interest in identifying composite DNA methylation patterns (sometimes referred to as the "methylome") that may predict sensitivity to therapeutic agents. Similarly, in the microRNA field individual microRNAs were found to be involved in the regulation of key genes associated with drug response. Composite microRNA expression patterns of a panel of cell lines showed significant correlations with compound potency patterns, suggesting that microRNAs may play a role in chemoresistance (55). Comprehensive microRNA profiling is therefore likely to generate valuable information on potential microRNA correlates of drug response. The speed at which microRNA profiling technologies develop largely depends on the extent of our knowledge of microRNA-coding sequences in the genome. Another important barrier to further technological advancement is the low specificity of microRNA detection, primarily due to the short length of the microRNA molecules and their high homology. As we describe the existing array platforms for DNA methylation and microRNA expression profiling, we critically analyze the technical issues that influence the development of these methodologies.

In summary, the main purpose of Chapter 2 is to provide sufficient background information on genomics technologies in the context of their applications in drug discovery. Those readers who are interested in a more detailed analysis of the established and newly developed methodologies are referred to several recent reviews (56–59). In particular, two excellent review articles (56, 57) comprehensively describe the main principles of CGH and discuss common technical issues related to CGH experiments. Since these reviews do not cover the use of single nucleotide polymorphism (SNP) genotyping arrays for copy number analysis, we dedicate more space in Chapter 2 to this methodology. The emerging microarray-based microRNA profiling methodologies are reviewed in a recent article (59), with a specific emphasis on microRNA detection in cancer. The current state of technologies for detection of promoter methylation is addressed in a recent review on methylation changes in cancer (58). Finally, the most mature of all the technologies presented in Chapter 2, gene expression microarrays, has been extensively covered in the literature, including several books.

The rest of the book is almost entirely devoted to the various applications of genomics technologies introduced in Figure 1.3. Biomarker discovery is a key area of genomics research and a rapidly growing field, as our understanding of the factors that determine response to most drugs is currently insufficient to rationally select patients for maximum therapeutic benefit. As we systematically review different types of **genomic biomarkers** and the methodologies for their identification in Chapter 3, we distinguish between two major types of biomarkers, based on their main application: (i) patient stratification markers, that is, markers that stratify the patient population into likely responders and nonresponders, and (ii) pharmacodynamic biomarkers, that is, markers that enable monitoring the drug effect and quantifying its benefits. In Chapter 3, we specifically focus on drug efficacy, as prediction and quantitation of drug toxicity are addressed in separate chapters devoted to toxicogenomics.

The term "genomic biomarkers" may be applied to any genomic feature (structural or functional) that can be used to predict or quantify the benefits or a drug. Therefore, we considered all types of genomic changes known to correlate with drug response, such as gene amplifications, mutations, polymorphisms, or gene expression changes. To align the biomarker chapter with the technologies chapter, we chose to classify genomic biomarkers into DNA- and RNA-based, with the former group further subdivided into **DNA copy number alterations, mutations**, and **epigenetic changes** and the latter group comprised of gene expression patterns. Since many SNPs have been demonstrated to correlate with drug response, it would be logical to include them in the DNA-based biomarker section. Although the distinction between normal genetic variation and abnormal changes in the DNA structure is often quite vague, the effects of interindividual genetic variation on drug response are commonly considered the subject of pharmacogenetics and pharmacogenomics. We therefore decided to cover the normal genetic variability separately rather than include it in the biomarker chapter.

In drug discovery, biomarkers are particularly useful when they are discovered early, because they can then be applied to prioritize therapeutic targets, optimize the lead compound in terms of both predicted efficacy and toxicity, and generate hypotheses about the patient subpopulations that are more likely to respond to the drug with minimal toxicity. Most importantly, if biomarker candidates are identified early in discovery rather than in clinical trials, they may be tested and validated early enough in the clinic to enable rational patient selection for late-stage clinical trials. Thus early incorporation of biomarker programs into drug discovery may facilitate prioritization of drug discovery resources and optimize the design of subsequent clinical trials. Therefore, throughout the biomarker chapter we will focus on opportunities for **early discovery of biomarker candidates**.

A separate and equally important task in drug discovery is early assessment of the toxicity of candidate compounds. As classical toxicity evaluation represents a major bottleneck in the discovery and development pipeline, the need arises for novel methodologies that would enable early prediction of compound toxicity. This need has prompted the development of a scientific discipline commonly referred to as **toxicogenomics**, which will be comprehensively covered in this book (Chapters 4, 5, and 6). The term "toxicogenomics" has been generally used to describe the application of the genomics technologies in the field of toxicology. In this book, we restrict the use of the term "toxicogenomics" to describe the use of transcriptomic data to detect, investigate, or characterize toxicological effects of chemical entities. Chapter 4 provides an overview of the basic principles of toxicogenomics and defines its place in drug discovery. To facilitate further adoption of the methodology in the pharmaceutical industry, the practical aspects of toxicogenomics are discussed. To enable toxicogenomics analysis in a setting with limited experimental capacity, existing toxicogenomics databases are introduced. Chapter 4 thus provides practical advice to aid the incorporation of toxicogenomics in drug discovery and development.

To illustrate the concepts introduced in Chapter 4, we describe in Chapter 5 several specific applications of gene expression analysis in predictive, diagnostic, and mechanistic toxicology. The chapter focuses on **in vivo applications of toxicogenomics** and reviews multiple successful examples of toxicological assessment of small molecule therapeutics in rats and mice. Overall, preclinical toxicology studies are relatively effective in predicting potential human toxicity. According to a retrospective evaluation of data for 150 drugs, 94% of human toxicities are detected preclinically (60). The focus of the in vivo toxicogenomics chapter is on the ability of toxicogenomics to predict potential toxicity in the clinic earlier, with less compound and lower costs, as compared to traditional toxicology assessment. To facilitate the understanding of toxicogenomics in the context of drug discovery, the key steps of preclinical toxicology assessment are described. Special attention is paid to the issue of relevance of toxicological findings in animals to possible adverse effects in the clinic. We discuss the applications of toxicogenomics to detect toxicity, using the following organs as illustrations: liver, kidney, gastrointestinal tract, and the male reproductive system. The ability of toxicogenomics to predict potential carcinogenicity of compounds is considered separately. Although the evidence accumulated to date does not indicate conclusively that the carcinogenic potential of compounds can be predicted early by studying gene expression, assessment of carcinogenicity is a particularly important application of toxicogenomics, because of the value it may generate by complementing or possibly even replacing some of the current long and resource-intensive in vivo carcinogenicity assays. Finally, we consider the use of hemogenomics, or application of toxicogenomics to blood. Blood represents a tissue with great potential for predictive toxicogenomics, because of its accessibility in clinical studies. Thus, if compound toxicity could be predicted by measuring gene expression in blood, the findings from preclinical toxicology studies would be potentially transferable to clinical trials. As evidence is being accumulated in favor of applying gene expression data from blood to predict disease outcome and treatment response, the idea of using blood in toxicogenomics as a surrogate to monitor potential toxicity in other organs becomes more and more realistic. The advantages of this approach are discussed in the section focused on hemogenomics.

In addition to the in vivo applications of toxicogenomics reviewed in Chapter 5, data from our laboratory as well as published findings also suggest that toxicogenomics represents a feasible approach to in vitro screening of compounds for toxicity. This prompted us to include a separate chapter on in vitro toxicogenomics. In Chapter 6 we analyze the value creation proposition of **in vitro toxicogenomics**, which stems from its early timing, low compound requirements, and potential to reduce the amount of animal testing required. Indeed, animal studies are lengthy and require gram amounts of compounds, which means that they cannot be routinely conducted for multiple compounds in an early series that are typically synthesized in small amounts by bench chemists. In addition, animal welfare issues represent a strong incentive to reduce testing in animals as much as possible. In contrast, in vitro toxicogenomics assessment can be conducted as early as at the lead optimization stage, thus providing an initial read on the compound's potential toxicity liabilities. Typically, it only requires microgram or milligram quantities of the test compound, obviating the necessity for large-scale synthesis and thus enabling early application.

In Chapter 6, we review existing in vitro toxicogenomics data from our laboratory and published reports, including attempts to predict hepatotoxicity, as well as several common toxic changes and end points, such as phospholipidosis, and genotoxicity. Throughout the chapter, we emphasize that in vitro alternatives to animal toxicology studies are only useful if they generate reliable data that can be used to infer the effects of compounds in humans. As more cost-effective and flexible analytical gene expression platforms with appropriate throughput are developed, toxicogenomics is likely to become more practical as an approach to early toxicological screening of compounds.

In contrast to predictive applications, **mechanistic toxicogenomics** is greatly facilitated by the use of in vitro systems. Hypotheses on the compound mechanism can be generated based on pathway deregulation data obtained from in vitro systems. While such systems, typically based on cultured cells, do not properly reflect the complexity of the tissue of interest, their advantage is that they can assess directly the effect of the compounds on the target cell type, without the added effects of the compound metabolism, pharmacokinetics, etc. that are always observed during in vivo studies. Furthermore, in mechanistic studies speed and cost are not as important as data content, since they would be used in a retroactive mode using limited sets of compounds. Chapter 6 analyzes in detail the existing mechanistic data generated through in vitro toxicogenomics approaches.

The interindividual variability in response to drugs is determined to a large extent by genetic polymorphisms. The latter have been a subject of **pharmacogenetics**, a scientific discipline dealing with the influence of *individual* germ line polymorphisms on drug response. More recently, the focus has been shifting toward studying the effects of genetic variation on a genome-wide scale, giving rise to **pharmacogenomics**. Given the increasing role of the concept of personalized medicine in today's drug discovery, we devoted an entire chapter to the study of **germ line polymorphisms affecting response to therapeutic agents**. A significant part of Chapter 7 represents a collection of case studies in several therapeutic areas, namely, oncology, inflammation, virology, and neuroscience. While most of these studies deal with approved therapeutic agents, we emphasize the key learnings from the perspective of drug discovery. Consistent with the main focus of this book on applications of genomics in drug discovery, we analyze potential strategies for *early* identification of polymorphisms as predictors of drug response. The overwhelming majority of the studies reviewed in Chapter 7 deal with SNPs, sequence variants that were until recently considered the dominant form of genetic variation in humans. However, a new form of variation in the human genome, copy number variants (CNVs), has recently been discovered. Because of the tranformational role of this discovery in pharmacogenomics, we devote a separate subsection in Chapter 7 to copy number variation in humans. While only very limited evidence is available in support of their role in drug response, it has already been demonstrated that CNVs encompass more sequence content than SNPs, implying their important role in interindividual variability. Therefore, we review several key studies on CNVs, highlighting polymorphisms in disease-related and drug-metabolizing genes. To conclude the chapter, we compare the candidate gene, genome-wide, and pathway-based approaches for discovery of polymorphisms, emphasizing the advantages and disadvantages of each of these from the standpoint of drug discovery and development.

A large volume of data has been accumulated on the **role of polymorphisms in drug-metabolizing enzymes in the pharmacokinetics (PK) of drugs**. This information is extensively used by the pharmaceutical industry to incorporate pharmacogenetic (PG) assays into the drug development process. This prompted us to include a separate chapter devoted to the overview and the practical aspects of the PG–PK studies. Chapter 8 outlines the pharmacogenetic aspects of drug disposition and comprehensively reviews the available information on the known drug-metabolizing enzymes and polymorphisms that affect their function. The utility of PG–PK studies in clinical studies is analyzed, with emphasis on their early incorporation in clinical development.

As part of their role in developing standards for drugs and diagnostics, the regulatory agencies are actively participating in the process of industrywide adoption of genomic technologies by issuing guidance documents and establishing procedures for submission and review of genomic data. Therefore, in Chapter 9 of this book we review the regulatory developments and initiatives related to the use of genomic data in drug discovery. For instance, **a guidance on the regulatory submission of pharmacogenomics data** issued by the U.S. Food and Drug Administration (FDA) is cited as a milestone document aimed at promoting the use of genomic technologies to improve the efficacy and safety of novel drugs. We discuss an initiative by the FDA to enable **voluntary genomic data submission (VGDS)**, an innovative approach to encourage the industry to share genomic data on new medicines without making them a part of the formal regulatory submission. However, certain types of genomic data are required to be included in the formal submission, and we cite in Chapter 9 the criteria set by the agency for recognizing these types. Since application of genomics technologies may result in the development of a genomic diagnostic test, we include a discussion of a draft concept paper by the FDA on codevelopment of drugs and diagnostics. Additionally, since pharmacogenomics tests are considered in vitro diagnostics (IVDs), and IVDs are regarded as medical devices by the FDA, we briefly review some regulations related to diagnostic tests. In a recent development, the FDA released a **Draft Guidance on in vitro Diagnostic Multivariate Index Assays (IVDMIAs)**, which is relevant to genomics tests because it specifically deals with composite scores combining input from multiple variables and intended to predict the outcome of the disease or guide therapeutic intervention, a definition that includes such tests as microarray-derived gene expression signatures. Finally, Chapter 9 discusses **classification of biomarkers** proposed by the FDA and cites recent papers by FDA scientists outlining a process for biomarker qualification.

In summary, this book represents an attempt to cover several critical and rapidly developing areas of drug discovery, namely, biomarker research, toxicogenomics, and pharmacogenomics. All three fields have been recognized as transformational for drug discovery and development, but despite a significant overlap in their applications, they have not yet been considered together in a single text. Here, we attempted to review the current state of the three areas, emphasizing the synergies between them. Indeed, as the development of modern genomic technologies enables routine screening of clinical samples for gene copy number abnormalities, mutations, gene expression, and germ line polymorphisms, concurrent application of these technologies in clinical trials will certainly facilitate the discovery of genomic patterns associated with better drug response and lower toxicity. These genomic patterns would then be used to rationally select subjects for treatment and individually tailor pharmacological intervention to appropriate populations, thus advancing the concept of personalized medicine for the benefit of the patients.

REFERENCES

1. Finishing the euchromatic sequence of the human genome. *Nature* 2004;431:931–45.
2. Lander ES, Linton LM, Birren B, Nusbaum C, Zody MC, Baldwin J, Devon K, Dewar K, Doyle M, FitzHugh W, et al. Initial sequencing and analysis of the human genome. *Nature* 2001;409:860–921.

3. Venter JC, Adams MD, Myers EW, Li PW, Mural RJ, Sutton GG, Smith HO, Yandell M, Evans CA, Holt RA, et al. The sequence of the human genome. *Science* 2001;291:1304–51.
4. Singh D, Febbo PG, Ross K, Jackson DG, Manola J, Ladd C, Tamayo P, Renshaw AA, D' Amico AV, Richie JP, Lander ES, Loda M, Kantoff PW, Golub TR, and Sellers WR. Gene expression correlates of clinical prostate cancer behavior. *Cancer Cell* 2002;1:203–9.
5. Shipp MA, Ross KN, Tamayo P, Weng AP, Kutok JL, Aguiar RC, Gaasenbeek M, Angelo M, Reich M, Pinkus GS, Ray TS, Koval MA, Last KW, Norton A, Lister TA, Mesirov J, Neuberg DS, Lander ES, Aster JC, and Golub TR. Diffuse large B-cell lymphoma outcome prediction by gene-expression profiling and supervised machine learning. *Nat Med* 2002;8:68–74.
6. Ramaswamy S, Tamayo P, Rifkin R, Mukherjee S, Yeang CH, Angelo M, Ladd C, Reich M, Latulippe E, Mesirov JP, Poggio T, Gerald W, Loda M, Lander ES, and Golub TR. Multiclass cancer diagnosis using tumor gene expression signatures. *Proc Natl Acad Sci USA* 2001;98:15149–54.
7. Pomeroy SL, Tamayo P, Gaasenbeek M, Sturla LM, Angelo M, McLaughlin ME, Kim JY, Goumnerova LC, Black PM, Lau C, et al. Prediction of central nervous system embryonal tumour outcome based on gene expression. *Nature* 2002;415:436–42.
8. Armstrong SA, Staunton JE, Silverman LB, Pieters R, den Boer ML, Minden MD, Sallan SE, Lander ES, Golub TR, and Korsmeyer SJ. MLL translocations specify a distinct gene expression profile that distinguishes a unique leukemia. *Nat Genet* 2002;30:41–7.
9. Golub TR, Slonim DK, Tamayo P, Huard C, Gaasenbeek M, Mesirov JP, Coller H, Loh ML, Downing JR, Caligiuri MA, Bloomfield CD, and Lander ES. Molecular classification of cancer: class discovery and class prediction by gene expression monitoring. *Science* 1999;286:531–7.
10. Perou CM, Sorlie T, Eisen MB, van de Rijn M, Jeffrey SS, Rees CA, Pollack JR, Ross DT, Johnsen H, Akslen LA, Fluge O, Pergamenschikov A, Williams C, Zhu SX, Lonning PE, Borresen-Dale AL, Brown PO, and Botstein D. Molecular portraits of human breast tumours. *Nature* 2000;406:747–52.
11. Potti A, Dressman HK, Bild A, Riedel RF, Chan G, Sayer R, Cragun J, Cottrill H, Kelley MJ, Petersen R, Harpole D, Marks J, Berchuck A, Ginsburg GS, Febbo P, Lancaster J, and Nevins JR. Genomic signatures to guide the use of chemotherapeutics. *Nat Med* 2006;12:1294–300.
12. Sorlie T, Perou CM, Tibshirani R, Aas T, Geisler S, Johnsen H, Hastie T, Eisen MB, van de Rijn M, Jeffrey SS, Thorsen T, Quist H, Matese JC, Brown PO, Botstein D, Eystein Lonning P, and Borresen-Dale AL. Gene expression patterns of breast carcinomas distinguish tumor subclasses with clinical implications. *Proc Natl Acad Sci USA* 2001;98:10869–74.
13. Bittner M, Meltzer P, Chen Y, Jiang Y, Seftor E, Hendrix M, Radmacher M, Simon R, Yakhini Z, Ben-Dor A, et al. Molecular classification of cutaneous malignant melanoma by gene expression profiling. *Nature* 2000;406:536–40.
14. van 't Veer LJ, Dai H, van de Vijver MJ, He YD, Hart AA, Mao M, Peterse HL, van der Kooy K, Marton MJ, Witteveen AT, Schreiber GJ, Kerkhoven RM, Roberts C, Linsley PS, Bernards R, and Friend SH. Gene expression profiling predicts clinical outcome of breast cancer. *Nature* 2002;415:530–6.
15. Alizadeh AA, Eisen MB, Davis RE, Ma C, Lossos IS, Rosenwald A, Boldrick JC, Sabet H, Tran T, Yu X, et al. Distinct types of diffuse large B-cell lymphoma identified by gene expression profiling. *Nature* 2000;403:503–11.
16. Bea S, Zettl A, Wright G, Salaverria I, Jehn P, Moreno V, Burek C, Ott G, Puig X, Yang L, et al. Diffuse large B-cell lymphoma subgroups have distinct genetic profiles that influence tumor biology and improve gene expression-based survival prediction. *Blood* 2005;106:3183–90.
17. Iqbal J, Sanger WG, Horsman DE, Rosenwald A, Pickering DL, Dave B, Dave S, Xiao L, Cao K, Zhu Q, et al. BCL2 translocation defines a unique tumor subset within the germinal center B-cell-like diffuse large B-cell lymphoma. *Am J Pathol* 2004;165:159–66.
18. Rosenwald A, Wright G, Chan WC, Connors JM, Campo E, Fisher RI, Gascoyne RD, Muller-Hermelink HK, Smeland EB, Giltnane JM, et al. The use of molecular profiling

to predict survival after chemotherapy for diffuse large-B-cell lymphoma. *N Engl J Med* 2002;346:1937–47.

19. Staudt LM, and Dave S. The biology of human lymphoid malignancies revealed by gene expression profiling. *Adv Immunol* 2005;87:163–208.
20. Wright G, Tan B, Rosenwald A, Hurt EH, Wiestner A, and Staudt LM. A gene expression-based method to diagnose clinically distinct subgroups of diffuse large B cell lymphoma. *Proc Natl Acad Sci USA* 2003;100:9991–6.
21. Staudt LM. Molecular diagnosis of the hematologic cancers. *N Engl J Med* 2003;348:1777–85.
22. Druker BJ, Talpaz M, Resta DJ, Peng B, Buchdunger E, Ford JM, Lydon NB, Kantarjian H, Capdeville R, Ohno-Jones S, and Sawyers CL. Efficacy and safety of a specific inhibitor of the BCR-ABL tyrosine kinase in chronic myeloid leukemia. *N Engl J Med* 2001;344:1031–7.
23. Druker BJ. Effects of a selective inhibitor of the Abl tyrosine kinase on the growth of Bcr-Abl positive cells. *Nat Med* 1996;2:561–6.
24. Rowley JD. The critical role of chromosome translocations in human leukemias. *Annu Rev Genet* 1998;32:495–519.
25. Vogel CL, Cobleigh MA, Tripathy D, Gutheil JC, Harris LN, Fehrenbacher L, Slamon DJ, Murphy M, Novotny WF, Burchmore M, Shak S, Stewart SJ, and Press M. Efficacy and safety of trastuzumab as a single agent in first-line treatment of HER2-overexpressing metastatic breast cancer. *J Clin Oncol* 2002;20:719–26.
26. Slamon DJ, Leyland-Jones B, Shak S, Fuchs H, Paton V, Bajamonde A, Fleming T, Eiermann W, Wolter J, Pegram M, Baselga J, and Norton L. Use of chemotherapy plus a monoclonal antibody against HER2 for metastatic breast cancer that overexpresses HER2. *N Engl J Med* 2001;344:783–92.
27. Cobleigh MA, Vogel CL, Tripathy D, Robert NJ, Scholl S, Fehrenbacher L, Wolter JM, Paton V, Shak S, Lieberman G, and Slamon DJ. Multinational study of the efficacy and safety of humanized anti-HER2 monoclonal antibody in women who have HER2-overexpressing metastatic breast cancer that has progressed after chemotherapy for metastatic disease. *J Clin Oncol* 1999;17:2639–48.
28. Kantarjian H, Sawyers C, Hochhaus A, Guilhot F, Schiffer C, Gambacorti-Passerini C, Niederwieser D, Resta D, Capdeville R, Zoellner U, et al. Hematologic and cytogenetic responses to imatinib mesylate in chronic myelogenous leukemia. *N Engl J Med* 2002;346:645–52.
29. Million RP. Impact of genetic diagnostics on drug development strategy. *Nat Rev Drug Discov* 2006;5:459–62.
30. Hudziak RM, Lewis GD, Winget M, Fendly BM, Shepard HM, and Ullrich A. p185HER2 monoclonal antibody has antiproliferative effects in vitro and sensitizes human breast tumor cells to tumor necrosis factor. *Mol Cell Biol* 1989;9:1165–72.
31. Armstrong SA, Kung AL, Mabon ME, Silverman LB, Stam RW, Den Boer ML, Pieters R, Kersey JH, Sallan SE, Fletcher JA, Golub TR, Griffin JD, and Korsmeyer SJ. Inhibition of FLT3 in MLL. Validation of a therapeutic target identified by gene expression based classification. *Cancer Cell* 2003;3:173–83.
32. Elenitoba-Johnson KS, Jenson SD, Abbott RT, Palais RA, Bohling SD, Lin Z, Tripp S, Shami PJ, Wang LY, Coupland RW, Buckstein R, Perez-Ordonez B, Perkins SL, Dube ID, and Lim MS. Involvement of multiple signaling pathways in follicular lymphoma transformation: p38-mitogen-activated protein kinase as a target for therapy. *Proc Natl Acad Sci USA* 2003;100:7259–64.
33. Welsh JB, Sapinoso LM, Su AI, Kern SG, Wang-Rodriguez J, Moskaluk CA, Frierson HF, Jr., and Hampton GM. Analysis of gene expression identifies candidate markers and pharmacological targets in prostate cancer. *Cancer Res* 2001;61:5974–8.
34. Lee JA, and Lupski JR. Genomic rearrangements and gene copy-number alterations as a cause of nervous system disorders. *Neuron* 2006;52:103–21.

35. Cheng KW, Lahad JP, Kuo WL, Lapuk A, Yamada K, Auersperg N, Liu J, Smith-McCune K, Lu KH, Fishman D, Gray JW, and Mills GB. The RAB25 small GTPase determines aggressiveness of ovarian and breast cancers. *Nat Med* 2004;10:1251–6.
36. Chin K, DeVries S, Fridlyand J, Spellman PT, Roydasgupta R, Kuo WL, Lapuk A, Neve RM, Qian Z, Ryder T, Chen F, Feiler H, Tokuyasu T, Kingsley C, Dairkee S, Meng Z, Chew K, Pinkel D, Jain A, Ljung BM, Esserman L, Albertson DG, Waldman FM, and Gray JW. Genomic and transcriptional aberrations linked to breast cancer pathophysiologies. *Cancer Cell* 2006;10:529–41.
37. Service RF. Surviving the blockbuster syndrome. *Science* 2004;303:1796–9.
38. Kola I, and Landis J. Can the pharmaceutical industry reduce attrition rates? *Nat Rev Drug Discov* 2004;3:711–5.
39. Glaser KB, Staver MJ, Waring JF, Stender J, Ulrich RG, and Davidsen SK. Gene expression profiling of multiple histone deacetylase (HDAC) inhibitors: defining a common gene set produced by HDAC inhibition in T24 and MDA carcinoma cell lines. *Mol Cancer Ther* 2003;2:151–63.
40. Lamb J. The Connectivity Map: a new tool for biomedical research. *Nat Rev Cancer* 2007;7:54–60.
41. Lamb J, Crawford ED, Peck D, Modell JW, Blat IC, Wrobel MJ, Lerner J, Brunet JP, Subramanian A, Ross KN, Reich M, Hieronymus H, Wei G, Armstrong SA, Haggarty SJ, Clemons PA, Wei R, Carr SA, Lander ES, and Golub TR. The Connectivity Map: using gene-expression signatures to connect small molecules, genes, and disease. *Science* 2006;313:1929–35.
42. Cheok MH, Yang W, Pui CH, Downing JR, Cheng C, Naeve CW, Relling MV, and Evans WE. Treatment-specific changes in gene expression discriminate in vivo drug response in human leukemia cells. *Nat Genet* 2003;34:85–90.
43. Bonham M, Arnold H, Montgomery B, and Nelson PS. Molecular effects of the herbal compound PC-SPES: identification of activity pathways in prostate carcinoma. *Cancer Res* 2002;62:3920–4.
44. Maier S, Dahlstroem C, Haefliger C, Plum A, and Piepenbrock C. Identifying DNA methylation biomarkers of cancer drug response. *Am J Pharmacogenomics* 2005;5:223–32.
45. Calin GA, and Croce CM. MicroRNA signatures in human cancers. *Nat Rev Cancer* 2006;6:857–66.
46. Lu J, Getz G, Miska EA, Alvarez-Saavedra E, Lamb J, Peck D, Sweet-Cordero A, Ebert BL, Mak RH, Ferrando AA, Downing JR, Jacks T, Horvitz HR, and Golub TR. MicroRNA expression profiles classify human cancers. *Nature* 2005;435:834–8.
47. Ruepp S, Boess F, Suter L, de Vera MC, Steiner G, Steele T, Weiser T, and Albertini S. Assessment of hepatotoxic liabilities by transcript profiling. *Toxicol Appl Pharmacol* 2005;207:161–70.
48. Foster WR, Chen SJ, He A, Truong A, Bhaskaran V, Nelson DM, Dambach DM, Lehman-McKeeman LD, and Car BD. A retrospective analysis of toxicogenomics in the safety assessment of drug candidates. *Toxicol Pathol* 2007;35:621–35.
49. Yang Y, Blomme EA, and Waring JF. Toxicogenomics in drug discovery: from preclinical studies to clinical trials. *Chem Biol Interact* 2004;150:71–85.
50. Searfoss GH, Ryan TP, and Jolly RA. The role of transcriptome analysis in pre-clinical toxicology. *Curr Mol Med* 2005;5:53–64.
51. Suter L, Babiss LE, and Wheeldon EB. Toxicogenomics in predictive toxicology in drug development. *Chem Biol* 2004;11:161–71.
52. Ransohoff DF. Bias as a threat to the validity of cancer molecular-marker research. *Nat Rev Cancer* 2005;5:142–9.
53. Monti S, Savage KJ, Kutok JL, Feuerhake F, Kurtin P, Mihm M, Wu B, Pasqualucci L, Neuberg D, Aguiar RC, Dal Cin P, Ladd C, Pinkus GS, Salles G, Harris NL, Dalla-Favera R, Habermann TM, Aster JC, Golub TR, and Shipp MA. Molecular profiling of diffuse large

B-cell lymphoma identifies robust subtypes including one characterized by host inflammatory response. *Blood* 2005;105:1851–61.

54. Paik S, Tang G, Shak S, Kim C, Baker J, Kim W, Cronin M, Baehner FL, Watson D, Bryant J, Costantino JP, Geyer CE, Jr., Wickerham DL, and Wolmark N. Gene expression and benefit of chemotherapy in women with node-negative, estrogen receptor-positive breast cancer. *J Clin Oncol* 2006;24:3726–34.
55. Blower PE, Verducci JS, Lin S, Zhou J, Chung JH, Dai Z, Liu CG, Reinhold W, Lorenzi PL, Kaldjian EP, Croce CM, Weinstein JN, and Sadee W. MicroRNA expression profiles for the NCI-60 cancer cell panel. *Mol Cancer Ther* 2007;6:1483–91.
56. Pinkel D, and Albertson DG. Comparative genomic hybridization. *Annu Rev Genomics Human Genet* 2005;6:331–54.
57. Pinkel D, and Albertson DG. Array comparative genomic hybridization and its applications in cancer. *Nat Genet* 2005;37 Suppl:S11–7.
58. Ushijima T. Detection and interpretation of altered methylation patterns in cancer cells. *Nat Rev Cancer* 2005;5:223–31.
59. Calin GA, and Croce CM. MicroRNA signatures in human cancers. *Nat Rev Cancer* 2006;6:857–66.
60. Olson H, Betton G, Robinson D, Thomas K, Monro A, Kolaja G, Lilly P, Sanders J, Sipes G, Bracken W, Dorato M, Van Deun K, Smith P, Berger B, and Heller A. Concordance of the toxicity of pharmaceuticals in humans and in animals. *Regul Toxicol Pharmacol* 2000;32:56–67.

Chapter 2

Genomics Technologies as Tools in Drug Discovery

2.1. INTRODUCTION TO GENOMICS TECHNOLOGIES

The term "genomics" refers to the study of the structure, variation, and function of the whole genome of an organism. Hence, the spectrum of technologies covered by the definition "genomic technologies" should include all methodologies that enable the analysis of either structure or function of the entire genome. As our knowledge of the human and other commonly studied genomes is very far from being complete, in most cases we limit ourselves to a portion of the genome that has been sufficiently characterized to justify development and commercialization of analysis tools. Despite this limitation, it is fair to define genomics technologies as methods that permit the study of the entire genome. Hence, one common characteristic of genomics technologies is their high-throughput multiplexed nature, which enables scaling of the analysis to include a substantial portion of the organism's genome. The majority of the existing genomic technologies utilize microarrays (discussed in detail below), but the growing spectrum of genome-wide methodologies also includes bead-based techniques, as well as numerous high-throughput assays for studying gene function, such as multiplexed siRNA-silencing assays and knock-in methodologies. This chapter largely focuses on microarray-based technologies that allow simultaneous quantitation of thousands of DNA or RNA molecules, while the rest of the book largely deals with their applications at different stages of drug discovery.

Fundamentally, all **microarray methodologies** quantify nucleic acids in a biological sample through monitoring their interaction with a specially designed library of molecular probes of known sequence. The probes are arranged in a predetermined order on a solid support, whereas the nucleic acids in the test

Genomics in Drug Discovery and Development, by Dimitri Semizarov and Eric Blomme

sample are preprocessed to ensure that they can be detected after hybridization. A microarray can be manufactured either through individual synthesis of a probe directly on the array support or by depositing presynthesized nucleic acid probes on the chip. The first in situ microarray fabrication protocol was reported in 1991 (1) and employed a photolithographic method, whereby the sequence of the oligonucleotide probes synthesized at different positions on the microarray is determined by photolithographic masks controlled by a computer. At each step of the oligonucleotide synthesis (i.e., at each nucleotide addition), the masks ensure that only selected positions on the array are illuminated to enable a photochemical coupling reaction. Thus the mask positions and the order of addition of chemical substrates ultimately determine the sequence and location of each oligonucleotide probe synthesized on the microarray. Two extremely important features of this method are its scalability and efficiency: the complete set of 4^n oligonucleotides (where n is the length) can be synthesized by performing only $4n$ chemical steps (2). The method was commercialized by Affymetrix Inc (Santa Clara, CA), initially to manufacture microarrays for gene expression analysis. A collection of 25-mer oligonucleotides was designed to hybridize a population of transcripts whose concentration was to be measured, and a protocol was developed to amplify, label, and hybridize the RNA from test samples.

An alternative in situ microarray fabrication protocol was developed at Protogene (Menlo Park, CA) and Agilent Technologies (Palo Alto, CA) in collaboration with Rosetta Inpharmatics (Seattle, WA). The protocol involved synthesis of oligonucleotides on a solid support through an ink-jet printing process and traditional phosphoramidite chemistry. The ink-jet technology enables rapid deposition of extremely small volumes of reagents at a very high accuracy, thus ensuring superior homogeneity of the spot. DNA microarrays can also be manufactured by using electrical addressing systems on a semiconductor chip. This methodology, commercialized by Combimatrix (Snoqualmie, WA), is based on connecting electrodes to addressable sites on the array. Oligonucleotide probes are synthesized one nucleotide at a time by adding an excess of a nucleotide and activating the electrode that corresponds to the site where this particular nucleotide is required for synthesis. The activated electrode initiates an electrochemical reaction that results in the attachment of the nucleotide to the oligonucleotide chain. Another approach (commercialized by Nimblegen Systems, LLC of Madison, WI) has been developed that employs so-called digital light processors, which represent arrays of very small mirrors attached to a computer chip. The mirrors are controlled by a computer to direct the light to particular positions on the microarray surface. The advantage of this approach over photolithography is its lower cost, as photolithographic masks are expensive.

While all in situ fabrication methodologies require sophisticated equipment, attachment of presynthesized DNA probes to a solid support can be performed in a smaller industrial or academic laboratory. Several types of robotic arrayers have been used to print various oligonucleotide-, cDNA-, and bacterial artificial chromosome-based probes (3–5). Typically, small volumes of oligonucleotide solutions are spotted on a slide pretreated with polylysine or polyamine, which facilitates absorption. As the density of the commercial microarrays increases and their cost drops, spotted arrays are used less and less frequently.

In this chapter, we describe the state-of-the-art **microarray methodologies for measuring gene expression, gene copy number, DNA methylation, and microRNA expression**. The order of the respective subchapters is not accidental: Gene expression microarrays are the most mature of the technologies, and comparative genomic hybridization (CGH) microarrays are becoming mainstream analysis tools, while DNA methylation and microRNA arrays are nascent technologies that require further development to enable their routine application in drug discovery. Special attention is paid to analysis of microarray data, as it is often the most complex part of the project. We have dedicated separate subchapters to bioinformatic aspects of gene expression and CGH microarray methodologies, where we review the relevant statistical approaches and describe the typical workflow of microarray data analysis.

2.2. GENE EXPRESSION MICROARRAYS: TECHNOLOGY

2.2.1. Standard Microarray Protocol

Generally speaking, **gene expression microarrays represent microchips containing thousands of DNA probes, which are used to analyze the abundance of multiple transcripts in a sample**. Based on the type of the probe, microarrays can be classified into oligonucleotide and cDNA arrays. While cDNA microarrays may potentially offer higher sensitivity of mRNA detection, difficulties in manufacturing and deposition of cloned and purified long DNA sequences have limited the use of cDNA arrays largely to academic laboratories. In this chapter, we focus on the more commonly used **oligonucleotide microarrays**. A glossary of microarray-related terms is presented in Table 2.1.

Figure 2.1 presents an outline of a typical **single-color microarray** experiment. Total RNA is isolated from the test sample by one of the established techniques (6, 7) and used as a template to synthesize cDNA. An oligo(dT) primer with an attached T7 sequence is used as a primer for reverse transcription. An enzyme called reverse transcriptase catalyzes the synthesis of cDNA, using the input RNA as a template. The resulting cDNA is then subjected to one round of DNA replication to generate double-stranded DNA. This reaction is catalyzed by a DNA polymerase. The resulting double-stranded DNA then serves as a template for T7 RNA polymerase, which recognizes the T7 sequences in the cDNA (8). The in vitro transcription is performed in the presence of biotinylated rNTPs to label the cRNA. This method of preparation of labeled amplified RNA was developed in 1990 and has been extensively validated since then (8–11).The cRNA synthesized through this procedure is fragmented and hybridized to the array.

After array hybridization, the array is washed to remove unbound molecules, stained with streptavidin-phycoerythrin and a bioinylated anti-streptavidin

Table 2.1 Glossary of Microarray-Related Terms

Probe	A DNA fragment attached to the microchip that is used to detect transcripts in the test sample
Reverse transcriptase	An enzyme capable of synthesizing DNA using RNA as a template
DNA polymerase	An enzyme capable of synthesizing DNA on a DNA template
cDNA	DNA synthesized off an mRNA template
cRNA	RNA synthesized in an in vitro transcription reaction using cDNA as a template
dNTP	A generic term referring to the four deoxyribonucleotides: dATP, dCTP, dGTP, and dTTP
rNTP	A generic term referring to the four ribonucleotides: ATP, CTP, GTP, and UTP
Oligonucleotide	A short synthetically prepared fragment of DNA or RNA
Ligase	An enzyme that catalyzes linkage of two DNA molecules
Restriction enzyme	An enzyme that catalyzes the cleavage of DNA at specific sites to produce discrete fragments
Transcription	Synthesis of RNA using DNA as a template
cDNA microarray	A microarray that uses cDNAs immobilized on a solid support as probes to interrogate nucleic acids in solution
Oligonucleotide microarray	A microarray that uses oligonucleotides immobilized on a solid support as probes to interrogate nucleic acids in solution

antibody, and scanned in a fluorescent scanner in order to quantify the signal for all the probes. After the acquisition of the image by the scanner, a specialized program overlays a grid onto the array to identify the spots and generates a table of signal intensities. A different program then processes the signal intensities for individual probes to generate the intensities for each individual gene, determine the background, and perform normalization. The signal intensity for each gene serves as a measure of the abundance of the corresponding transcript in the initial sample.

The quantity of the test RNA sample is an important factor in microarray analysis. The RNA polymerase synthesizes multiple copies of cRNA from each cDNA molecule, and the target preparation protocol results in an amplification of the original sample. For RNA quantities $>1\ \mu g$, the above-described protocol typically produces enough cRNA for at least one or two array hybridizations. Another important consideration is the integrity of the sample. If the total RNA is degraded, the reverse transcriptase will not be able to synthesize sufficiently long cDNAs and the cRNA products will not hybridize to all the probes for the transcript. Although $3'$ bias is an important consideration in microarray probe design, many probes on a microarray are removed by hundreds of nucleotides from the $3'$ terminus of the transcript. Probes that are distant from the $3'$ terminus will not hybridize if the cRNA products are of insufficient length. Similarly, suboptimal functioning of the reverse transcriptase may lead to shortened cDNAs and cRNAs and hence will result in an underestimation of the abundance of the

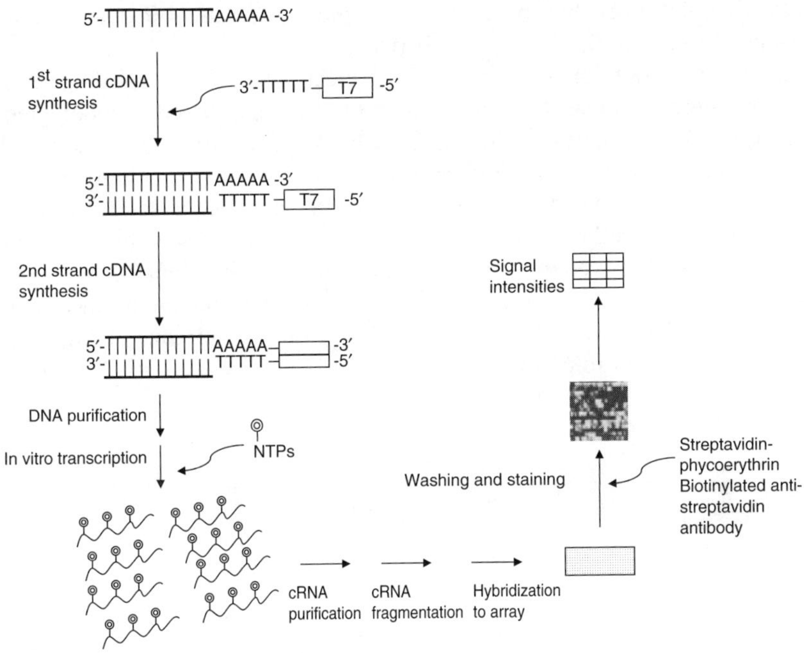

Figure 2.1 Gene expression profiling using DNA microarrays. The RNA sample is reverse transcribed to yield cDNA, which is then converted into double-stranded cDNA. The double-stranded cDNA is purified and used as a template for in vitro transcription in the presence of biotin-labeled rubonucleotides to produce labeled cRNA. The cRNA is then fragmented and hybridized to a microarray. After hybridization, the microarray is washed, stained to attach a fluorescent label to the biotin residues, and scanned to generate signal intensities, which are used to estimate the relative abundance of the corresponding mRNA in the initial sample.

transcripts. As microarray applications expand, more different sample types will need to be analyzed. Below we consider two special situations with regard to sample type.

2.2.2. Monitoring the Quality of Input RNA for Microarray Experiments

Several procedures can be used to determine the quality of a total RNA sample before a microarray experiment. Traditionally, RNA integrity has been evaluated with an agarose electrophoresis gel stained with ethidium bromide, followed by assessment of the resulting banding pattern. Gel images of intact human total RNA show two bands corresponding to the 28S and 18S ribosomal RNA (rRNA) species. RNA is considered of high quality when the ratio of 28S to 18S bands is close to 2. Since this approach relies on visual interpretation of gel images, it is subjective, not amenable to automation, digital processing, or quantitation, and does not permit interlaboratory comparisons.

Therefore the traditional denaturing gel electrophoresis is rapidly being replaced by more efficient and rapid microfluidics chip-based technologies, such as Agilent's Bioanalyzer chips (12). The Bioanalyzer is an automated bioanalytical device using microfluidics technology that provides eletrophoretic separations in an automated manner. Very small samples of total RNA or mRNA are separated in the channels of microfabricated chips according to their molecular weight, followed by detection via laser-induced fluorescence. The result is visualized as an electrophoretogram where the amount of measured fluorescence correlates with the amount of RNA of a given size. Figure 2.2 presents Bioanalyzer profiles of an intact and a degraded total RNA sample.

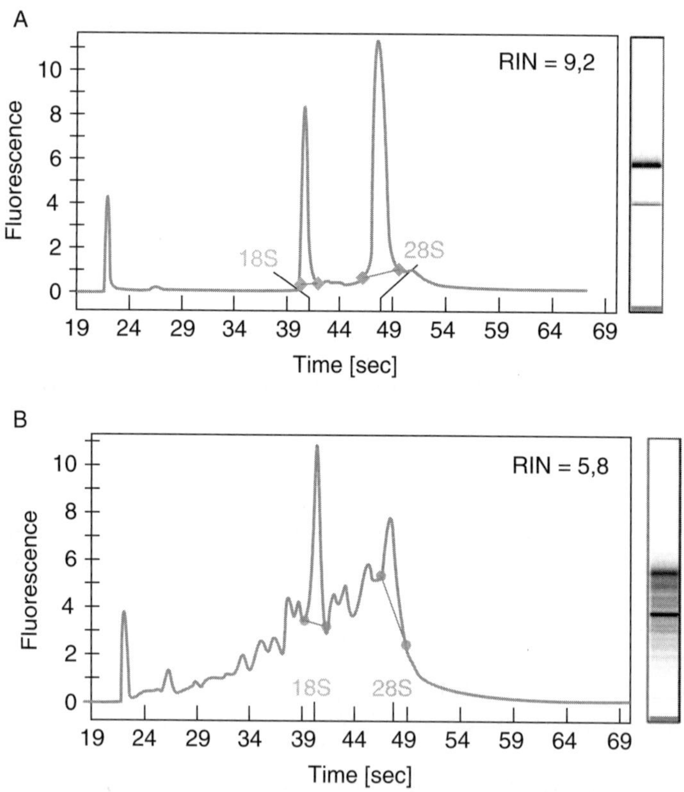

Figure 2.2 Examples of Bioanalyzer RNA profiles for intact (A) and degraded (B) human total RNA. The pictures are adapted from the manufacturers's manual, which can be found at www.agilent.com. An intact total RNA sample yields a flat baseline and two well-defined peaks corresponding to 18S and 28S ribosomal RNAs. A degraded total RNA sample typically produces wide peaks corresponding to accumulated degradation products, whereas the 18S and 28S peaks are poorly defined.

Clearly, even a small degree of RNA degradation can be easily detected with this technology, as it would result in the flattening of the 18S and 28S ribosomal RNA peaks and the appearance of the signal in the low-molecular-weight part of the graph.

Since RNA quality data are generated in a digital format, they can be reprocessed to enable additional calculations and quantification of RNA degradation. Historically, the first measure of RNA integrity was the ratio of the 28S to 18S ribosomal RNAs. When first released, the Bioanalyzer software calculated the ratio of the two ribosomal bands, following the commonly used approach for RNA integrity assessment. However, the use of ribosomal ratio for RNA quality assessment has several disadvantages. In many cases, ribosomal ratios showed only weak correlation with RNA integrity (13), and proper analysis of the electrophoretic traces requires substantial expertise from the user and is of a subjective nature. To provide a tool for standardization of RNA quality control, a user-independent automated procedure was recently developed (12). It is based on the calculation of an RNA Integrity Number (RIN). The algorithm was developed by using methods from information theory to rank features according to their information content and using a Bayesian approach to train and select a prediction model on the basis of artificial neural networks. The resulting algorithm is a user-independent, automated, and reliable procedure for standardization of RNA quality control that yields an RIN.

In summary, the advantages of the microchip-based technologies for RNA assessment include the short duration of the protocol, the small amounts of the RNA sample required, and the easy quantitation of the results. All this makes the chip-based technologies most appropriate for quality control of RNA samples used in drug discovery applications, particularly because of the need to standardize the quality control procedures for regulatory submissions.

2.2.3. Specialized Microarray Protocols for Archived and Small Samples

The standard expression microarray protocol requires high-quality intact RNA. However, as microarrays became important tools in biomarker research, the scientific community started looking for ways to apply the microarray technology to retrospectively analyze archived human tissue samples. Freezing a sample immediately after surgical resection typically preserves RNA. Therefore such samples can be analyzed with a standard microarray protocol. It is critical that the sample be frozen immediately, because even quick manipulation of tissue results in changes in gene expression (14, 15) that may be mistaken for true characteristics of the sample. Even more importantly, prolonged manipulation of a tissue specimen at room temperature results in RNA degradation, leading to synthesis of truncated cRNA fragments during target preparation and subsequently to underestimation of transcript abundance and missed gene calls.

The most difficult challenge is presented by **formalin-fixed paraffin-embedded (FFPE) samples**, as formalin fixation results in irreversible modification and degradation of RNA (16, 17). It is noteworthy that formalin fixation results in a wide spectrum of RNA modifications, including cross-linking, addition of monomethylol residues to the nucleic bases, and adenine dimerization (16). A specialized microarray for analysis of FFPE samples was developed (18). It contains mostly probe sets that are directed against the three hundred 3′-terminal nucleotides of the transcripts (instead of the 600-nucleotide limit set for regular microarrays). The increased 3′-bias is intended to facilitate binding of shortened cRNAs synthesized off truncated cDNAs. However, a solution remains to be found for analysis of highly modified RNA, as RNA with modified bases has a limited capacity to produce cDNA in the reverse transcription reaction. The chip is designed for use with a reagent system (19), which enables RNA isolation from FFPE tissues as well as its amplification and labeling. Another option for profiling FFPE tissue samples is a so-called tiling array, i.e., an microarray providing unbiased assessment of transcriptional activity (20). An example of such microarray is an exon junction microarray (21). The rationale behind using tiling arrays for analysis of FFPE samples is based on the fact that the detection of truncated and cross-linked mRNA fragments typically found in FFPE samples is optimized by uninterrupted coverage of the transcriptionally active areas of the genome.

Significant progress has also been made in analysis of small samples. The progress in this area was fueled by the introduction of a tissue dissection technology called **laser capture microdissection** (14, 22). The technique involves placing a transparent film over a tissue section and selectively adhering the cells of interest to the film with a fixed-position, short-duration, focused pulse from an infrared laser (Fig. 2.3). During the procedure, the tissue is visualized microscopically. The film with the procured tissue is then removed from the section

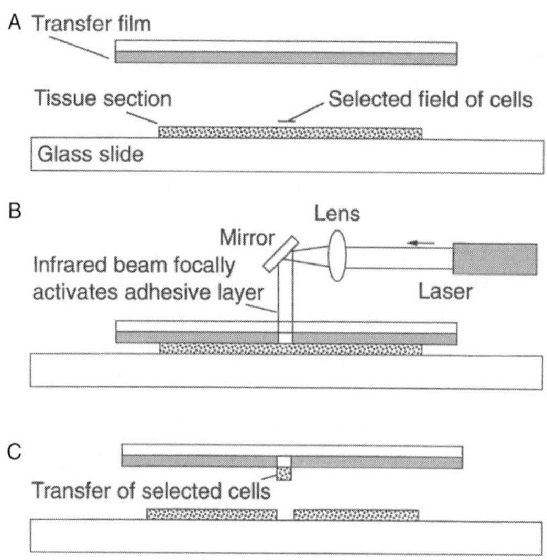

Figure 2.3 Laser capture microdissection protocol. A) A tissue section is mounted on a microscope slide and covered with transparent film. Cells of interest are selected visually under a microscope. B) A laser beam focused on the cells of interest is activated, causing the film to adhere to the selected cells. C) The film is removed together with the attached cells. At this point, the cells of interest can be lysed and further processed. Reproduced with permission from Emmert-Buck et al. (1996) *Science* 274: 998–1001 (ref. 22).

and used to isolate DNA or RNA (22). As laser capture microdissection of tissue samples became common, an urgent need arose for a microarray protocol suitable for analysis of samples ranging from 100 to 1000 cells. The problem was solved by introducing an additional round of RNA amplification. In a two-round amplification protocol (Fig. 2.4), the first round is performed with regular rNTPs, while the second round uses labeled rNTPs as in the single-round amplification protocol described in the previous subsection. Today, protocols involving laser capture microdissection and RNA isolation from single cells followed by gene expression analysis have become routine. They have greatly facilitated analysis of pure tumor cells and comparison with adjacent normal tissue (23–25).

2.2.4. Quality of Microarray Data and Technical Parameters of Microarrays

Sensitivity, specificity, and accuracy of measuring the transcript concentrations are the most commonly considered technical parameters of gene expression microarrays. The **sensitivity threshold of a microarray** can be defined as the lowest mRNA concentration that can be measured in an experiment. The sensitivity of a microarray is typically measured by spiking experiments (26, 27). Generally, the limit of detection for existing expression microarray platforms is in the range of one to 10 mRNA copies per cell (28).

The sensitivity of a microarray is partly determined by the length of the probes. Longer probes generally provide higher sensitivity (28). Comparisons of oligonucleotide microarrays with different probe lengths indicate that 30-mers provide a 10-fold greater sensitivity than 25-mers (26). As a significant fraction of the human transcriptome is represented by low-abundance mRNAs, it may seem beneficial to use microarrays with longer probes to maximize the spectrum of detectable transcripts. However, the trade-off to higher sensitivity is a decrease in specificity. The **specificity of a microarray probe** can be defined as its ability to discriminate between its perfect-match complementary sequence and any possible mismatched sequences. Longer oligonucleotides have an increased capacity to bind nontargeted sequences, and this decreases the specificity of the microarray. An increase in the oligonucleotide length from 25 to 30 nucleotides resulted in a decrease in the specificity, as defined by the relative signal intensity for the perfect match versus a mismatched sequence (26).

The **accuracy of mRNA quantitation** is an important parameter that reflects the precision with which a microarray determines the absolute concentration of a transcript in a sample or its relative concentration in different samples (expression ratio). The more commonly used single-color oligonucleotide platforms can be used to measure either the absolute or the relative mRNA concentrations, but absolute measurements require calibration with known amounts of the transcripts. Moreover, the accuracy of measurement for absolute concentrations is lower, as different probes hybridizing to different regions of the same mRNA produce widely different signal intensities (29). However, when the same probes

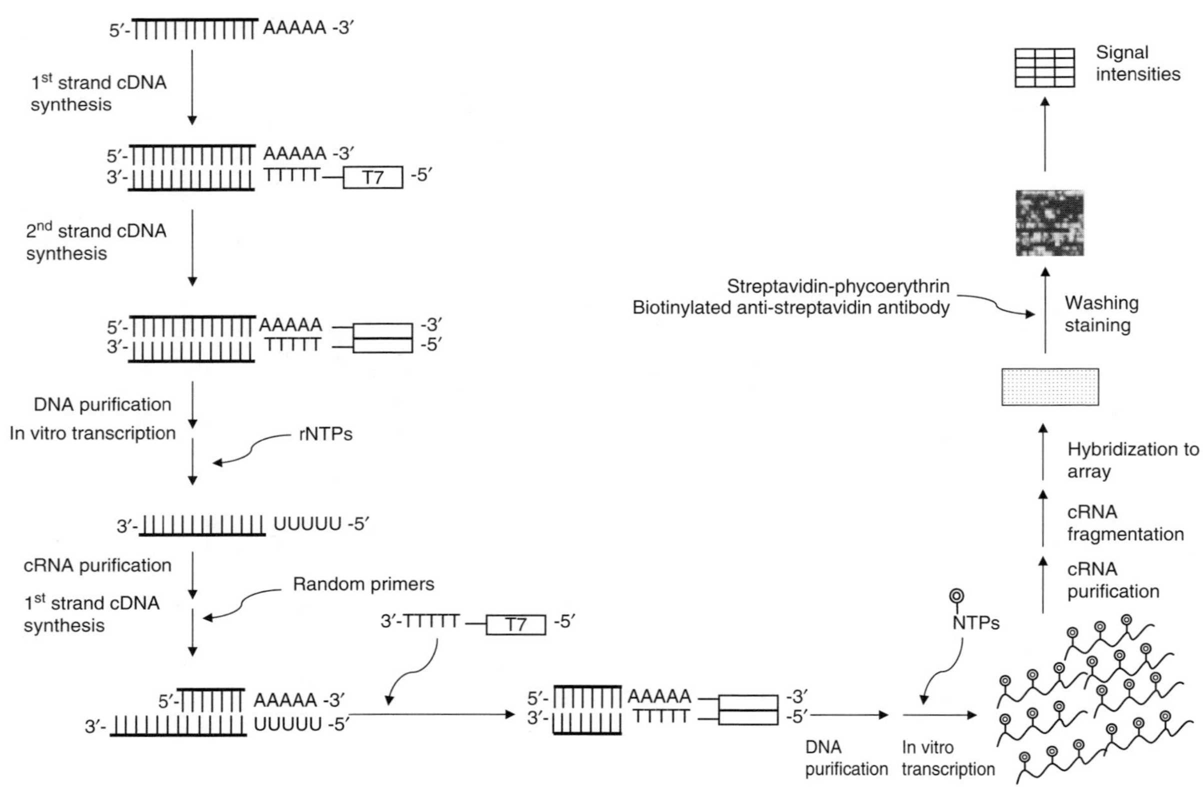

Figure 2.4 Gene expression profiling of small samples using DNA microarrays. Unlike the standard protocol outlined in Fig. 2.1, this procedure involves two rounds of RNA amplification. The first round is performed in the presence of regular rNTPs, while the second round involves biotin-modified rNTPs to label the cRNA product.

are used to measure the ratio of the transcript concentrations in two samples, the accuracy of measurement increases dramatically (28). Fortunately, most biological problems addressed by gene expression profiling require the measurement of gene expression ratios between two or more samples rather than the absolute amount of mRNA per cell. Therefore, the vast majority of the accumulated gene expression microarray data deals with expression ratios rather than absolute mRNA concentrations.

The most common DNA microarray platforms, such as the GeneChip® by Affymetrix, accurately detect the existence and direction of gene expression changes for 85–90% of the genes measured (28). Although several problems, such as systematic compression of gene expression ratios, have been reported (30), we believe that the accuracy of the existing commercial microarray platforms is sufficient for drug discovery applications, such as target identification, compound optimization, and biomarker discovery. Indeed, as we demonstrate in the subsequent chapters, multiple microarray data sets generated in test sample sets were subsequently validated with independent validation sample sets and, in some cases, validated in independent studies, thus ultimately demonstrating the applicability and value of the technology in drug discovery.

2.2.5. Reproducibility of Expression Microarrays and Cross-Platform Comparisons

The growing role of genomics information in U.S. Food and Drug Administration (FDA) submissions imposes a number of requirements on microarray data. Most importantly, one needs to be able to compare data sets obtained on different platforms in different laboratories. Several gene expression microarray platforms are available commercially (31). Because of the differences in microarray design, probe length and sequence, normalization methods, and analytic software, one would expect variation in transcript measurement in data sets obtained on these platforms. Generally speaking, comparisons between microarray data sets from different platforms can be made in several dimensions. First, one can compare the lists of genes detected by each platform. Obviously, microarrays with broader genome coverage have the potential to detect more transcripts in a sample. However, the sensitivity of a microarray would be the limiting factor for detecting low-abundance transcripts. Therefore, between two array types with equivalent genome coverage, the more sensitive array is likely to produce a longer list of expressed genes for the same sample. Second, one can compare the lists of differentially expressed genes for the same pair of samples obtained on two different microarray platforms. In this case, any difference would reflect the accuracy and the dynamic range of the microarrays. Two array types may detect a gene as expressed in both samples in a pair, but one array type may detect a difference between its transcript concentrations in the two samples and the other may not. This may happen, for example, if the variability between the signals for different probes for the same gene is high relative to the difference between the average probe intensities for each gene. In this case, the analytical software will likely call the gene in both samples as expressed (as the probes all show intensities higher than the background), but will not detect the difference between its expression levels in the two samples (as the noise level in measurements is too high relative to the true difference between the transcript concentrations). Finally, one can compare the fold changes for each differentially expressed gene obtained with two different platforms. Because of various technical factors, different arrays may detect a gene as differentially expressed but misrepresent the fold change. In particular, compression of gene expression ratios is a known problem in microarray analysis of differential gene expression (30).

As the issue of microarray cross-platform correlation was addressed by various laboratories, awareness grew of potential measurement inconsistencies. A group of U.S. National Institutes of Health (NIH) scientists profiled the same cell line with microarrays from three major commercially available platforms and used unsupervised clustering and principal component analysis (PCA) to determine the correlation between the data sets (32). The results suggested that the largest source of variation between the data sets is attributable to the differences contributed by the platforms themselves. The Pearson linear correlation for gene expression measurements across platforms was in the range of 0.48 to 0.60, indicating significant differences between the datasets. Differential gene expression between two samples was measured with the same platforms. Dichotomous classification of the genes for differential expression with a Bonferroni-corrected alpha yielded practically no overlap between the lists of differentially expressed genes. This is a particularly worrisome finding because determination of changes in gene expression is the most common application of microarrays in drug discovery applications. Under less stringent cutoffs, a larger overlap was detected, but the majority of differentially expressed genes were still unique to each technology. The authors concluded that continued refinement of microarray technology and careful verification of microarray data are required to enable reliable gene expression measurements.

Other cross-platform studies have also detected substantial discordance between gene expression measurements (33, 34). However, a number of recently published studies claim a high degree of cross-platform correlation. A comparison of three short oligonuicleotide microarray platforms revealed a high degree of overlap between the gene sets found to be differentially expressed (35). Generally speaking, the authors have found that the main source of variation between data sets is biological rather than technological. A recent comparative study of the Affymetrix and Illumina gene expression microarray platforms indicated that the two platforms generate highly comparable data, especially for genes predicted to be differentially expressed (36).

As expression microarray technology continues to mature, several possible solutions can be envisioned for the problem of cross-platform variation. First, for high-impact decision-critical studies (such as drug toxicity or genomic biomarker validation), the sponsors may be required to provide data sets obtained on more than one platform. For example, if a gene expression signature is claimed to be predictive of toxicity or drug response, the authors may be required to validate the signature on different microarray platforms, in addition to the proposed "diagnostic" platform to be used in the decision making process on the drug candidate. In this case, the transcript abundance measurements for the genes constituting the signature should lie within a predefined range for all microarray platforms. This validation can be done by independent microarray service providers specializing in certain microarray types, as most drug discovery organizations run a single microarray platform.

Another possible solution to the problem of discordance between microarray platforms is **incorporation of a standard measurement** into all gene expression microarray data sets published in scientific journals and submitted to public databases. The use of a standard RNA sample and a predefined normalization and analysis algorithm would thus provide a calibration measurement that can be used to assess the technical parameters of the microarray platform and determine whether comparison of data sets is appropriate. In May 2003, an organization called the External RNA Controls Consortium or ERCC was created to develop commonly agreed-upon and tested external controls for gene expression microarrays.

Scientists from over 50 biotechnology, pharmaceutical, clinical, and government and academic research organizations are participating in the Consortium with the goal of designing a universal industry standard. The specific goals of the Consortium are to:

1. Design a Certified Reference Material (CRM) that would contain clones that could be used to generate approximately 100 unique RNA sequences from several species (mouse, rat, *Drosophila*, etc.) not present in human databases.
2. Create a public repository of the selected clones and provide access to the repository to all interested gene expression microarray users.
3. Publish a standard protocol for preparing external RNA controls from the clones in the repository.
4. Provide all users with bioinformatics algorithms and programs to analyze the signals for the external RNA controls.

The Consortium has prepared a specification document for external RNA controls available at www.nist.gov. It is proposed that the external controls be introduced into the gene expression microarray protocol at the cDNA synthesis step, along with total RNA from the test samples. Thus, the external control RNA would go through the same target preparation and array hybridization procedures as the test samples and therefore would serve as a true calibration tool for the entire microarray protocol. At the time of preparation of this chapter, the Consortium is continuing to develop tools for evaluating microarray data. The active participation of key industrial and academic players provides the necessary support for this effort and raises substantial expectations for increased standardization and usability of microarray data generated on different technological platforms.

Another important effort in this area, a Microarray Quality Control Project (MAQC), was initiated in 2005. It is particularly relevant to the applications of microarrays in drug discovery because it is sponsored and led by researchers from the FDA. A detailed description of the mission, goals, and approaches of the Microarray Quality Control Project can be found on the following website: www.fda.gov/nctr/science/centers/toxicoinformatics/maqc/index.htm. The MAQC project involves six FDA Centers, all major providers of microarray platforms, and Environmental Protection Agency (EPA), National Institutes of Standards and Testing (NIST), and a number of academic laboratories. Its goal is to establish QC metrics and thresholds for objective evaluation of the performance achievable by various microarray platforms and assessment of the applicability of various microarray data analysis methods. The current approach of the MAQC project is to select two RNA samples for three species (human, rat, and mouse) and calibrate differential gene expression levels between the two samples with microarrays and other technologies (e.g., QRT-PCR). The resulting microarray data sets are used for assessing the precision and cross-platform/laboratory comparability of microarrays, while the QRT-PCR data sets enable evaluation of the nature and magnitude of any systematic biases that may exist between microarrays and QRT-PCR. The availability of the calibrated RNA samples combined with the resulting microarray and QRT-PCR data sets, which will be made readily accessible to the microarray community, will enable individual laboratories to identify and correct for experimental failures.

In summary, the genomics community now fully recognizes the potential consequences of variability in microarray data in the context of the increasing role of genomics data in drug discovery research and particularly in regulatory submissions. This has prompted the FDA and representatives of the industry to initiate a number of joint projects, which are likely to produce a common set of standards for microarray data evaluation and comparison. These efforts endorse an optimistic view of the developments in the microarray community in the next 5–10 years. We envision that within this time frame standardized calibration experiments will become a common practice, leading to more reproducible and accurate results. Researchers currently using custom spotted arrays will likely switch to commercial platforms; the number of commercial platforms will likely be reduced to two or three, reflecting the growing effort to optimize performance and standardize results. The spectrum of microarray applications will widen in terms of both information generated (gene copy number, epigenetic modifications, etc.) and the types of samples amenable to microarray analysis (archived tissues, very small samples, etc.).

2.2.6. Microarray Databases and Annotation of Microarray Data

The value of microarray data increases substantially with the increase in the size of the database and the number of relevant samples profiled. For example, identification of therapeutic targets, molecular classification of disease, compound assessment and optimization, and biomarker discovery are all examples of tasks facilitated by large depositories of microarray data.

2.2.6.1. Target Identification

Identification of genes consistently overexpressed in the target diseased tissue is a common and logical strategy in identification of therapeutic targets. If a database contains a critical number of gene expression profiles for the disease under study, the data sets can be mined to identify genes of potential relevance to the disease. Historically the first and the simplest method to be used for this purpose was unsupervised two-dimensional clustering. Figure 2.5 presents an example of a two-dimensional cluster, where each row represents a gene and each column represents a tumor sample. The data are subjected to unsupervised hierarchical agglomerative clustering to identify patterns within the data set (a detailed discussion of clustering algorithms and sample classification can be found in the data analysis section of this chapter). Clearly, by revealing clusters of upregulated genes, this analysis facilitates identification of genes overexpressed in multiple samples representing the same disease. A critical requirement for this type of analysis is a sufficient number of experiments in the database. Typically, 50–100 samples representing the disease of interest need to be considered to identify candidate targets by studying gene expression. One significant advantage of this type of analysis is that it permits identification of clusters of genes, i.e., patterns of coexpression, as opposed to single genes. This facilitates detection of intergene relationships and functional networks that characterize the diseased tissue. It is noteworthy, however, that mining microarray databases and selection of candidate targets *in silico* is only the first step, to be followed by multiple validation experiments to establish the functional involvement of the gene products in the disease process.

2.2.6.2. Disease Classification

Traditional histopathological classification of cancer does not reflect its genetic heterogeneity. Molecular classification, i.e., classification based on the genomic

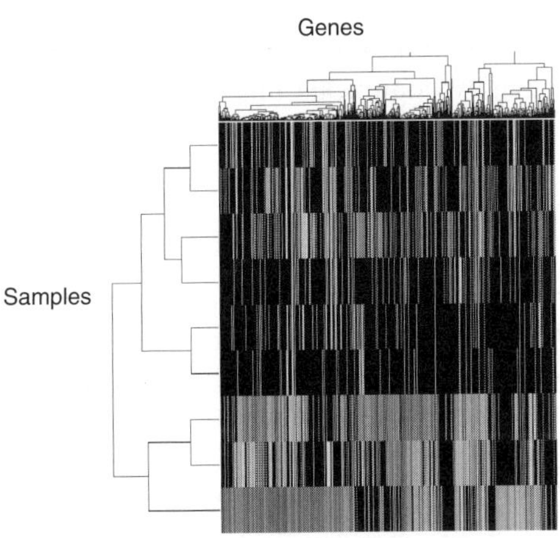

Figure 2.5 An example of a heat map obtained by hierarchical two-dimensional clustering of nine samples. Each row represents a sample, and each column represents a gene. Red color is used for upregulation and blue color is used for downregulation of genes, with black color reserved for unaffected genes. The dendrogram on the left illustrates the degree of relatedness between the expression profiles of the samples, and the dendrogram on the top reflects the similarity of the expression levels for each gene across the samples See color insert.

profiles, may identify patient subgroups likely to have the same outcome and treatment response and thus help optimize treatment decisions. Genomic classification of cancer has already facilitated patient stratification for treatment (37–39). This approach is particularly important for development of novel therapeutic agents. Molecular classification of disease can be facilitated by databases containing gene expression signatures of multiple patient samples. Gene expression profiles for a large number (usually >100) of patient samples representing multiple experiments from different laboratories can be downloaded from the database and subjected to one of the established classification procedures (reviewed in the data analysis subchapter). Again, the simplest and historically the first method to be used for disease classification was unsupervised two-dimensional clustering. In the case of cancer, various clustering procedures have revealed multiple subtypes of the disease based on their gene expression signatures (see Fig. 2.6 for an example of breast cancer classification (40)). Undoubtedly, large microarray databases will prove to be invaluable tools in molecular classification of disease, because the statistical power of such classification improves substantially with the number of samples profiled. In a typical cancer classification study, 50 to 150 patient samples are profiled (for examples, see (41–44)). Pooling multiple microarray data sets for the same disease would greatly improve the robustness of the classification, as each subtype would likely be represented by a larger number of cases. Even more importantly, a large database containing multiple data sets would facilitate validation of gene expression classifiers. Specifically, as a single microarray data set is subjected to unsupervised two-dimensional clustering to identify candidate molecular subtypes of the disease, these subtypes are characterized by subtype-specific gene expression signatures. However, since these signatures are generated from a single microarray study using a single population of patients (often from the same clinical site), there is a significant potential for bias (45). However, if additional microarray data sets are available in the database, the candidate gene expression signatures can be validated in an independent set of patient samples. This validation can be performed by exporting the signatures from the original experimental cluster and testing their predictive power in an independent data set. If the signatures can reliably predict the disease subtype in an unrelated data set, their validity is obviously much higher. However, interlaboratory and interplatform variability in microarray measurements, a widely recognized problem, will undoubtedly complicate this type of analysis. This problem was comprehensively addressed above in this chapter.

2.2.6.3. Compound Assessment

If a microarray database contains a sufficient number of gene expression signatures for a spectrum of compounds profiled in a relevant model system, it can be used to assess and optimize a therapeutic candidate, in terms of both efficacy and toxicity. This type of analysis is schematically illustrated in Fig. 2.7. Once a therapeutic candidate is profiled in a model system, its gene expression signature is compared with the signatures of known compounds in the database to

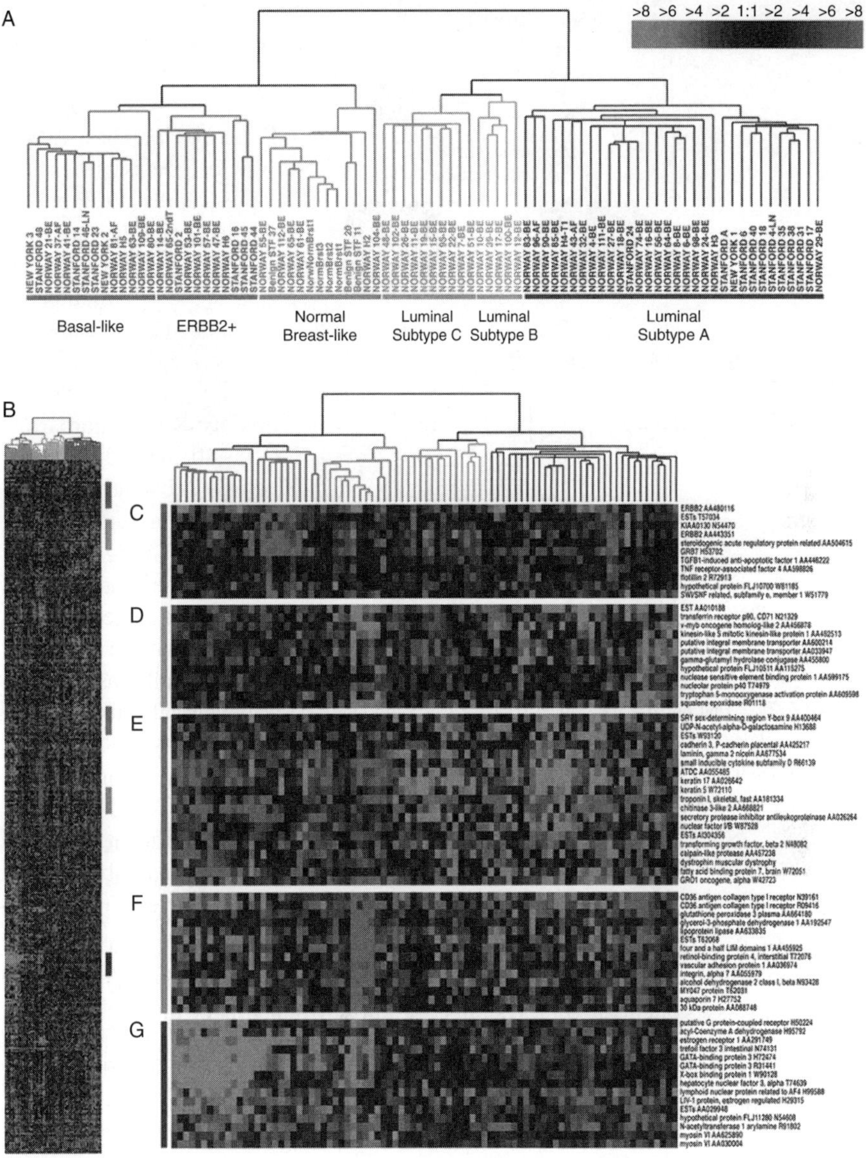

Figure 2.6 Gene expression patterns of 85 samples representing 78 breast carcinomas, three benign tumors, and four normal tissues, analyzed by hierarchical clustering. *A*) The tumor specimens were divided into five (or six) subtypes based on differences in gene expression. The cluster dendrogram showing the six subtypes of tumors are colored as luminal subtype A, dark blue; luminal subtype B, yellow; luminal subtype C, light blue; normal breast-like, green; basal-like, red; and ERBB2+, pink. *B*) The full cluster diagram scaled down. The colored bars on the right represent the inserts presented in *C*–*G*. *C*) *ERBB2* amplicon cluster. *D*) Novel unknown cluster. *E*) Basal epithelial cell-enriched cluster. *F*) Normal breast-like cluster. *G*) Luminal epithelial gene cluster containing *ER*. Figure is reproduced with permission from Sorlie et al. (2001), *Proc Natl Acad Sci USA* 98: 10869–10874 (ref. 40) See color insert.

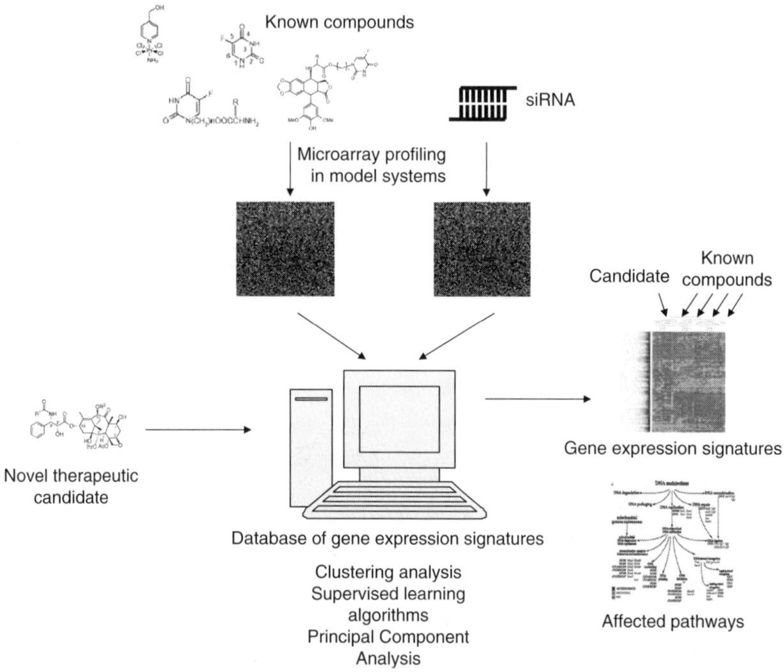

Figure 2.7 Creation of a genomics database for compound selection and optimization. Gene expression signatures of target inhibition and elimination are obtained with known inhibitors and siRNA, respectively, for multiple targets. Novel compounds synthesized to inhibit the target are profiled with the same microarray and their gene expression signatures are utilized to identify the affected pathways and targets. See color insert.

formulate hypotheses about its mechanism of efficacy or toxicity. For efficacy evaluation, the model system needs to be relevant to the target tissue (for example, lung cancer cell lines could serve as a model system for lung cancer), whereas for toxicity assessment, model systems are chosen that reflect the toxicological changes that occur in liver, kidney, or other organs. As discussed in Chapter 6, the most common in vitro toxicogenomics model system is hepatocytes. The application of microarray databases to predict and characterize compound toxicity is comprehensively analyzed in the chapters devoted to toxicogenomics. A critical requirement for the utility of a microarray database for compound assessment is a sufficient number of known compounds to cover a wide spectrum of on-target events or toxicological changes. Since compound assessment is a highly specialized task that requires accumulation of a significant number of known compounds and a substantial amount of chemigenomics expertise, specialized subscription databases are typically used for this purpose. Examples of such databases are reviewed in Chapter 4 of this book.

Substantial efforts have been made to create public microarray data repositories. For example, **Gene Expression Omnibus (GEO)** at the National

Center for Biotechnology Information (NCBI), the largest public repository for high-throughput gene expression data, contains over 180,000 samples (as of late 2007). The GEO database is available to the scientific community at the following URL: http://www.ncbi.nlm.nih.gov/geo.

The principal architecture of the GEO database is based on a three-level organization of the data sets: by platform, sample, and series (46, 47). The platform contains a description of the probes (oligonucleotides, cDNA, etc.) used to interrogate the samples. The sample provides information on the transcript abundance for each gene in each experimental sample, while referencing the platform used in the experiment. The series summarizes all the information on the experiment (samples profiled, experimental conditions, treatments, time courses, etc.) and often contains summaries and tables representing the analysis performed by the submitter. Typically, these summaries identify the genes significantly regulated in the experiment.

The data in GEO can be queried with two NCBI Entrez databases, Entrez GEO-DataSets and Entrez GEO-Profiles. Entrez GEO-DataSets presents an experiment-centric view of the data in GEO. Experiments or data sets of interest can be found by searching for attributes such as text keywords, organism studied, microarray platform, author of the publication, and experimental variable information. For example, searching the data sets for "NCI60" produces an entry related to the NCI60 panel of cancer cell lines. When a relevant DataSet is identified, the experiment can be further queried for gene expression profiles of interest by using the supplementary tools provided on the DataSet record.

Entrez GEO-Profiles provides a gene-centric view of the data in the database. Gene expression profiles of interest can be located by running a search for such categories as gene name, GenBank accession number, GEO accession number, and gene description. For example, a search for "Bcl-2" produces a list of entries for multiple gene expression data sets where the Bcl-2 gene was interrogated.

Several useful tools are available to manipulate the gene expression profiles in the database. For example, a selection of hierarchical and K-means clustering algorithms are provided. Two-dimensional clusters can be downloaded, plotted as line charts, or linked directly to Entrez GEO-Profiles. The *Query subset A vs. B* tool can be used to identify genes that display marked differences in expression level between two specified sets of samples within a DataSet, as calculated with t-tests or fold differences. Genes that meet the user-defined criteria are presented in Entrez GEO-Profiles. The *Subset effects* tool retrieves all gene expression profiles that are marked as displaying significant effects with respect to a specific experimental variable, for example, "cancer type" or "cell line."

With tens of thousands of microarray experiments and specialized data analysis tools, GEO is well-suited for addressing such complex problems as gene-gene and compound-compound interactions, analysis of pathways activated by a compound, and comparison of the effects of different compounds on a pharmacologically relevant system. Therefore, even with the emergence of multiple subscription databases, this public repository remains an extremely valuable tool for drug discovery scientists.

Another commonly used microarray database that may be of interest for drug discovery scientists is the **ArrayExpress database** run by the European Bioinformatics Institute. The database can be accessed through the following Internet address: http://www.ebi.ac.uk/microarray.

The ArrayExpress suite of databases and applications consists of the following (48):

- MIAMExpress, a web-based tool for microarray data submission
- ArrayExpress data repository that permits open and password-protected access to the submitted data
- A query-optimized data warehouse containing a curated subset of normalized data
- Expression Profiler, an integrated online visualization and analysis tool for microarray data

All the software in the ArrayExpress suite is open-source. The highest level of organization in the ArrayExpress database is the "Experiment," which may include one or more microarray hybridizations. Usually experiments are linked to publications. The ArrayExpress query interface enables searching for specific experiments, protocols, and array designs by various attributes, such as species, authors, or array platforms. When an experiment has been selected, the data set can be downloaded locally for further analysis. Alternatively, the data can be visualized online and analyzed with the function called Expression Profiler. Password-protected access to prepublication data is provided for submitters and reviewers.

The ArrayExpress data warehouse based on the BioMart technology (49) enables queries for genes based on such attributes as gene names, Gene Ontology (GO) annotations, or sample properties. Gene expression values can be retrieved and visualized for multiple experiments. For example, running a search for sample property "lung cancer" and for the gene name "Bcl-w" would retrieve all the experiments that profiled samples related to lung cancer and used microarrays that interrogate the bcl-w gene. The data for the genes retrieved can be visualized with line plots across different experiments. Data can be selected and labeled for further analysis. The pages presenting the data provide links to data annotation and supporting raw data supplied by the submitting authors.

Several other public microarray repositories are available to the genomics community (Stanford and Yale databases, available at http://genome-www5.stanford.edu and http://www.med.yale.edu/microarray, respectively) that contain unique data not available in other databases. Additionally, a number of databases tailored to specific therapeutic areas are available on a subscription basis. A standout in the oncology area is the Oncomine database, available at http://www.oncomine.org. It contains over 20,000 individual microarray profiles for 40 cancer types (as of late 2007). Oncomine is built similarly to the databases described above, with the two key modules used to search for data: "GENE" and "STUDY." A distinguishing feature of the Oncomine database is its special focus on therapeutic targets. It provides a platform to examine the expression of all known therapeutic targets in various cancers. The spectrum of targets is not limited to those in cancer, because the authors hypothesized that their database might reveal novel drug target/cancer type associations, possibly implying novel applications of currently used therapeutic agents. A set of 148 known drug targets and their respective drugs was compiled by querying the Therapeutic Target Database (50) and by automated PubMed searches. Sixty-five of these targets were found to be significantly overexpressed in at least one differential expression analysis. Within the *STUDY* module, the user can apply the therapeutic target filter to identify the targets most overexpressed in a particular differential expression analysis.

A number of other specialized gene expression databases are available on a subscription basis, for example, ToxExpress® (Gene Logic) or DrugMatrix™ (Iconix Pharmaceuticals). These databases are focused on a particular task (such as prediction of toxicity of compounds), contain abundant descriptions of the compounds, treatments, and species profiles, and therefore are well-suited for drug discovery applications. Chapter 4 discusses toxicogenomics databases in more detail.

Many scientific journals now require that authors using microarrays in their studies submit raw microarray data as supplementary information or deposit the data into a public repository, such as the databases described above. Many leading genomics laboratories post raw microarray data on their websites. Regrettably, however, the available microarray data sets are published in different formats and are supplemented with only insufficient array annotation information. This often complicates or prevents data comparisons and pooling of data sets.

An important prerequisite for exchanging microarray data and creating public microarray databases is the **development of a common format and common content for data submissions**. The most successful attempt in this area has been the project titled **Minimum Information About a Microarray Experiment (MIAME)**, which was initiated in 2001 by a group of microarray researchers (51). Conceptually, any gene expression microarray data set can be represented by a model consisting of three parts: (i) gene expression matrix, in which each element contains information about the expression level of the respective gene in a particular sample, (ii) gene annotation, which can be represented by a link to EntrezGene, and (iii) sample annotation (Fig. 2.8). The latter part is the most problematic one, as there is no common format for presenting the information about a sample, a treatment, or a species involved. Not surprisingly, most publicly available gene expression microarray data sets lack sample annotation information sufficient to reproduce the experiment.

The MIAME document proposes a common standard for supplying the minimum required information about microarray data. The authors of the document suggest that the sample annotation should be sufficient to (i) interpret the experiment, (ii) permit comparisons to other experiments, and (iii) reproduce the experiment. At the same time the experimental annotation should be structured in such a way that automated querying and efficient mining would be

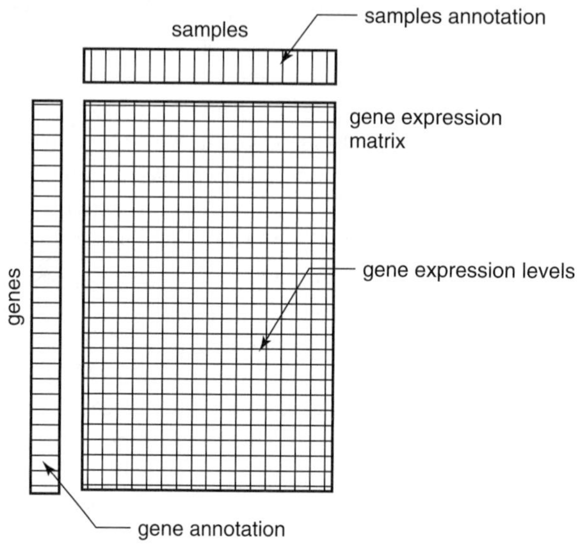

Figure 2.8 Conceptual view of gene expression data. The model has three parts: (i) gene annotation, which may be given as links to gene sequence databases, (ii) sample annotation, for which there currently are no public external databases (except the species taxonomy), and (iii) the gene expression matrix, in which each position contains information characterizing the expression of a particular gene in a particular sample. Reproduced with permission from Brazma, A. et al. (2001) *Nat Genet* 29(4): 365–371.

possible. Briefly, the minimum information about a published microarray experiment should contain six main parts (51):

1. Description of the experimental design (replication, treatments, time course, etc.)
2. Microarray design (probe length and sequence, array layout, slide type, surface, manufacturer, etc.)
3. Description of the samples (species, strain, nucleic acid isolation protocol, labeling protocol, etc.)
4. Hybridization protocol (quantity of nucleic acid hybridized, hybridization buffer, hybridization time, volume, and temperature, etc.)
5. Signal measurement (scanner, scanning protocol, signal quantification matrix, etc.). It was proposed that raw image files be included in this part. This is the most difficult part to control, because raw image files are very large and their storage would significantly increase the size of the database.
6. Normalization controls (housekeeping genes, spiking, or whole array normalization, normalization algorithm used, etc.)

In summary, while the current repositories and publically available individual data sets continue to vary in terms of experimental annotation and platform information, the continued effort in unification of the microarray information and the increased acceptance of the MIAME standards by the microarray community are likely to facilitate pooling of microarray data sets and thus enhance their value for drug discovery applications.

2.3. GENE EXPRESSION MICROARRAYS: DATA ANALYSIS

2.3.1. Identification of Significant Gene Expression Changes

Most microarray experiments involve either a comparison between a treatment and the baseline or a comparison between the test sample and a reference. Therefore, the first level of data analysis almost inevitably involves building gene expression ratios, i.e., calculating the ratios between the intensity values for the same gene from two different chips. A t-test is typically used to determine the significance of the difference between the control and the test values for each gene. The data can then be filtered to remove insignificant changes. Methods based on conventional t-tests provide the probability (P) that a difference in gene expression occurred by chance. It is common to set up a significance threshold at P value $\leq$0.01. Although $P\ value = 0.01$ is a reasonably stringent cutoff for experiments designed to evaluate small numbers of genes, a microarray experiment measuring the expression of 15,000 genes (such as an experiment using Affymetrix U133A arrays) would identify 150 genes as differentially regulated by chance.

To reduce the number of false positives, **significance analysis of microarrays (SAM)** (52) can be used. This method identifies genes with statistically significant changes in expression by assimilating a set of gene-specific t-tests. Each gene is assigned a score on the basis of its change in gene expression relative to the standard deviation of repeated measurements for that gene. Genes with scores greater than a threshold are deemed potentially significant. The percentage of such genes identified by chance is the false discovery rate (FDR). To estimate the FDR, nonsense genes are identified by analyzing permutations of the measurements. The threshold can be adjusted to identify smaller or larger sets of genes, and FDRs are calculated for each set. Other false discovery analysis methods have recently been introduced, some of which include analysis of false negatives (53, 54).

To improve the robustness of analysis, multiple replicates of the same sample are typically run. The commonly accepted minimum is two replicates; however, use of triplicates minimizes the false positive rate (55). A microarray experiment consists of multiple steps, and each step represents a potential source of variation. The **variation of the measured gene expression data** can be categorized into two generic sources: biological and technical variations. The biological variation in measured gene expression comes from different animals or different cell lines or tissues. It reflects the variability in gene expression between the different biological samples used in the experiment. Biological variation can be assessed only by using independent biological replicates. If all biological samples are pooled the biological variation is minimized, but the potentially useful information on the variability in gene expression between different animals or cells is lost. Technical variation accounts for the variation associated with the use of microarray techniques unrelated to the biological samples. The biological, technical, and residual variations are independent from each other. The variation in a measured intensity is the sum of these variations.

The contributions of the technical and biological variability into the overall variability have been extensively studied. It has been established that the biological variation is the main component of the variation between microarray experiments (56, 57). Therefore, biological replicates (multiple plates of cells, multiple animals, etc.) rather than technical replicates (multiple arrays run for the same sample) should be used whenever possible. Once the data are filtered with respect to the statistical significance, an additional filter is usually set up to remove genes with a small fold change, which are less likely to be biologically relevant. As the robustness of microarrays improved with time, the fold change threshold was lowered; it is now commonly set at 1.5.

Simple lists of genes regulated as a result of a biological process provide limited information. As applications of expression microarrays have widened and the numbers of genes analyzed have increased, analysis methods have become more and more complex.

2.3.2. Sample Classification and Class Prediction with Expression Microarrays

One of the most common tasks in microarray data analysis is identification of common patterns of gene regulation in a population of samples. An example of such a task would be identification of genes coinduced in a series of treatments or discovery of genes associated with a particular biological characteristic of the samples (disease category, tissue type, etc.). Problems of this type are commonly solved by two-dimensional clustering, a statistical procedure whereby samples (each represented by one or more microarrays) are aggregated into clusters based on the similarity of their expression "signatures," while the genes are simultaneously clustered based on the similarity of their expression levels across the samples. The rationale behind clustering samples according to their expression profiles is simple: Samples with similar gene expression "signatures" are more likely to have common biological characteristics. Similarly, genes coregulated in a series of samples are more likely to be part of a common biological pathway activated in the samples under consideration. Thus two-dimensional clustering may provide very useful information on the degree of relatedness between samples and reveals the genes potentially relevant to the classification. Clustering results can be conveniently visualized with a gene expression matrix, or a heatmap, in which each column represents an experiment and each row represents a gene (Fig. 2.5 in the previous subsection of this chapter). Each element of the heatmap is colored based on the expression level, thus providing a convenient visual representation of the gene expression patterns across all the experiments. One of the most notable applications of clustering is in genomics-based cancer classification, which was pioneered in the late 1990s (for examples see refs. (44, 58–60)).

If clustering is done without any a priori introduced sample classification, it is referred to as **unsupervised clustering**. Because of its unbiased nature, unsupervised clustering is often used to identify patterns in previously unclassified complex data sets. Several unsupervised clustering algorithms are used for microarray data analysis (reviewed in (61, 62)). Hierarchical clustering is the most common algorithm. It uses an agglomerative approach, whereby expression profiles are successively joined to form groups based on similarity between them, thus forming a hierarchical tree, or a dendrogram (5). The latter presents a convenient visualization option and is often presented together with a heatmap (Fig. 2.5 in the previous subsection of this chapter). An alternative algorithm is k-means clustering, a divisive approach based on partitioning the data set into a predefined number (k) of clusters (63). Obviously, this requires some a priori knowledge of the biology of the data set so that the number of clusters could be preset. When the researcher can specify in advance not only the number of clusters, but also the relationships between them, self-organizing maps (SOMs) can be used, which organize the clusters into a "map" in which similar clusters are close to each other (64).

Unsupervised algorithms can find novel patterns in data sets, but they are not designed to classify data according to known classes. In contrast, **supervised clustering** approaches, such as support vector machines (SVMs) (65), take known classes and create rules for assigning genes or experiments into these classes. The user initially runs microarrays for a training set with known class labels and enters the gene expression profiles together with the classification information into the algorithm. This "trains" the algorithm, or teaches it to associate certain gene expression patterns with the predefined sample class labels. The next step is to profile samples from a new set of samples, the test set, and input the gene expression data into the algorithm. The latter will then classify the samples using the knowledge on class—expression pattern associations learned from the training set. SVMs have been used to identify genes with similar expression patterns, but their most powerful application is in classification of samples. They have been extensively used in cancer classification and in some cases proved to be more reliable than the traditional diagnostic methods (44, 66).

2.3.3. Pathway Analysis with Gene Expression Microarrays

At the early stages of adoption of the microarray technologies, the most common approach was to focus on the genes showing the greatest difference between two samples under study (between diseased and normal tissue or between treated and untreated cells, etc.). This approach, however, has a limited value in drug discovery for the following reasons. First, disease-related intracellular processes often affect the expression of many genes in a coordinated fashion. However, the changes in the expression levels of each individual gene may not be significant. In many cases, they will not meet the threshold of significance set by the researcher (typically 1.5 to 2-fold) and therefore will be overlooked. However, if these changes are mapped to intracellular signaling pathways, they may reveal significant perturbations in the cellular state and point at important pathways. Second, the problem of interlaboratory and interplatform variability complicates comparison and combined analysis of data sets for the same disease produced in different laboratories. However, conversion of gene expression data into pathway activation data may facilitate data comparison by increasing focus on the biologically relevant changes and removing noise. Finally, if stringent cutoffs using multiple hypothesis testing are used, in many cases no individual genes will meet the threshold of statistical significance and the relevant biological processes will thus be masked by the noise.

The gene expression pattern of a cell reflects its phenotype and may provide information on the intracellular signaling pathways functioning in the cell. In in vitro experiments, studying the gene induction patterns caused by a treatment may help identify the pathways activated or repressed by the treatment. A prerequisite for all these applications is the ability of the researcher to map gene expression profiles to signaling pathways, i.e., to identify the associations between the affected genes and the known pathways.

There are several programs that facilitate association of gene expression patterns with predefined biological classes (67–75). They use one of the existing gene classification systems, such as Gene Ontology (GO) (76, 77), Biocarta (78), or KEGG (79), to determine the enrichment of an expression signature in a certain motif, such as "cell cycle control," "DNA biosynthesis," etc. Gene Ontology is the most commonly used annotation system, which classifies a significant fraction of the genome (~15,000 genes) according to their involvement in a biological process or molecular function or their cellular localization. It is built hierarchically and involves a parent-child relationship between its terms. Programs such as MappFinder allow the researcher to identify the GO terms that show correlated gene expression changes in a microarray experiment. The affected GO terms can then be rank-ordered based on the Z-score, a statistic that reflects the number of genes in the term meeting the criteria for fold change in the microarray experiment (67). MappFinder was one of the first GO-based programs designed for analysis of gene expression data. It has since been used to study the effects of various factors on intracellular pathways in vitro (80, 81) and in vivo (82). Combined use of siRNA-mediated gene silencing and MappFinder analysis of expression signatures has been suggested as an approach to pathway profiling (80). Other pathway analysis programs have been developed in the past several years, which allow convenient visualization of the pathway analysis results (70–75). One disadvantage of Gene Ontology-based analysis is that most genes are represented in multiple nodes and that many Gene Function branches overlap in function, creating redundancy and complicating the interpretation of the results.

An alternative approach to pathway analysis of microarray data is to use manually curated pathways instead of Gene Ontology. This approach often permits greater focus and lower redundancy, especially when studying specific disease-related pathways. A method called **Gene Set Enrichment Analysis**

(GSEA) was developed in 2005 to examine microarray data at the level of gene sets (83). Gene sets are created based on common function, common patterns of regulation in previous experiments, close chromosomal location, or common association with published biological pathways. The method analyzes gene expression signatures from samples belonging to two classes, labeled 1 or 2. Genes are ranked based on the correlation between their expression and the class distinction by using any suitable metric. A predefined gene set S is created, which contains genes with common characteristics. The GSEA method determines whether genes from the set S are found at the top or bottom of the ranked list L derived from the experiment. This would mean that the group S is affected in the experiment (the pathway is activated or suppressed, if S represents a pathway). Initially, an enrichment score (ES) is calculated, which reflects the degree to which a gene set S is overrepresented at the top or bottom of the ranked experimental list L. The score is calculated by walking down the list L, increasing a running-sum statistic when a gene in S is encountered, and decreasing it when genes not in S are encountered. The magnitude of the increment depends on the correlation of the gene with the phenotype. To determine the statistical significance (P value) of the ES, the phenotype labels are permuted and the ES of the gene set is recomputed for the permuted data to generate a null distribution for the ES. The empirical, nominal P value of the observed ES is then calculated relative to this null distribution (83).

The GSEA method has been applied to identify pathways activated in skeletal muscle of diabetes patients (84). DNA microarrays were used to profile expression of over 22,000 genes in skeletal muscle biopsy samples from 43 age-matched males [17 with normal glucose tolerance (NGT), 8 with impaired glucose tolerance (IGT), and 18 with type 2 diabetes]. When assessed with traditional methods that take into account the multiple comparisons implicit in microarray analysis, no individual genes had a significant difference in expression between the patient groups. The GSEA method was applied to the microarray data, using 149 predefined gene sets. Of these gene sets, 113 had been grouped according to their involvement in metabolic pathways (derived from public or local curation) and 36 consisted of gene clusters that are coregulated in a mouse expression atlas of 46 tissues. The gene sets were selected without regard to the results of the microarray data from the affected individuals. The GSEA procedure yielded the maximal ES score for a manually curated gene set corresponding to oxidative phosphorylation pathways. Importantly, although for each individual gene in the set the decrease in expression was only ~20% (below the threshold of significance for a typical microarray study), the decrease was consistent across the set: 89% (94 of 106) of the genes were underexpressed in patients with type 2 diabetes relative to those with NGT. This study clearly demonstrates the advantage of systematic analysis of microarray data in the context of carefully defined pathways, because this type of analysis enables the identification of significant changes at the level of biological processes when no significant changes are apparent at the level of individual genes.

The GSEA method has also been applied to characterize newly identified subtypes of diffuse large B-cell lymphoma (85). To address the difficult question of biological heterogeneity of the disease, samples from 176 patients were profiled on expression microarrays. The resulting gene expression signatures were subjected to multiple clustering methods and comprehensive genetic analyses to identify discrete subsets of tumors. Three different clustering algorithms [hierarchical clustering (HC), self-organizing maps (SOMs), and probabilistic clustering (PC)] were used, and the top 5% of genes with the highest reproducibility across duplicate samples and largest variation across patient tumors were considered. With all three clustering algorithms, the most robust substructure included three discrete clusters. To interpret the clustering pattern and differentiate the molecular profiles based on biological pathways, the GSEA method was used. The same predefined gene sets were fed into the algorithm to determine which pathways are activated in the three putative subtypes of diffuse large B-cell lymphoma. A total of 281 gene sets were used from four independent sources: (i) Biocarta (an Internet resource that includes 169 biological pathways involved in adhesion, apoptosis, cell activation, cell cycle regulation, cell signaling, cytokines/chemokines, developmental biology, hematopoiesis, immunology, metabolism, and neuroscience; available at www.biocarta.com); (ii) GenMAPP (Gene MicroArray Pathway Profiler, a set of web-accessible pathways and gene families including 45 gene sets involved in metabolic and cell signaling processes; available at www.genmapp.org); (iii) 64 manually curated pathways involved in mitochondrial function and metabolism that are coregulated in normal murine tissues; and (iv) three previously described gene sets coregulated in diffuse large B-cell lymphoma (86).

The first putative subtype of the disease was characterized by upregulation of genes involved in oxidative phosphorylation, mitochondrial function, and the electron transport chain. Comprehensive analysis of this cluster revealed upregulation of the genes comprising the nicotinamide adenine dinucleotide dehydrogenase (NADH) complex and cytochrome c/cytochrome c oxidase (COX) complex. Importantly, the tumors of this subtype had an increased expression of the *Bcl-2* family member, BFL-1/A1, a protein with known antiapoptotic functions. These findings enabled by GSEA analysis are consistent with the function of Bcl-2 members in regulation of mitochondrial membrane potential and cytochrome *c* release.

The second subtype revealed increased expression of genes involved in cell cycle regulation, including members of the CDK2 and MCM (minichromosome maintenance-deficient) families. This DLBCL cluster was also characterized by upregulation of several members of the B-cell receptor (BCR) signaling cascade (CD19, Ig, CD79a, BLK, SYK, PLC-2, and MAP4K) and pro-proliferation transcription factors (MYC, PAX5, OBF-1, E2A, *BCL6*, and STAT6). In contrast to the first two clusters, the third putative subtype of diffuse large B-cell lymphoma revealed a signature that was mapped by GSEA to T-cell–mediated immune responses and the classical complement pathway. The tumors of the third subtype also had relatively higher levels of coregulated mediators of inflammation and components of the connective tissue. Thus GSEA enabled identification of a subtype that is characterized primarily by the host response rather than the tumor itself. This study represents an elegant application of GSEA to characterization of novel disease subtypes revealed by unsupervised clustering-based microarray analysis methods.

Thus the GSEA method has been established as a powerful technique for converting gene expression profiling data into higher-order structures, such as biological functions and intracellular signaling pathways. This makes the method particularly valuable for drug discovery applications, such as compound characterization and optimization, early assessment of toxicity, as well as biomarker discovery. One can anticipate that studies such as the lymphoma profiling and classification work described above will become an integral part of the translational biology programs in pharmaceutical research and development. Genomic characterization of the disease of interest, followed by biomarker discovery in the respective groups and patient stratification based on genomic markers, will produce molecular data that will be fed back into early in vitro discovery process to drive compound selection and optimize the in vitro disease models.

A distinct approach has recently been developed that makes use of compendia of gene expression signatures generated by *artificial* activation of certain intracellular pathways. A pioneering study in this area (87) involved analysis of regulatory pathways controlled by the following genes: *HRAS* (Harvey rat sarcoma viral oncogene homolog), *MYC* (myelocytomatosis viral oncogene homolog) and *E2F1*, *E2F2*, and *E2F3* (encoding E2F transcription factors 1, 2 and 3, respectively). Recombinant adenoviruses were used to express Ras, Myc, or E2F proteins in quiescent primary mouse embryo fibroblasts. The resulting transfectants were profiled on gene expression microarrays to generate signatures of Ras, Myc, and E2F activation. The data were used to derive so-called "metagenes" or linear combinations of individual gene expression values. These metagenes were then validated by testing their ability to predict the activation of their respective pathways in a normal, physiological setting. Mouse embryonic fibroblasts (MEFs) were stimulated to proliferate by serum stimulation and profiled on gene expression microarrays at different time points after serum addition. Both the Ras and Myc metagenes predicted activity at the early time points, consistent with the known kinetics of accumulation of Myc and Ras activity. In contrast, the E2F metagenes predicted activity at much later time points (15–20 hours), consistent with the timing for accumulation of E2F activity. Most importantly, the metagenes were used to predict the Myc and Ras state in a series of mammary tumors that developed in transgenic mice expressing either Myc or Ras from the MMTV enhancer. Normal mammary tissue samples as well as mammary tumors induced by expression of the oncogene *ERBB2* (*NEU* or *HER-2*) were used as controls. The metagenes accurately predicted each of the Ras tumors and separated them from the controls, but did not distinguish between the Ras tumors and the Myc tumors. The fact that the Myc tumors are predicted by both the Myc and the Ras metagenes is in agreement with the previously demonstrated activation of Ras in both sets of tumors and demonstrates the potential of the metagene analysis approach in microarray-based pathway analysis.

A comprehensive study that followed in 2006 sought to establish **microarray-based pathway signatures** as a guide for development of targeted therapies (88). Primary human mammary epithelial cell cultures (HMECs) were used to develop a series of "pathway signatures." Individual oncogenic pathways were activated in these cells by transfection with recombinant adenoviruses, followed by microarray-based gene expression profiling and application of supervised classification methods to the resulting gene expression signatures. This analysis identified a set of genes whose expression was most highly correlated with the oncogene-activated status (relative to the green fluorescent protein, or GFP, control). Similar to the previous study, this analysis identified the dominant principal components from these sets of genes to define corresponding phenotype-related metagenes.

The predictive capacity of these microarray-derived pathway signatures was validated in a number of mouse cancer models. The human metagenes for Myc, Ras, and E2F3 were converted into mouse metagenes by keeping the genes common to both human and mouse data and eliminating the uniquely human genes. The pathway predictions made with the metagenes correlated closely with the molecular basis for tumor induction. For example, mouse (MMTV)-*MYC* tumors revealed the highest probability of Myc pathway deregulation, while Rb-null tumors showed the highest degree of E2F3 deregulation. The probability of Ras pathway upregulation was the highest in the MMTV-*HRAS* mice. In addition to predicting the status of individual pathways, this approach is capable of identifying the patterns of pathway deregulation (88), thus providing a framework for comprehensive analysis of human tumors and model systems for drug screening. With hierarchical clustering of pathways, human adenocarcinomas were distinguished from squamous cell carcinomas based primarily on their Ras pathway status. Additionally, coordinated deregulation of Ras, β-catenin, Src, and Myc pathways defined a population of patients with poor survival (a median survival of 19.7 months vs. 51.3 months for all other clusters). Thus the results of this study validate the concept of using oncogenic pathway signatures derived in model systems to predict the pathway status in tumors.

The real power of the pathway signature approach in drug discovery is in its ability to predict sensitivity of a model system (and potentially, a tumor) to therapeutic agents targeted to specific intracellular pathways. To investigate this possibility the authors of the approach coupled screening of candidate therapeutics in breast cancer cell lines with microarray-based analysis of pathway deregulation in these cells. The drug sensitivity testing was performed with standard colorimetric assays. The Ras pathway activity was correlated with the sensitivity of cells to two inhibitors of the Ras pathway, a farnesyl transferase inhibitor (L-744,832) and a farnesylthiosalicylic acid (FTS). The activity of the Src pathway was correlated with the targeted pathway inhibitor SU6656. In all three cases, a close correlation was observed between the probability of Ras and Src pathway deregulation based on the microarray-based signatures and the extent of growth inhibition by the corresponding agents. It is noteworthy that no correlation was found between the Ras pathway activity and sensitivity to the Src-targeting agent, and vice versa. These exciting findings lay the foundation of a new approach to stratification of cancers and prediction of response to therapeutic agents: microarray-based pathway profiling using predefined pathway signatures.

Prediction of the deregulation of various pathways in a diseased human tissue sample through microarray-based pathway analysis offers tremendous opportunities in patient stratification and biomarker identification. As new pathway-specific biomarkers are identified in clinical samples, they can be used to stratify the patient population for targeted therapeutics. The biomarkers can also be used to select appropriate model systems for drug screening that would match the pathway profiles of the target patient populations. In cell-based drug screening systems and murine disease models, microarray-based pathway profiling can be used to study the effects of the compound. Indeed, activation of various pathways by a therapeutic candidate may reveal the compound mechanism and thus aid in compound optimization in terms of efficacy. Additionally, if the new pathway analysis paradigm is adopted by the toxicology community and specific toxic pathway gene sets are generated and validated, the approach will aid in early assessment of compound toxicity. This is discussed in detail in Chapters 4, 5, and 6. Finally, and perhaps most importantly, if the concept of drug sensitivity prediction using pathway signatures is validated for multiple pathways, the method will offer a potential basis for guiding the use of targeted therapeutic agents. As this information is accumulated, the value of the approach will increase, as it will be used to guide combination therapies, i.e., select combination of drugs to target all pathways found to be activated in the disease subtype under consideration.

Given the large amounts of gene expression data accumulated in the literature, integrative analysis of multiple data sets related to the same disease represents a very attractive idea. The precedent for such analysis was established when so-called **meta-analysis** was performed for four different gene expression data sets for prostate cancer (89). The authors identified a molecular signature common to the data sets, thus generating a robust signature of the disease. The signature was then mapped to KEGG pathways (79) to reveal a common biological motif, activation of polyamine biosynthesis (89). Other studies identified common gene signatures in different breast and lung cancer data sets (90–92). The existence of common motifs in data sets from different laboratories despite the well-publicized problem of interplatform variability presents strong evidence in favor of microarrays as tools for identification of drug targets and biomarkers. As the amount of information derived from microarrays continues to increase, new and more complex data analysis procedures will emerge that will facilitate current and future applications of the technology.

2.3.4. Common Problems Affecting the Validity of Microarray Studies

Common issues in validating microarray results in a clinical biomarker setting have received a lot of attention in the recent years, as microarray technology took a central place in biomarker discovery. Several recent reviews (45, 93–96) have addressed common issues that affect the validity of microarray data in clinical research. One of the most common issues is **overfitting**. Overfitting occurs when a composite biomarker, such as a gene expression signature, is discovered to discriminate between two subtypes of disease, two groups of patients, etc., but it is designed to perfectly fit the initial data set used in its discovery. Because of the large number of predictors in a composite signature (i.e., individual genes whose expression is measured), "a perfect discriminator" can be designed for the initial data set purely by chance. Consequently, such a composite biomarker has no predictive value in any data sets other than the initial "training" data set used to construct the model. Testing for overfitting is not difficult: The composite biomarker needs to be evaluated in a separate, independent set of samples. The use of a separate validation data set is becoming common. The key consideration here is complete independence: The training data set cannot be included in the validation set. In the context of biomarker discovery, this presents a special problem, as samples of microarray quality are often hard to obtain. However, as microarray-based biomarkers and diagnostics become part of regulatory submissions, the problem of overfitting will need to be resolved by appropriate validation studies. The potential effects of this and other common data analysis problems in the context of regulatory submissions are analyzed in Section 2.8 of this chapter.

A separate and common problem in designing microarray studies is **bias**. Generally speaking, bias may be defined as an erroneous association of a characteristic with a group in a way that distorts a comparison with another group (45). In an example from the biomarker discovery field, two sample sets representing two subtypes of a cancer (A and B) are used to identify a genomic biomarker that would be able to distinguish between the two subtypes, but if the RNA samples from cancers A and B were prepared with different protocols, the study might be affected by bias. Indeed, different laboratories often use tissue fixation protocols that vary in their ability to conserve RNA. Hence, if the RNA from sample set B is more degraded than the RNA from sample set A, the resulting bias will lead to identification of a "biomarker" that will be associated with the RNA quality rather than with the cancer subtype. The bias resulting from inappropriate selection of initial samples is often referred to as **bias of inequality at baseline**. In clinical trials, this type of bias is always addressed by randomizing the initial patient populations. It is noteworthy that a demonstration of the reproducibility of microarray measurements may prevent overfitting as it would eliminate chance as the possible cause of discrimination between the sample groups, but it does not address the problem of bias. The problem of bias can be addressed by several important procedures: randomization, blinding of investigators, and uniform handling of specimens and data (45).

Generalizability, often referred to as "external validity," is a separate but related problem, which concerns the population to which a biomarker can be applied. The generalizability of a biomarker is determined by the selection of patients for the biomarker discovery study, in terms of their disease type, age, symptoms, and other parameters. A biomarker will have external validity if the training and validation data sets used in development are representative of the target population.

Today, the aforementioned problems are widely ignored by investigators involved in microarray research (45). In many cases, these problems could be avoided by simple and straightforward processes, such as ensuring baseline equality of characteristics of individuals and uniformity in sample collection, storage, and handling. As the conceptual frameworks are developed for assessing the validity of complex genomic biomarkers, such as gene expression signatures, it is important for the investigators engaged in microarray research to carefully consider possible sources of bias.

2.4. COMPARATIVE GENOMIC HYBRIDIZATION: TECHNOLOGY

Chromosomal aberrations are detrimental events associated with a number of developmental diseases and cancer. Amplifications and deletions of chromosomal regions occurring in somatic cells are believed to be one of the main factors leading to cancer. Although fluorescent in situ hybridization (FISH; for a recent review, see (97)) has been effectively applied to analyze known genetic aberrations for decades, until recently there was no method for detecting gene copy number alterations on a whole-genome scale. **Comparative genomic hybridization (CGH)**, a technique that enables genome-wide analysis of chromosomal aberrations, was first described by Kallioniemi and colleagues in 1992 (98). The method involves hybridization of the test DNA (sometimes mixed with reference DNA) to a complete representation of the genome attached to a solid support. Originally, CGH was performed on metaphase chromosome spreads, but in the past decade microarray-based CGH has become dominant (reviewed in (99, 100)). A glossary of terms related to gene copy number measurement is provided in Table 2.2.

BOX 2-1 *Detection of Copy Number Alterations by CGH*

Chromosomal Aberrations

Changes in the chromosome structure and number that may include exchange of material between chromosomes, gain or loss of restricted regions of the genome, as well as gain or loss of individual chromosomes or portions thereof and altered ploidy.

Chromosomal aberrations may be balanced (no gain or loss of genetic material) or unbalanced.

Copy Number Abnormalities

Gains or losses of chromosomal material ranging in size from whole chromosomes to fragments involving single genes or portions of genes. Copy number gains may result in just one extra copy or dozens of additional copies of the locus, while losses may include heterozygous or homozygous deletions. Amplified genomic material can be organized as repeated units at a single locus or as extrachromosomal DNA, or be scattered throughout the chromosome. Gene copy number alterations frequently result in changes in gene expression.

Copy Number Changes in Human Disease

Gains and losses of genomic loci in germ line DNA are associated with neurodevelopmental syndromes and multiple congenital abnormalities. Somatic gene copy number abnormalities are fundamental to oncogenesis. Gains of oncogenes and losses of tumor suppressor genes are well known to be early and fundamental events in the development of most cancers.

Copy Number Variation in Human Population

Large-scale copy number polymorphism has recently been discovered in humans. Copy number polymorphisms are microscopic and submicroscopic variants (deletions, duplications, insertions, and inversions) that cover millions of nucleotides within the human genome. They may be responsible for a significant portion of interindividual variability in humans.

Array-Based Comparative Genomic Hybridization

A high-throughput microarray technology for genome-wide detection of gene copy number abnormalities. Digested and labeled genomic DNA is hybridized to a representation of the genome (such as BACs or oligonucleotides) to obtain fluorescent signals indicative of the relative abundance of each genomic locus interrogated. The ratio of the signal for the test sample to that of the normal control (copy number of two) is used to determine the copy number of the locus in the test sample. The copy number profile is then mapped to the genome to identify genes located in the gained/lost regions.

Types of CGH Arrays

CGH arrays may carry BACs, cDNAs, or oligonucleotides. While most early CGH arrays were BAC-based, recently oligonucleotide arrays have become dominant. The main advantage of oligonucleotides is that, unlike BACs, they can be produced synthetically and deposited in a highly controlled fashion, thus facilitating quality control. The number of oligonucleotides and hence the number of loci interrogated is determined only by the maximum density of the microarray. Therefore, the genomic resolution of an oligonucleotide array is theoretically unlimited, while the resolution of BAC arrays is limited by the size of the clones.

The original array CGH protocols employed a two-color hybridization scheme, whereby the test DNA is labeled with a red fluorescent dye while the reference normal DNA is labeled green (Fig. 2.9). Although genomic DNA can be labeled and hybridized directly, some CGH protocols involve a PCR-based amplification step. Once the DNA is labeled, the test and the reference samples are mixed and hybridized to the array. Cot-1 DNA is typically added to suppress the hybridization of repetitive sequences. After hybridization, the array is washed and scanned to generate the red and green fluorescence intensity values for each probe on the array. The data are then normalized and presented as ratios of test to normal (usually on a log scale). An example of the CGH output for one chromosome in one sample is shown in Fig. 2.9. Ratios between the test and reference samples for multiple positions on a chromosome provide information on the copy number for each region measured. The copy number profile of a sample typically consists of a series of plateaus corresponding to regions with a constant copy number, flanked by abrupt transitions.

An important limitation on the use of CGH is that it can measure changes in copy number but it cannot detect certain chromosomal translocations or changes in ploidy. Indeed, many chromosomal translocations occur without gain or loss of chromosomal material (balanced translocations). Ploidy changes that result in a similar increase in the number of each chromosome (such as a chromosome number of 4n) cannot be detected by CGH because normalization during CGH is based on the total amount of DNA loaded on a chip, so an increase in ploidy will only result in a lower requirement for the number of cells required to run one chip and will not affect the signal. Balanced DNA translocations cannot be detected by CGH because they rearrange the chromosomes but do not affect the copy

Table 2.2 Glossary of Terms Related to Gene Copy Number Measurement*

Term	Definition
Bacterial artificial chromosome (BAC)	An artificially created chromosome in which medium-sized segments of foreign DNA (100,000 to 300,000 bases in length from another species) are cloned into bacteria. Once the foreign DNA has been cloned into the bacterial chromosome, many copies of it can be made and sequenced.
Fluorescent in situ hybridization (FISH)	A physical mapping approach that uses fluorescein tags to detect hybridization of probes with metaphase chromosomes and with the less condensed somatic interphase chromatin
Ploidy	The number of sets of chromosomes in a cell
Single nucleotide polymorphism (SNP)	A DNA sequence variation that occurs when a single nucleotide in the genome sequence is altered. Each individual has many single nucleotide polymorphisms that together create a unique DNA pattern for that person.
Heterozygous deletion	Loss of one copy of a DNA segment
Homozygous deletion	Loss of both copies of a DNA segment

*For general microarray terminology, see Table 2.1.

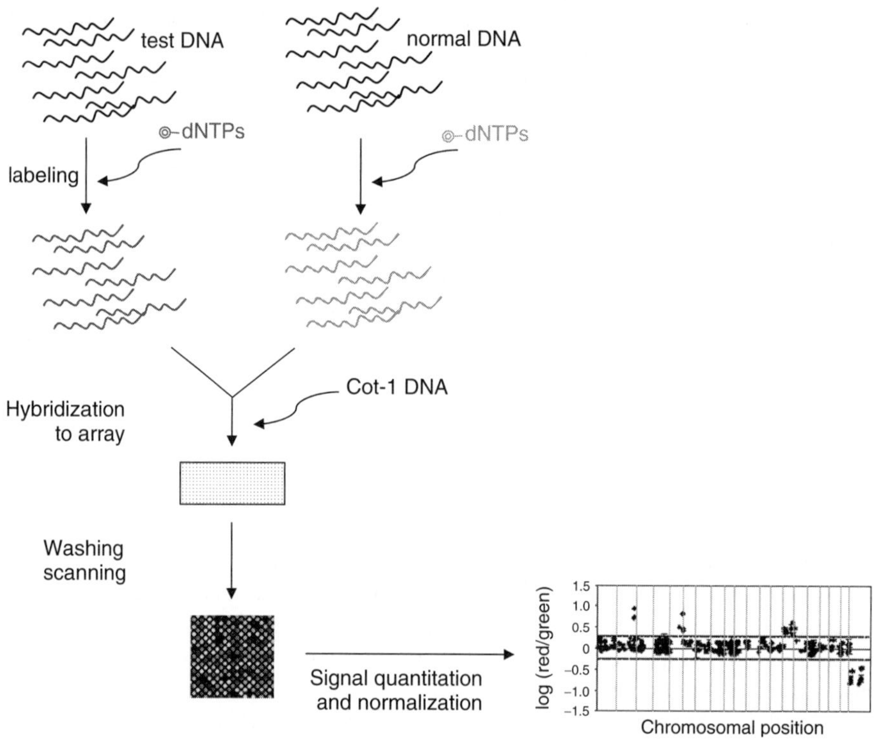

Figure 2.9 Two-color procedure for comparative genomic hybridization (CGH). The test gDNA and the reference normal gDNA are labeled with two different fluorophores. The gDNA samples are mixed and hybridized to a CGH array. Cot-1 DNA is added to eliminate the signal from repetitive sequences. After hybridization, the array is washed and scanned to generate signal intensities for all regions of interest. See color insert.

number of the sequences involved. However, unbalanced translocations, as well as all types of amplifications and deletions, are detectable because they result in a gain or loss of chromosomal material. Figure 2.10 illustrates possible types of chromosomal aberrations and indicates the abnormalities that can be detected by CGH.

The value of a CGH array increases with an improvement in genome coverage, resolution, and reproducibility. Several types of array platforms are currently used for CGH. Historically, the genome was represented on CGH arrays as a collection of bacterial artificial chromosomes (BACs). However, direct use of BACs as microarray probes is technically very difficult, because (*i*) BACs are single-copy vectors, (*ii*) the yield of DNA from BAC cultures is low, and (*iii*) high-molecular-weight DNA is difficult to spot at concentrations sufficient

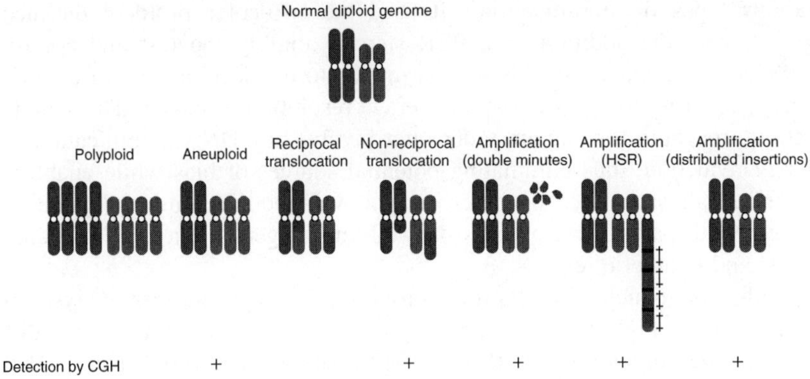

Figure 2.10 Types of chromosomal aberrations that can be detected by CGH. Reproduced with permission from Albertson et al. (2003) *Nat Genet* 34: 369–376 (ref. 100).

to obtain a good ratio of signal to noise in the hybridizations. Therefore, representations of BACs rather than BACs themselves were generated and spotted on an early CGH microarray (3, 101). Thousands of BACs were propagated and used as templates to generate PCR products, which were purified and deposited on a microarray. Highly reproducible measurements were obtained for a number of human cancer cell lines over a wide dynamic range (from homozygous deletions to high-level amplifications). The arrays provided sufficiently high sensitivity to detect single copy amplifications and deletions (3). The main drawback of the earlier BAC arrays was low resolution. Spotting 2460 BAC clones in triplicate provided an average resolution of 1.4 Mb across the genome (3). Continued efforts to increase the genomic resolution of BAC arrays resulted in the development in 2004 of a high-density BAC array that contained approximately 32,000 clones arranged in a tiling fashion and covering the entire genome (102). The array provided a significantly higher resolution and made possible detecting amplifications as small as 300 kb and deletions as small as 240 kb. Microarrays containing cDNAs have been extensively used for CGH (4, 103–105). The advantages of cDNA arrays include higher reproducibility, easier manufacturing, and better representation of the genome (100). However, multiple probes are required to detect small copy number changes, and more sample needs to be used (several micrograms), because of the lower sensitivity of the array (100).

A significant breakthrough was achieved in 2004 when two **oligonucleotide-based platforms** were developed for CGH (106, 107). One of these was a microarray containing 60-mer oligonucleotide probes synthesized in situ by an ink-jet technology (106). The array provided a significant improvement in resolution and was shown to reliably detect single-copy losses, homozygous deletions, and various types of amplifications. It used the two-color protocol outlined in Fig. 2.9, with the addition of a PCR step to amplify the test and control DNAs. Subsequent updates of this microarray led to the development of a 244K microarray, covering the genome at an average resolution of 6.4 kb. The sample preparation procedure for this array does not involve any DNA amplification or complexity reduction, thus eliminating potential sources of bias while enabling analysis of small samples. This microarray covers both coding and noncoding regions, with particular emphasis on well-known genes, microRNA-coding sequences, and telomeric regions.

The other oligonucleotide platform proposed for copy number analysis in 2004 represented a high-density microarray originally designed for detection of single nucleotide polymorphisms (SNPs) (107). The array covered over 10,000 SNPs distributed across the genome. Each SNP was interrogated by multiple 25-mers synthesized *in situ* by a photolithographic method. Unlike the two-color CGH protocol presented in Fig. 2.9, the SNP array protocol involves labeling DNA by incorporation of biotinylated dNTPs (Fig. 2.11). Briefly, the protocol starts with 250 ng of genomic DNA per array. Digestion with a restriction enzyme is used to reduce the complexity of DNA, followed by ligation of adapters to facilitate the subsequent PCR step. The complexity of DNA is further reduced by a PCR procedure optimized to yield fragments of a specified size range. The PCR products are purified to remove primers and nucleotides, fragmented, end-labeled, and hybridized to a microarray (108). After array hybridization, the array is stained with streptavidin-phycoerythrin and a bioinylated anti-streptavidin antibody. Signal intensities from individual SNP measurements are smoothed across a user-defined smoothing window with a specialized algorithm. The resulting values are then compared to a preloaded reference data set for normal DNA to produce an estimate of the copy number in the experimental sample. The array was used to evaluate chromosomal aberrations on a genome-wide scale in a number of cancer cells. It reliably detected chromosomal amplifications as well as homozygous and hemizygous deletions simultaneously with loss of heterozygosity (LOH) detection (107). The arrays produced results generally comparable with those obtained on BAC and cDNA arrays, but the authors reported a substantially lower noise level and a much higher resolution, averaging approximately 300 kb (107).

The next generation of the SNP microarray has an increased SNP coverage (approximately 114,000 SNPs), which corresponds to a resolution of <100 kb. The original manufacturer's software for data analysis smoothed the signals for all the SNPs and compared the data with the internal reference data set for over 100 individuals. New software programs for copy number analysis on 100K arrays enable paired analysis, i.e., estimation of the copy number in the test sample relative to a reference sample run in the same experiment, as well as unpaired analysis, which involves a preloaded set of reference samples. Additional smoothing options as well as more advanced analysis methodologies, such as Hidden Markov Models analysis, are also available. These are considered

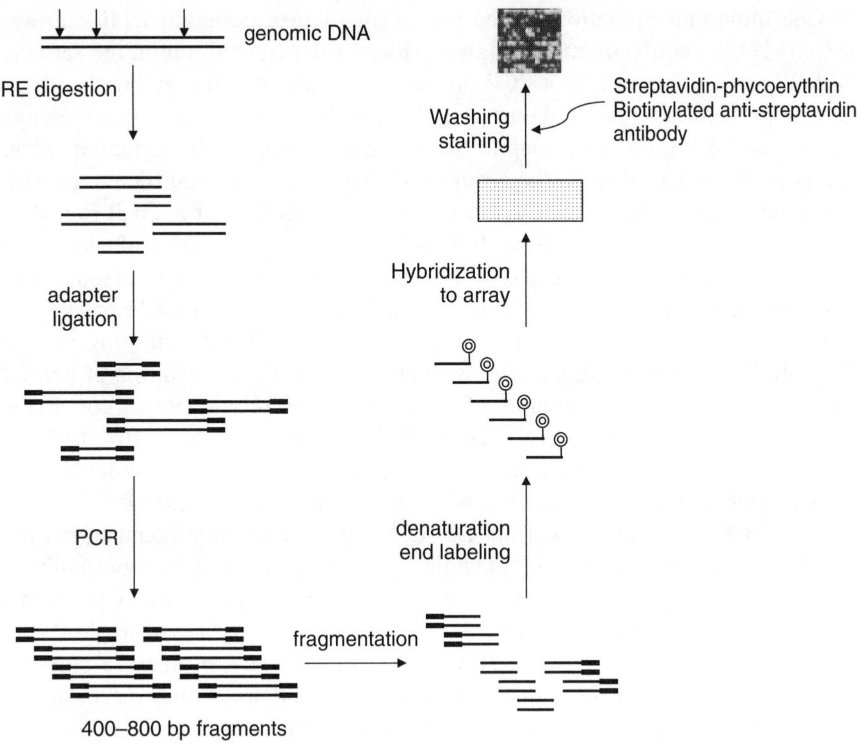

Figure 2.11 Gene copy number analysis using SNP genotyping microarrays. The genomic DNA sample is digested with a restriction endonuclease for complexity reduction. Adapters are ligated to the restriction fragments, and the fragments are amplified by PCR. The PCR products are denatured, labeled, and hybridized to a microrarray. The array is washed and stained to produce signal intensities for each probe. The data for multiple SNPs are smoothed and compared to an internal control to generate an estimate of the copy number for the chromosomal region.

in the subsection devoted to CGH bioinformatics. The output of the analysis is the absolute copy number for each SNP position on the chip. The new 100K microarray has been validated in a large study of 101 lung carcinoma samples (tumors and cell lines) (109) and subsequently in other studies. The increased resolution of the array enabled identification of several small amplifications and homozygous deletions that had not been previously detected by other CGH protocols (109). The trend for increased density of genome coverage continued with a 500K SNP genotyping array set introduced in 2006. The set consists of two microarrays, Nsp and Sty, that interrogate approximately 262,000 and 238,000 SNPs. The 500K array set utilizes the same protocol as the previous generations of SNP genotyping microarrays (depicted in Fig. 2.11).

One important limitation on the use of high-density oligonucleotide arrays for CGH is the **requirement for high purity of the target tissue** in the sample. The SNP array protocol uses a PCR amplification step, which may produce nonlinear amplification effects and thus diminish the ability of the procedure to detect deletions and even amplifications in heterogeneous samples. In particular, Zhao et al. (107) reported a drop in the accuracy of scoring homozygous deletions with decreasing purity of tumor DNA. For example, the deletion of a small region in a cell line was detectable only with ≥90% purity of tumor DNA. Deletion of another region was only detected at 100% purity of tumor DNA. In our laboratory, we have studied the effects of sample heterogeneity on CGH profiles by using tissues with varying tumor-to-normal ratio, as well as by titrating normal DNA into DNA from tumor samples with known aberrations. Addition of normal DNA results in a gradual flattening of the copy number profile. No abrupt effects were observed as the fraction of normal DNA was increased from 10% to 50%. While high-level amplifications were still present in all samples, low-level copy number gains and heterozygous deletions were essentially undetectable in samples with more than 30% of normal DNA. An increase in the amount of normal DNA had similar effects on short and long regions of chromosomal abnormalities. These findings suggest that CGH on oligonucleotide-based microarray platforms may still be a useful method for heterogeneous samples, as it will likely detect high-level copy number changes. However, if one intends to obtain a comprehensive genomic profile of a diseased tissue, it is important that the tissue be carefully dissected to remove the contaminating normal cells.

As novel CGH microarray platforms are developed and existing platforms are improved to increase genome coverage and improve sensitivity and other parameters, the choice of the platform becomes a matter of prioritizing the required performance characteristics for a particular project. For example, identification of genomic biomarkers in preclinical model systems imposes the requirements of high resolution and reproducibility on the CGH platform. Biomarker discovery in clinical samples requires that the array protocol be flexible with respect to the sample type (i.e., it should be amenable to processing of diseased samples containing normal tissue and be adaptable to analysis of formalin-fixed paraffin-embedded samples). Pharmacogenetic studies aimed at detecting copy number polymorphisms in germ line DNA rely on the ability of the CGH microarray to detect very subtle variation in copy number over short genomic regions (such as a single copy gain or loss for a region of 20–50 kb), emphasizing the sensitivity and high genomic resolution of the array. If allele-specific copy number needs to be measured, the array platform needs to be able to detect loss of heterozygosity in addition to measuring the overall copy number.

When resolution of a CGH microarray is considered, the main questions asked by the researcher are "What is the minimal length of an aberration that can be reliably detected?" and "How precisely can the boundaries of an aberration be defined?" It has been proposed that the performance of a CGH array be described by the probability of detecting any alteration of a given size (reviewed in ref. (110)). The probability of detecting a copy number alteration can be determined for different alteration sizes by calculating the percentage of alterations of a given size that fall between adjacent probes on the microarray. An algorithm named ResCalc has been designed to apply this methodology in evaluating different array platforms (110), and the results of this analysis for several widely used platforms have been reported. It is noteworthy that the optimal performance calculated theoretically based on the probe spacing is not attainable in practice, primarily because of experimental noise. Although high-level amplifications and homozygous deletions are readily detected in most cases given sufficient genomic coverage, single copy gains and losses are deemphasized in the presence of experimental noise. Additionally, high variability and noise complicate delineation of the borders of an aberration (Fig. 2.12A). One possible solution to reduce the noise is to probe the same locus with multiple probes and then estimate the copy number by pooling the signal from the different probes, while eliminating outliers. Various smoothing and segmentation algorithms have also been employed for this purpose (discussed in the CGH Bioinformatics subchapter).

Generally, when high genomic resolution is a priority, oligonucleotide-based CGH array platforms have a significant advantage over BAC-based microarrays. The resolution of BAC arrays is limited by the size of BAC clones, while oligonucleotides can theoretically be designed to interrogate any locus in the genome, with no limitation on the density of the coverage. Spotting BACs in a tiling pattern may help overcome this limitation. However, the deciding factor in drug discovery research is often the reproducibility of the technology. Oligonucleotide array platforms are fundamentally more reproducible than BAC-based ones, because oligonucleotides can be easily manufactured through chemical synthesis and precisely deposited on a solid support or synthesized directly on the array. To the contrary, BACs represent large clones, which need to be maintained and propagated biologically, complicating the quality control process.

Among the existing oligonucleotide platforms, arrays designed specifically for copy number analysis (such as the 244K CGH array by Agilent) have the advantage of more even probe coverage. They are typically gene-centric, as they are intentionally designed to interrogate gene sequences, as opposed to SNPs. Chromosomal regions that are known to be frequently amplified in cancer and loci that contain genes important in human disease typically receive denser probe coverage. At the same time, SNP genotyping arrays have the advantage of capturing loss of heterozygosity events. Therefore, they should be used when determination of the allele-specific copy number is important, for example, in some pharmacogenomic studies. The use of genotyping microarrays for copy number analysis has limitations, however. For example, SNPs at a primer locus or a restriction site would affect the signal intensity, resulting in incorrect readings of the copy number at the locus. These artifacts would be individual-specific, because they are caused by individual variants.

Oligonucleotide–based protocols typically require less genomic DNA (as little as 250 ng) and are potentially less sensitive to the degradation of genomic

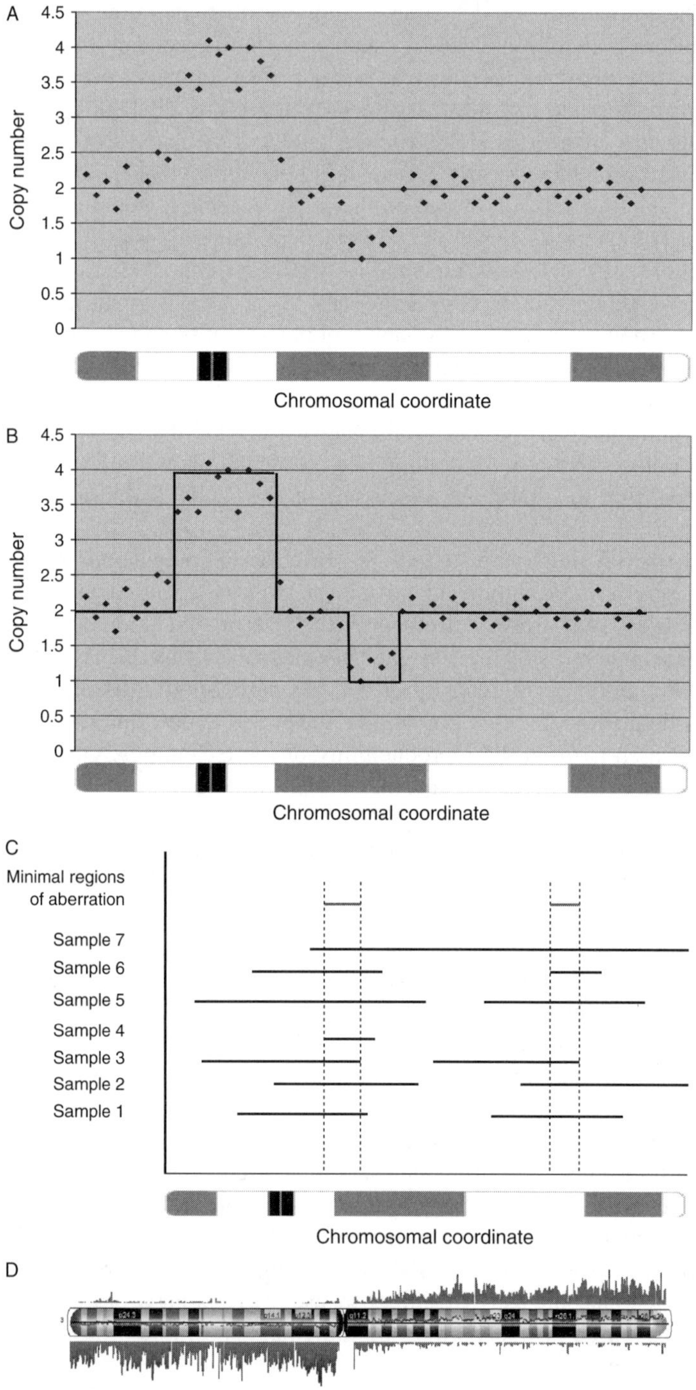

Figure 2.12 (Continued)

Figure 2.12 A) Measurement noise in CGH data complicates the identification of copy number abnormalities, because it (i) results in noninteger copy number values and (ii) blurs transitions between segments with a constant copy number. B) A typical copy number profile consists of regions of constant copy number flanked by abrupt transitions. C) Determination of recurrently aberrant genomic regions in a population of samples. D) A plot used to illustrate recurrent copy number aberrations in a CGH data set (often referred to as a frequency plot). One chromosome is shown with the gray bars facing upward corresponding to the percentages of samples containing a copy number gain (above a preset threshold, such as 3 copies), and with the blue bars facing downward indicating the percentage of samples having a loss at the corresponding locus. This type of graph facilitates identification of recurrent copy number abnormalities present in a data set.

DNA frequently seen in archived samples. Therefore, oligonucleotide arrays are preferred tools for biomarker discovery studies that utilize clinical samples. When the quantity of DNA required for the labeling protocol (250–500 ng, equivalent to approximately 50,000–100,000 cells) is limiting, amplification can be considered. Numerous DNA amplification methods have been suggested for CGH on small samples (111–117). The sample type is also an important consideration in array CGH. The easiest type of sample to work with is cell culture, because the isolation of high-quality DNA is routine and the cell population is homogeneous. Analysis of frozen tissue samples presents more difficulties because of the potential sample heterogeneity. For example, tumor samples often contain significant amounts of normal tissue, and this dilutes the signal obtained for aberrations in the tumor. Profiling of archived FFPE samples by CGH presents the greatest challenges, because of the poor quality of DNA isolated from such samples. Fixation protocols used in hospitals typically result in a number of known alterations in DNA, including degradation, cross-linking, and modification or loss of bases (118–120). The average fragment size of DNA decreases with increasing fixation time (121). The concentration of formalin used for tissue fixation and the age of the sample also affect the quality of the genomic DNA preparation (122).

The feasibility of profiling of FFPE samples for copy number abnormalities and LOH has recently been demonstrated (123). The 500K SNP genotyping array was used to assess the copy number changes in preselected samples that had undergone quality control testing to eliminate severely degraded genomic DNA. Several extraction methods were evaluated to identify a protocol that would provide DNA most suitable for array analysis. It was shown that PCR-based assessment of DNA quality predicted the downstream performance of FFPE DNA samples better than their age. More specifically, random amplified polymorphic DNA-PCR or RAPD-PCR (124) was used to evaluate the quality of genomic DNA samples and qualify them for CGH analysis.

Before estimating the copy number, concordance of genotype calls between paired FFPE and fresh frozen ovarian tumor DNA samples was examined. The average overall concordance between FFPE and fresh frozen samples from the same tumor was 93.6%, implying adequate processing of the FFPE DNA by the protocol. As the copy number profiles for FFPE samples were analyzed, the effect of fragment size on the copy number estimate was considered. Larger restriction fragments are not efficiently amplified during the PCR step, and therefore the probes mapping to these fragments yield incorrect estimates of the copy number (123). Exclusion of the SNPs that map to larger restriction fragments significantly increases the amplitude of copy number changes and at the same time reduces the measurement noise for the FFPE samples. Therefore a filter was proposed to eliminate SNPs corresponding to larger restriction fragments, resulting in the reduction of the total number of probesets contributing to copy number analysis to 308,788. Signal intensities from the remaining SNPs mapping to smaller restriction fragments yielded copy number estimates for FFPE samples consistent with those from the matching fresh frozen samples. In addition to filtering based on the fragment size, compensation against fragment size bias was performed to remove bias due to the amplicon size. Quadratic regression was applied to eliminate fluctuation in copy number, while exclusion of SNPs mapping to amplicons >700 bp before regression effectively reduced the fragment size bias.

Thus guidelines have been outlined for copy number analysis of archived FFPE samples on SNP genotyping arrays, enabling rational sample selection and selective analysis of most informative probesets on the array. Archived tissue samples represent an invaluable resource for genetic analysis because the existence of large banks of FFPE tissues with clinical annotation makes possible retrospective analysis of correlation between the genomic profile of the disease and the outcome or response to treatment. This goal undoubtedly justifies the amount of effort devoted to the optimization of FFPE CGH protocols. It is important to note that the task of genomic analysis of archived samples would be significantly facilitated if the fixation protocols used by hospitals were standardized, thus eliminating the variation in the DNA quality. Obviously, a protocol minimizing DNA degradation would be preferred as the common standard.

In summary, the choice between the existing CGH microarray platforms should be dictated by the relative importance of the following factors: detection sensitivity, genomic resolution, accuracy of breakpoint determination, signal reproducibility, requirement for LOH detection, and type of test sample. In the past several years, oligonucleotide-based CGH platforms have become dominant for most applications, including cancer gene discovery, biomarker identification, pharmacogenetic studies of copy number variation, and others. Their merits include easy quality control and standardization of manufacturing processes, flexibility in genomic content, high density of coverage, and high genomic resolution. Importantly, their content can be easily modified as the human genome sequence is updated. As many genomics projects in drug discovery eventually target regulatory submissions, the interlaboratory reproducibility and data standardization issues take a central place in experimental design. In this context, it is anticipated that several commercial oligonucleotide-based microarrays will become platforms of choice for most drug discovery applications.

2.5. COMPARATIVE GENOMIC HYBRIDIZATION: DATA ANALYSIS

Generally speaking, the computational and statistical manipulations of the CGH data may be divided into three major steps:

- Processing of signals to generate copy number estimates for each locus interrogated in a given sample
- Generation of the copy number profile for a single sample
- Multisample analysis, i.e., analysis of recurrent events.

Once a CGH microarray has been hybridized, washed, and scanned, the fluorescent signals on the array need to be processed **to obtain an estimate of the copy number at the respective loci**. The processing involves several steps typical for microarray data, namely, image gridding, spot segmentation, spot quantification, signal normalization to correct for spatial bias, and pooling of signals from replicate probes. These procedures were originally developed and optimized for expression microarray analysis. Their adaptation to CGH is generally straightforward; in fact, most commercial CGH arrays use the same solid support and same scanning equipment as gene expression microarrays. Manipulations specific to copy number assessment thus begin as the signal intensities for individual loci in the genome are transformed into estimates of the copy number for these loci. To obtain a copy number estimate, ratios are generated between the signals for the test sample and the reference sample, with the assumption that the reference DNA contains two copies of each locus interrogated. Normalization is applied to set the median ratio to a standard value, assuming that the majority of the genome in the test sample is normal. However, this is difficult when the test genome is significantly modified. Additionally, in two-color protocols, the pooling of ratios is often performed by using a dye reversal.

In the ideal situation of unbiased measurement, the ratios would always equal $n/2$, where n is the copy number for the experimental sample, i.e., a heterozygous deletion would yield a ratio of 0.5, while a gain of one copy would produce a ratio of 1.5. However, in practice microarray experiments are always associated with some level of variation resulting from differences in dye incorporation (for two-color protocols) or differences in overall array intensity (for single-color protocols), as well as variation in probe hybridization efficiency at different loci. Additionally, experimental samples are often heterogeneous and contain cells with different copy numbers of the same chromosomal regions. All these factors contribute to the formation of experimental noise, which complicates the next step of CGH data analysis, namely, the definition of regions of copy number change.

In a sample carrying chromosomal gains and losses, a typical copy number profile consists of plateaus of constant copy number flanked by transitions (Fig. 2.12B). Generally speaking, the main challenges in analyzing data from CGH microarrays for a single sample are to identify the copy number value and the statistical significance for all the plateau regions (genomic segments with a constant altered copy number) and to precisely determine the boundaries of each of them. Before this can be done, however, it is often necessary to filter out outlier probes that do not fit the regional copy number pattern. This is not straightforward, as many copy number abnormalities are very small (possibly including just one probe), and filtering procedures that are too stringent may eliminate true aberrations. It is noteworthy that as the density of CGH microarrays is increased, smaller aberrations are more likely to receive better probe coverage, and the risk of losing true aberrations during filtering decreases. The most common filters remove probes that show nonuniformity, saturation, or a high error standard deviation for replicates. It is also customary to remove aberrations that are smaller than a certain number of probes (for example <3 probes).

A number of statistical methods have been applied to **define genomic segments with different copy numbers** (125, 126). Broadly, existing methodologies for delineations of copy number alterations can be divided into two categories: smoothing based and segmentation based. Smoothing-based approaches operate under the assumption that true data points in a segment considered are smoother than any kind of experimental noise superimposed on them. They decrease noise by adjusting individual data points based on comparisons with adjacent data points. By reducing signal variation within genomic segments of constant copy number, smoothing methods decrease the number of false positives in a data set, but they also blur the borders between segments, and thus affect the accuracy of their delineation. The segmentation-based methodologies transform copy number data points for a chromosome into a series of segments. The unknown boundaries and unknown heights are estimated from the raw data points by utilizing an optimization criterion. In comparison with the smoothing approaches, segmentation methods yield less reliable copy number segments (relatively more false positives), but better identify the breakpoints (boundaries between segments).

In early CGH analyses, simple procedures were proposed that involved smoothing the test-to-normal ratios and setting thresholds for calling gains and losses (127). Data were smoothed by using the symmetric five-nearest neighbors moving average, while copy number alterations were determined by measuring the deviation of the ratios from the control "normal *versus* normal" ratios. Another early approach was to fit Gaussian distributions to the copy number plots and use the standard deviation of the central Gaussian distribution to define segments with an altered copy number (128). Regions were considered to have a significantly increased or decreased copy number if the test-to-normal ratio was at least three standard deviations away from the mean of the central distribution.

Currently, a number of **smoothing algorithms** are used. A software package for copy number analysis on SNP genotyping arrays (129) uses Gaussian smoothing to increase the signal-to-noise ratio in the data and improve the discrimination between genomic segments with different copy numbers (130). A Gaussian kernel-smoothing average is used to average the copy number and the *P* value of individual data points over a used-preset genomic interval. The procedure reduces the random noise within a given region and thus minimizes the false positive rate. The kernel-smoothing methodology works by accentuating genomic segments in which consecutive copy number measurements have the same type of alteration (gain or loss). Alternative smoothing methods include quantile smoothing (131) and wavelet analysis (132). The quantile smoothing methodology is based on penalized quantile regression and clearly defined boundaries between genomic segments with a constant copy number. Denoising by wavelets is a nonparametric technique that represents copy number data as a linear combination of wavelets. Each wavelet is characterized by two independent variables: one for the location of the change and the other for the copy number aberration size. This method is also meant to emphasize boundaries between genomic segments.

Segmentation methodologies, such as Hidden Markov Models, are ideally suited for analysis of CGH data as they partition data points along the chromosome into sets with the same copy number (133, 134). As this methodology is applied to CGH data, it is assumed that the underlying copy numbers are the hidden states with certain transition probabilities. It partitions observations $y_1, y_2, \ldots, y_n$ made at locations $t_1, t_2, \ldots, t_n$ into K (hidden) states, where $K < n$. In the context of copy number profiles, the observations represent either normalized test-to-normal signal ratios associated with probes (oligonucleotides or BACs) arranged on a chromosome. For each probe, $t_1, t_2, \ldots, t_n$ are the genomic coordinates. The maximum number of states K is limited by the number of probes on the chromosome and the computation time; it is typically limited to <5. Segmentation by Hidden Markov Models has been applied to CGH data from various platforms (oligonucleotides, BACs) and incorporated into commercially available software packages for analysis of copy number data from SNP genotyping arrays (Partek Genomics Suite available at www.partek.com and Copy Number Analysis Tool or CNAT available at www.affymetrix.com). Other segmentation procedures have been applied to copy number analysis, including binary segmentation adapted to allow splits into two or three segments (135). An approach has also been proposed that estimates the number of genomic segments in an adaptive manner (136).

In summary, a wide spectrum of methodologies is available for generating copy number profiles from CGH data. Their relative performance is likely to be dependent on multiple factors, including the noise level in the raw data, the spacing between probes on the array, the frequency of high-level and low-level aberrations in the data set, and the size of aberrant regions. It has been suggested that smoothing methods work well in noisy data sets, although their output may be more difficult to interpret (126). A combination of smoothing and segmentation procedures has been proposed for improved performance (126). It is important to consider the spacing between the probes on the array, because, generally speaking, longer distances between probes make it less likely that they belong to the same copy number segment, and it is not clear how different algorithms take this into account. The uniformity of probe coverage should be considered as well.

Once reliable copy number profiles are obtained for individual samples with the procedures described above, the researcher may focus on the next step, the combined analysis of the entire population of samples profiled. The goals at this step may include **identification of recurrent individual aberrations** or patterns thereof and discovery of **abnormalities associated with a phenotype**, such as a certain disease outcome or drug response. Identification of recurrent copy number abnormalities is an important first step in target identification and biomarker discovery, particularly in the field of oncology (comprehensively reviewed in 3.2.1). Discovery of copy number alterations associated with drug response is a promising strategy in the identification of patient stratification biomarkers. Additionally, accumulating evidence suggests that classification of tumors based on their gene copy number profiles may facilitate selection of preclinical tumor models. Therefore, methodologies for analysis of multiple samples are actively being developed to complement the existing spectrum of single-sample analysis procedures.

Recurrent copy number abnormalities in a group of samples can be identified by setting a simple frequency threshold (percentage of samples carrying an aberration) and a copy number threshold and determining the minimal regions of aberration, i.e., the overlaps between copy number aberrations in different samples (Fig. 2.12C). This procedure can be illustrated by a frequency plot, also occasionally referred to as a penetrance graph (Fig. 2.12D). The minimal regions of recurrent aberrations can be determined manually by analyzing the borders of the minimal regions on the frequency plots. Several software packages have been developed for this purpose. For example, a program called STAC (Significance Testing for Aberrant Copy Number) detects regions of copy number gain or loss that occur across an entire sample population or within a subset of samples more often than would be expected under a reasonable null model (137). The algorithm provides a rigorous mechanism for localizing regions of significance and can accept data from any of the existing array platforms. Currently, the main deficiency of STAC is its inability to factor in the amplitude of copy number change. Since high-level amplifications and homozygous deletions typically receive more consideration in copy number analysis, it would be logical to weigh copy number alterations according to the amplitude of change. The statistics used in the STAC algorithm are being modified to account for weighted intervals, where the weight of an interval is reflective of its degree of gain/loss (137).

A number of available software programs provide the option of plotting the frequency of copy number change and detecting recurrent abnormalities (Table 2.3). For example, the Partek Genomics Suite can import data from Affymetrix SNP genotyping arrays (100K, 500K, or SNP6.0), produce copy number profiles for individual samples, and summarize the data in a frequency plot. A freely available program called DNA-Chip Analyzer (dCHIP) can import

Table 2.3 Selected Software Programs for Analysis of Multisample CGH Data

Program	Functionalities	Access
Partek Genomics Suite	• Multisample display of CGH data • Multiple smoothing algorithms and HMM tranformation • Frequency plots • Association of aberrations with class labels (Fisher's exact test) • Hierarchical clustering	www.partek.com
dCHIP	• Multisample display of CGH data • Frequency plots	www.dchip.org
CGH Analyzer	• Multisample display of CGH data • Penetrance plots • Clustering	http://www.genomics.upenn.edu/people/faculty/weberb/CGH/html/software.htm
STAC	• Multisample display of CGH data • Frequency plots • Specialized statistical procedure to detect recurrent aberrations	http://cbil.upenn.edu/STAC
CGH Analytics	• Data filtering and normalization • Multisample display of CGH data • Adjustment of the baseline (normal copy number) • Frequency plots	www.agilent.com

microarray data from different platforms and plot copy number data for multiple samples, enabling frequency calculations and identification of minimal common regions of aberration. The CGH Analytics program is designed for analysis of CGH data from the Agilent platform and enables data filtering and identification of frequent aberrations with a penetrance plot. An additional benefit of this software package is the ability to perform combined analysis of copy number and gene expression data. The combined analysis tool uses the aberration calls detected from copy number analysis as a starting point to select genes in the aberrant regions that also show expression changes in the samples of interest. The CGH and gene expression data sets are then visualized together on a heat map along with plots of statistical relevance.

A separate and difficult problem in the analysis of CGH data is the identification of copy number abnormalities associated with a predefined phenotype, such as poor disease outcome or drug resistance. In the field of gene expression analysis, a number of statistical approaches have been used to determine genes associated with a predefined class label (for an example, see ref. (138)). The main difficulty in adapting these approaches to CGH data stems from the fact that copy number profiles do not represent a composite of discrete biologically meaningful units (unlike gene expression signatures, which are composed of genes). Instead, each sample in a set contains distinct regions of aberration, and their borders are different in each sample. Therefore, association of individual regions of copy number change cannot be done in a systematic manner. As we developed a software package for multisample copy number analysis, we approached the problem by separately testing the copy number values for each individual probeset on the microarray for association with the desired phenotype, followed by combining successive associated probes with a set of rules that defined altered genomic segments (139). More specifically, our program tested for statistical significance, using Fisher's Exact test to determine whether a probeset shows preferential gain/loss in one of the two predefined sample subsets (sensitive or resistant cell lines). The copy number thresholds for gains and losses were set at 2.8 and 1.5, respectively. For each probe set, a 2×2 contingency table was constructed for testing the significance of an increase or decrease in copy number in the two groups. The program calculated Fisher's Exact test P values for each probe and assembled contiguous significant probes into genomic segments. A minimum of three contiguous SNPs meeting the P-value threshold were considered to be one region. The algorithm developed was used in a study that identified a genomic marker of sensitivity of lung cancer cells to a targeted cancer drug (139).

Another important task in analysis of copy number profiling data is **identification of subclasses of samples** within a population. This procedure may be applied, for example, in evaluation of preclinical cancer models, as a large population of samples for the tumor type of interest is profiled to identify genomically characterized subgroups with the goal of matching a preclinical model to each subtype. The problem of pattern identification in copy number data has been addressed in a number of recent studies that used unsupervised classification by nonnegative matrix factorization (NMF) (140, 141). The NMF methodology is based on decomposition by parts that can reduce the dimension of microarray data from thousands of signals to a handful of metasignals. Originally it was introduced as a method to decompose images; the method yielded a decomposition of human faces into components that resembled facial parts, such as nose, chin, eyes, etc. (142). The method was then applied to gene expression data and elucidated meaningful structures in several complex data sets (143). A modification of NMF adapted to analysis of CGH data (140, 141) performs the following successive manipulations:

- The dimensionality of the CGH data set is reduced by eliminating probes that have identical copy number values in all samples in the data set.
- The negative log ratios associated with copy number losses are converted to positive values by introducing two dimensions, "gain" ($\log_2$ ratio >0) and "loss" ($\log_2$ ratio <0).

- The number of clusters is specified by the user, and NMF analysis is applied to the resulting data set.
- The stability of the resulting clusters is tested by removing a fraction of samples randomly, repeating the procedure, and comparing the clustering pattern with that for the entire data set.
- Fisher's Exact test is used to determine genomic regions that show statistically significant differences in copy number between clusters.
- The NMF procedure is repeated for a larger number of clusters until the clusters become unstable. The greatest number of stable clusters is likely to reflect the inherent structure of the data set.

As the methodology was applied to a data set for 67 multiple myeloma samples, the initial data reduction step reduced the number of elements from 16,084 probes to 942 segmented genomic values with constant copy numbers. The resulting 67×942 matrix was converted into nonnegative values by introducing two dimensions to each of the segments ("gain" and "loss"), to yield a nonnegative 67×1884 matrix. As the matrix was subjected to the NMF procedure, ranks $K = 2$, 3, and 4 produced stable clusters, implying the existence of up to four subgroups in the population of multiple myeloma samples. To test for stability of clusters, the classification procedure was rerun after randomly removing 10% of the samples in each iteration. For two major clusters of multiple myeloma samples that showed significant correlation with the known hyperdiploid versus nonhyperdiploid subclasses of the disease, Fisher's exact test was then applied to determine genomic segments that show statistically significant differences in copy number alterations between the clusters. Contingency tables were created for each of the 942 genomic segments, classified as gained, normal, or lost. Fisher's Exact test P values were then calculated for cluster 1 *versus* cluster 2 for gains and losses. The regions with the lowest P values were thus the main distinguishing copy number features between the two subtypes of the disease.

The NMF methodology has significant advantages over hierarchical clustering, frequently used in analysis of gene expression data. Unlike clustering, it reveals substructures within sample populations, but does not force subclassification when it is not biologically relevant (143). The NMF procedure shows significant promise in the multisample analysis of copy number data, specifically when the goal is to identify biologically relevant subgroups of samples.

In summary, we have reviewed the procedures and algorithms used at the three steps of CGH data analysis: microarray signal processing, generation of copy number profiles for a single sample, and multisample analysis. While the manipulations at the first step are very similar to the well-established procedures for gene expression analysis and therefore are routine, the second step of the CGH data analysis is potentially more challenging, as a number of issues complicate the definition of genomic segments with an altered copy number. Finally, at the third step, the researcher faces the highest level of uncertainty, as few software programs exist for comprehensive analysis of multiple samples. However, as the

CGH becomes a mainstream technology in drug target identification, biomarker discovery, and preclinical model selection, it is likely that existing statistical algorithms will be adapted to process CGH data, thus enabling development of comprehensive software solutions to extract meaningful information on copy number alterations from multiple biological samples.

2.6. MICROARRAY-BASED DNA METHYLATION PROFILING

Over the past decade, the study of epigenetic DNA modifications, in particular DNA methylation, has become an important area for drug discovery, primarily in oncology. It is widely recognized that the methyl group at the fifth position of the cytosine ring, which is present in approximately 80% of CpG dinucleotide fragments in the human genome, represents an important functional element of the genome, prompting references to the "fifth base" in the human genome. The most important functional role currently assigned to the methylated regions in the genome is in regulation of gene expression. Specifically, aberrant promoter methylation patterns have been detected in different human cancers (reviewed in ref. (144)). The observed diversity of promoter methylation patterns in various cancers and their functional association with tumorigenic phenotypes have prompted significant interest in correlating DNA methylation profiles to tumor type and drug response. The potential uses of promoter methylation patterns as composite genomic biomarkers are reviewed in detail in Chapter 3 on biomarkers. Discovery of DNA methylation biomarkers is dependent on the availability of a robust high-throughput technology for detecting methylation on a genome-wide scale. The progress in the development of such methodologies will be the main subject of this subsection.

BOX 2-2 *DNA Methylation and Its Detection*

Occurrence of DNA methylation

The term "DNA methylation" refers to addition of a methyl group at the fifth position of the cytosine ring, resulting in formation of methyl-cytosine, which may be considered the fifth base in the human genome. DNA methylation typically occurs at so-called CpG islands which are relatively infrequent in the human genome.

Significance of DNA Methylation

Methylation alters the chromosome structure and defines regions for transcriptional regulation. Methylation of CpG islands in the promoter regions causes transcriptional silencing of the respective genes. Transcriptional activity of a significant portion of human genesis is regulated by methylation. Changes in methylation patterns resulting in transcriptional deregulation are believed to be early events in tumorigenesis.

Methylation Patterns as Biomarkers

In addition to being tissue-specific, DNA methylation patterns may be used to characterize subgroups of tumors within a population of cancers of the same histopathological origin and, as such, may be explored as biomarkers predictive of disease outcome and drug sensitivity.

Detection by Use of Methylation-Sensitive Restriction Enzymes

This methodology takes advantage of the existence of restriction enzymes that do not cut DNA containing methyl-cytosine. The test and control samples are digested with a methylation-insensitive restriction enzyme, followed by ligation of adapters. The resulting fragments are then treated with a methylation-sensitive restriction enzyme, followed by PCR amplification. Methylated fragments remain intact and amplify at the next step, while unmethylated sequences are digested and become unusable as templates for PCR. The PCR products are hybridized to specially designed microarrays, and the ratio of the fluorescence intensities for the known sites present on the array are used to determine the methylation status.

Detection Through Bisulfite Conversion of DNA

This methodology relies on selective chemical modification of unmodified cytosines to differentiate between methylated and unmethylated sites. Genomic DNA from the test and control samples is treated with bisulfite, resulting in conversion of cytosine residues to uracils. As the DNA is then amplified by PCR, the DNA polymerase incorporates adenosine residues opposite the uracils in the template, resulting in formation of DNA duplexes with AT pairs at the position of interest. Meanwhile, the methyl-cytosine residues in the methylated sample are not affected by bisulfite and the DNA polymerase enters a guanosine residue opposite the methyl-cytosine, thus conserving the GC pair at the position of interest. As the two samples are hybridized to specialized microarrays designed to interrogate the genomic positions in question, the ratio of the fluorescent signals can be used to determine the methylation status at the position. See color insert.

Generally speaking, most methods for detecting methylated sites are based either on cutting the genomic DNA with methylation-sensitive restriction enzymes (Fig. 2.13A) or on selective bisulfite conversion of unmethylated cytosine into uracil (Fig. 2.13B) (145, 146). It is noteworthy that detection of aberrant methylation patterns in diseased samples should involve a comparison with relevant precursor cells (as opposed to a general reference sample of normal tissue as in mutation analysis or CGH). This is necessary because methylation patterns are tissue-specific (147, 148). In the past decade, a number of different microarray-based approaches have been proposed for methylation analysis. The

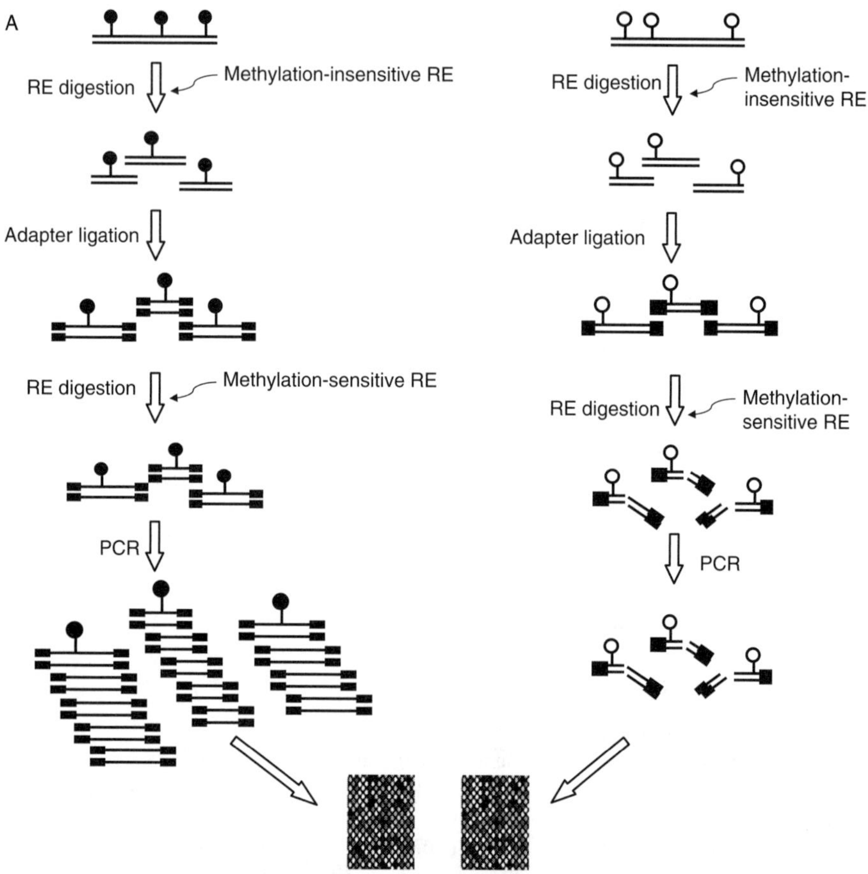

Figure 2.13 A) Microarray methodology for methylation profiling based on the use of methylation-sensitive restriction enzymes. Methylated cytosines are designated by black circles, while unmethylated cytosines are shown as open circles. The test and control samples are digested with a methylation-insensitive restriction enzyme, followed by ligation of adapters. The resulting fragments are then subjected to a methylation-sensitive restriction enzyme. Methylated fragments remain intact and amplify at the next step, while unmethylated sequences are digested and become unusable as templates for PCR. The post-PCR samples are then hybridized to specialized microarrays, and the ratio of the fluorescent intensities for the known sites present on the array is used to determine the methylation status. B) Bisulfite-based methodology for methylation profiling. Genomic DNA from the test and control samples is subjected to bisulfite treatment, resulting in conversion of cytosine residues to uracils. As the DNA is then amplified in a PCR reaction, the DNA polymerase enters an adenosine residue when it encounters an uracil in the template, resulting in formation of nascent DNA chains with adenosine and thymidine residues at the position of interest. At the same time, in the methylated sample, the methyl-cytosine residue is not affected by bisulfite and the DNA polymerase enters a guanosine residue opposite the methyl-cytosine, leading to the conservation of the GC pair at the position of interest. As the two samples are hybridized to specialized microarrays designed to interrogate the position in question, the ratio of the fluorescent signals can be used to determine the methylation status at the position. See color insert.

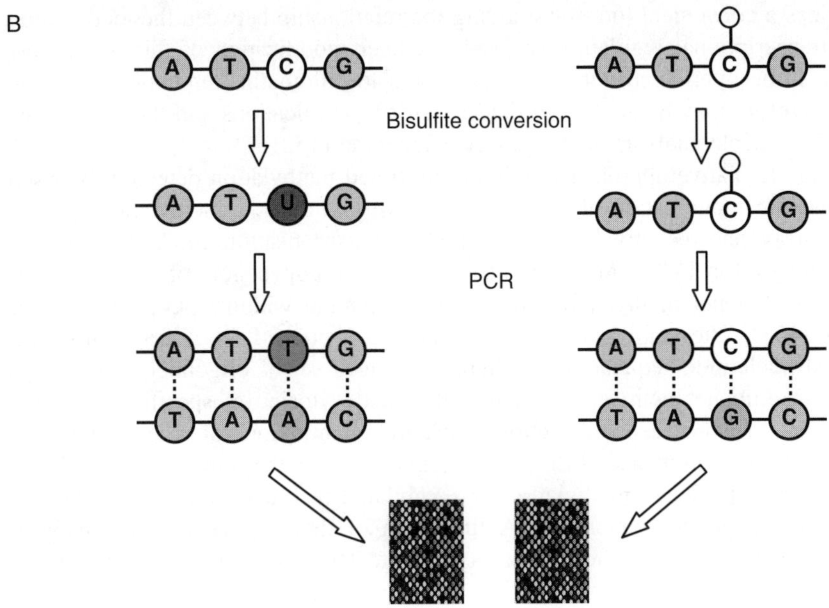

Figure 2.13 (Continued)

first methylation microarray (differential methylation hybridization or DMH array) interrogated 276 known CGI sequences and was hybridized with a probe that contained methylated CpG-rich DNA fragments (149). This probe was prepared by digesting genomic DNA with a methylation-insensitive (*Mse*I) and a methylation-sensitive (*Bst*UI) restriction enzyme. Methylated fragments were PCR-amplified with an adaptor for *Mse*I. Regions of DNA that contained an unmethylated CpG island did not amplify, because they were cleaved by *Bst*UI. Meanwhile, methylated regions were not digested by *Bst*UI, and therefore were amplified by PCR. The post-PCR product was hybridized to the microarray, and the differences in hybridization signals were used to identify differentially methylated CpG islands in the test and control samples.

This microarray was later expanded to interrogate 1104 CpG islands and applied to study aberrant methylation patterns in breast cancer (150). Furthermore, the DMH methodology was refined to enable combined assessment of methylation and gene expression (151). To this end, a microarray containing 1162 expressed CpG island sequence tags (ECISTs) was developed, and the GC-rich portions of the ECISTs were used to determine the methylation status, while the exon-containing regions were utilized for measuring gene expression. This ECIST panel was used to analyze the breast cancer cell line MDA-MB-231, which was treated with a demethylating agent. Microarray profile analysis identified 30 methylation-silenced genes, the expression of which could be directly reactivated by demethylation. Thus the dual methylation/expression microarray provides a convenient tool for studying the relationship between the methylation and transcriptional regulation of genes. A later modification of this microarray can also be hybridized with DNA subjected to chromatin immunoprecipitation and therefore may be applied to detect histone modifications and thus represents a tool for triple analysis of the cancer epigenome (152).

An alternative approach to microarray-based methylation detection is based on bisulfite modification of genomic DNA. In particular, a methylation-specific microarray has recently been developed that uses bisulfite-modified DNA as a template for PCR amplification, resulting in conversion of unmethylated cytosine, but not methylated cytosine, into thymine within CpG islands under study (153). The amplified product containing a pool of DNA fragments with altered nucleotide sequences was hybridized to a set of oligonucleotide arrays that discriminate methylated and unmethylated cytosine at specific nucleotide positions. Fluorescence detection enabled quantitation of the differential hybridization and hence differential methylation in the samples. The ability of the array to detect methylation was validated in breast cancer cell lines by bisulfite nucleotide sequencing. A limitation of this approach is the inability of the methylation-specific microarray to detect methylation changes at single CpG sites because of cross-reactivity (153). This limitation was overcome by designing probes to include two or more CpG sites, as it is usually not necessary to assess the overall methylation status of a given CpG island by analyzing every CpG site within the locus.

In conclusion, genome-wide methylation profiling technologies hold great promise for biomarker discovery. Some of the most pressing technology issues today are related to the probe selection and optimization of microarray design. Probe selection is hindered by the incomplete knowledge of the functional consequences of methylation. Additionally, the instability of the methylation patterns in culture (154) complicates biomarker identification in cell lines and necessitates validation in clinical samples. Further development of the microarray-based methylation profiling methodologies will undoubtedly help advance drug discovery by facilitating genomic stratification of human diseases and identification of biomarkers.

2.7. MICROARRAY-BASED MICRORNA PROFILING

Discovered in the mid-1990s, microRNAs (miRNAs) are now recognized as one of the major regulatory gene families in eukaryotes. Hundreds of miRNAs have already been discovered in humans, mice, plants, and viruses, and some miRNAs are conserved from *Caenorhabditis elegans* to *Homo sapiens*. Generally speaking, miRNAs are small noncoding RNA molecules 20–22 nucleotides long, generated through processing of 60- to 110-nucleotide-long precursor RNA molecules (for reviews see refs. (155–157)). MicroRNAs have been implicated in the control of several crucial biological processes, including proliferation, differentiation, apoptosis, and development (155), and have been shown to exert their effects by imperfect pairing with target mRNAs and transcriptional or posttranscriptional regulation of their expression (158–160). The effects of miRNAs are widespread: According to various computational and modeling studies, miRNAs may regulate the expression of up to 30% of all human genes (160, 161). Because an individual miRNA molecule is capable of targeting numerous transcripts, often in combination with other miRNAs, gene regulation by miRNAs is mediated by extremely complex regulatory networks. It has recently been demonstrated that miRNA expression profiles may serve as unique molecular "signatures" of human tumors, thus facilitating tumor classification (157, 162). In this section, we review the current status of miRNA profiling technologies, with a particular focus on emerging high-throughput microarray-based methodologies.

BOX 2-3 *MicroRNAs and Their Detection*

MicroRNA Structure and Function

MicroRNAs (miRNAs) are small single-stranded noncoding endogenous RNAs that play an important role in gene regulation. MicroRNA represents short (19–30 nucleotides) single-stranded RNA molecules. Mature miRNAs regulate gene expression by forming ribonucleoprotein complexes, which interact with the target mRNA, resulting in cleavage of the target or suppression of translation. Translation inhibition mediated by miRNAs is a combinatorial process, whereby each miRNA is capable of binding to more than one mRNA and each mRNA is bound by multiple miRNAs for repression.

MicroRNA Expression and Abundance in Cells

MicroRNAs are transcribed as long pri-miRNAs, which are then converted to shorter (70–90 nucleotides) hairpin progenitors, pre-miRNAs, and finally, to mature miRNAs. MicroRNA expression levels in the cell span four orders of magnitude.

MicroRNA Importance in Cancer

By interfering with gene expression, miRNAs may control cell proliferation and differentiation. Aberrant patterns of miRNA expression have been detected in breast, colon, and other cancers, as well as in leukemias and lymphomas. miRNA signatures have been associated with specific clinical and pathological features of the tumor. Sequences coding for known miRNAs have been mapped to regions frequently gained and lost in human cancer.

MicroRNAs as Biomarkers

Because certain miRNA expression patterns are associated with clinicopathological subtypes of tumors, it is logical to suggest that the levels of individual miRNAs or composite miRNA signatures may correlate with the clinical behavior of the tumor and its response to therapeutics, similar to gene expression or DNA copy number signatures. More research is needed to identify specific miRNA-based biomarkers predictive of disease outcome or drug response.

MicroRNA Detection Methodologies

Several microarray- and bead-based methods are currently used in high-throughput detection of miRNAs. Oligonucleotide microarrays have been designed to detect hundreds of miRNAs, but the specificity of detection remains a significant challenge, because of the short length of miRNAs, the high sequence homology between miRNA molecules, and the presence of miRNA progenitors common to many miRNAs.

As the task of comprehensive miRNA profiling of tissues is seemingly similar to the well-established microarray approaches to gene expression profiling, it is logical to apply oligonucleotide microarrays to facilitate high-throughput miRNA detection. However, several factors specific to miRNA need to be taken into account for microarray design. First, because miRNAs are generated by processing of larger precursor molecules, the detection methodology needs to discriminate between the mature miRNAs and the precursors to enable an accurate assessment of the unique miRNA profile of the tissue under study. One possible approach is to reduce the complexity of the sample by separating low-molecular-weight miRNA from the more abundant high-molecular-weight RNA species before array hybridization (Fig. 2.14). Second, miRNAs are very short molecules relative to mRNAs, and many of them display a high degree of homology, increasing the requirement for detection specificity. Third, the short length of miRNAs limits the possibilities for optimizing the probe hybridization conditions, as the GC content may vary significantly. Consequently, a significant challenge in the design of miRNA profiling microarrays is to maintain the uniformity of hybridization conditions. Finally, since miRNA concentrations in the cell can span four orders of magnitude (163), miRNA profiling microarrays need to be highly sensitive and are required to have a large linear dynamic range.

An early methodology based on oligonucleotide microarrays was developed in 2004 to interrogate 245 human and mouse miRNAs (164). The array contained 368 probes 40 nucleotides in length, spotted in triplicate. To ensure specificity, the following rules were used in probe design: (i) $<$75% global cross-homology and $<$20 bases in any 100% alignment to the relevant organism, (ii) $<$16 bases in alignments to repetitive elements, (iii) $<$16 bases of low complexity, homopolymeric stretches of no more than 6 bases, (iv) G + C content between 30% and 70%, (v) no more than 11 windows of size 10 with G + C content outside 30–70%, and (vi) no self 5-mers. A total of 5 μg of total RNA was used for hybridization, which is comparable to the requirement for gene expression microarrays. To evaluate the ability of the microarray to distinguish between mature miRNAs and their high-molecular-weight precursors, two different probes were designed for 76 miRNAs, one specific to the active

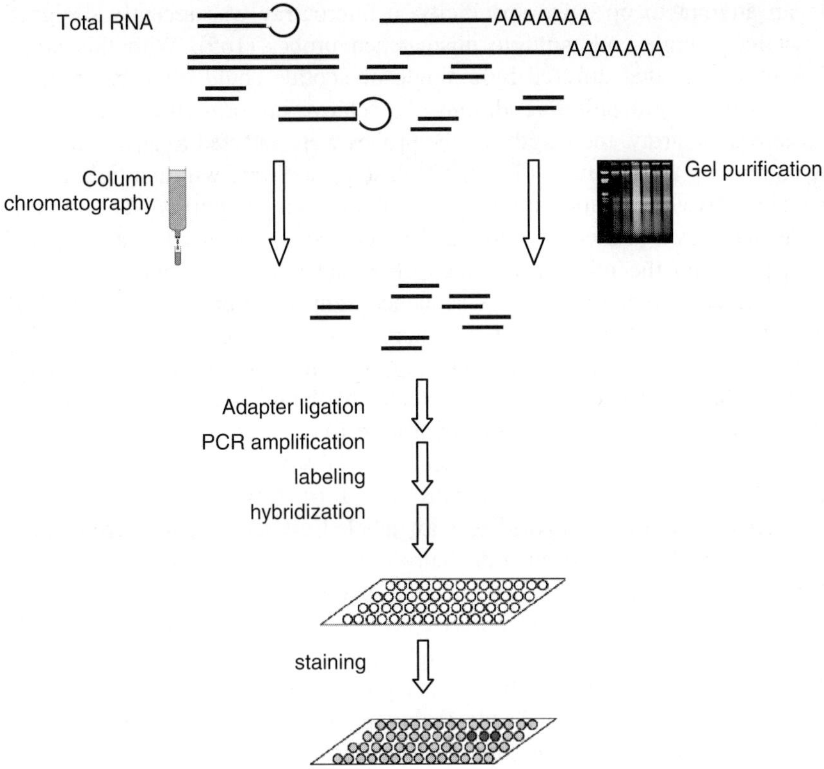

Figure 2.14 High-throughput microRNA profiling using microarrays. Total RNA from the sample of interest is fractionated to separate mature miRNAs from messenger RNAs, hairpin miRNA precursors, long double-stranded RNA, and other molecules. The miRNA-enriched fraction is then ligated with adapters, reverse transcribed, amplified through polymerase chain reaction, and labeled, with procedures typical for microarray protocols. The labeled miRNA is hybridized to a microarray, followed by microarray staining, scanning, and quantitation of the miRNA molecules interrogated.

sequence and the other one to the precursor. The microarray was used to analyze a panel of 20 normal human tissues, including 18 of adult origin and two of fetal origin. Different tissues displayed distinct patterns of miRNA expression, with each tissue presenting a specific signature. Because an unsupervised hierarchical clustering algorithm was applied, the same tissue types from different individuals clustered together, which is consistent with the reported tissue specificity of miRNA expression (155–157). The possibility of detecting both the mature miRNAs and their high-molecular-weight precursors increased the utility of the tool. This design of this microarray has served well to establish the feasibility of microarray-based high-throughput miRNA profiling.

In an attempt to optimize specificity, a microarray was recently designed that contained single and multiple mismatched probes (165). With this array, miRNA sequences that differed by a single nucleotide could often be resolved under stringent hybridization conditions. To achieve uniform stringency for all probes across the array, the lengths of the probes were selected to ensure that the predicted melting temperatures (T_m) for each sequence were within a tight range. To enable relative quantitation of each miRNA in a sample, normalization to reference nucleotides was used. A synthetic reference oligonucleotide was prepared for each probe on the microarray. Each oligonucleotide represented the sense strand of the corresponding miRNA sequence flanked by primer sites for PCR amplification. To test the ability of the microarray to correctly measure relative miRNA abundance, it was used to analyze cloned total RNA from mixed-stage *C. elegans*. The relative concentrations detected by the microarray correlated well with the cloning frequencies of the known miRNAs.

A microarray was recently designed for miRNA profiling that contained probes with length exceeding that of miRNAs (166). It was shown that 60-mer oligonucleotide probes can hybridize with labeled cRNA produced from miRNAs, but not with their precursors, hairpin RNAs. The microarray protocol involved size fractionation and amplification of total RNA. The signal intensity depended on the location of the miRNA sequences within the oligonucleotide probe, with the 5′-region yielding the highest signal and the 3′-region producing the lowest signal. Accordingly, 60-mer probes harboring one miRNA copy at the 5′-terminus yielded signals of similar intensity to probes containing two or three miRNA copies. It was also demonstrated that mismatches within the miRNA sequence substantially decrease or eliminate the signal, implying that the observed signals accurately reflect the abundance of matching miRNAs in the labeled cRNA. Hybridization signals obtained from cRNA prepared from size-fractionated RNA smaller than 140 nucleotides were derived from miRNAs rather than their high-molecular-weight precursors, possibly because the hairpin structure of the precursors makes them less available for hybridization. Since the oligonucleotide 60-mer probes complementary to the hairpin precursors are also expected to form hairpin structures, the finding that miRNAs are able to effectively hybridize to these probes suggests that miRNAs are able to efficiently displace self-annealed strands. As the microarray was used to study the expression of known human miRNAs in several tissues, it was found that miRNAs that differ by only a few nucleotides often display very different expression patterns in tissues. For example, miRNAs let-7A and let-7B, which are different from each other by two nucleotides, have a very similar expression pattern, while let-7c, which is one nucleotide different from both let-7A and let-7B, has a different expression profile with significantly lower expression in placenta and brain, but not in the other tissues. The similar expression pattern of let-7A and let-7B is probably due to the fact that they are both coded by the same cluster on chromosome 22. Thus the data reported support the specificity of a microarray carrying 60-mer oligonucleotide probes and suggest that longer probes are acceptable for miRNA detection.

As the probe design methodologies were further refined to address the challenges of detecting miRNAs, a multiplexed profiling assay was developed that enabled the creation of a commercial miRNA profiling microarray (167). Similar to the protocols described above, the assay was designed to discriminate between mature miRNAs and their high-molecular-weight precursors. The distinguishing feature of this assay is that it does not involve a size fractionation step, enabling a significant reduction in the amount of the input total RNA (to 120 ng). This became possible owing to a specialized probe design strategy aimed at selective detection of mature miRNAs. The probes on the microarray represented unmodified oligonucleotides of varying length containing several novel design features. First, a guanosine was added to the 5′ terminus of each probe sequence to pair with the 3′ cytosine of the corresponding miRNA introduced during labeling. This additional G-C pair increased the stability of the target miRNA–probe duplex relative to duplexes formed by homologous RNAs. As a result, for most mature miRNAs interrogated by the microarray the calculated melting temperatures were above 55°C under the hybridization conditions selected. The miRNAs with melting temperatures exceeding 57.5°C were destabilized by truncating the probe at the 5′ terminus of the miRNA sequence. Since most of the sequence homology among miRNAs tends to occur at the 5′ terminus, removal of bases from the 5′ terminus has little effect on the detection specificity. Second, to facilitate discrimination between mature miRNA and unintended potential targets, a hairpin structure was incorporated into the 5′ terminus of the oligonucleotide probe, adjacent to the 3′ end of the hybridizing miRNA sequence. The function of the hairpin is to destabilize potential duplexes formed by untargeted high-molecular-weight precursor RNAs and to stabilize the target miRNA–probe duplex. Finally, the optimal probe length was empirically selected as follows. For each miRNA interrogated by the microarray, all possible probe sequences with a calculated T_m between 50°C and 60°C were synthesized on microarrays both with and without the 5′ hairpin. The miRNAs were then hybridized to the microarrays at several temperatures within the 50–65°C range. Two sequences that melted just below and just above 57.5°C were selected for the microarray. The sequence specificity of the probes was then verified experimentally by hybridizing 19 synthetic human miRNAs with high sequence homology to other miRNAs. Very little cross-hybridization was observed for miRNAs that differed by more than one nucleotide (less than 1%). With several exceptions, miRNAs hybridized only to their specific probes. Higher cross-hybridization was more commonly detected for single nucleotide purine-to-purine or pyrimidine-to-pyrimidine substitutions. Further optimization of the probe melting temperatures should improve specificity, although it is unlikely to completely eliminate cross-hybridization, given the complexity of the RNA mixture applied to the microarray (167). One way to validate the ability of a microarray to specifically detect mature miRNA sequences is to compare the hybridization patterns for size-fractionated total RNA (selected for small RNAs) and unfractionated RNA (containing miRNA precursors and non-miRNAs). For the majority of miRNAs, a very high degree of correlation was observed between the hybridization patterns from total and purified small

RNAs, implying that optimization of probe design may eliminate the need for size fractionation (167).

The sensitivity and dynamic range are important considerations in miRNA microarray design, because miRNA concentrations in the cell may vary by four orders of magnitude (163). As a mixture of miRNAs was hybridized at concentrations ranging from 0.2 amol to 2 fmol, the hybridization signals were proportional to the concentrations of input miRNA over the entire range tested, suggesting that the detection limit of the assay is lower than 0.2 amol, and that the linear dynamic range of the assay exceeds four orders of magnitude.

A bead-based approach to miRNA profiling has recently been developed that addressed the issue of detection specificity (162). Oligonucleotide probes designed to interrogate known human miRNAs were attached to carboxylated 5-μm polystyrene beads impregnated with mixtures of two fluorescent dyes. Changing the relative concentration of the two dyes resulted in creation of up to 100 colors, thus enabling detection of up to 100 individual miRNAs. Total RNA from the samples of interest is subjected to adaptor ligation, which uses both the 5′-phosphate and the 3′-hydroxyl groups of miRNA. The molecules are then reverse transcribed, amplified by polymerase chain reaction using a common biotinylated primer, hybridized to the capture beads, and stained with streptavidin-phycoerythrin. The miRNA-carrying beads are then subjected to flow cytometry to detect the bead color (reflecting miRNA identity) and phycoerythrin intensity (reflecting miRNA abundance). The advantage of bead-based hybridization is that it approximates more closely hybridization in solution and thus may provide higher specificity relative to methodologies that employ a solid surface (162).

In summary, the sensitive and relatively specific microarrays described above may serve as a foundation for further development of high-throughput genome-wide methods for assessment of the relative abundance of miRNA in human cells. Besides improving accuracy and further validation, the main areas for further development are improved coverage, better discrimination between homologous molecules, and adaptation of the assay to archived samples.

2.8. TECHNICAL ISSUES IN GENOMICS EXPERIMENTS AND REGULATORY SUBMISSIONS OF MICROARRAY DATA

Application of genomics technologies offers an enormous potential to improve drug discovery and development, and may offer critical insights into the mechanisms of the drug's efficacy and toxicity and aid in biomarker identification. However, inclusion of microarray data in a regulatory submission presents a number of serious challenges, both for the sponsors and the evaluators of the microarray data. In this section we briefly review the potential technical issues in the context of regulatory submissions of microarray data. A more general and comprehensive discussion of pharmacogenomics submissions to the FDA is presented in Chapter 9.

In a document titled "The Critical Path to New Medical Products" (168), FDA experts identified genomics technologies as being crucial in advancing medical product development and personalized medicine. The regulatory agencies are working with the microarray community on designing standards for interpretation and submission of microarray data for regulatory decision making. These joint efforts have already produced a number of valuable documents that can be used as guidance in preparing microarray data packages for regulatory submissions. For example, a recent concept article addressed the issues arising in codevelopment of therapeutic agents and genomics-based tests used to stratify the patient population for treatment (169). A number of guidance documents specifically addressed microarray technologies and the use of gene expression patterns in drug discovery (170, 171).

Experts from FDA have participated in various educational workshops aimed at improving the understanding of the current issues related to microarray data analysis and submission. For example, a joint workshop, "Microarrays in Transcriptional Profiling," was organized by the FDA, Johns Hopkins University, and the Pharmaceutical Research and Manufacturers of America (PhRMA) in July 2005. Technical issues addressed at the workshop included:

- Reproducibility of microarray data from different laboratories and different technological platforms
- Sources of variation in microarray data
- Use of proper controls in gene expression microarray experiments
- Proper data normalization
- Microarray profiling of different types of clinical samples
- Clinical validation of microarray data

The specific technical issues that need to be addressed are determined by the stage of the drug discovery and development and by applications. Our review of these issues is organized by application, starting with the earliest, hypothesis-generating studies typically run at the early discovery stage and ending with biomarker discovery validation tasks and pharmacogenetic applications, commonly tackled in clinical development.

2.8.1. Study of a Drug's Mechanism of Action by Gene Expression Profiling

Greater understanding of the compound mechanism can be achieved by profiling the compound in cultured cells representative of the target disease or expressing the drug target followed by pathway analysis of the resulting gene expression signatures. The signatures may help identify previously unknown pathways activated by the compound or provide additional information on the activation of the known pathways (time course, concentrations required, etc.). Overall, this approach may provide clues and help formulate hypothesis, but a definitive study would require verification by an independent method that would explore protein phosphorylation, transcription factor translocation or activation, etc. Therefore, microarray data may be submitted as part of a larger data set if it is critical to prove that the compound acts through a certain mechanism or affects certain cellular processes to exhibit its on-target activity. Off-target effects of a compound can also be explored with DNA microarrays, but studies of this type are more commonly performed in the context of a compound's toxicity or undesired effects. In summary, submission of efficacy-related microarray data may provide useful information on the compound mechanism of action and give rise to hypotheses on its efficacy and toxicity, but it has little value alone, if not supported by confirmatory studies using other methods.

According to the FDA guidance for pharmacogenomics data submissions (170), genomics data from exploratory studies or that is research data, such as from general gene expression analyses in cells/animals/humans, or single nucleotide polymorphism (SNP) analysis of trial participants, does not have to be submitted. However, the FDA welcomes voluntary submission of such data under the Voluntary Genomic Data Submission (VGDS) program. Data from voluntary genomic data submissions will not be used for regulatory decision making. However, voluntary submissions can benefit both the industry and the FDA in a general way by providing a means for sponsors to ensure that regulatory scientists are familiar with and prepared to appropriately evaluate future genomic submissions. Both FDA and industry scientists would benefit from an enhanced understanding of relevant scientific and technical issues.

As gene expression microarrays are used to characterize compound efficacy and mechanism, the main technical issue that may arise is experimental variability. Most importantly, interplatform and interuser variability in microarray data will result in differences in the gene expression signatures produced for the same compound under the same conditions, but on different arrays in different laboratories. This variability may yield so-called false positives unrelated to the compound's mechanism of action and in turn lead to false hypotheses about the pathways activated by the compound and its mechanism. Alternatively, if very stringent statistical cutoffs are used, increased variability may lead to a high level of false negatives, i.e., genes affected by the compound but not included in the analysis. The problem of cross-platform and interuser variability in microarray experiments is comprehensively addressed in the previous subsections of this chapter and the possible solutions reviewed there can be applied.

2.8.2. Early Assessment of Drug Toxicity in Model Systems

This application is closely related to the mechanistic studies outlined above, but the focus here is primarily on the toxic mechanism of the compound. Various model systems related to the organs, targets of toxicity, are used to profile the compound. The resulting gene expression signatures are then related to the signatures of compounds with known toxicity mechanisms to make conclusions on the toxic profile of the compound under study. This approach, which represents the main focus of an emerging scientific field called toxicogenomics, is comprehensively discussed in Chapters 4, 5, and 6.

If toxicogenomics data sets for a compound are intended for an FDA submission, the technical issues will be similar to those outlined in the previous subsections. If the variability between microarray measurements is high, the toxicogenomic signatures will not be sufficiently robust to properly predict the toxic mechanism of the compound. Interlaboratory variability is a particularly important factor in toxicogenomics, as gene expression signatures for the test compound are typically generated in the sponsor's laboratory while the toxicogenomic signatures for the reference compounds are often provided by specialized laboratories or taken from commercial toxicogenomics databases. Interplatform variability is an even more significant hurdle, if the test and reference compounds have been profiled on different microarray platforms. If the gene sets represented on the two platforms do not fully overlap, the unique gene sets will have to be excluded from consideration, resulting in incomplete assessment of the toxic mechanism. Even more significantly, different microarray platforms often have different thresholds for calling gene expression, resulting in variability in assessing the toxicological changes. As the technology matures, one could anticipate that specialized arrays will be developed for toxicogenomics studies by a limited number of manufacturers, thus limiting the interplatform component of microarray variability. In the meantime, voluntary submissions of genomic data should be able to educate the sponsors and the FDA scientists and allow them to formulate the relevant requirements and standards.

2.8.3. Biomarker Identification in Discovery and Early Development

Early discovery of pharmacogenomic and patient stratification biomarkers may significantly improve the efficiency of clinical drug development by allowing rational dose optimization and patient selection. Therefore, biomarker programs are now often incorporated into early discovery projects using disease models. Gene expression may play a key role in biomarker discovery, if they are used to (i) determine the pathway activation and target inhibition status after drug treatment, facilitating identification of markers of target inhibition; or (ii) identify basal patterns of gene expression in untreated cells associated with sensitivity to the drug, thus enabling rational patient selection.

According to the recent FDA guidance on pharmacogenomics data submissions (170), all genomics data must be submitted to the Investigational New Drug Application (IND) if any of the following apply:

1. The test results are used for making decisions pertaining to a specific clinical trial, or in an animal trial used to support safety (e.g., the results

will affect dose and dose schedule selection, entry criteria into a clinical trial safety monitoring, or subject stratification).

2. A sponsor is using the test results to support scientific arguments pertaining to, for example, the pharmacological mechanism of action, the selection of drug dosing and dosing schedule, or the safety and effectiveness of a drug.
3. Test results constitute a known valid biomarker for physiological, pathophysiological, pharmacological, toxicological, or clinical states or outcomes in humans, or the test is a known valid biomarker for a safety outcome in animal studies.

Thus, if the discovery-stage biomarker identification program results in any product affecting clinical development, and the data become part of the IND, the sponsors need to pay attention to the technical issues from the onset of the program. In addition to the potential problems outlined in the previous subsections, potential bias in study design may affect the validity of the resulting biomarker in the patient population. In particular, overfitting of the data often occurs when multivariate models are used for a large number of potential predictors while trying to discriminate with a relatively small number of patients (45). The bias may affect the validity of a biomarker if appropriate rules are not followed when designing the preclinical biomarker identification studies. A special subsection of the bioinformatics section covers the problem of bias and contains recommendations on study design. In particular, randomization of patients at baseline, uniformity in sample collection and processing, and blinding are recommended as procedures to address the problem of bias.

2.8.4. Patient Stratification in Clinical Trials with Gene Expression Signatures

If preclinical or early clinical studies yield a candidate genomic biomarker for patient stratification, the next logical step is to validate its predictive power in a large phase III study. Moreover, if a sufficient amount of supportive data has been accumulated at the discovery and phase I and II clinical stages, a candidate biomarker can be used to select patients for phase III clinical trials. The latter situation presents an ideal scenario in drug development, as rational selection of patients for phase III trials based in their predicted sensitivity to the drug will increase the response rates, shorten the duration, and cut the cost of the phase III development. The concept of using gene expression signatures as composite biomarkers for predicting disease outcome and patient response has been proven in a number of recent pioneering studies (43, 60, 66, 172–179).

If microarray-derived gene expression signatures are used to select for responders in the patient population, the data must be submitted to the FDA (170). Moreover, if patients are included in or excluded from a clinical trial based on a composite biomarker, such as a gene expression signature, FDA recommends codevelopment of the drug and the pharmacogenomic tests, if they are not currently available, and submission of complete information on the test/drug combination to the agency (169). The FDA plans to issue further guidance on codevelopment of pharmacogenomic tests and drugs.

Because of the complexity of measuring composite biomarkers and the stringent requirements for clinical in vitro diagnostics, a number of technology and bioinformatics issues must be addressed before this application becomes common. First, the inclusion of a common RNA control (such as those developed by the Microarray Quality Control Project; see the previous subsection) is required if a microarray-based test is to be run by different clinical laboratories. Such a control would facilitate normalization and allow laboratories to assess the reproducibility of their microarray readouts. Additionally, no reliable method currently exists for microaray-based gene expression profiling of formalin-fixed paraffin-embedded (FFPE) tissue samples. All existing microarray protocols require that total RNA be isolated from flash-frozen tissue, whereas this sample type is not commonly available from clinical trials. The additional complications associated with RNA isolation from FFPE samples and microarray profiling of this RNA (addressed in an earlier subchapter) will necessitate validation of a newly discovered microarray-derived composite biomarker discovered in preclinical studies. In particular, the degree of RNA degradation and nucleic base modification characteristic of archived FFPE samples varies depending on the tissue fixation procedure, the sample age, and the tissue type. This means that the expression levels for the components of the composite biomarker signature will depend on these technical parameters. Multiple data analysis issues, such as data overfitting to a particular sample set and sample selection bias, will also affect the validity of the measurements for composite biomarkers derived from clinical trial samples. All these considerations will need to be addressed before microarray-derived composite biomarkers will become routine tools for patient stratification for treatment.

2.8.5. Genotyping of Patients in Clinical Studies to Predict Drug Response

Interindividual variation in drug metabolism is an important aspect of drug development. This variability in drug response and toxicity could be due to genetic factors such as single nucleotide polymorphisms (SNPs) in genes encoding drug-metabolizing enzymes, drug transporters, and DNA repair enzymes. Such SNPs are common in the general population and are known to influence drug pharmacokinetics. For example, cytochrome *P*-450 (CYP) proteins are heme-containing enzymes that are responsible for the oxidative metabolism of various endogenous and exogenous compounds (180). A member of the CYP3A subfamily, the CYP3A5 gene, exhibits high interpopulational variability in allele frequency and haplotype structure (180). This gene is known to play an important role in the metabolism of many prescription drugs. In some patients, it may have a severely reduced activity. Hence, it is likely that it may contribute to the differential drug responses among different individuals (181). Because genetic polymorphisms may cause variability in gene expression and hence variable drug response, microarray-based detection of SNPs in relevant drug processing genes could improve the drug development process by enabling personalized dose selection, elimination of nonresponders, and prevention of toxic effects.

As mentioned above, the FDA mandates submission in an IND of any pharmacogenetic test results that are used for making decisions pertaining to a specific clinical trial, or in an animal trial used to support safety (e.g., the results will affect dose and dose schedule selection, entry criteria into a clinical trial safety monitoring, or subject stratification) (170). If a microarray-based genotyping test is used to stratify patients based on drug metabolism or to select the dose, a number of technology-related issues will have to be addressed. These issues will be specific to the genotyping arrays. While the genotyping arrays generally have a higher reproducibility than gene expression arrays (because of the specifics of the output, namely, discrete calls as opposed to continuous expression values), they are prone to reduced call rates. If the experimental procedures permit cross-contamination of DNA samples or introduction of foreign DNA molecules, the SNP call rates will be low, rendering the genotype unusable. The current validated protocols for microarray-based SNP genotyping use frozen tissue. Adapting the method to archived FFPE samples will require modifications of the DNA processing protocol. Additionally, investigators designing microarray genotyping studies need to consider the issues of overfitting, bias, and generalizability outlined in the subsection devoted to common problems affecting the validity of microarray studies.

2.9. CONCLUSION

The intention of this chapter was to introduce several key genomics technologies applied in the discovery and development of new therapeutic agents. Four key microarray-based technologies were covered: gene expression microarrays, comparative genomic hybridization, methylation microarrays, and microRNA arrays. Gene expression microarrays represent a mature technology, which is now widely used at all stages of drug discovery and is currently being adapted to clinical diagnostics applications. They are applied in target identification and validation, compound characterization, and biomarker discovery in most therapeutic areas, including oncology, neurosciences, immunology, and metabolic disease. Array-based comparative genomic hybridization has only recently been commercialized but is already becoming an indispensable tool in oncology drug discovery, primarily because of the importance of DNA copy number alterations in tumorigenesis. As substantial evidence is accumulated for a significant role of copy number variation in neural system disorders and psychiatric diseases, array CGH will undoubtedly become a useful tool in neuroscience drug discovery.

Methylation and microRNA arrays are currently being developed for identification of tissue-specific signatures, which may serve as biomarkers for patient stratification. These tools are primarily being explored in oncology, but it is likely that tissue-specific methylation and miRNA expression patterns will prove to be useful biomarkers in other disease areas. The following chapters will analyze the established and developing applications of these technologies at different stages of drug discovery.

REFERENCES

1. Fodor SP, Read JL, Pirrung MC, Stryer L, Lu AT, and Solas D. Light-directed, spatially addressable parallel chemical synthesis. *Science* 1991;251:767–73.
2. Fodor SP, Rava RP, Huang XC, Pease AC, Holmes CP, and Adams CL. Multiplexed biochemical assays with biological chips. *Nature* 1993;364:555–6.
3. Snijders AM, Nowak N, Segraves R, Blackwood S, Brown N, Conroy J, Hamilton G, Hindle AK, Huey B, Kimura K, Law S, Myambo K, Palmer J, Ylstra B, Yue JP, Gray JW, Jain AN, Pinkel D, and Albertson DG. Assembly of microarrays for genome-wide measurement of DNA copy number. *Nat Genet* 2001;29:263–4.
4. Pollack JR, Perou CM, Alizadeh AA, Eisen MB, Pergamenschikov A, Williams CF, Jeffrey SS, Botstein D, and Brown PO. Genome-wide analysis of DNA copy-number changes using cDNA microarrays. *Nat Genet* 1999;23:41–6.
5. Eisen MB and Brown PO. DNA arrays for analysis of gene expression. *Methods Enzymol* 1999;303:179–205.
6. Chomczynski P and Sacchi N. Single-step method of RNA isolation by acid guanidinium thiocyanate-phenol-chloroform extraction. *Anal Biochem* 1987;162:156–9.
7. Chomczynski P. A reagent for the single-step simultaneous isolation of RNA, DNA and proteins from cell and tissue samples. *Biotechniques* 1993;15:532–4, 6–7.
8. Van Gelder RN, von Zastrow ME, Yool A, Dement WC, Barchas JD, and Eberwine JH. Amplified RNA synthesized from limited quantities of heterogeneous cDNA. *Proc Natl Acad Sci USA* 1990;87:1663–7.
9. Eberwine J, Spencer C, Miyashiro K, Mackler S, and Finnell R. Complementary DNA synthesis in situ: methods and applications. *Methods Enzymol* 1992;216:80–100.
10. Eberwine J, Yeh H, Miyashiro K, Cao Y, Nair S, Finnell R, Zettel M, and Coleman P. Analysis of gene expression in single live neurons. *Proc Natl Acad Sci USA* 1992;89:3010–4.
11. Phillips J and Eberwine JH. Antisense RNA amplification: a linear amplification method for analyzing the mRNA population from single living cells. *Methods* 1996;10:283–8.
12. Schroeder A, Mueller O, Stocker S, Salowsky R, Leiber M, Gassmann M, Lightfoot S, Menzel W, Granzow M, and Ragg T. The RIN: an RNA integrity number for assigning integrity values to RNA measurements. *BMC Mol Biol* 2006;7:3.
13. Imbeaud S, Graudens E, Boulanger V, Barlet X, Zaborski P, Eveno E, Mueller O, Schroeder A, and Auffray C. Towards standardization of RNA quality assessment using user-independent classifiers of microcapillary electrophoresis traces. *Nucleic Acids Res* 2005;33:e56.
14. Perou CM, Sorlie T, Eisen MB, van de Rijn M, Jeffrey SS, Rees CA, Pollack JR, Ross DT, Johnsen H, Akslen LA, Fluge O, Pergamenschikov A, Williams C, Zhu SX, Lonning PE, Borresen-Dale AL, Brown PO, and Botstein D. Molecular portraits of human breast tumours. *Nature* 2000;406:747–52.

15. Huang J, Qi R, Quackenbush J, Dauway E, Lazaridis E, and Yeatman T. Effects of ischemia on gene expression. *J Surg Res* 2001;99:222–7.
16. Masuda N, Ohnishi T, Kawamoto S, Monden M, and Okubo K. Analysis of chemical modification of RNA from formalin-fixed samples and optimization of molecular biology applications for such samples. *Nucleic Acids Res* 1999;27:4436–43.
17. Cronin M, Pho M, Dutta D, Stephans JC, Shak S, Kiefer MC, Esteban JM, and Baker JB. Measurement of gene expression in archival paraffin-embedded tissues: development and performance of a 92-gene reverse transcriptase-polymerase chain reaction assay. *Am J Pathol* 2004;164:35–42.
18. Erlander MG, Ellis SP, Olson JM, Wittliff B, Smith B, Younger J, and Sgroi DC. Global gene expression profiling from formalin-fixed paraffin-embedded (FFPE) tissues: implications for retrospective and prospective clinical studies. *Proc Am Soc Clin Oncol* 2003;22:125.
19. http://www.arctur.com/research_portal/products/paradise_main.htm.
20. Mockler TC, Chan S, Sundaresan A, Chen H, Jacobsen SE, and Ecker JR. Applications of DNA tiling arrays for whole-genome analysis. *Genomics* 2005;85:1–15.
21. Johnson JM, Castle J, Garrett-Engele P, Kan Z, Loerch PM, Armour CD, Santos R, Schadt EE, Stoughton R, and Shoemaker DD. Genome-wide survey of human alternative pre-mRNA splicing with exon junction microarrays. *Science* 2003;302:2141–4.
22. Emmert-Buck MR, Bonner RF, Smith PD, Chuaqui RF, Zhuang Z, Goldstein SR, Weiss RA, and Liotta LA. Laser capture microdissection. *Science* 1996;274:998–1001.
23. Ma XJ, Wang Z, Ryan PD, Isakoff SJ, Barmettler A, Fuller A, Muir B, Mohapatra G, Salunga R, Tuggle JT, et al. A two-gene expression ratio predicts clinical outcome in breast cancer patients treated with tamoxifen. *Cancer Cell* 2004;5:607–16.
24. Fuller AP, Palmer-Toy D, Erlander MG, and Sgroi DC. Laser capture microdissection and advanced molecular analysis of human breast cancer. *J Mammary Gland Biol Neoplasia* 2003;8:335–45.
25. Ma XJ, Salunga R, Tuggle JT, Gaudet J, Enright E, McQuary P, Payette T, Pistone M, Stecker K, Zhang BM, Zhou YX, Varnholt H, Smith B, Gadd M, Chatfield E, Kessler J, Baer TM, Erlander MG, and Sgroi DC. Gene expression profiles of human breast cancer progression. *Proc Natl Acad Sci USA* 2003;100:5974–9.
26. Shippy R, Sendera TJ, Lockner R, Palaniappan C, Kaysser-Kranich T, Watts G, and Alsobrook J. Performance evaluation of commercial short-oligonucleotide microarrays and the impact of noise in making cross-platform correlations. *BMC Genomics* 2004;5:61.
27. Ramakrishnan R, Dorris D, Lublinsky A, Nguyen A, Domanus M, Prokhorova A, Gieser L, Touma E, Lockner R, Tata M, Zhu X, Patterson M, Shippy R, Sendera TJ, and Mazumder A. An assessment of Motorola CodeLink microarray performance for gene expression profiling applications. *Nucleic Acids Res* 2002;30:e30.
28. Draghici S, Khatri P, Eklund AC, and Szallasi Z. Reliability and reproducibility issues in DNA microarray measurements. *Trends Genet* 2006;22:101–9.
29. Draghici S. *Data Analysis Tools for DNA Microarrays*. CRC Press, 2003.
30. Dorris D, Nguyen A, Gieser L, Lockner R, Lublinsky A, Patterson M, Touma E, Sendera T, Elghanian R, and Mazumder A. Oligodeoxyribonucleotide probe accessibility on a three-dimensional DNA microarray surface and the effect of hybridization time on the accuracy of expression ratios. *BMC Biotechnol* 2003;3:6.
31. Holloway AJ, Van Laar RK, Tothill RW, and Bowtell DD. Options available—from start to finish—for obtaining data from DNA microarrays II. *Nat Genet* 2002;32 Suppl:481–9.
32. Tan PK, Downey TJ, Spitznagel EL, Jr., Xu P, Fu D, Dimitrov DS, Lempicki RA, Raaka BM, and Cam MC. Evaluation of gene expression measurements from commercial microarray platforms. *Nucleic Acids Res* 2003;31:5676–84.

33. Kuo WP, Jenssen TK, Butte AJ, Ohno-Machado L, and Kohane IS. Analysis of matched mRNA measurements from two different microarray technologies. *Bioinformatics* 2002;18:405–12.
34. Kothapalli R, Yoder SJ, Mane S, and Loughran TP, Jr. Microarray results: how accurate are they? *BMC Bioinformatics* 2002;3:22.
35. Yauk CL, Berndt ML, Williams A, and Douglas GR. Comprehensive comparison of six microarray technologies. *Nucleic Acids Res* 2004;32:e124.
36. Barnes M, Freudenberg J, Thompson S, Aronow B, and Pavlidis P. Experimental comparison and cross-validation of the Affymetrix and Illumina gene expression analysis platforms. *Nucleic Acids Res* 2005;33:5914–23.
37. Paik S, Tang G, Shak S, Kim C, Baker J, Kim W, Cronin M, Baehner FL, Watson D, Bryant J, Costantino JP, Geyer CE, Jr., Wickerham DL, and Wolmark N. Gene expression and benefit of chemotherapy in women with node-negative, estrogen receptor-positive breast cancer. *J Clin Oncol* 2006;24:3726–34.
38. Paik S, Shak S, Tang G, Kim C, Baker J, Cronin M, Baehner FL, Walker MG, Watson D, Park T, Hiller W, Fisher ER, Wickerham DL, Bryant J, and Wolmark N. A multigene assay to predict recurrence of tamoxifen-treated, node-negative breast cancer. *N Engl J Med* 2004;351:2817–26.
39. Glas AM, Floore A, Delahaye LJ, Witteveen AT, Pover RC, Bakx N, Lahti-Domenici JS, Bruinsma TJ, Warmoes MO, Bernards R, Wessels LF, and Van't Veer LJ. Converting a breast cancer microarray signature into a high-throughput diagnostic test. *BMC Genomics* 2006;7:278.
40. Sorlie T, Perou CM, Tibshirani R, Aas T, Geisler S, Johnsen H, Hastie T, Eisen MB, van de Rijn M, Jeffrey SS, Thorsen T, Quist H, Matese JC, Brown PO, Botstein D, Eystein Lonning P, and Borresen-Dale AL. Gene expression patterns of breast carcinomas distinguish tumor subclasses with clinical implications. *Proc Natl Acad Sci USA* 2001;98:10869–74.
41. Armstrong SA, Staunton JE, Silverman LB, Pieters R, den Boer ML, Minden MD, Sallan SE, Lander ES, Golub TR, and Korsmeyer SJ. MLL translocations specify a distinct gene expression profile that distinguishes a unique leukemia. *Nat Genet* 2002;30:41–7.
42. Bea S, Zettl A, Wright G, Salaverria I, Jehn P, Moreno V, Burek C, Ott G, Puig X, Yang L, et al. Diffuse large B-cell lymphoma subgroups have distinct genetic profiles that influence tumor Biology and improve gene expression-based survival prediction. *Blood* 2005;106:31813–90.
43. Bohen SP, Troyanskaya OG, Alter O, Warnke R, Botstein D, Brown PO, and Levy R. Variation in gene expression patterns in follicular lymphoma and the response to rituximab. *Proc Natl Acad Sci USA* 2003;100:1926–30.
44. Golub TR, Slonim DK, Tamayo P, Huard C, Gaasenbeek M, Mesirov JP, Coller H, Loh ML, Downing JR, Caligiuri MA, Bloomfield CD, and Lander ES. Molecular classification of cancer: class discovery and class prediction by gene expression monitoring. *Science* 1999;286:531–7.
45. Ransohoff DF. Bias as a threat to the validity of cancer molecular-marker research. *Nat Rev Cancer* 2005;5:142–9.
46. Barrett T, Suzek TO, Troup DB, Wilhite SE, Ngau WC, Ledoux P, Rudnev D, Lash AE, Fujibuchi W, and Edgar R. NCBI GEO: mining millions of expression profiles—database and tools. *Nucleic Acids Res* 2005;33:D562–6.
47. Edgar R, Domrachev M, and Lash AE. Gene Expression Omnibus: NCBI gene expression and hybridization array data repository. *Nucleic Acids Res* 2002;30:207–10.
48. Parkinson H, Sarkans U, Shojatalab M, Abeygunawardena N, Contrino S, Coulson R, Farne A, Garcia Lara G, Holloway E, Kapushesky M, Lilja P, Mukherjee G, Oezcimen A, Rayner T, Rocca-Serra P, Sharma A, Sansone S, and Brazma A. ArrayExpress—a public repository for microarray gene expression data at the EBI. *Nucleic Acids Res* 2005;33:D553–5.

49. Mulder NJ, Apweiler R, Attwood TK, Bairoch A, Barrell D, Bateman A, Binns D, Biswas M, Bradley P, Bork P, et al. The InterPro Database, 2003 brings increased coverage and new features. *Nucleic Acids Res* 2003;31:315–8.
50. Chen X, Ji ZL, and Chen YZ. TTD: Therapeutic Target Database. *Nucleic Acids Res* 2002; 30:412–5.
51. Brazma A, Hingamp P, Quackenbush J, Sherlock G, Spellman P, Stoeckert C, Aach J, Ansorge W, Ball CA, Causton HC, Gaasterland T, Glenisson P, Holstege FC, Kim IF, Markowitz V, Matese JC, Parkinson H, Robinson A, Sarkans U, Schulze-Kremer S, Stewart J, Taylor R, Vilo J, and Vingron M. Minimum information about a microarray experiment (MIAME)—toward standards for microarray data. *Nat Genet* 2001;29:365–71.
52. Tusher VG, Tibshirani R, and Chu G. Significance analysis of microarrays applied to the ionizing radiation response. *Proc Natl Acad Sci USA* 2001;98:5116–21.
53. Tsai CA, Hsueh HM, and Chen JJ. Estimation of false discovery rates in multiple testing: application to gene microarray data. *Biometrics* 2003;59:1071–81.
54. Delongchamp RR, Bowyer JF, Chen JJ, and Kodell RL. Multiple-testing strategy for analyzing cDNA array data on gene expression. *Biometrics* 2004;60:774–82.
55. Lee ML, Kuo FC, Whitmore GA, and Sklar J. Importance of replication in microarray gene expression studies: statistical methods and evidence from repetitive cDNA hybridizations. *Proc Natl Acad Sci USA* 2000;97:9834–9.
56. Novak JP, Sladek R, and Hudson TJ. Characterization of variability in large-scale gene expression data: implications for study design. *Genomics* 2002;79:104–13.
57. Chen JJ, Delongchamp RR, Tsai CA, Hsueh HM, Sistare F, Thompson KL, Desai VG, and Fuscoe JC. Analysis of variance components in gene expression data. *Bioinformatics* 2004;20:1436–46.
58. Ramaswamy S, Tamayo P, Rifkin R, Mukherjee S, Yeang CH, Angelo M, Ladd C, Reich M, Latulippe E, Mesirov JP, Poggio T, Gerald W, Loda M, Lander ES, and Golub TR. Multiclass cancer diagnosis using tumor gene expression signatures. *Proc Natl Acad Sci USA* 2001;98:15149–54.
59. Bittner M, Meltzer P, Chen Y, Jiang Y, Seftor E, Hendrix M, Radmacher M, Simon R, Yakhini Z, Ben-Dor A, et al. Molecular classification of cutaneous malignant melanoma by gene expression profiling. *Nature* 2000;406:536–40.
60. van 't Veer LJ, Dai H, van de Vijver MJ, He YD, Hart AA, Mao M, Peterse HL, van der Kooy K, Marton MJ, Witteveen AT, Schreiber GJ, Kerkhoven RM, Roberts C, Linsley PS, Bernards R, and Friend SH. Gene expression profiling predicts clinical outcome of breast cancer. *Nature* 2002;415:530–6.
61. Quackenbush J. Computational analysis of microarray data. *Nat Rev Genet* 2001;2:418–27.
62. Raychaudhuri S, Sutphin PD, Chang JT, and Altman RB. Basic microarray analysis: grouping and feature reduction. *Trends Biotechnol* 2001;19:189–93.
63. Tavazoie S, Hughes JD, Campbell MJ, Cho RJ, and Church GM. Systematic determination of genetic network architecture. *Nat Genet* 1999;22:281–5.
64. Toronen P, Kolehmainen M, Wong G, and Castren E. Analysis of gene expression data using self-organizing maps. *FEBS Lett* 1999;451:142–6.
65. Brown MP, Grundy WN, Lin D, Cristianini N, Sugnet CW, Furey TS, Ares M, Jr., and Haussler D. Knowledge-based analysis of microarray gene expression data by using support vector machines. *Proc Natl Acad Sci USA* 2000;97:262–7.
66. Nutt CL, Mani DR, Betensky RA, Tamayo P, Cairncross JG, Ladd C, Pohl U, Hartmann C, McLaughlin ME, Batchelor TT, Black PM, von Deimling A, Pomeroy SL, Golub TR, and Louis DN. Gene expression-based classification of malignant gliomas correlates better with survival than histological classification. *Cancer Res* 2003;63:1602–7.

67. Doniger SW, Salomonis N, Dahlquist KD, Vranizan K, Lawlor SC, and Conklin BR. MAPPFinder: using Gene Ontology and GenMAPP to create a global gene-expression profile from microarray data. *Genome Biol* 2003;4:R7.
68. Volinia S, Evangelisti R, Francioso F, Arcelli D, Carella M, and Gasparini P. GOAL: automated Gene Ontology analysis of expression profiles. *Nucleic Acids Res* 2004;32:W492–9.
69. http://david.niaid.nih.gov/david/.
70. Vaquerizas JM, Conde L, Yankilevich P, Cabezon A, Minguez P, Diaz-Uriarte R, Al-Shahrour F, Herrero J, and Dopazo J. GEPAS, an experiment-oriented pipeline for the analysis of microarray gene expression data. *Nucleic Acids Res* 2005;33:W616–20.
71. Al-Shahrour F, Minguez P, Vaquerizas JM, Conde L, and Dopazo J. BABELOMICS: a suite of web tools for functional annotation and analysis of groups of genes in high-throughput experiments. *Nucleic Acids Res* 2005;33:W460–4.
72. Al-Shahrour F, Diaz-Uriarte R, and Dopazo J. FatiGO: a web tool for finding significant associations of Gene Ontology terms with groups of genes. *Bioinformatics* 2004;20:578–80.
73. Mlecnik B, Scheideler M, Hackl H, Hartler J, Sanchez-Cabo F, and Trajanoski Z. PathwayExplorer: web service for visualizing high-throughput expression data on biological pathways. *Nucleic Acids Res* 2005;33:W633–7.
74. www.ingenuity.com.
75. www.genego.com.
76. Harris MA, Clark J, Ireland A, Lomax J, Ashburner M, Foulger R, Eilbeck K, Lewis S, Marshall B, Mungall C, et al. The Gene Ontology (GO) database and informatics resource. *Nucleic Acids Res* 2004;32:D258–61.
77. Ashburner M, Ball CA, Blake JA, Botstein D, Butler H, Cherry JM, Davis AP, Dolinski K, Dwight SS, Eppig JT, Harris MA, Hill DP, Issel-Tarver L, Kasarskis A, Lewis S, Matese JC, Richardson JE, Ringwald M, Rubin GM, and Sherlock G. Gene ontology: tool for the unification of biology. The Gene Ontology Consortium. *Nat Genet* 2000;25:25–9.
78. www.biocarta.com.
79. Kanehisa M, Goto S, Kawashima S, Okuno Y, and Hattori M. The KEGG resource for deciphering the genome. *Nucleic Acids Res* 2004;32:D277–80.
80. Semizarov D, Kroeger P, and Fesik S. siRNA-mediated gene silencing: a global genome view. *Nucleic Acids Res* 2004;32:3836–45.
81. Zucchini C, Bianchini M, Valvassori L, Perdichizzi S, Benini S, Manara MC, Solmi R, Strippoli P, Picci P, Carinci P, and Scotlandi K. Identification of candidate genes involved in the reversal of malignant phenotype of osteosarcoma cells transfected with the liver/bone/kidney alkaline phosphatase gene. *Bone* 2004;34:672–9.
82. Meyer MH, Dulde E, and Meyer RA, Jr. The genomic response of the mouse kidney to low-phosphate diet is altered in X-linked hypophosphatemia. *Physiol Genomics* 2004;18:4–11.
83. Subramanian A, Tamayo P, Mootha VK, Mukherjee S, Ebert BL, Gillette MA, Paulovich A, Pomeroy SL, Golub TR, Lander ES, and Mesirov JP. Gene set enrichment analysis: a knowledge-based approach for interpreting genome-wide expression profiles. *Proc Natl Acad Sci USA* 2005;102:15545–50.
84. Mootha VK, Lindgren CM, Eriksson KF, Subramanian A, Sihag S, Lehar J, Puigserver P, Carlsson E, Ridderstrale M, Laurila E, Houstis N, Daly MJ, Patterson N, Mesirov JP, Golub TR, Tamayo P, Spiegelman B, Lander ES, Hirschhorn JN, Altshuler D, and Groop LC. PGC-1alpha-responsive genes involved in oxidative phosphorylation are coordinately downregulated in human diabetes. *Nat Genet* 2003;34:267–73.
85. Monti S, Savage KJ, Kutok JL, Feuerhake F, Kurtin P, Mihm M, Wu B, Pasqualucci L, Neuberg D, Aguiar RC, Dal Cin P, Ladd C, Pinkus GS, Salles G, Harris NL, Dalla-Favera R, Habermann TM, Aster JC, Golub TR, and Shipp MA. Molecular profiling of diffuse large

B-cell lymphoma identifies robust subtypes including one characterized by host inflammatory response. *Blood* 2005;105:1851–61.

86. Rosenwald A, Wright G, Chan WC, Connors JM, Campo E, Fisher RI, Gascoyne RD, Muller-Hermelink HK, Smeland EB, Giltnane JM, et al. The use of molecular profiling to predict survival after chemotherapy for diffuse large-B-cell lymphoma. *N Engl J Med* 2002;346:1937–47.
87. Huang E, Ishida S, Pittman J, Dressman H, Bild A, Kloos M, D' Amico M, Pestell RG, West M, and Nevins Jr. Gene expression phenotypic models that predict the activity of oncogenic pathways. *Nat Genet* 2003;34:226–30.
88. Bild AH, Yao G, Chang JT, Wang Q, Potti A, Chasse D, Joshi MB, Harpole D, Lancaster JM, Berchuck A, Olson JA, Jr., Marks JR, Dressman HK, West M, and Nevins JR. Oncogenic pathway signatures in human cancers as a guide to targeted therapies. *Nature* 2006;439:353–7.
89. Rhodes DR, Barrette TR, Rubin MA, Ghosh D, and Chinnaiyan AM. Meta-analysis of microarrays: interstudy validation of gene expression profiles reveals pathway dysregulation in prostate cancer. *Cancer Res* 2002;62:4427–33.
90. Jiang H, Deng Y, Chen HS, Tao L, Sha Q, Chen J, Tsai CJ, and Zhang S. Joint analysis of two microarray gene-expression data sets to select lung adenocarcinoma marker genes. *BMC Bioinformatics* 2004;5:81.
91. Sorlie T, Tibshirani R, Parker J, Hastie T, Marron JS, Nobel A, Deng S, Johnsen H, Pesich R, Geisler S, Demeter J, Perou CM, Lonning PE, Brown PO, Borresen-Dale AL, and Botstein D. Repeated observation of breast tumor subtypes in independent gene expression data sets. *Proc Natl Acad Sci USA* 2003;100:8418–23.
92. Shen R, Ghosh D, and Chinnaiyan AM. Prognostic meta-signature of breast cancer developed by two-stage mixture modeling of microarray data. *BMC Genomics* 2004;5:94.
93. Ntzani EE, and Ioannidis JP. Predictive ability of DNA microarrays for cancer outcomes and correlates: an empirical assessment. *Lancet* 2003;362:1439–44.
94. Simon R, Radmacher MD, and Dobbin K. Design of studies using DNA microarrays. *Genet Epidemiol* 2002;23:21–36.
95. Simon R. Using DNA microarrays for diagnostic and prognostic prediction. *Expert Rev Mol Diagn* 2003;3:587–95.
96. Simon R, Radmacher MD, Dobbin K, and McShane LM. Pitfalls in the use of DNA microarray data for diagnostic and prognostic classification. *J Natl Cancer Inst* 2003;95:14–8.
97. Levsky JM and Singer RH. Fluorescence in situ hybridization: past, present and future. *J Cell Sci* 2003;116:2833–8.
98. Kallioniemi A, Kallioniemi OP, Sudar D, Rutovitz D, Gray JW, Waldman F, and Pinkel D. Comparative genomic hybridization for molecular cytogenetic analysis of solid tumors. *Science* 1992;258:818–21.
99. Pinkel D and Albertson DG. Array comparative genomic hybridization and its applications in cancer. *Nat Genet* 2005;37 Suppl:S11–7.
100. Albertson DG and Pinkel D. Genomic microarrays in human genetic disease and cancer. *Hum Mol Genet* 2003;12 Spec No 2:R145–52.
101. Pinkel D, Segraves R, Sudar D, Clark S, Poole I, Kowbel D, Collins C, Kuo WL, Chen C, Zhai Y, Dairkee SH, Ljung BM, Gray JW, and Albertson DG. High resolution analysis of DNA copy number variation using comparative genomic hybridization to microarrays. *Nat Genet* 1998;20:207–11.
102. Ishkanian AS, Malloff CA, Watson SK, deLeeuw RJ, Chi B, Coe BP, Snijders A, Albertson DG, Pinkel D, Marra MA, Ling V, MacAulay C, and Lam WL. A tiling resolution DNA microarray with complete coverage of the human genome. *Nat Genet* 2004;36:299–303.

103. Clark J, Edwards S, Feber A, Flohr P, John M, Giddings I, Crossland S, Stratton MR, Wooster R, Campbell C, and Cooper CS. Genome-wide screening for complete genetic loss in prostate cancer by comparative hybridization onto cDNA microarrays. *Oncogene* 2003;22:1247–52.

104. Pollack JR, Sorlie T, Perou CM, Rees CA, Jeffrey SS, Lonning PE, Tibshirani R, Botstein D, Borresen-Dale AL, and Brown PO. Microarray analysis reveals a major direct role of DNA copy number alteration in the transcriptional program of human breast tumors. *Proc Natl Acad Sci USA* 2002;99:12963–8.

105. Hyman E, Kauraniemi P, Hautaniemi S, Wolf M, Mousses S, Rozenblum E, Ringner M, Sauter G, Monni O, Elkahloun A, Kallioniemi OP, and Kallioniemi A. Impact of DNA amplification on gene expression patterns in breast cancer. *Cancer Res* 2002;62:6240–5.

106. Barrett MT, Scheffer A, Ben-Dor A, Sampas N, Lipson D, Kincaid R, Tsang P, Curry B, Baird K, Meltzer PS, Yakhini Z, Bruhn L, and Laderman S. Comparative genomic hybridization using oligonucleotide microarrays and total genomic DNA. *Proc Natl Acad Sci USA* 2004;101:17765–70.

107. Zhao X, Li C, Paez JG, Chin K, Janne PA, Chen TH, Girard L, Minna J, Christiani D, Leo C, Gray JW, Sellers WR, and Meyerson M. An integrated view of copy number and allelic alterations in the cancer genome using single nucleotide polymorphism arrays. *Cancer Res* 2004;64:3060–71.

108. Kennedy GC, Matsuzaki H, Dong S, Liu W-m, Huang J, Liu G, Su X, Cao M, Chen W, Zhang J, Liu W, Yang G, Di X, Ryder T, He Z, Surti U, Phillips MS, Boyce-Jacino MT, Fodor SP, and Jones KW. Large-scale genotyping of complex DNA. *Nat Biotechnol* 2003;21:1233–7.

109. Zhao X, Weir BA, LaFramboise T, Lin M, Beroukhim R, Garraway L, Beheshti J, Lee JC, Naoki K, Richards WG, Sugarbaker D, Chen F, Rubin MA, Janne PA, Girard L, Minna J, Christiani D, Li C, Sellers WR, and Meyerson M. Homozygous deletions and chromosome amplifications in human lung carcinomas revealed by single nucleotide polymorphism array analysis. *Cancer Res* 2005;65:5561–70.

110. Coe BP, Ylstra B, Carvalho B, Meijer GA, MacAulay C, and Lam WL. Resolving the resolution of array CGH. *Genomics* 2007;89:647–53.

111. Hosono S, Faruqi AF, Dean FB, Du Y, Sun Z, Wu X, Du J, Kingsmore SF, Egholm M, and Lasken RS. Unbiased whole-genome amplification directly from clinical samples. *Genome Res* 2003;13:954–64.

112. Devries S, Nyante S, Korkola J, Segraves R, Nakao K, Moore D, Bae H, Wilhelm M, Hwang S, and Waldman F. Array-based comparative genomic hybridization from formalin-fixed, paraffin-embedded breast tumors. *J Mol Diagn* 2005;7:65–71.

113. Lage JM, Leamon JH, Pejovic T, Hamann S, Lacey M, Dillon D, Segraves R, Vossbrinck B, Gonzalez A, Pinkel D, Albertson DG, Costa J, and Lizardi PM. Whole genome analysis of genetic alterations in small DNA samples using hyperbranched strand displacement amplification and array-CGH. *Genome Res* 2003;13:294–307.

114. Daigo Y, Chin SF, Gorringe KL, Bobrow LG, Ponder BA, Pharoah PD, and Caldas C. Degenerate oligonucleotide primed-polymerase chain reaction-based array comparative genomic hybridization for extensive amplicon profiling of breast cancers: a new approach for the molecular analysis of paraffin-embedded cancer tissue. *Am J Pathol* 2001;158:1623–31.

115. Wang G, Brennan C, Rook M, Wolfe JL, Leo C, Chin L, Pan H, Liu WH, Price B, and Makrigiorgos GM. Balanced-PCR amplification allows unbiased identification of genomic copy changes in minute cell and tissue samples. *Nucleic Acids Res* 2004;32:e76.

116. Tanabe C, Aoyagi K, Sakiyama T, Kohno T, Yanagitani N, Akimoto S, Sakamoto M, Sakamoto H, Yokota J, Ohki M, Terada M, Yoshida T, and Sasaki H. Evaluation of a whole-genome

amplification method based on adaptor-ligation PCR of randomly sheared genomic DNA. *Genes Chromosomes Cancer* 2003;38:168–76.

117. Guillaud-Bataille M, Valent A, Soularue P, Perot C, Inda MM, Receveur A, Smaili S, Crollius HR, Benard J, Bernheim A, Gidrol X, and Danglot G. Detecting single DNA copy number variations in complex genomes using one nanogram of starting DNA and BAC-array CGH. *Nucleic Acids Res* 2004;32:e112.
118. Giannella C, Zito FA, Colonna F, Paradiso A, Marzullo F, Alaibac M, and Schittulli F. Comparison of formalin, ethanol, and Histochoice fixation on the PCR amplification from paraffin-embedded breast cancer tissue. *Eur J Clin Chem Clin Biochem* 1997;35:633–5.
119. Koshiba M, Ogawa K, Hamazaki S, Sugiyama T, Ogawa O, and Kitajima T. The effect of formalin fixation on DNA and the extraction of high-molecular-weight DNA from fixed and embedded tissues. *Pathol Res Pract* 1993;189:66–72.
120. Tbakhi A, Totos G, Pettay JD, Myles J, and Tubbs RR. The effect of fixation on detection of B-cell clonality by polymerase chain reaction. *Mod Pathol* 1999;12:272–8.
121. Ghazvini S, Char DH, Kroll S, Waldman FM, and Pinkel D. Comparative genomic hybridization analysis of archival formalin-fixed paraffin-embedded uveal melanomas. *Cancer Genet Cytogenet* 1996;90:95–101.
122. Semizarov D. unpublished observations.
123. Jacobs S, Thompson ER, Nannya Y, Yamamoto G, Pillai R, Ogawa S, Bailey DK, and Campbell IG. Genome-wide, high-resolution detection of copy number, loss of heterozygosity, and genotypes from formalin-fixed, paraffin-embedded tumor tissue using microarrays. *Cancer Res* 2007;67:2544–51.
124. Williams JG, Kubelik AR, Livak KJ, Rafalski JA, and Tingey SV. DNA polymorphisms amplified by arbitrary primers are useful as genetic markers. *Nucleic Acids Res* 1990;18:6531–5.
125. Willenbrock H, and Fridlyand J. A comparison study: applying segmentation to array CGH data for downstream analyses. *Bioinformatics* 2005;21:4084–91.
126. Lai WR, Johnson MD, Kucherlapati R, and Park PJ. Comparative analysis of algorithms for identifying amplifications and deletions in array CGH data. *Bioinformatics* 2005;21:3763–70.
127. Pollack JR, Sorlie T, Perou CM, Rees CA, Jeffrey SS, Lonning PE, Tibshirani R, Botstein D, Borresen-Dale A-L, and Brown PO. Microarray analysis reveals a major direct role of DNA copy number alteration in the transcriptional program of human breast tumors. *Proc Natl Acad Sci USA* 2002;99:12963–8.
128. Hodgson G, Hager JH, Volik S, Hariono S, Wernick M, Moore D, Albertson DG, Pinkel D, Collins C, Hanahan D, and Gray JW. Genome scanning with array CGH delineates regional alterations in mouse islet carcinomas. *Nat Genet* 2001;29:459–64.
129. Huang J, Wei W, Zhang J, Liu G, Bignell GR, Stratton MR, Futreal PA, Wooster R, Jones KW, and Shapero MH. Whole genome DNA copy number changes identified by high density oligonucleotide arrays. *Hum Genomics* 2004;1:287–99.
130. Slater HR, Bailey DK, Ren H, Cao M, Bell K, Nasioulas S, Henke R, Choo KHA, and Kennedy GC. High-resolution identification of chromosomal abnormalities using oligonucleotide arrays containing 116,204 SNPs. *Am J Hum Genet* 2005;77:709–26.
131. Eilers PHC, and de Menezes RX. Quantile smoothing of array CGH data. *Bioinformatics* 2005;21:1146–53.
132. Hsu L, Self SG, Grove D, Randolph T, Wang K, Delrow JJ, Loo L, and Porter P. Denoising array-based comparative genomic hybridization data using wavelets. *Biostatistics* 2005; 6:211–26.
133. Colella S, Yau C, Taylor JM, Mirza G, Butler H, Clouston P, Bassett AS, Seller A, Holmes CC, and Ragoussis J. QuantiSNP: an Objective Bayes Hidden-Markov Model to detect and accurately map copy number variation using SNP genotyping data. *Nucleic Acids Res* 2007;35:2013–25.

134. Shah SP, Xuan X, DeLeeuw RJ, Khojasteh M, Lam WL, Ng R, and Murphy KP. Integrating copy number polymorphisms into array CGH analysis using a robust HMM. *Bioinformatics* 2006;22:e431–9.

135. Olshen AB, Venkatraman ES, Lucito R, and Wigler M. Circular binary segmentation for the analysis of array-based DNA copy number data. *Biostatistics* 2004;5:557–72.

136. Picard F, Robin S, Lavielle M, Vaisse C, and Daudin J-J. A statistical approach for array CGH data analysis. *BMC Bioinformatics* 2005;6:27.

137. Diskin SJ, Eck T, Greshock J, Mosse YP, Naylor T, Stoeckert CJ, Jr., Weber BL, Maris JM, and Grant GR. STAC: A method for testing the significance of DNA copy number aberrations across multiple array-CGH experiments. *Genome Res* 2006;16:1149–58.

138. Shipp MA, Ross KN, Tamayo P, Weng AP, Kutok JL, Aguiar RCT, Gaasenbeek M, Angelo M, Reich M, Pinkus GS, Ray TS, Koval MA, Last KW, Norton A, Lister TA, Mesirov J, Neuberg DS, Lander ES, Aster JC, and Golub TR. Diffuse large B-cell lymphoma outcome prediction by gene-expression profiling and supervised machine learning. *Nat Med* 2002;8:68–74.

139. Olejniczak ET, Van Sant C, Anderson MG, Wang G, Tahir SK, Sauter G, Lesniewski R, and Semizarov D. Integrative genomic analysis of small-cell lung carcinoma reveals correlates of sensitivity to bcl-2 antagonists and uncovers novel chromosomal gains. *Mol Cancer Res* 2007;5:331–9.

140. Maher EA, Brennan C, Wen PY, Durso L, Ligon KL, Richardson A, Khatry D, Feng B, Sinha R, Louis DN, Quackenbush J, Black PM, Chin L, and DePinho RA. Marked genomic differences characterize primary and secondary glioblastoma subtypes and identify two distinct molecular and clinical secondary glioblastoma entities. *Cancer Res* 2006;66:11502–13.

141. Carrasco DR, Tonon G, Huang Y, Zhang Y, Sinha R, Feng B, Stewart JP, Zhan F, Khatry D, Protopopova M, Protopopov A, Sukhdeo K, Hanamura I, Stephens O, Barlogie B, Anderson KC, Chin L, Shaughnessy JD, Jr., Brennan C, and Depinho RA. High-resolution genomic profiles define distinct clinico-pathogenetic subgroups of multiple myeloma patients. *Cancer Cell* 2006;9:313–25.

142. Lee DD and Seung HS. Learning the parts of objects by non-negative matrix factorization. *Nature* 1999;401:788–91.

143. Brunet JP, Tamayo P, Golub TR, and Mesirov JP. Metagenes and molecular pattern discovery using matrix factorization. *Proc Natl Acad Sci USA* 2004;101:4164–9.

144. Ting AH, McGarvey KM, and Baylin SB. The cancer epigenome—components and functional correlates. *Genes Dev* 2006;20:3215–31.

145. Ushijima T. Detection and interpretation of altered methylation patterns in cancer cells. *Nat Rev Cancer* 2005;5:223–31.

146. Adorjan P, Distler J, Lipscher E, Model F, Muller J, Pelet C, Braun A, Florl AR, Gutig D, Grabs G, et al. Tumour class prediction and discovery by microarray-based DNA methylation analysis. *Nucleic Acids Res* 2002;30:e21.

147. Ohgane J, Hattori N, Oda M, Tanaka S, and Shiota K. Differentiation of trophoblast lineage is associated with DNA methylation and demethylation. *Biochem Biophys Res Commun* 2002;290:701–6.

148. Futscher BW, Oshiro MM, Wozniak RJ, Holtan N, Hanigan CL, Duan H, and Domann FE. Role for DNA methylation in the control of cell type specific maspin expression. *Nat Genet* 2002;31:175–9.

149. Huang TH, Perry MR, and Laux DE. Methylation profiling of CpG islands in human breast cancer cells. *Hum Mol Genet* 1999;8:459–70.

150. Yan PS, Perry MR, Laux DE, Asare AL, Caldwell CW, and Huang TH. CpG island arrays: an application toward deciphering epigenetic signatures of breast cancer. *Clin Cancer Res* 2000;6:1432–8.

151. Shi H, Yan PS, Chen CM, Rahmatpanah F, Lofton-Day C, Caldwell CW, and Huang TH. Expressed CpG island sequence tag microarray for dual screening of DNA hypermethylation and gene silencing in cancer cells. *Cancer Res* 2002;62:3214–20.
152. Shi H, Wei SH, Leu YW, Rahmatpanah F, Liu JC, Yan PS, Nephew KP, and Huang TH. Triple analysis of the cancer epigenome: an integrated microarray system for assessing gene expression, DNA methylation, and histone acetylation. *Cancer Res* 2003;63:2164–71.
153. Gitan RS, Shi H, Chen CM, Yan PS, and Huang TH. Methylation-specific oligonucleotide microarray: a new potential for high-throughput methylation analysis. *Genome Res* 2002; 12:158–64.
154. Kaneda A, Kaminishi M, Yanagihara K, Sugimura T, and Ushijima T. Identification of silencing of nine genes in human gastric cancers. *Cancer Res* 2002;62:6645–50.
155. Bartel DP. MicroRNAs: genomics, biogenesis, mechanism, and function. *Cell* 2004;116:281–97.
156. Zamore PD, and Haley B. Ribo-gnome: the big world of small RNAs. *Science* 2005;309:1519–24.
157. Calin GA and Croce CM. MicroRNA signatures in human cancers. *Nat Rev Cancer* 2006;6:857–66.
158. Rajewsky N. microRNA target predictions in animals. *Nat Genet* 2006;38 Suppl:S8–13.
159. Bartel DP and Chen CZ. Micromanagers of gene expression: the potentially widespread influence of metazoan microRNAs. *Nat Rev Genet* 2004;5:396–400.
160. Lim LP, Lau NC, Garrett-Engele P, Grimson A, Schelter JM, Castle J, Bartel DP, Linsley PS, and Johnson JM. Microarray analysis shows that some microRNAs downregulate large numbers of target mRNAs. *Nature* 2005;433:769–73.
161. Lewis BP, Burge CB, and Bartel DP. Conserved seed pairing, often flanked by adenosines, indicates that thousands of human genes are microRNA targets. *Cell* 2005;120:15–20.
162. Lu J, Getz G, Miska EA, Alvarez-Saavedra E, Lamb J, Peck D, Sweet-Cordero A, Ebert BL, Mak RH, Ferrando AA, Downing JR, Jacks T, Horvitz HR, and Golub TR. MicroRNA expression profiles classify human cancers. *Nature* 2005;435:834–8.
163. Cummins JM, He Y, Leary RJ, Pagliarini R, Diaz LA, Jr., Sjoblom T, Barad O, Bentwich Z, Szafranska AE, Labourier E, Raymond CK, Roberts BS, Juhl H, Kinzler KW, Vogelstein B, and Velculescu VE. The colorectal microRNAome. *Proc Natl Acad Sci USA* 2006;103:3687–92.
164. Liu C-G, Calin GA, Meloon B, Gamliel N, Sevignani C, Ferracin M, Dumitru CD, Shimizu M, Zupo S, Dono M, Alder H, Bullrich F, Negrini M, and Croce CM. An oligonucleotide microchip for genome-wide microRNA profiling in human and mouse tissues. *Proc Natl Acad Sci USA* 2004;101:9740–4.
165. Baskerville S and Bartel DP. Microarray profiling of microRNAs reveals frequent coexpression with neighboring miRNAs and host genes. *RNA* 2005;11:241–7.
166. Barad O, Meiri E, Avniel A, Aharonov R, Barzilai A, Bentwich I, Einav U, Gilad S, Hurban P, Karov Y, Lobenhofer EK, Sharon E, Shiboleth YM, Shtutman M, Bentwich Z, and Einat P. MicroRNA expression detected by oligonucleotide microarrays: system establishment and expression profiling in human tissues. *Genome Res* 2004;14:2486–94.
167. Wang H, Ach RA, and Curry B. Direct and sensitive miRNA profiling from low-input total RNA. *RNA* 2007;13:151–9.
168. FDA White Paper: "Innovation or stagnation, challenge and opportunity on the critical path to new medical products," available at http://www.fda.gov/oc/initiatives/criticalpath/whitepaper.html.
169. FDA White Paper: "Drug-device co-development concept paper (draft)," available at http://www.fda.gov/cder/genomics/pharmacoconceptfn.pdf.
170. FDA Guidance for industry: "Pharmacogenomics data submissions," available at http://www.fda.gov/cber/gdlns/pharmdtasub.pdf.

171. FDA Draft Guidance for Industry and FDA Staff - Pharmacogenetic Tests and Genetic Tests for Heritable Markers, available at http://www.fda.gov/cdrh/oivd/guidance/1549.html.
172. Oshita F, Ikehara M, Sekiyama A, Hamanaka N, Saito H, Yamada K, Noda K, Kameda Y, and Miyagi Y. Genomic-wide cDNA microarray screening to correlate gene expression profile with chemoresistance in patients with advanced lung cancer. *J Exp Ther Oncol* 2004;4:155–60.
173. Ramaswamy S and Golub TR. DNA microarrays in clinical oncology. *J Clin Oncol* 2002;20:1932–41.
174. Singh D, Febbo PG, Ross K, Jackson DG, Manola J, Ladd C, Tamayo P, Renshaw AA, D' Amico AV, Richie JP, Lander ES, Loda M, Kantoff PW, Golub TR, and Sellers WR. Gene expression correlates of clinical prostate cancer behavior. *Cancer Cell* 2002;1:203–9.
175. Shipp MA, Ross KN, Tamayo P, Weng AP, Kutok JL, Aguiar RC, Gaasenbeek M, Angelo M, Reich M, Pinkus GS, Ray TS, Koval MA, Last KW, Norton A, Lister TA, Mesirov J, Neuberg DS, Lander ES, Aster JC, and Golub TR. Diffuse large B-cell lymphoma outcome prediction by gene-expression profiling and supervised machine learning. *Nat Med* 2002;8:68–74.
176. Weiss MM, Kuipers EJ, Postma C, Snijders AM, Siccama I, Pinkel D, Westerga J, Meuwissen SG, Albertson DG, and Meijer GA. Genomic profiling of gastric cancer predicts lymph node status and survival. *Oncogene* 2003;22:1872–9.
177. Weiss MM, Kuipers EJ, Postma C, Snijders AM, Pinkel D, Meuwissen SG, Albertson D, and Meijer GA. Genomic alterations in primary gastric adenocarcinomas correlate with clinicopathological characteristics and survival. *Cell Oncol* 2004;26:307–17.
178. Wreesmann VB, Sieczka EM, Socci ND, Hezel M, Belbin TJ, Childs G, Patel SG, Patel KN, Tallini G, Prystowsky M, Shaha AR, Kraus D, Shah JP, Rao PH, Ghossein R, and Singh B. Genome-wide profiling of papillary thyroid cancer identifies MUC1 as an independent prognostic marker. *Cancer Res* 2004;64:3780–9.
179. Yeoh EJ, Ross ME, Shurtleff SA, Williams WK, Patel D, Mahfouz R, Behm FG, Raimondi SC, Relling MV, Patel A, Cheng C, Campana D, Wilkins D, Zhou X, Li J, Liu H, Pui CH, Evans WE, Naeve C, Wong L, and Downing JR. Classification, subtype discovery, and prediction of outcome in pediatric acute lymphoblastic leukemia by gene expression profiling. *Cancer Cell* 2002;1:133–43.
180. Katz DA, Murray B, Bhathena A, and Sahelijo L. Defining drug disposition determinants: a pharmacogenetic—pharmacokinetic strategy. *Nature Reviews Drug Discovery* 2008;7:293–305.
181. Lamba JK, Lin YS, Schuetz EG, and Thummel KE. Genetic contribution to variable human CYP3A-mediated metabolism. *Adv Drug Deliv Rev* 2002;54:1271–94.

Chapter 3

Genomic Biomarkers

3.1. INTRODUCTION TO GENOMIC BIOMARKERS

One of the key concepts of today's biomedical sciences is that of **personalized medicine**, i.e., the clinical use of individualized, genome-based pharmacotherapy. Most known diseases represent extremely complex biological phenotypes that reflect a complex of genetic and environmental factors. Understanding of the genomic factors that determine a patient's response to a drug facilitates targeted development and application of drugs to maximize the therapeutic benefit. Currently this represents a major challenge for most diseases, as the existing array of disease markers is not sufficient to derive a relationship between the patient's genotype and the molecular type of the disease on one hand and the response to existing or novel therapies on the other.

Although traditional biomarkers have been used in medicine throughout most of the twentieth century, the last decade has witnessed an exponential growth in investment in the biomarker area, in both human and financial resources. This increase is primarily driven by the realization of the potential economic value of biomarkers in the context of rapidly rising drug development costs and enabled by the recent successes in the study of the human genome. The progress of modern genomic technologies (such as gene expression microarrays, array-based comparative genomic hybridization, genotyping microarrays, and others) has facilitated discovery of a principally novel type of biomarkers, often referred to as genomic or molecular biomarkers. As the focus of this book is on the applications of genomics in drug discovery, we restrict this chapter to genomic biomarkers, i.e., biomarkers that may be discovered with genomic technologies. Today, genomic biomarker discovery represents an area of significant focus for most pharmaceutical companies. The biggest challenge in genomic biomarker discovery is in correlating the complex patterns of DNA or RNA with the biological event of interest. Because of the enormous complexity of the DNA and RNA patterns,

Genomics in Drug Discovery and Development, by Dimitri Semizarov and Eric Blomme

advanced statistical procedures are typically required to discover and validate a genomic biomarker. In Chapter 2 devoted to genomics technologies, we review multiple statistical procedures used in the analysis of expression microarray and array CGH data.

The increased interest in biomarkers led to the creation of various organizations such as the Biomarkers Definitions Working Group (1). The group has suggested the following definition of a biomarker that has subsequently been widely accepted by the community: "a characteristic that is objectively measured and evaluated as an indicator of normal biologic processes, pathogenic processes, or pharmacologic responses to a therapeutic intervention." The major applications of biomarkers in the context of pharmacological intervention are in (i) **selecting patients for therapy** and (ii) **determining the benefits of therapy**. The former relates to so-called patient stratification biomarkers, or markers that stratify the patient population into likely responders and nonresponders. A commonly cited example of the genomic biomarkers of this type is an amplification of the HER2/neu gene in breast cancer. Women carrying this amplification are selected as candidates for therapy with Herceptin®, a drug targeting the HER2 protein. Clinical trials have demonstrated that Herceptin® is efficacious in women with a HER2/neu amplification, but not in patients with a normal HER2/neu copy number (2, 3). Additionally, a genomic biomarker can be used to select patients who are less likely to exhibit a toxic response to the therapeutic under consideration. For instance, a toxic response to the oncology drug irinotecan can be predicted from the patient's genotype. Specifically, the dose-limiting toxicities of irinotecan, namely, diarrhea and leukopenia, are associated with a genetic variant in the UGT1A1 gene. The majority of people have six TA repeats in the promoter of this gene, but some have seven, and this increase results in lower expression levels of UGT1A1, resulting in increased accumulation of irinotecan and subsequent toxic responses, such as diarrhea and leukopenia (4–6). Selecting patients with the optimal window between efficacy and toxicity is the major focus of pharmacogenetics and pharmacogenomics, two scientific disciplines that offer the potential to predict occurrence of a disease, determine the potential variability in drug response and toxicity in all potential subpopulations of patients, and optimize the treatment regimen for these subgroups.

The second application of biomarkers is in monitoring the drug effect and quantifying its benefits. As changes in RNA expression patterns correlate with the expression and secretion of the protein products, early identification of an RNA biomarker predicting the on-target efficacy may facilitate subsequent characterization and optimization of lead compounds directed at the target of interest. As the compound progresses toward clinical trials, a biomarker suitable for assessment of clinical benefit may be derived from the initial RNA marker. For instance, if the protein product of the marker gene is secreted and can be measured in serum, it may be used to quantify the benefit of the drug in humans. Alternatively, measurements of the protein product in the diseased tissue may be considered, although tissue availability may limit the utility of this biomarker. Similarly, if early toxicogenomics tests yield an RNA marker of toxicity, the latter can then be used to monitor toxicity in preclinical and clinical studies. However, today, the major application of toxicogenomics is in compound selection and optimization, as toxicogenomic signatures represent complex composite biomarkers that are typically measured in the target organs of toxicity.

It is noteworthy that other biomarker classifications exist, such as the classification developed by the Biomarkers and Surrogate Endpoints Working Group (1). The group classifies biomarkers into type 0, type I, and type II. Type 0 biomarkers are markers of the natural history of the disease, which correlate longitudinally with known clinical indices of the disease. Type I biomarkers reflect the effects of a therapeutic intervention and are related to the mechanism of action of the drug. Type II markers are also referred to as surrogate end points, as they predict clinical benefit. For the sake of consistency and greater focus on drug discovery, in this chapter we classify biomarkers into patient stratification biomarkers and monitoring biomarkers, including biomarkers of efficacy and toxicity.

BOX 3-1 *Terminology Developed by the Biomarker Definitions Working Group*

Biomarker

A characteristic that is objectively measured and evaluated as an indicator of normal biological processes, pathogenic processes, or pharmacological responses to a therapeutic intervention

Clinical End Point

A characteristic or variable that reflects how a patient feels, functions, or survives

Surrogate End Point

A biomarker that is intended to substitute for a clinical end point. A surrogate end point is expected to predict clinical benefit (or harm or lack of benefit or harm) based on epidemiological, therapeutic, pathophysiological, or other scientific evidence.

Biomarker discovery is closely related to the other applications of genomics covered in this book, such as characterization of compound efficacy and toxicity and candidate compound optimization. For instance, mechanism of action studies using microarrays often result in identification of efficacy biomarkers, either directly (by pointing at significant upregulation of a transcript) or indirectly (by pointing at activation or repression of a signaling pathway that has known biomarkers). Early characterization of compound toxicity may reveal specific markers (single gene or composite) that may serve as early predictors of toxicity. Therefore the application of genomic technologies in the discovery of biomarkers of efficacy or toxicity or patient stratification biomarkers is an integral part of this book. At the same time, the identification of diagnostic markers unrelated to pharmacological intervention is a separate field that will not be covered in this book.

Biomarkers are especially valuable if they can be used early in discovery to prioritize therapeutic targets, optimize the lead compound both in terms of predicted efficacy and toxicity, and generate hypotheses about the patient subpopulations that are more likely to respond to the drug with minimal toxicity. Biomarker programs can be initiated as early as target validation and compound optimization stages and can be applied to the existing in vitro and in vivo model systems. The biomarker candidates identified in the models systems can then be tested and validated in patients during early-stage clinical trials. The aforementioned example of HER2 is an intriguing illustration of the feasibility of such a transition. It is known that a number of breast cancer cell lines carry an amplification of the HER2/neu gene and consequently overexpress the HER2 protein. When Herceptin® was tested in cell-based assays it displayed selective potency toward cell lines with the amplification, and this correlation between the amplification and response was validated in humans. When the HER2 amplification marker was used to select patients for Herceptin® therapy, statistically significant response rates were achieved that would not be attainable in the unselected population (2, 3). Thus early incorporation of biomarker programs into drug discovery may facilitate prioritization of drug discovery resources and optimize the design of subsequent clinical trials. Therefore a major focus of this chapter will be on *early* identification of biomarkers in drug discovery.

Various structural modifications of the genome and differences in gene expression have been considered as possible genomic biomarkers. This chapter is organized by the type of nucleic acid measured and the type of modification detected (Table 3.1). The list of **DNA modifications** that can be reliably measured today by the existing technologies includes copy number changes, mutations, single nucleotide polymorphisms (SNPs), and methylation. Known **RNA-based biomarkers** include single transcripts and composite gene expression "signatures." The latter have received the most attention in the past decade, mostly because of the early adoption and higher degree of maturity of the gene expression microarray and QPCR technologies. In the past year, a number of intriguing studies have been conducted that associated microRNA (miRNA) profiles of cells with oncogenic processes in some cancers, thus raising the possibility of using miRNA as a stratification biomarker as the field matures. This prompted us to include miRNAs in this chapter.

For each biomarker type, we comprehensively review the discovery process and the applications of the existing markers to stratify the patient population and monitor the drug effects. Before genomic biomarkers can become an integral part of clinical practice, a number of issues must be addressed related to their analytical and clinical validity and utility. Therefore, special attention is paid to the validation of genomic biomarkers and demonstration of their clinical utility.

Table 3.1 Types of Genomic Biomarkers Covered in This Chapter

Genomic Biomarker Type		Detection Technology
DNA	Copy number changes	Comparative genomic hybridization (CGH); quantitative PCR (QPCR); fluorescent *in situ* hybridization (FISH)
	SNPs	Genotyping microarrays, sequencing, bead-based assays
	Mutations	Sequencing, microarrays, indirect methods (PCR amplification followed by HPLC or cleavage assays)
	Methylation	Methylation microarrays
RNA	Single transcript expression	Quantitative RT-PCR (Q-RT-PCR)
	Gene expression signatures	Gene expression microarrays; multiplex Q-RT-PCR
	miRNA	miRNA microarrays

3.2. DNA BIOMARKERS

As the genetic changes in cells often represent the driving force behind the pathogenic process in a diseased tissue, identification of the exact DNA modification events that produce the pathological phenotype and determine its responsiveness to therapies represents a logical and common strategy in biomarker discovery. However, although the sequence of the human genome was published more than 5 years ago (7, 8), only a handful of DNA modifications have been reliably and quantitatively associated with responsiveness to drugs or susceptibility to their toxic effects. The challenge is to establish a quantitative relationship between the marker and the response, demonstrate the statistical significance, and achieve the necessary level of validation in a relevant patient population.

So far, DNA biomarkers capable of predicting response to therapy have been derived from several sources. The first one, DNA copy number changes, is particularly important in cancer, as oncogenic transformation is often driven by amplifications of oncogenes and loss of tumor suppressor genes. Copy number changes have also been detected in a number of other diseases, such as neurodevelopmental syndromes and multiple congenital abnormalities, but they have not yet been transformed into useful biomarkers in these settings. In this chapter, we comprehensively cover the use of genomic technologies in the discovery of DNA copy number abnormalities as biomarkers in the field of cancer.

Other important types of DNA alterations that may serve as stratification markers are mutations and germ line polymorphisms. Mutations in the coding sequence of drug targets may result in the altered affinity of the drug to the target and consequently higher sensitivity of the cells to the therapeutic. Similarly, the presence of single nucleotide polymorphisms (SNPs) in the gene coding for the drug target is an important consideration in drug discovery, as they may result in variability in drug sensitivity in different populations and therefore serve as patient stratification markers. Mutations that can potentially be used as biomarkers are considered in this chapter, while germ line polymorphisms are reviewed in Chapter 7.

Finally, epigenetic modifications of DNA, primarily DNA methylation, have been shown to be an important factor in tumorigenesis. Methylation of the promoter regions causes silencing of genes and is known to deregulate the function of genes involved in malignant transformation and maintenance of the malignant phenotype. The variety of epigenetic modifications contributes to the genomic heterogeneity of cancer and therefore represents an opportunity for patient stratification. Although most of the work related to identification of epigenetic modifications for cancer classification has been of an exploratory nature, the significant potential of this epigenetic marker in patient stratification and the relative maturity of the appropriate genomic technologies have prompted us to include a discussion of this marker type in this chapter.

3.2.1. DNA Copy Number Alterations

Alterations at the DNA level are fundamental events underlying the initiation and progression of various human diseases. Chromosomal aberrations are detrimental events associated with a number of developmental diseases, such as neurodevelopmental syndromes and multiple congenital abnormalities. Amplifications and deletions of chromosomal regions occurring in somatic cells are believed to be one of the main factors leading to cancer. To date, gene copy number changes have received the most study in cancer. Most of this subsection is therefore devoted to the review of copy number alterations as biomarkers in cancer, and a brief summary of the current state of relevant research in neuroscience is given at the end.

3.2.1.1. DNA Copy Number Alterations in Cancer

Historically, the role of chromosomal aberrations was first recognized and investigated in leukemias (for a comprehensive review, see (9)). Recurrent structural aberrations of chromosomes are extremely frequent events in blood cancers. A common aberration in leukemia is a balanced chromosomal translocation, an event involving fusion of two chromosomes or balanced exchange of genetic material between two chromosomal regions. One of the most common abnormalities is the t(9; 22) translocation in chronic myelogenous leukemia (CML) (9, 10). This translocation results in the synthesis of a fusion protein, a BCR-ABL tyrosine kinase, whose oncogenic role has been firmly established (11). A drug targeting the fusion protein kinase has recently been developed and shown to be safe and efficacious in patients with CML (12). However, the critical role of chromosomal aberrations is not limited to leukemias. It has more recently been demonstrated that multiple types of chromosomal abnormalities are involved in the pathogenesis of solid tumors (for a review, see (13)). A broad range of chromosomal aberrations has been detected in solid tumors, including changes in ploidy, gains or losses of individual chromosomes, balanced and unbalanced translocations, and gains and losses of chromosomal fragments. Several efforts to catalog recurrent chromosomal abnormalities in cancers are under way, resulting in a number of useful public databases (Table 3.2).

Because the focus of this book is on the application of high-throughput genomics technologies in biomarker discovery, we limit the scope of the present chapter to **unbalanced chromosomal aberrations**, events that are accompanied by gene copy number changes and therefore are amenable to genome-wide detection by microarray-based methodologies. Amplifications of known oncogenes and losses of tumor suppressor genes are proven to play an important role in tumorigenesis. In many cases it is not easy to identify genes driving tumorigenesis based on information on recurrent amplifications, as amplified chromosomal regions often contain multiple genes, including potential drivers of oncogenic processes. However, the involvement of individual genes in an amplicon in tumorigenesis may be clarified by additional studies, such as analysis of gene expression, targeted gene silencing, and various functional assays (14). It has been demonstrated that DNA copy number gains often cause gene overexpression (15–17).

Several technologies have been used successfully to identify gene copy number changes. Although fluorescent *in situ* hybridization (FISH; for a recent review, see (18)) has been effectively applied to analyze known genetic aberrations for decades, until recently there was no method for detecting gene copy number alterations on a whole-genome scale. Comparative genomic hybridization (CGH), a technique that enables genome-wide analysis of chromosomal aberrations, is rapidly gaining acceptance in the biomarker community as the microarray-based CGH platforms mature. A detailed description of the technical aspects of CGH is given in Chapter 2 devoted to genomic technologies.

Perhaps the most successful example of a DNA copy number abnormality as a biomarker is the HER2 amplification in breast cancer. Because the discovery of the HER2 marker is a prototypical story of genomic biomarker identification from the early discovery of the HER2-targeting drug Herceptin® to regulatory

Table 3.2 Public Databases of Chromosomal Aberrations in Human Cancer

Database	URL
National Cancer Institute Cancer Genome Anatomy Project (CGAP)	http://cgap.nci.nih.gov
Mitelman Database of Chromosome Aberrations in Cancer	http://cgap.nci.nih.gov/Chromosomes/Mitelman
Charité Hospital online CGH tumor database	http://amba.charite.de/~ksch/cghdatabase/index.htm
Progenetix CGH database	www.progenetix.de
SKY, M-FISH, and CGH database at NCI	http://www.ncbi.nlm.nih.gov/sky/
Atlas of Genetics and Cytogenetics in Oncology and Haematology	http://atlasgeneticsoncology.org/index.html

approval and successful application in the clinic, we believe it deserves a detailed description in this chapter. The HER2 gene product is a transmembrane growth factor receptor normally expressed in secretory epithelia. It controls an intracellular signaling pathway that regulates growth and development (19). In a normal cell that expresses HER2, there are two copies of the gene and about 50,000 molecules of the protein on the cell surface. However, 20–30% of breast cancers carry an amplification of the HER2 gene, which leads to significant overexpression of the protein. The relationship between HER2 amplification and breast cancer clinical course and survival was described as early as 1987 (20). The amplification and overexpression of HER2 in breast cancer correlates with lymph node metastasis, high nuclear grade, relatively higher cell proliferation rate, negative hormone receptor status, and an overall more aggressive clinical course (19, 20).

As a known mediator of pro-oncogenic processes in the cell, HER2 received early attention as a therapeutic target for breast cancer. A humanized monoclonal antibody was promptly developed to target the extracellular portion of the receptor molecule (21). The antibody, named trastuzumab (Herceptin®), acts by binding to the HER2 receptor and causing its internalization and the blockage of signal transduction (2, 3, 21, 22). It was first tested on cultured breast cancer cells (23). Of six mammary carcinoma cell lines tested, only the lines with HER2 amplification (SK-BR3, MBA-MB-175, and MDA-MD-361) were sensitive to the antibody. The established correlation between HER2 amplification and sensitivity to Herceptin was later used in the clinical development of the drug (2, 3). The antibody has been shown to be effective in HER2-amplified patients when used as a single agent (3) or in combination with traditional chemotherapeutic agents (24, 25).

Both HER2 gene amplification and protein overexpression can be detected in formalin-fixed paraffin-embedded (FFPE) clinical samples (26). Initially, the immunohistochemistry-based detection test was the method of choice, but in the past decade evidence has been accumulated that points to a higher accuracy and a higher prediction power of the FISH-based detection of the HER2 gene amplification (27). Currently, HER2-targeted FISH diagnostic tests are available commercially, and the label for trastuzumab (Herceptin®) specifies that the drug should only be used in HER2-amplified patients. The HER2 story clearly represents a hallmark in the history of personalized medicine, as one of the first examples of successful patient stratification strategy based on a genomic biomarker. It is noteworthy that the correlation between HER2 amplification and sensitivity to trastuzumab was first detected in preclinical model systems and then reproduced in clinical trials, thus setting a precedent for early identification of genomics biomarkers in the drug discovery process.

Another example of a known gene copy number abnormality correlating with the sensitivity to a drug is an amplification of the EGFR gene. The EGFR copy number has recently been shown to predict response to gefitinib, a drug targeting EFGR tyrosine kinase and originally approved for treatment of lung cancer (28–30). Although no predictive molecular markers had been identified at the time of approval, somatic mutations in the kinase domain of *EGFR* have been subsequently shown to correlate to gefitinib sensitivity, and more recently, an increase in EFGR gene copy number was linked to gefitinib response (28–30). More specifically, in one study the time to progression was significantly longer in patients with increased *EGFR* copy numbers (median, 3.0 vs. 1.4 months; log-rank P value = 0.021) (28). In another study, tumors from 102 non-small-cell lung carcinoma patients treated with gefitinib were evaluated for EGFR status by FISH and immunohistochemistry. Amplification or high polysomy of the EGFR gene (seen in 33 of 102 patients) and high protein expression (seen in 58 of 98 patients) were statistically significantly associated with better response (36% vs. 3%, mean *difference* = 34%, 95% CI = 16.6–50.3; P value < 0.001) (29).

In addition to these advanced examples of gene copy number as stratification biomarker, multiple earlier proof-of-concept studies have been conducted to identify molecular subgroups of various cancers and to correlate copy number abnormalities with the disease outcome and susceptibility to treatments. As the CGH technology matures, it becomes the method of choice for initial genome-wide screening of cancer patient samples. For example, neuroblastoma, a childhood tumor derived from neural crest cells, displays remarkable genetic heterogeneity, with several well-known recurrent aberrations, such as MYCN amplification, 17q gains, and 1p losses (31, 32). The presence of these aberrations correlates with the outcome (31, 32), raising a possibility of patient stratification in the development of therapeutics against the disease.

Recently, a highly annotated panel of 101 prospectively collected diagnostic neuroblastoma primary tumors was profiled on oligonucleotide-based microarrays (33). Gene copy number at the prognostically relevant loci 1p36, 2p24 (*MYCN*), 11q23, and 17q23 was determined by quantitative PCR and found to be aberrant in 26, 20, 40, and 38 cases, respectively. In addition, 72 diagnostic neuroblastoma primary tumors assayed in a different laboratory were used as an independent validation set. Unsupervised hierarchical clustering showed that gene expression was highly correlated with genomic alterations and clinical markers of tumor behavior. The vast majority of samples with *MYCN* amplification and 1p36 loss of heterozygosity (LOH) clustered together on a terminal node of the sample dendrogram, whereas the majority of samples with 11q deletion clustered separately and both of these were largely distinct from the copy number-neutral group of tumors. The accumulated body of data on gene copy number profiles of neuroblastoma suggests that neuroblastoma drug development programs would benefit from a stratification scheme, whereby the model system is screened for genetic abnormalities and the response to the agent is correlated with the copy number profile.

In the past few years, CGH has been used to discover patient stratification biomarkers for several types of cancer. In archived tumors from 64 prostate cancer patients, 32 of whom had recurred postoperatively, a loss at 8p23.2 was identified (34). The loss was associated with advanced disease, but most importantly, a chromosomal gain was discovered at 11q13.1 that was predictive of postoperative recurrence, independent of the stage of the disease. One gene (MEN1) coding for the nuclear protein menin was mapped to the amplified region, and its expression was correlated with disease recurrence, thus establishing the gene as a biomarker of the aggressive recurrent disease. In another study, 35 gastric carcinomas were profiled on ~2400 – element BAC CGH arrays, and the patterns of chromosomal aberrations were correlated with the clinical history of the patients (35). Hierarchical clustering of the CGH profiles revealed three predominant groups. Membership in these groups correlated with lymph node status and survival. Patients from cluster 3 had a significantly better prognosis than patients from clusters 1 and 2. Although no commonly amplified genes were identified, this study clearly demonstrates the power of genome-wide CGH in patient stratification and biomarker discovery.

In breast cancer, analysis of copy number alterations suggests the existence of several genomic subgroups of the disease that differ in clinical features and prognosis. Recently, 305 unselected primary invasive breast tumors were profiled by CGH to identify recurrent copy number alterations (36). Abnormalities were detected in >90% of the tumors and involved all chromosomes, most frequently 1q (gains in 55% of the tumors), 8q (losses in 41%), 16p (losses in 40%), 17q (losses in 28%), 13q (losses in 27%), 16q (losses in 22%), 20q (losses in 19%), 8p (losses in 18%), and 11q (losses in 16%). When these copy number abnormalities (CNAs) were tested for association with Fisher's Exact test, 18 pairs of CNAs were found to be significantly associated. The most frequent pairs were 8p loss/8q loss, 17q loss/20q loss, and 4q loss/13q loss. To identify subgroups within the population, principal component analysis and distance-based tree modeling were applied. Three distinct composite patterns were observed. Group A was defined by losses at 1q, 16p, and 16q, group B by losses at 11q, 20q, 17q, and 13q, and group C by losses at 8p and 8q. These alterations correlated to positive estrogen receptor (ER) and progesterone receptor (PgR) status (P value < 0.001 and <0.05, respectively). Molecular groups B and C were associated with DNA ploidy changes (P value < 0.001 and <0.05, respectively), high histological grade, and lymph node positivity (P value < 0.05). Tumors in group B also had a high proliferation rate, large size (P value < 0.001), and negative PgR status (P value < 0.05). Patients with type A tumors only had a significantly higher breast cancer survival rate than all other patients. The worst survival was observed in group C. The 5-year survival rates varied from 96% in group A to 56% in group C. These correlations were independent of node status, tumor size, and PgR status in a multivariate analysis. This study establishes a relationship between the genomic portrait of a tumor and its histopathological and clinical features and clearly demonstrates the power of CGH in cancer classification.

The genetic heterogeneity of breast cancer has also been addressed by other CGH studies. A panel of 148 primary tumors was screened with CGH arrays containing 287 BAC clones representing cancer-related loci to obtain molecular portraits of this tumor type (37). Chromosomal gains were detected in 136 tumors (91.9%) and losses in 123 tumors (83.1%). Recurrent amplifications were observed at 8q11–qtel, 1q21–qtel, 17q11–q12, and 11q13, and frequent losses were observed at 16q12–qtel, 11ptel–p15.5, 1p36–ptel, 17p11.2–p12, and 8ptel–p22. The obtained genomic profiles were correlated with the survival data. Patients with tumors carrying less than 5% total copy number changes had a better overall survival (log-rank test: P value = 0.0417). Unsupervised hierarchical clustering of the copy number profiles revealed four clusters in the panel. Poor prognosis was associated with the cluster characterized by amplification of three regions containing known breast oncogenes (11q13, 17q12, and 20q13). The tumors in the good prognosis group [Nottingham Prognostic Index (NPI) $\leq$3.4] displayed frequent losses of 16q24–qtel. This study illustrates the potential of CGH in the development of a molecular taxonomy of breast cancer and providing a path for identification of genomic biomarkers for different subgroups of the disease. However, the limited genome coverage of the array used limits the utility of the classification.

Another recent CGH study of breast cancer attempted to identify genomic features that would differentiate between the known histological subtypes of the disease, invasive ductal carcinoma and invasive lobular carcinoma (38). These two types differ with regard to presentation, metastatic spread, and epidemiological features. Bacterial artificial chromosome (BAC) CGH microarrays with an average resolution of 0.5 Mb were used to profile 40 invasive breast cancers. The data were subjected to classification and unsupervised hierarchical cluster analyses. The best discrimination between the two subtypes was achieved when combining the copy number patterns of chromosome arms 1q and 16p, which were significantly more often gained in invasive lobular carcinoma. These regions were further condensed to subregions 1q24.2–25.1, 1q25.3–q31.3, and 16p11.2. Located within the candidate gains on 1q are two genes, *FMO2* and *PTGS2*, known to be overexpressed in invasive lobular carcinoma relative to invasive ductal carcinoma. Unsupervised hierarchical cluster analysis identified three molecular subgroups that are characterized by different aberration patterns, in particular concerning gain of *MYC* (8q24) and the identified candidate regions on 1q24.2–25.1, 1q25.3–q31.3, and 16p11.2. These genetic subgroups differed with regard to histology, tumor grading, frequency of alterations, and estrogen receptor (ER) expression. Thus copy number abnormalities on 1q and 16p were identified as significant classifiers of histological and molecular subgroups in breast cancer. It remains to be demonstrated whether these abnormalities will correlate with the sensitivity of cancers to therapy.

Correlation between gene copy number abnormalities and response to therapy in breast cancer was explored in a recent study of ER-positive breast cancer samples (39). A significant proportion of ER-positive breast cancers recur after tamoxifen therapy, which severely worsens the prognosis. Array CGH on microarrays containing 1440 human BACs was used to assess copy number changes in 28 fresh-frozen ER-positive breast cancer tissues. All of the patients included had received at least 1 year of tamoxifen treatment. Nine patients had distant recurrence within 5 years (recurrence group) of diagnosis, and 19 patients were alive without disease at least 5 years after diagnosis (nonrecurrence group). In an unsupervised clustering analysis, samples from each group were well separated. The average frequency of copy number changes was similar between the two groups. The most significant chromosomal alterations found more often in the recurrence group by two different statistical methods were loss of 11p15.5–p15.4, 1p36.33, 11q13.1, and 11p11.2 (adjusted *P values* < 0.001). In subgroup analysis according to lymph node status, loss of 11p15 and 1p36 were found more often in the recurrence group, with borderline significance within the lymph node-positive patients (adjusted P value $= 0.052$). In the Kaplan–Meier analysis of the 28 patients, both loss at 11p15.5 and gain at 8q21.13 were significantly associated with distant metastasis-free survival (P value < 0.001 and P value $= 0.006$, respectively). The loss of 1p36.33 was also a significant prognostic factor for distant recurrence (P value < 0.001). Multivariate analysis using the Cox proportional hazard model with parameters including age, T stage, lymph node status, nuclear grade, and progesterone receptor revealed that loss of 11p15.5 was the most significant factor among the different variables [hazard ratio 12.3 (95% confidence interval: 2.7–55.4)]. It was thus demonstrated that the patterns of genomic alterations detected by array CGH are different between ER-positive breast cancer patients in the recurrence and nonrecurrence groups after surgery and tamoxifen treatment. The loss of 11p15 and 1p36 and gain of 8q21 were significantly associated with distant recurrence of the disease within 5 years of diagnosis. These copy number alterations can thus be regarded as candidate markers for tamoxifen resistance in ER-positive breast cancers. It is noteworthy that the gene copy number alterations implicated in this study as markers of tamoxifen resistance had also been identified in other studies as recurrent events in breast cancer. This confirms the clinical relevance of the aberrations and supports their further examination.

A number of proof-of-concept studies have been performed to identify genes amplified in specific tumors. For example, a CGH screen of ovarian and breast cancers using array CGH revealed a recurrent amplification on chromosome 1q22 (14). The amplification was centered on the gene coding for a small GTPase called RAB25. The aberration was associated with decreased survival in both types of cancer. RAB25 was then shown to increase anchorage-independent cell proliferation and suppress apoptosis and anoikis. The authors concluded that RAB25 represents an attractive therapeutic target. Because of the association of RAB25 with survival and proliferation, it is tempting to speculate that this recurrent amplification may present a genomic biomarker that defines the aggressiveness of the tumor and its susceptibility to therapies.

Arrays based on BACs have been used to characterize the genomic profiles of malignant fibrous histiocytoma, an aggressive tumor that shows no distinct line of differentiation (40). The 6q23 band was found to be frequently amplified in this type of cancer. The authors characterized the genes residing in the amplified region and identified ASK1 (MAP3K5) as a candidate target for therapeutic intervention. In another study, 44 lung adenocarcinomas and small-cell carcinomas were profiled to determine the most frequent aberrations in these tumors (41). Two genomic regions, 8p12 and 20q11, were found to be frequently amplified in both types of cancer. Several genes residing in these regions were identified, and their overexpression was confirmed by quantitative RT-PCR. Both CGH and gene expression microarrays have been applied to study uveal melanoma (42). It was shown that the gain of chromosome 8 correlates most strongly with the expression of DDEF1, a gene located at 8q24. It was also found that overexpression of DDEF1 results in increased cell motility. The authors concluded that DDEF1 represents an attractive therapeutic target.

Diffuse large B-cell lymphoma is a highly heterogeneous disease that displays significant diversity with respect to clinical presentation and outcome. Different subtypes of the disease require different therapeutic approaches. A recent report (43) described CGH profiling of 224 diffuse large B-cell lymphomas previously examined by expression microarrays. The authors reported a high degree of correlation between the gene copy number and the expression data and identified a chromosomal region (3p11–p12) that provided prognostic information that is statistically independent of the previously built gene expression-based model.

Several less common tumors showing significant variability in outcome have also been studied by CGH. Papillary thyroid cancer (PTC) has two distinct variants, the more aggressive tall-cell thyroid cancer (TCV) and conventional thyroid cancer (cPTC). A panel of 25 TCVs and 45 cPTCs was profiled by CGH and gene expression microarrays to identify genomic features that would distinguish between the two subtypes of the disease (44). Significant differences were identified in the patterns of chromosomal gains and losses. One gene, MUC1, was of particular interest because it was both amplified and overexpressed in TCV. Its overexpression was confirmed by immunohistochemistry on independent TCV samples. Multivariate analysis showed a significant correlation between MUC1 expression level and the outcome of treatment, establishing the gene as a prognostic marker in PTC.

Copy number-based predictors of disease outcome have recently been determined for a rare tumor called giant-cell tumor of bone (GCTb), a disease for which no predictive markers currently exist (45). Array CGH was used to profile 20 frozen samples of GCTb. A separate subset of 59 GCTb samples with outcome data was used for validation. A gain of a 1-Mb region at 20q11.1 was identified as the most frequent copy number change in the initial data set. In the validation set of 59 cases the minimal common region of copy number gain at 20q11.1 was observed in 54% of the samples. The minimal region contained genes TPX2 and BCL2L1 (an antiapoptotic Bcl-2 family member also known as Bcl-xL). Confirmatory Southern blot analysis for these two genes identified TPX2 as the gene with the highest copy number. Immunohistochemistry for TPX2 expression correlated with amplification, while BCL2L1 expression was not detected. Kaplan–Meier analysis for progression-free survival showed a statistically significant difference between the groups with and without the 20q11.1 amplification ($P = 0.0001$). Multivariate analysis identified the 20q11.1 amplification ($P = 0.001$) as the only marker for poor outcome that reached statistical significance. Thus the 20q11.1 amplification can be used as a marker of prognostic importance in GCTb. *TPX2* was proposed as a candidate oncogene in the core-amplified region at 20q11.1.

First steps have been made toward developing a diagnostic tool based on a CGH array. A novel CGH chip has been designed that can detect recurrent chromosomal abnormalities in B-cell chronic lymphocytic leukemia (B-CLL) (46). Availability of an accurate diagnostic tool is particularly important in this disease, because it is characterized by a highly variable clinical course ranging from rapid progression with fatal outcome to normal life expectancy. Since the presence of specific chromosomal aberrations has been clearly associated with prognosis, a targeted CGH array would present an opportunity to simultaneously assess multiple prognostic markers for improved outcome prediction. The chip contained a total of 644 DNA elements and covered all the known regions frequently altered in B-CLL, as well as some other B-cell neoplasms. This chip was validated in a series of 106 B-CLL patients in a blinded fashion by correlating the CGH profiles to interphase cytogenetic data obtained by FISH. The chip reliably detected known aberrations and CLL and revealed a number of novel abnormalities, thus providing a platform for future diagnostic development.

In summary, high-throughput detection of gene copy number aberrations in cancer has already enabled a new cancer classification, whereby patterns of abnormalities define a cancer subtype. Application of CGH to several tumor types has resulted in identification of new recurrent regions of amplification, yielding new potential therapeutic targets. Further development and commercialization of CGH technology will likely yield a comprehensive picture of recurrent chromosomal aberrations in different tumors and thus will result in a new taxonomy of cancer and improved patient stratification strategies.

3.2.1.2. DNA Copy Number Alterations in Other Diseases

While structural aberrations in the genome have been causally associated with cancer for many years, it is only recently that gene copy number alterations were implicated in other diseases. It is now well known that a number of neurodegenerative and neurodevelopmental disorders are associated with genomic rearrangements in germ line DNA (47). The list of such disorders includes peripheral and central nervous system neuropathies, well-recognized syndromes with characteristic behavioral or neurocognitive phenotypes, as well as several psychiatric illnesses. Moreover, indisputable evidence has been accumulated that points to a role for copy number aberrations in the pathogenesis of neurodevelopmental and neurodegenerative disorders.

One of the best-studied examples of copy number change in neuroscience is a deletion of a 1.6-Mb region on chromosome 7q11.23 in Williams–Beuren syndrome. The region contains 28 genes, some of which play a role in language skills and visuospatial proficiency. Consistent with this observation, patients with William-Beuren syndrome exhibit learning disability or mental retardation, featuring severe visuospatial construction deficits, as well as attention-deficit hyperactivity disorder (47).

A known copy number abnormality associated with neurodegenerative disease is a germ line duplication of a region on chromosome 21q21 containing the gene encoding the amyloid precursor protein (APP). This duplication ranges in size from 580 kb to 6.37 Mb and causes autosomal dominant early-onset Alzheimer disease, which is accompanied by dementia and cerebral amyloid angiopathy (48). Consistent with this, increased expression of APP in a trisomic mouse model of Down syndrome has been shown to disrupt nerve growth factor transport and induce degeneration of cholinergic neurons (49). A number of other chromosomal regions, such as 8q22.1–24.1 and 15q.14, have been linked to psychiatric disorders (47).

The main implication of these findings for drug discovery is that nervous system disorders caused by gene copy number abnormalities may be treated by correcting the gene dosage. One of the possible therapeutic strategies is to correct the abnormal gene expression resulting from the change in the gene copy number. A number of approaches are currently being explored in various in vivo disease models (47). Once candidate therapeutics are identified, we anticipate that their development will involve biomarker programs directed at identification of copy number correlates of disease outcome and drug response.

3.2.1.3. Identification of DNA Copy Number Biomarkers in Drug Discovery

Although the bulk of the studies on gene copy number abnormalities have been of an exploratory nature, one can expect that comprehensive copy number profiling of clinical samples will become routine in the near future. Once the recurrent aberrations are identified in the patient population of interest, the population can be stratified based on the presence or absence of a specific aberration or combination of aberrations. If a correlation is found between a specific genomic profile (such as a HER2 amplification described in the beginning of this subchapter) and the response to an existing drug or a therapeutic agent in development, then the profile can be explored as a patient stratification marker (Fig. 3.1A). If a correlation is found, a FISH probe can be developed for use as a companion diagnostic tool for the therapeutic agent under consideration.

An alternative, proactive approach involves application of comparative genomic hybridization early in the discovery process to analyze the copy number profiles of the model systems used to screen candidate compounds (Fig. 3.1B). Once a copy number abnormality (or a pattern of abnormalities) is found that correlates with the sensitivity of the model system to the candidate compound, it is further explored in in vivo model systems, such as human tumor xenografts in mice. At this stage, additional studies can be done to establish functional relationships between the gene amplification and the sensitivity to the agent. For example, the genes in the amplified region can be knocked down with an siRNA and the effects of the knockdown on the cell phenotype can be analyzed. If the preliminary data look promising, a knockout mouse can be created to further ascertain the mechanism of drug sensitivity related to the gene copy number abnormality. At this point, the copy number abnormality is considered a candidate stratification biomarker. If compound optimization continues at

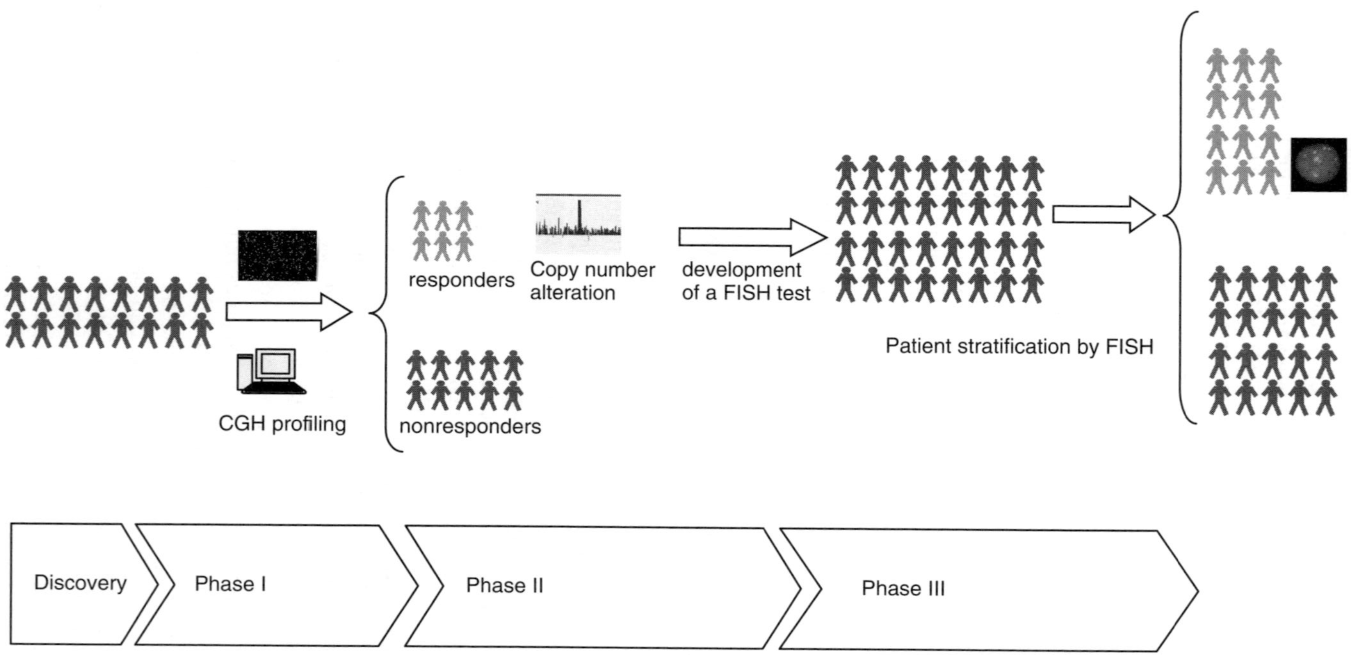

Figure 3.1 Applications of gene copy number profiling in the discovery of patient stratification biomarkers. A) Approach based on CGH profiling of patients enrolled in clinical trials. Tumor samples are profiled by CGH to identify gene copy number abnormalities associated with drug response. FISH probes for the marker regions are then designed and validated in a larger patient population (phase III trials or additional studies designed to validate the diagnostic).
B) Proactive approach based on early identification of drug sensitivity markers in preclinical model systems. Cell lines and xenografts used to screen the drug candidate are profiled to determine gene copy number abnormalities associated with drug sensitivity. FISH probes are then developed for the sensitivity marker regions and applied to analyze the marker in the patients enrolled in phase II and phase III clinical trials. See color insert.

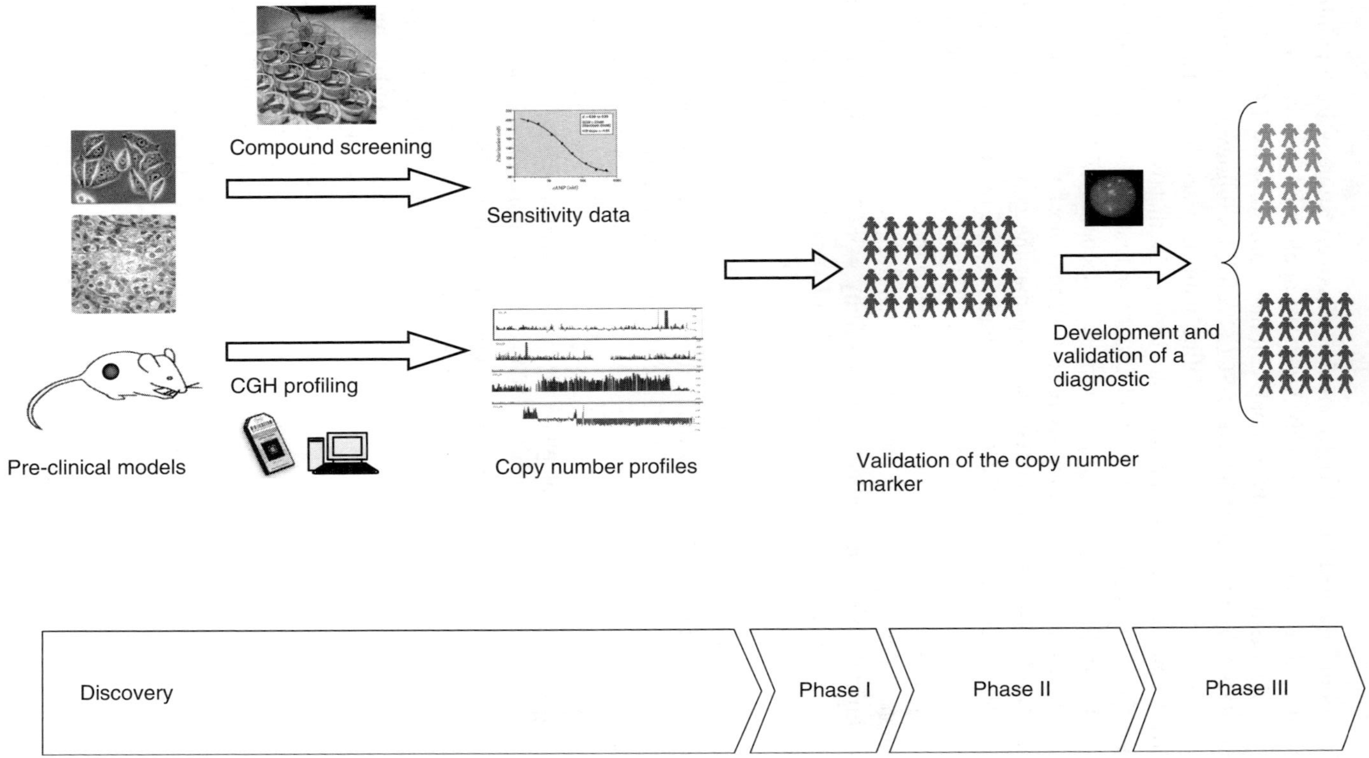

Figure 3.1 (Continued)

this stage, it is particularly important to establish the correlation for additional agents.

We have implemented this early biomarker discovery strategy for a number of our oncology drug candidates. For example, we applied integrative genomics to identify predictors of sensitivity of small-cell lung carcinoma (SCLC) to Bcl-2 family inhibitors (50). Our CGH screen of SCLC cell lines followed by unbiased genome-wide analysis of aberrations correlating with drug sensitivity has identified a novel amplification on chromosome 18q21–23 that is associated with sensitivity of cells to ABT-737 and ABT-263, two first-in-class Bcl-2 family inhibitors. This marker region had the strongest association with sensitivity to the agents and harbored two apoptosis-related genes, most notably Bcl-2, the target of the drugs. Microarray analysis showed that the genes residing in the region are also overexpressed in the sensitive lines relative to the resistant lines. To establish the clinical relevance of the 18q.21–23 gain, we studied the copy number of the Bcl-2 gene in SCLC tumors and showed that it is amplified in 48% of the patients. Taken together, our data suggest that the 18q21–23 copy number may be a clinically relevant predictor for sensitivity of SCLC to Bcl-2 antagonists as they are tested in clinical trials. In light of our findings, the ABT-263 phase I/II clinical protocol was amended to include measurement of the Bcl-2 gene copy number by FISH.

If a clear relationship is established between the presence of the stratification biomarker and the sensitivity to the optimized compound, patient selection should be considered for clinical trials. Phase II of the clinical trials is an appropriate time to apply the patient selection strategy. At this time a diagnostic test needs to be developed and optimized. If the phase II results support the stratification strategy, then patients for phase III can be selected exclusively based on the presence of the genomic marker. If the stratification scheme proves to be successful, drug labeling will need to be considered. If patients are selected for phase III trials based on the positive signal from a diagnostic test, then the use of this diagnostic test must be mandated on the drug label. In phase III of clinical development, the diagnostic test for the genomic marker must undergo clinical validation. This includes the establishment of clinical sensitivity and specificity, as well as other relevant performance characteristics in the chosen patient population. Biomarker performance characteristics and clinical utility considerations are covered in a separate subsection. The specifics of trial design for biomarker-driven clinical development are also considered separately. At the end of a successful phase III, a regulatory approval would need to be obtained for the diagnostic test. The FDA would expect many of these diagnostic tests—in particular those with high-risk profiles—to be processed as class III products subject to premarket review under the premarket application approval (PMA) process (http://www.fda.gov/cdrh/pmapage.html). The key steps during codevelopment of drugs and diagnostics have been outlined in a related FDA concept document (51).

In conclusion, DNA copy number abnormalities have proven their utility as patient stratification biomarkers, as demonstrated by the example of HER2 amplification in breast cancer. Numerous correlations between genomic abnormalities and clinical behavior have been established for a variety of human cancers. The wealth of the existing evidence as well as the rapidly growing understanding of the potential value of DNA biomarkers will undoubtedly lead in the next 5–10 years to discovery and validation of multiple DNA copy number alterations as stratification markers for targeted cancer therapies.

3.2.2. Mutations

Mutations in disease-relevant genes and genes involved in toxicological mechanisms can serve as genomic biomarkers of efficacy and toxicity, respectively. Any changes in the nucleotide sequence of disease-causing genes, in particularly those representing drug targets, are often considered as stratification biomarker candidates, as they may affect the progression of the disease and sensitivity to therapy (this point is illustrated in Fig. 3.2). Here we consider examples of mutations as stratification markers predictive of drug efficacy. This section is heavily biased toward oncology as analysis of mutations and their correlation with sensitivity to therapy have been particularly fruitful in the field of cancer. Mutations in coding regions of oncogenes and tumor suppressor genes have been extensively studied as predictors of outcome. Historically the first genes to be subjected to mutation analysis were the best-known cancer genes, such as p53 and the Rb (retinoblastoma) gene, KRAS, and APC (the adenomatous polyposis) gene (reviewed in (52, 53)). As they have a clear role in the oncogenic processes driving the disease progression, they are naturally regarded as mechanistic markers for patient stratification. We will begin by reviewing the existing evidence for the common cancer genes as biomarkers and by analyzing their prognostic potential. The emphasis will be on the progression from an early association with the disease biology to correlation with the disease outcome, and finally to use as predictors of response to therapy. Later in the chapter we discuss more recent exciting findings for novel targeted cancer drugs, such as EGFR and ABL inhibitors, emphasizing the mechanistic basis for the use of the biomarkers under consideration.

Genomics technologies have greatly facilitated mutational analysis and enabled high-throughput analysis of hundreds of samples. In particular, specialized microarrays have been designed to facilitate high-throughput mutation detection in p53 (54–58). For example, high-throughput mutational analysis of the p53 gene in hundreds of formalin-fixed paraffin-embedded tumor samples has been successfully conducted (54), thus validating the technology for biomarker discovery applications. A detailed discussion of microarray-based technology for mutation identification is given in Chapter 2 devoted to genomics technologies.

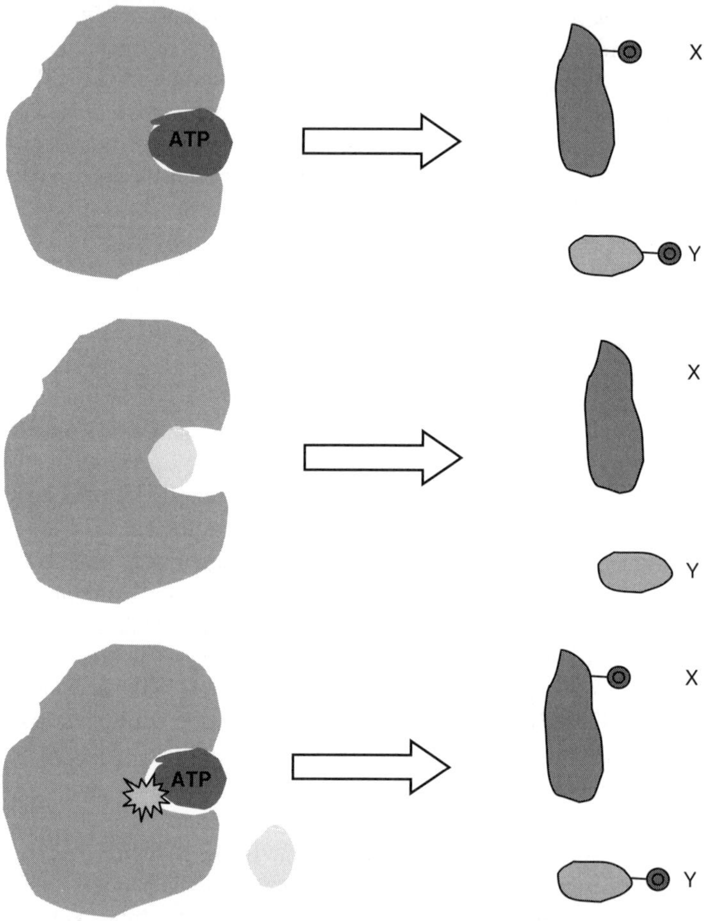

Figure 3.2 Possible effects of mutations in the drug target on the drug efficacy. In this example, the drug target is a protein kinase, and the drug is a small-molecule protein kinase inhibitor, which phosphorylates substrates X and Y. The wild-type enzyme is efficiently inhibited by the drug, resulting in abrogation of substrate phosphorylation. The enzyme that carries a mutation does not bind the drug as efficiently, as the phosphorylation continues to occur in the presence of the drug. See color insert.

3.2.2.1. p53 Mutations

As one of the first important cancer genes identified, p53 has been subjected to extensive mutational analysis (5259–62). For example, p53 mutation status was related to prognosis and outcome of adjuvant therapy in breast cancer by analyzing the sequence of the complete coding region of the gene in 316 patients, of which 97 were lymph node positive and 206 were lymph node negative (59). A total of 69 individual mutations were detected in the coding sequence of the gene.

In node-positive and node-negative patients, the mutation sites were different. The presence of mutations in the evolutionarily conserved regions II and V was associated with poor prognosis. Most importantly, the specific mutations found were explored as potential predictors of the outcome of adjuvant systemic therapy, in particular with tamoxifen, along with radiotherapy. It was found that the therapy was of less benefit to patients with p53 mutations and lymph node-positive tumors.

In a separate study involving 90 breast cancer patients (60), the presence of a p53 gene mutation was found to be the single most powerful predictor of disease recurrence and death. It is noteworthy that it was shown that genomic detection of p53 mutations is a more reliable method for predicting outcome than immunohistochemical detection of the p53 protein overexpression. Since lymph node-positive patients are typically selected for adjuvant chemotherapy, it was of particular interest to determine whether p53 gene mutations are reliable prognosis indicators in two separate groups, lymph node-negative and lymph node-positive patients. A relationship between p53 mutations and expected outcome of chemotherapy or radiation therapy would have a strong mechanistic foundation, because anticancer therapy yields defects in the p53-dependent apoptotic pathway (63, 64). Hence it is logical to hypothesize that a mutation in the p53 gene would reduce the potential therapeutic benefits of a chemotherapeutic agent. Indeed, in the sample of 53 node-negative patients examined in the above-mentioned study (60), of 18 patients with a p53 gene mutation, eight relapsed and four died. Meanwhile, of 35 node-positive patients without mutation, seven relapsed and only one died (P value $=$ 0.0025 and 0.0003, respectively). Interestingly, in the lymph node-negative group, the proportion of relapses in patients with a p53 gene mutation was similar, implying that the predictive power of p53 mutations is not affected by lymph node status. Although larger studies are needed for individual chemotherapeutic drugs to validate p53 mutations as stratification biomarkers, it is clear from the existing data that mutational analysis of p53 may enable identification of subpopulations of patients who are more likely to respond to existing and experimental therapeutics. Therefore, one may already argue that the mutational status of p53 needs to be routinely determined and considered when making patient stratification decisions in oncology clinical trials.

3.2.2.2. K-ras Mutations

Another example of extensively studied mutation patterns in cancer is mutations in the K-ras oncogene in lung carcinomas. Initially, K-ras mutations were identified and associated with survival and disease progression, but it was not until much later, when targeted therapies for lung cancer were invented, that these mutations were used to stratify patients for treatment. This example is very illustrative as one can follow the entire process of biomarker discovery, from early identification of the genetic change to its validation as a predictor of sensitivity to a drug. As early as in the 1980s, K-ras mutations were reported in non-small-cell lung carcinoma (NSCLC), particularly in adenocarcinoma (65). The majority of these mutations were in codon 12 of K-*ras;* G to T transversions were most common (66). It has been reported that approximately 30% of adenocarcinomas of the lung carry K-*ras* mutations at codon 12 (67, 68). An association between smoking and the presence of *ras* mutations has been observed (67).

As the nature of K-ras mutations and their frequency in lung cancer have been established, researchers have begun to address their potential application as a prognostic marker, initially in patients receiving standard treatment. A study of 69 completely resected patients with adenocarcinoma of the lung (69) detected point mutations in codon 12 of K-*ras* in 28% of the patients. With a median follow-up period of 3 years, 12 of the 19 patients with mutations died, compared with 16 of 50 patients without mutations (P value $= 0.002$). Thus the presence of K-*ras* mutations was associated with shortened disease-free and overall survival (69). A later study explored the association between codon 12 K-ras mutations and survival in early-stage NSCLC (70). Samples from 260 patients with stage I and stage II NSCLC were analyzed for codon 12 K-ras mutations. Mutations were detected in 35 of 213 assessable specimens (16.4%); G to T transversions were the most common, with 18 of 35 cysteine (51%) and seven of 35 valine (20%) mutations detected. For patients with stage II disease, the median survival was 13 months for those with K-*ras* mutations ($n = 13$) and 38 months for those without K-*ras* mutations ($n = 49$) (P value $= 0.03$).

As evidence had accumulated in favor of patient stratification by K-ras mutational status, an attempt was made to correlate mutational activation of K-ras with the response to standard chemotherapeutic treatment in advanced adenocarcinoma of the lung (71). A course of mesna, ifosfamide, carboplatin, and etoposide (MICE) was administered to 62 patients whose K-ras mutational status had been determined by analyzing archived biopsy samples. All clinical characteristics associated with sensitivity to chemotherapy, such as survival, response, or pattern of metastasis, were similar between the mutation-positive and mutation-negative groups. Thus K-ras mutations were not predictive of response to untargeted cytotoxic therapy. However, in a recent study employing targeted agents, K-ras mutations were shown to predict drug response (72). A systematic evaluation for the presence of mutations in exon 2 of K-ras and exons 18 through 21 of EGFR was performed for 60 lung adenocarcinomas from patients with known responses to two small-molecule EGFR inhibitors, gefitinib and erlotinib. It is noteworthy that ras oncogenes encode GTPases that participate in intracellular signaling downstream of ERBB receptors, including EGFR, and therefore the search for biomarkers in the ras family was a logical strategy. Additionally, K-ras and EGFR mutations are rarely present in the same patients, implying that their functions in tumorigenesis are redundant (73). Analysis of the 60 patients showed that mutations in exon 2 of K-ras correlated with a lack of response to gefitinib and erlotinib ($n = 9$, 0% response rate, 95% confidence interval of 0–30%) (72). None of the patients sensitive to the drugs had K-ras exon 2 mutations ($n = 21$). These data suggest that K-ras mutations are associated with primary resistance to gefitinib and erlotinib and offer the possibility of using the mutational status of K-ras as a predictive marker of response to EGFR inhibitors in patients known to carry EGFR mutations. However, the comparison of the drug response rates for wild-type EGFR patients with K-ras mutations (5 of 22) and wild-type EGFR patients with no K-ras mutations (0 of 9) did not reach statistical significance (P value = 0.29) (72), indicating that further work is required to define mutations that may stratify patients with wild-type EGFR. Overall, K-ras mutations represent a convincing example of the genomic biomarker discovery process, from initial observations to correlative studies for disease outcome, and finally to establishment as a stratification marker for therapy.

3.2.2.3. EGFR Mutations

More recently, there has been significant focus on mutations in the genes representing drug targets. Perhaps the most intriguing correlation between drug efficacy and the mutational status of the drug target has recently been discovered for two novel drugs for NSCLC, gefitinib and erlotinib. Previous examples described the use of known markers (K-ras mutations) to stratify for treatment with these small-molecule agents. However, the approval of these novel agents has inspired a wave of targeted studies that attempted to identify stratification markers within the target of the drugs, EGFR, a gene not previously explored for genomic alterations in lung cancer.

Clinical trials have revealed significant variability in the response to gefitinib among ethnic groups, with significantly higher responses seen in Japanese patients than in a population of European origin (27.5% vs. 10.4%, in a multi-institutional phase II trial) (74). This prompted researchers to address potential differences in the mutation patterns in the receptor tyrosine kinase genes in NSCLC tissues from two cohorts of patients from Japan and the United States (75). Somatic mutations of the EGFR gene were detected in 15 of 58 unselected tumors from Japan and only 1 of 61 from the United States. Additionally, EGFR mutations were found in a separate set of lung cancer samples from U.S. patients who responded to gefitinib therapy. These results suggested that EGFR mutations predict sensitivity to gefitinib.

This observation in clinical samples has prompted the authors to investigate the relationship between EGFR mutations and cell response to gefitinib in vitro. The mutation status and response to gefitinib were studied for four lung adenocarcinoma and bronchioloalveolar carcinoma cell lines. The H3255 cell line, originally derived from a Caucasian female nonsmoker with lung adenocarcinoma, was 50-fold more sensitive to gefitinib than the other three lines, as established by an IC_{50} of 40 nM for cell survival in a 72-hour assay. In this hypersensitive cell line, treatment with 100 nM gefitinib completely inhibited EGFR autophosphorylation and suppressed the phosphorylation of known downstream targets of EGFR such as the extracellular signal-regulated kinase 1/2 (ERK1/2) and the v-akt murine thymoma viral oncogene homolog (AKT kinase). In contrast, the other three cell lines revealed comparable levels of inhibition of target protein phosphorylation only when gefitinib was used at concentrations approximately 100 times as high. It was shown that the hypersensitive line H3255 carries an EGFR mutation, whereas the other three cell lines contain wild-type EGFR, suggesting that EGFR mutations correlate with gefitinib sensitivity in vitro. This study thus presents an example of translational research that was initiated by identifying differences in responders and nonresponders and then transferred in vitro to examine in more detail the mechanism whereby the proposed genomic biomarker affects cancer cell response to the drug.

An independent study searched for mutations in the *EGFR* gene in primary tumors from patients with non–small-cell lung cancer who had a response to gefitinib, those who did not have a response, and those who had not been exposed to gefitinib (76). It was hypothesized that NSCLC patients who responded to the agent had somatic mutations in the EGFR gene that would indicate the essential role of the EGFR signaling pathway in tumorigenesis. To search for such mutations, rearrangements within the extracellular domain of EGFR were initially considered because they had been found in gliomas. Since no such rearrangements were found, the entire coding region of the EGFR gene was sequenced. Heterozygous mutations were observed in eight of nine responders, all of which were localized in the region coding for the tyrosine kinase domain of EGFR. Four responders were shown to carry in-frame deletions that eliminated amino acids 746 through 750 (delE746–A750) in Patient 1, 747 through 751 (delL747–T751insS) in Patient 2, and 747 through 753 (delL747–P753insS) in Patients 3 and 4. It is of particular interest that all these deletions overlapped, with the minimal deleted region encoding four amino acids within exon 19. Three other responders carried amino acid substitutions within exon 21: leucine to arginine at codon 858 (L858R) in Patients 5 and 6 and leucine to glutamine at codon 861 (L861Q) in Patient 7. Another missense mutation in the tyrosine kinase domain resulted in the substitution of cysteine for glycine at codon 719 within exon 18 (G719C) in Patient 8. Matched normal tissue for several of the patients revealed wild-type EGFR, indicating that the mutations were of somatic nature. In contrast, no mutations were detected in seven nonresponder patients with NSCLC (Fisher's exact test P value < 0.001).

Similarly to the previous work, this study made use of the patient-derived genomic marker data to investigate the mechanistic consequences of EGFR mutations. The receptor with the L747–P753insS deletion and the receptor with the L858R missense mutation were expressed in cultured cells. In the absence of relevant growth factors, neither wild-type nor mutant EGFR exhibited any level of autophosphorylation. However, the addition of EGF resulted in a two- to threefold higher activation level of both mutant EGFRs, as compared with the activation of the wild-type receptor. Furthermore, the two mutant receptors demonstrated continued activation for up to 3 hours, while the activation of normal EGFR was downregulated after 15 minutes.

The natural next step was to test the sensitivity of mutant EGFR receptors to gefitinib in vitro. It is noteworthy that seven of the eight detected EGFR mutations are localized near the ATP binding site, which is targeted by gefitinib. Autophosphorylation of EGFR was measured after ligand treatment of cells pretreated with various concentrations of gefitinib. Complete inhibition of wild-type

EGFR was achieved at 2.0 μM gefitinib, whereas the complete inhibition of the two mutants required only 0.2 μM. In light of the above-described findings, the authors postulated that the common EGFR mutations result in repositioning of the critical residues in the ATP-binding site of EGFR, stabilizing their interaction with both ATP and its competitive inhibitor gefitinib.

Another study confirmed and extended the observations described above by identifying EGFR gene mutations in a cohort of patients treated with gefitinib and erlotinib, another approved tyrosine kinase inhibitor (77). Similar types of mutations were detected in seven of 10 gefitinib-sensitive tumors, whereas no mutations were found in eight gefitinib-refractory tumors (P value = 0.004). Five of seven tumors that responded to erlotinib had analogous somatic mutations in the EGFR gene, while all of the 10 screened erlotinib-resistant tumors had a wild-type EGFR gene. Because the majority of the mutation-positive samples represented adenocarcinomas from nonsmokers, EGFR exons 2–28 were screened in 15 adenocarcinomas from untreated never-smokers. Seven tumors were shown to carry tyrosine kinase domain mutations, compared with only four of 81 NSCLC samples resected from former or current smokers (P value = 0.0001). Based on these data, the authors suggested that adenocarcinomas from never-smokers comprise a distinct subset of lung cancers, because they frequently contain mutations within the tyrosine kinase domain of EGFR and are sensitive to gefitinib and erlotinib.

The breakthrough studies summarized above (75, 76) have fully demonstrated the utility of mutations as predictors of sensitivity of tumors to targeted agents. Additionally, they linked the marker mutation patterns to the mechanisms of the disease. It should be mentioned that these studies were conducted after gefitinib and erlotinib had been approved for the treatment of lung cancer patients. As with gene copy number-based biomarkers, an alternative strategy for mutation markers is to proactively identify possible genomic correlates of drug sensitivity in preclinical model systems. This would enable biomarker-guided patient selection for clinical trials and consequently would increase response rates, decrease the duration and cost of trials, and accelerate drug approval.

3.2.2.4. Bcr-abl and KIT Mutations

Mutations can also serve as markers of acquired resistance. In this case, they cannot be used to predict drug response at diagnosis, but they can be used to monitor the patient during therapy and, when a mutation is confirmed, to make a decision on discontinuation of the current therapy. A recent example of such biomarker is a mutation of the bcl-abl gene as a marker of resistance of chronic myeloid leukemia (CML) to imatinib, an approved therapeutic agent targeting the bcr-abl kinase. CML is a pluripotent hematopoietic stem cell cancer driven by the expression of the *BCR-ABL* fusion gene, which codes for a cytoplasmic protein with constitutive tyrosine kinase activity (78, 79). In clinical trials, imatinib induced remissions in patients in both the chronic phase and the blast crisis. Responses in chronic phase have been durable, but remissions observed in blast crisis patients only lasted 2–6 months, despite continued drug treatment (80, 81).

A pioneering study employing molecular analysis of nine clinical samples established that resistance to imatinib is caused by reactivation of BCR-ABL signal transduction in all cases examined (82). In six of nine cases, resistance was associated with a single amino acid substitution in a threonine residue of the Abl kinase domain, which forms a critical hydrogen bond with the drug. A confirmatory experiment in vitro showed that substitution of threonine with isoleucine confers imatinib resistance. In the three remaining patients, resistance to imatinib was associated with a genomic gain of bcr-abl. Analogously to CML, imatinib has proven to be efficacious in gastrointestinal stromal tumors (GIST), but resistance has been an issue. These tumors are initiated by aberrant signaling from the KIT receptor tyrosine kinase (83). As a selective inhibitor of KIT, bcr-abl, and PDGFR, imatinib has been tested in GIST and proven to be the only drug efficacious in this disease (84). Unfortunately, imatinib resistance began to emerge, raising the question of whether mutations in the KIT gene could serve as markers of resistance, similarly to mutations in bcr-abl in CML. To identify markers of resistance to imatinib in GIST, 12 patients with initial near-complete response to imatinib were studied (85). Seven of these patients were shown to carry mutations in exon 11 of KIT, and the remaining five harbored mutations in exon 9. Within 31 months of treatment, five patients developed imatinib-resistant metastases, whereas in seven patients quiescent residual GIST persisted. All six imatinib-resistant patients who rapidly progressed had the same missense mutation in KIT that resulted in an amino acid substitution in the tyrosine kinase domain 1. Thus this study has identified candidate markers for imatinib resistance in GIST.

A separate study addressed the issue of acquired resistance of GIST to imatinib (86). Molecular genotyping has been performed for 31 patients who were treated with imatinib, followed by surgical resection of the tumor. Among these patients, 13 were considered responders, three were initially resistant to the drug, and 15 acquired resistance after initial benefit from the drug. In the responder and primary resistance groups, no secondary mutations were detected in the KIT or PDGFR genes. In contrast, such mutations were found in seven of 15 (46%) patients with acquired resistance, each of whom also had a primary mutation in exon 11 of KIT. Most of the secondary mutations were localized in exon 17 of KIT. The results of this study imply that **mutation patterns** rather than individual mutations should be considered as candidate patient stratification biomarkers. As these and other mutations are evaluated as markers of resistance, their importance in the disease mechanism will continue to be studied. At this point, it is becoming clear that molecular genotyping of GIST patients, as well as CML patients, is a necessary step in determining the optimal therapeutic regimen.

The evidence accumulated so far indicates that the majority of somatic mutations in cancer can be mapped to a relatively small number of codons. Based on this observation, it was suggested that a limited number of carefully designed sequencing assays should be able to effectively interrogate a large proportion of known cancer-related mutations (87). In particular, it was shown that for known RAS, EGFR, and BRAF mutations, 16–44 genetic assays per gene captured

90–99% of the mutation spectrum. This prompted the creation of a large-scale **high-throughput cancer mutation profiling project**, which focused primarily on known critical or targetable cancer mutations (87). Accordingly, 245 genotyping assays were designed that interrogated 238 known somatic mutations within 17 oncogenes. As the assays were applied to over 1000 cancer samples, priority was given to mutations with high prevalence (such as RAS family mutations), proven clinical implications (for example, KIT and EGFR), and strong association with response to therapeutic agents (for example, BRAF). Of 17 oncogenes analyzed, 14 were shown to contain at least one mutation, and 298 samples (30%) were mutated at least once. This proof-of-concept high-throughput mutation profiling study has revealed previously unknown oncogene mutations in several tumor types and uncovered an unexpectedly high number of co-occurring mutations. This approach offers a new dimension in tumor genetics by enabling simultaneous real-time detection of a large spectrum of cancer-related mutations and thus facilitating cancer classification and biomarker discovery.

In conclusion, therapies targeting proteins mutated in cancer offer enormous potential for in the treatment of cancer. Companion diagnostics are essential for realizing this potential, as they would enable selection of patients who carry the target mutations. Although concurrent development of companion diagnostics and therapeutics would significantly increase the cost and the complexity of drug discovery and clinical trials, these additional efforts are clearly justified by the potential benefits realized by early identification of tumors susceptible to the drug and more rational clinical trial design. If the drug and the diagnostic test receive approval for a specific cancer, other tumor types can be chosen for clinical development based on the presence of the target mutation. This approach, namely, **targeting molecular lesions rather than histopathological cancer types**, provides substantial clinical and economic advantages compared with the current practice of random selection of tumor types and enrollment of unselected patient populations and therefore will likely become the dominant approach in the near future.

3.2.3. Epigenetic Markers

Epigenetic inheritance can be defined as biological information, other than the DNA sequence, that controls gene expression and is inherited during cell division. In the past decades it has become evident that epigenetic inheritance plays an important role in many normal and pathological processes in humans. The phenomenon of epigenetic regulation contributes to the differences between growing, senescent, and immortal cells and tumor and normal cells, and is important in cell differentiation and aging (88, 89). Three major types of epigenetic modification exist: **DNA methylation, histone modification**, and **genetic imprinting** (Table 3.3). All these modifications are amenable to genome-wide study and can be associated with gene expression. The study of epigenetic modifications on a whole-genome scale is the subject of a new discipline referred to as **epigenomics** (90). In Chapter 2 we review the

Table 3.3 Known Types of Epigenetic Modification in Humans

Type of Epigenetic Modification	Mechanism
DNA methylation	A covalent modification of DNA, whereby a methyl group is transferred from *S*-adenosylmethionine to the C-5 position of cytosine by a family of cytosine (DNA-5)-methyltransferases. DNA methylation is observed almost exclusively at cytosine residues that are followed by a guanine (CpG). It controls expression of many genes and silences repeat elements in the genome.
Histone modification	Occurs through acetylation, methylation, or phosphorylation. Controls transcription and is inherited during cell division. These modifications are regulated by proteins that are also involved in DNA methylating complexes. The precise mechanisms of histone modification are not yet known.
Genomic imprinting	Relative silencing of one parental allele compared with the other parental allele that occurs through differential methylation of regions within or near imprinted genes.

modern approaches to genome-wide profiling for epigenetic modifications. Rapid development of genome-wide approaches has enabled the creation of centers of excellence dedicated to study of gene regulation by epigenetic modifications. In the US, the Center for the Epigenetics of Common Human Disease at Johns Hopkins University (http://www.hopkinsmedicine.org/epigenetics/) is developing high-throughput methodologies for detecting epigenetic changes and linking known epigenetic modifications to the mechanisms of human disease. In Europe, the Human Epigenome Project (HEP; www.epigenome.org) is a collaborative effort, established in order to identify and interpret genome-wide DNA methylation patterns for all human genes in all major tissues. A substantial amount of data has been accumulated on the epigenetic patterns in human tissues and their association with disease. In this subsection, we review epigenetic modifications that have demonstrated the potential to serve as biomarkers.

Abnormal DNA methylation patterns have been associated with various tumors and therefore have been extensively studied in the past decade. DNA methylation almost always occurs at C nucleotides followed by G, or at so-called **CpG sites**. These sites are present in the genome at a relatively low frequency; their clusters are referred to as CpG islands. One of the challenges in studying CpG island methylation is distinguishing them from repetitive DNA sequences, because the latter are also heavily methylated (91). Strict criteria have therefore been proposed to define CpG islands in the genome (92). Particularly important is methylation of CpG islands in the promoter regions of genes, because it has been shown to cause gene silencing (93). Approximately half of human genes are estimated to contain CpG islands in their promoters (94).

DNA methylation changes occur with aging (95) and have been associated with disease (96). In cancer, tumor cells frequently reveal altered methylation patterns, such as global demethylation and promoter-localized hypermethylation (Fig. 3.3) (88). These patterns alter the structure and function of DNA by causing unwanted transcription of repeat elements and aberrant activation of genes and contributing to the phenomenon of genomic instability by interfering with the replication machinery (88). Particularly important in cancer is aberrant silencing of genes involved in the initiation and progression of tumors, such as known tumor suppressor genes, developmental transcription factors, tissue remodeling genes, as well as DNA repair, cell cycle control, and antiapoptotic genes (88). In fact, any single cancer may have all such genes silenced simultaneously (97).

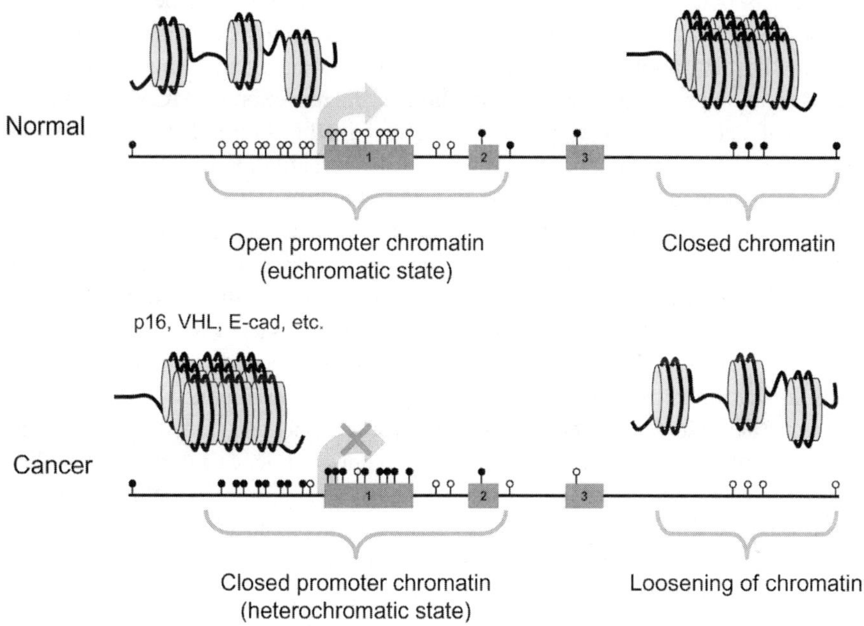

Figure 3.3 Epigenetic changes associated with cancer. In normal cells, CpG islands in proximal gene promoter regions (a three-exon gene is shown, with each exon marked in blue and numbered) are protected from DNA methylation (cytosines are shown as open circles) and reside in restricted regions of open chromatin, or euchromatic states, that permit gene transcription (blue arrow). In contrast, for most regions of the genome, such as in the bodies of many genes and areas outside genes, the cytosines in CpG dinucleotides are methylated (closed circles). This DNA methylation is characteristic of the bulk of the human genome, which is packaged as closed chromatin that does not permit transcription. In cancerous cells, genes may be abnormally silenced by methylation. Proximal promoter CpG islands are hypermethylated and reside in a closed chromatin, which does not favor transcription. In other regions of the genome, cytosines in CpG dinucleotides are hypomethylated and are associated with aberrantly loose chromatin. Reproduced with permission from Ting et al. (2006) *Genes Dev* 20: 3215 (88).

In some cases, tumor-specific loss of gene function has more epigenetic than genetic causes (98). As more information is accumulated on aberrant DNA methylation in tumors, it is becoming increasingly evident that epigenetic mechanisms may contribute to the early phases of tumorigenesis, rather than occur in mature tumors as a secondary effect (88). Because of the importance of epigenetic modification in cancer and the wide variety of epigenetic patterns observed in different tumors, methylation patterns can and need to be explored as genomic biomarkers. In this subsection we review the existing evidence for possible **association of DNA methylation profiles with response to therapeutic agents**.

An important class of anticancer therapeutics is represented by DNA damaging agents, which suppress tumorigenesis by interfering with DNA synthesis and ultimately causing cell death. Because the mechanism of action of these drugs is based on hindering DNA replication, it is logical that their efficacy has been associated with the expression of DNA repair enzymes. In particular, one such enzyme, *O*-6-methylguanine—DNA methyltransferase (MGMT), reverses the alkylation of DNA caused by alkylating agents, such as carmustine. Because approximately 30% of gliomas do not express MGMT (99), the efficacy of carmustine is of particular interest in this tumor type. Since DNA methylation is one of the potential mechanisms of regulation of gene expression, a study was conducted to determine whether methylation of the MGMT gene can serve as a marker predictive of patient response to carmustine in glioma (100). The methylation status of the gene was assessed in 47 glioma patients by methylation-specific PCR after bisulfite conversion of the genomic DNA. It was shown that the promoter region of MGMT is methylated in 19 (40%) of the samples. In this subpopulation, 12 (63%) patients responded to the drug. Meanwhile, in the unmethylated population, only one of 28 patients (4%) responded to carmustine. Moreover, the lack of methylation was associated with a much higher risk of death (hazard ratio, 9.5; 95 percent confidence interval, 3.0–42.7; P value < 0.001). In univariate analysis, the MGMT promoter methylation was the only factor that had a statistically significant association with survival. The median time to disease progression was 21 months for methylated gliomas and 8 months for unmethylated gliomas (P value < 0.001).

Hypermethylation of the MGMT promoter has also been evaluated as a predictive marker of response of gliomas to another commonly used alkylating agent, temozolomide (101). Using a methylation-specific PCR approach, we assessed the methylation status of the CpG island of MGMT in 92 glioma patients who received temozolomide as first-line chemotherapy or as a treatment for recurrent disease. Hypermethylation status was associated with the response to temozolomide in patients receiving the drug as a first-line therapy ($n = 40$), as eight of 12 patients with MGMT-methylated tumors (66.7%) had a complete or a partial response, compared with 7 of 28 patients with unmethylated tumors (25%; P value $= 0.03$). However, in patients receiving temozolomide as a treatment for recurrent disease, the MGMT promoter hypermethylation status in the initial glioma tumor did not correlate with the drug response (P value $= 0.729$).

These promising correlative data have prompted an examination of MGMT promoter methylation status as a predictive marker in a prospective clinical trial setting (102). The epigenetic marker was assessed by methylation-specific PCR in the tumor biopsies from 38 patients who had undergone tumor resection for newly diagnosed glioblastoma and were subsequently enrolled in a prospective phase II clinical trial to evaluate concomitant and adjuvant temozolomide and radiation. In contrast to the studies described above, the patient population in this trial was homogeneous. It was shown that inactivation of the *MGMT* gene by promoter methylation is associated with longer survival (P value $= 0.0051$). At 18 months, the survival was 62% (16 of 26) for patients with a methylated *MGMT* promoter but reached only 8% (1 of 12) for patients with no methylation (Fisher's Exact test P value $= 0.002$). In the presence of other clinically relevant factors, methylation of the MGMT promoter remains the only significant predictor (P value $= 0.017$; Cox regression). *MGMT* methylation status was thus demonstrated to be an independent predictor for glioblastoma patients treated with temozolomide.

The MGMT gene has also been shown to be transcriptionally silenced in a fraction of diffuse large B-cell lymphomas (DLBCL) (103). An attempt was made to correlate the methylation status of the MGMT gene with the outcome and drug response in this disease (104). In a retrospective cohort study, methylation-specific PCR was applied to analyze the MGMT promoter methylation status in tumors from DLBCL patients receiving cyclophosphamide as part of multidrug regimens. Promoter methylation was detected in 30 (36%) of 84 individuals. In this subgroup, 23 patients (77%) experienced complete remission, four (13%) experienced partial remission, and three (10%) did not respond. Meanwhile, among the patients with no promoter methylation ($n = 54$), 34 patients (63%) experienced complete remission, eight (15%) experienced partial remission, and 12 (22%) showed no response. The higher response rate in patients with tumors carrying a methylated MGMT promoter was not statistically significant (P value $= 0.3$). However, the methylation of the MGMT promoter was associated with a statistically significant increase in overall survival [hazard ratio for time to death for nonmethylation vs. *methylation* $= 2.8$; 95% confidence interval (CI) $= 1.2$–7.5; P value $= 0.01$] and progression-free survival (hazard ratio for time to progression for nonmethylation versus *methylation* $= 2.6$; 95% $CI = 1.3$–5.8; P value $= 0.02$). Hypermethylation of the MGMT promoter was an independent prognostic factor, stronger than the established prognostic factors, such as age or disease stage. Thus methylation status of the MGMT gene may serve as a useful genomic marker for predicting survival in DLBCL patients treated with cyclophosphamide–based combination therapies.

Thus the studies on MGMT promoter methylation provide support to the hypothesis that the response to alkylating agents may be associated with the methylation status of DNA repair-related genes. For other classes of anticancer agents, attempts were also made to find epigenetic correlates of efficacy. In particular, the CHFR gene has emerged as a promising candidate for epigenetic studies. The gene encodes a protein containing forkhead-associated and RING finger domains as well as ubiquitin–ligase activity and regulating the metaphase entry checkpoint of the cell cycle. It has been shown that the expression of CHFR is silenced through CpG methylation in 45% of cancer cell lines, 40% of primary colorectal cancers, 53% of colorectal adenomas, and 30% of primary head and neck cancers (105). The expression of the gene was reliably associated with both CpG methylation and deacetylation of histones H3 and H4 in the CpG-rich regulatory region. Furthermore, it was clearly demonstrated that CpG methylation and thus silencing of *CHFR* depend on the activities of two DNA methyltransferases, DNMT1 and DNMT3b, as their genetic inactivation restored *CHFR* expression, thus revealing a role for methylation in the mitotic checkpoint (105). The functional consequences of this epigenetic inactivation of CHFR in gastric cancers were investigated with a panel of 20 gastric cancer cell lines (106). It was found that the CHFR gene is silenced by DNA methylation of the 5′ region of the gene in 20% of the gastric cancer cell lines tested and in 39% of primary gastric cancers. Additionally, histones H3 and H4 were deacetylated in cell lines showing aberrant methylation, implying a role for histone deacetylation in CHFR gene silencing. Those cell lines in which CHFR was silenced exhibited impaired checkpoint function, which resulted in nuclear localization of cyclin B1 after treatment with the microtubule inhibitors docetaxel or paclitaxel. Thus CHFR inactivation likely plays a key role in tumorigenesis in gastric cancer by altering the mitotic checkpoint function. The results of this study suggest that the hypermethylated status of the CHFR promoter is associated with sensitivity of cells to docetaxel and paclitaxel (106). While more validation is clearly needed for the CHFR promoter as a marker of drug response, the data described above and the clear connection between CHFR and the mechanism of tubulin polymerization inhibitors warrant significant enthusiasm for development of epigenetic biomarkers predictive of response of gastric cancers to docetaxel and paclitaxel.

In addition to the promising data reviewed above, a number of in vitro studies have addressed correlation between the methylation status of selected genes and the sensitivity of cells to anticancer drugs. Genes with an established connection to cancer phenotypes or drug resistance have received the most study (107, 108). This illustrates one possible approach to discovery of DNA methylation biomarkers, whereby the methylation status is evaluated for genes a priori related to the disease mechanism or drug sensitivity. Using an analogy from the pharmacogenetic field, we suggest using the term “candidate gene approach” when referring to this methodology. An alternative approach may involve genome-wide profiling of patient tumors or cell lines with methylation microarrays followed by statistical analysis of the resulting profiles to identify patterns of methylation associated with drug sensitivity. Development of novel high-throughput microarray-based technologies for methylation profiling will enable this genome-wide methodology, thus facilitating biomarker discovery in situations where little or no prior knowledge exists on the epigenetic regulation of genes involved in the drug’s mechanism of action. More generally, DNA methylation patterns can be used to classify tumors, alone or in combination with other genomic profiles such as gene copy number or gene expression signatures. For this application, DNA methylation does not necessarily have to cause gene silencing; it would simply provide a signature specific to the subpopulation of tumors, a "molecular portrait," that may be associated with clinically important information. More studies are required to explore the feasibility of this use of methylation profiles. If epigenetic silencing is indeed as important in gene regulation as gene copy number changes and other mechanisms, it is likely that continued research in this field, together with further technological advances in microarray-based methylation analysis, will generate useful composite biomarkers for drug discovery and development.

3.3. RNA BIOMARKERS

A cancer phenotype reflects somatic mutations in the tumor cells, the inherited genotype of the patient, and the influence of the tumor environment. The interaction of these three factors yields enormous complexity and heterogeneity of the phenotypes, resulting in variation in outcomes and response to drugs. Since gene expression is affected by all of the factors mentioned above, expression signatures of tumor cells have the potential to capture the complexity of tumor phenotypes and thus may be used as genomic correlates of disease outcome and drug response. Indeed, in this subchapter we demonstrate how microarray-based gene expression analysis of tumor tissue has been used to classify cancers, discover new subtypes of the disease, and predict disease outcome and sensitivity to therapeutic agents.

Patterns of gene expression, often referred to as **"gene expression signatures,"** have been extensively explored as genomic biomarkers, both as predictors of efficacy and toxicity. This has been largely a result of the rapid development in high-throughput microarray technologies for measuring gene expression on a genome-wide scale. Development and commercialization of high-density gene expression microarrays has preceded the introduction of viable microarray platforms for CGH or epigenetic analysis. Therefore, a significantly larger body of data has been accumulated in this area, permitting discovery and validation of multiple gene expression signatures as biomarkers of disease outcome and drug response. Several advanced studies have resulted in development and commercialization of diagnostic tests that enable patient stratification for therapy. We begin this subsection by reviewing the most advanced examples of the development of gene expression biomarkers and then analyze other studies that are, in our opinion, most likely to make impact on biomarker development and drug discovery in general. Because gene expression microarrays, unlike CGH arrays, are often used to examine the effects of drug treatments on an organism or on cells, it is important to distinguish these studies from the experimental design that will be frequently cited in this subsection, namely, generation of *baseline* gene expression profiles *before the treatment* and then identification of gene expression signatures that are associated with drug response.

3.3.1. Gene Expression Biomarkers Validated as Diagnostic Tests

Breast cancer is a heterogeneous disease that shows significant variability in outcome and response to treatment. Patients with the same disease stage and other clinical parameters often have markedly different prognoses and display variable responses to standard treatments. The best nongenomic predictors of metastasis are histological grade of the tumor and lymph node status, but they fail to accurately predict the outcome and classify patients for therapy (109). Chemotherapy decreases the risk of metastases by ~30%, but 70–80% of patients who receive chemotherapy would have survived without it (110, 111). Overtreatment of breast cancer patients is associated with unnecessary toxicity and represents a substantial socioeconomic problem. Additionally, it has been demonstrated that not all patients derive equal benefit from adjuvant chemotherapy (112, 113). Correct classification of patients for therapy should thus involve assessment of the baseline risk of metastases as well as prediction of response to chemotherapy. As genetic heterogeneity may explain the clinical variability of the disease, a significant amount of effort has been devoted to the examination of the genomic profiles of breast cancers and discovery of genomic biomarkers predictive of the disease outcome. The development of predictive biomarkers for breast cancer outcome and response to adjuvant therapy has become possible because of several pioneering studies that have provided a critical mass of information on the genomic profiles of breast cancers.

As early as 2000, a large-scale study of gene expression was conducted in 65 breast cancer specimens (114). Sets of coexpressed genes were discovered, for which specific genomic features were associated with physiological variation. Strikingly, hierarchical clustering of the gene expression patterns revealed two clearly defined subgroups of tumors. This first-level genomics-based classification largely coincided with the previous assignment of the tumors into ER-negative and ER-positive groups. Finer classification was also attempted, as gene expression patterns were identified that correlate with HER2 status. The next major development was the discovery of a gene expression pattern strongly predictive of distant metastases in lymph node-negative patients (115). Two groups of patients were profiled on gene expression microarrays: 34 patients who had developed distant metastases within 5 years of diagnosis and 44 individuals who remained metastasis-free for 5 years or longer (median follow-up 8.3 years). To identify composite markers of poor outcome, a multistep supervised classification method was applied to the resulting gene expression profiles. First, approximately 5000 genes were selected from the total of 25,000 genes measured, based on their regulation in at least three tumors. Correlation coefficients were measured for each gene to assess the correlation between their expression levels and poor outcome; 231 genes were found to be significantly associated with outcome (correlation coefficient >0.3 or <-0.3) and were stack-ranked by their correlation coefficients. The subsets of these genes were evaluated for correct classification of tumors by the leave-one-out cross-validation method. This procedure yielded a set of 70 genes with the greatest predictive power for outcome prediction. This classifier set correctly predicted the outcome for 65 of 78 patients (83%), with five poor-prognosis and eight good-prognosis patients misclassified. Functional annotation of the genes in the marker set revealed their association with metastasis and the oncogenic phenotypes. Specifically, genes associated with cell cycle, invasion, angiogenesis, and signal transduction were significantly overexpressed in the poor-prognosis signature, providing insight into the mechanisms that underlie the more aggressive phenotype of the poor-outcome tumors. Most importantly, however, this study laid the foundation for development of biomarkers that may stratify patients for adjuvant chemotherapy based on the expected disease outcome. In a follow-up report, the 70-gene predictive signature was evaluated in a set of 295 patients with primary breast carcinomas, which included the majority of the initial set of 78 tumors (116). Importantly, this population contained both lymph node-negative and lymph node-positive patients (151 and 144 individuals, respectively). The 70-gene composite marker was evaluated for predictive power by univariable and multivariable statistical analyses and shown to be a strong predictor of the development of distant metastases in the lymph node-negative as well as the lymph node-positive patient subpopulations. Among the 295 individuals examined, 180 had a poor prognosis signature and 115 had a good prognosis signature, while the mean overall 10-year survival rates were $54.6 \pm 4.4\%$ and $94.5 \pm 2.6\%$, respectively. Multivariable Cox regression analysis demonstrated that the prognosis marker is a strong independent predictor of disease outcome. It was a stronger predictor than the commonly used clinical and histological criteria. An interesting conclusion from the findings of these two studies is that the metastatic capacity of a breast tumor is an early and inherent genetically programmed property of the cancer. This conclusion was intriguing, as it argued against the widely accepted notion that the metastatic potential of a tumor is acquired late in the process of tumorigenesis (116). An important practical implication of this new hypothesis for drug discovery is that early prognostic assessments can be made and patients can be stratified for therapy based on their potential to develop distant metastases.

The gene sets generated in these and several other (117, 118) studies were used to develop a predictive multigene assay suitable for clinical use (119). Initially, a high-throughput real-time RT-PCR assay was designed to enable quantitation of gene expression in sections of fixed paraffin-embedded tissue (120). The nest step was to design a composite biomarker using candidate genes that had been associated by earlier studies with the outcome of breast cancer. A total of 250 genes were selected from previously published studies and relevant genomic databases and tested for associations with the recurrence of breast cancer by analyzing three independent clinical studies that involved 447 patients. Based on this analysis, the initial list of 250 genes was triaged to select a panel of 16 cancer-related genes and five reference genes that could serve as a composite biomarker predictive of breast cancer recurrence. To facilitate the predictions, an algorithm was designed that can compute a recurrence score for the test sample based on the expression levels of the selected 21 genes. The test was then validated with archived paraffin-embedded tumor samples from patients with lymph node-negative tamoxifen-treated breast cancer. The expression levels of the 21 genes were fed into the algorithm developed to calculate a recurrence score for each tumor and classify the patient into a risk group (low, intermediate, and high). In the examined population of 668 patients, 51%, 22%, and 27% of women were categorized as having a low, intermediate, and high risk, respectively. The Kaplan–Meier estimates of the rates of disease recurrence in 10 years in the low-risk, intermediate-risk, and high-risk groups were 6.8% (95% confidence interval, 4.0–9.6), 14.3% (95% confidence interval, 8.3–20.3), and 30.5% (95% confidence interval, 23.6–37.4). The distant recurrence rate in the low-risk group was significantly lower than that in the high-risk group (P value < 0.001). In a multivariate Cox model, the recurrence score provided significant predictive power that was independent of age and tumor size (P value < 0.001). The recurrence score was also predictive of overall survival (P value < 0.001) and could be used as a continuous function to predict distant recurrence in individual patients.

BOX 3-2 *Groups of Genes in the 21-Gene Signature Predictive of Breast Cancer Recurrence and Benefit from Chemotherapy (119)*

Proliferation	**HER2**	**Estrogen**
Ki67	GRB7	ER
STK15	HER2	PGR
Survivin		BCL2
CCNB1 (cyclin B1)	**Others**	SCUBE2
MYBL2	GSTM1	
	CD68	**Reference**
Invasion	BAG1	ACTB (β-actin)
MMP11 (stromolysin 3)		GAPDH
CTSL2 (cathepsin L2)		RPLPO
		GUS
Estrogen		TFRC
ER		
PGR		
BCL2		
SCUBE2		

The major significance of this study is that the recurrence score obtained with the 21-gene composite biomarker represented the **first gene expression-based biomarker validated for clinical use**. It is now available commercially as as a laboratory reference test (Oncotype DX®). A cost-utility analysis was performed for the Oncotype DX® test with a decision analytic model (121). The analysis was performed from a societal perspective, considering survival, quality of life, and relevant costs. In a hypothetical cohort of 100 patients, the recurrence score predicted more accurately than current guidelines the risk of recurrence in lymph-node negative ER-positive patients with early-stage breast cancer. The test was predicted on average to increase quality-adjusted survival by 8.6 years and reduce overall costs by $202,828.

From the drug discovery perspective, the value of a genomic marker increases significantly if it can predict drug response as opposed to just predicting disease outcome.

As the 21-gene composite marker was established as a reliable predictor of disease recurrence, the next important question was whether it could be used to predict the response to chemotherapy. Indeed, the gene set included genes involved in cell proliferation and hormonal response, motifs that are generally associated with response to chemotherapy. A retrospective study was therefore designed to explore the association between the Oncotype Recurrence Score™ and benefit of chemotherapy. The study was conducted with archived tissue blocks previously collected from a clinical trial, which tested the worth of adding chemotherapy to tamoxifen in the treatment of lymph node-negative, ER-positive patients (122). In a population of 651 patients (227 randomly assigned to tamoxifen and 424 randomly assigned to tamoxifen plus chemotherapy), the interaction between chemotherapy treatment and the Recurrence Score was statistically significant (P value $= 0.038$). Patients with high Recurrence Scores (≥ 31) derived a substantial benefit from chemotherapy, whereas women with low Recurrence Scores (<18) had minimal, if any, benefit from the chemotherapeutic regimen. Patients in the intermediate recurrence group did not appear to have a large benefit, but the uncertainty in the estimate cannot exclude a clinically important benefit. Thus this retrospective study clearly demonstrated that the 21-gene composite biomarker not only quantifies the likelihood of breast cancer recurrence in women with node-negative, ER-positive breast cancer but also predicts the magnitude of chemotherapy benefit in this patient population. These results have significant clinical implications for patients with low or relatively high Recurrence Scores (122). Specifically, for women with a low risk of recurrence, the anticipated benefit of adding chemotherapy to hormonal therapy may not exceed the risks, while for patients in the high recurrence risk category, the anticipated benefit of chemotherapy appears to outweigh the risks. The implications for drug discovery are less obvious, but we would argue that, given the clear relationship between the measured parameter and the anticipated benefit from the currently used chemotherapy, it would be instructive to examine the performance of the Oncotype DX® test in clinical development of novel agents for breast cancer. A relatively good availability of archived paraffin-embedded tissue samples from past clinical trials facilitates retrospective analysis of possible correlations of the marker with drug response.

Another recently developed diagnostic test for recurrence of breast cancer is not only derived from microarray data but actually based on a microarray platform. A customized microarray, named MammaPrint® (123), was designed to interrogate the genes in a 70-gene signature previously shown to predict disease outcome in young breast cancer patients (115). The original signature was generated with microarrays containing 60-mer oligonucleotide probes corresponding to 25,000 transcripts, a platform not readily amenable for clinical diagnostics. To facilitate the application of this composite 70-gene marker in a high-throughput diagnostic setting, a custom eight-pack miniarray was designed, which enabled eight individual simultaneous hybridizations on a single microarray slide. To validate the predictive power of the new microarray-based diagnostic, 162 patient samples from the previous study were profiled, and the patient classification results from the new microarray were compared with those from the original study. The prognosis prediction correlated extremely well for the two platforms (P value < 0.0001), implying that the custom diagnostic array can be used as a tool to predict disease outcome in young breast cancer patients. MammaPrint® was the first microarray-based tool to be validated as a diagnostic test and also the **first microarray to receive market clearance by the FDA**.

The FDA's focus on the emerging field of genomic biomarkers has been reinforced by the approval of MammaPrint® and reflected in a recent "Draft Guidance on in vitro Diagnostic Multivariate Assays." It remains to be seen whether the existing gene expression diagnostic tools, such as Oncotype® and MammaPrint®, will also prove useful for novel therapeutic agents, in addition to the current standard chemotherapy regimens. Clearly, additional clinical studies are necessary to expand the applications of these tests. However, retrospective analysis has been successfully used and remains an option for clinical validation of genomic biomarkers. Moreover, the patient stratification methodology used in the development of the biomarkers described above can be applied in discovery and development of novel cancer therapeutics, thus enabling early identification of the target population of patients who are likely to respond to the drug.

3.3.2. Other Examples of Gene Expression Biomarkers

The two advanced examples of predictive gene sets reviewed above clearly demonstrate that composite RNA biomarkers can guide prognosis and treatment of breast cancer. Not surprisingly, the past several years have witnessed an increased interest in the discovery of predictive gene signatures in other cancer types, such as non-small-cell lung carcinoma (NSCLC), ovarian carcinoma, and various lymphomas.

For NSCLC, clinical trials revealed that patients with stage IB, II, and IIIA, but not stage IA, benefit from adjuvant chemotherapy. However, approximately 25% of individuals with stage IA disease relapse after the surgery (124), implying that further stratification is necessary for this group. In a attempt to stratify early-stage NSCLC patients, gene expression signatures were identified that are associated with the risk of recurrence in a population of 89 patients (125). The signature was then evaluated for predictive power in two independent populations of patients (25 and 84 individuals). The **composite biomarker was shown to predict recurrence of NSCLC** better than the currently used clinical prognostic factors (predictive accuracy >70%), suggesting that it could be used to change a clinical decision in early-stage NSCLC. The biomarker would need to be validated in a prospective, randomized, phase III clinical trial, whereby it would be applied to identify patients with a high risk of recurrence. The high-risk patients would then be randomly assigned after surgery to two groups: the observation group (current standard of care) and the chemotherapy group (125). Such a study would provide clinical validation, the necessary next step to establish the clinical utility of the genomic biomarker.

As the inadequacy of the current staging system for NSCLC was widely recognized, more attempts were made to develop a genomic biomarker that would better predict the outcome of early-stage disease. The association between gene expression signatures and survival was also studied in a set of 125 patients who had undergone surgical resection of NSCLC (126). Risk score and decision tree analyses were then applied to develop a gene expression-based model for predicting the outcome of treatment for NSCLC. Sixteen genes associated with survival were identified by analyzing the microarray profiles and risk scores. Five genes (*DUSP6, MMD, STAT1, ERBB3*, and *LCK*) were selected for RT-PCR validation and decision tree analysis. The presence of a high-risk five-gene signature in the tumor was associated with an elevated risk of recurrence and reduced overall survival. It was clearly demonstrated that the five-gene marker is an independent predictor of relapse-free and overall survival. The marker was then validated in an independent cohort of 60 patients with NSCLC and a separate set of published microarray data for 86 NSCLC samples. While prospective, large-scale, multicenter studies are still necessary to clinically validate this biomarker, analysis of the existing data on genomic biomarkers in NSCLC warrants an optimistic view of genomic stratification of this heterogeneous disease. Just as the Oncotype® and MammaPrint® tests have improved the diagnosis and treatment of breast cancer, the newly emerging genomic biomarkers for NSCLC are likely to advance the management of lung cancer if they are proven to have sufficient clinical utility.

Another instructive example of the utility of genomic classification of disease is related to non-Hodgkin's lymphoma, a highly heterogeneous cancer of the immune system. Diffuse large B-cell lymphoma (DLBCL), the most common subtype of non-Hodgkin's lymphoma, is clinically heterogeneous: 40% of patients respond well to current chemotherapy and survive, whereas the remainder relapse and die of the disease (127). In the past several years, a major gene expression profiling effort has revealed that DLBCL is comprised of at least two genomically and clinically distinct diseases (128–130). The first subgroup of DLBCL, named germinal center B cell-like (GCB) DLBCL, is characterized by the expression of genes typical for normal germinal center B cells, while the second DLBCL subgroup, termed activated B cell-like (ABC) DLBCL, does not express germinal center B cell-restricted genes and instead expresses genes that are upregulated in blood B cells during mitogenic stimulation. More recently, a third subgroup of DLBCL has been discovered by genomic profiling and named primary mediastinal B cell lymphoma (PMBL) (131). At the molecular level, these three subgroups are regarded as separate diseases, because they arise from

B cells at different stages of their differentiation and activate different oncogenic pathways. At the clinical level, these three groups of DLBCL manifest different outcomes: The 5-year survival rates for GCB-DLBCL, ABC-DLBCL, and PMBCL are 59%, 30%, and 64%, respectively (129, 130, 132).

The main utility of genomic classification of lymphoma is in guiding the application of existing and novel therapeutics. For the currently used anthracycline-based chemotherapy regimens, a **molecular predictor of lymphoma patient survival** has been developed with gene expression microarrays (130). Biopsy samples from 240 DLBCL patients were profiled with gene expression microarrays, and the resulting expression signatures were subjected to hierarchical clustering. A composite biomarker predictive of outcome was constructed with the expression patterns associated with survival in a test set of 160 patients and then validated in a separate set of 80 patients. The three genomic subgroups of the disease were delineated. Patients in the GCB-DLBCL group had the highest rate of 5-year survival. However, the classification into the three subtypes explained some, but not all, of the variation in the survival rates of these patients. The Cox proportional hazards model was used to refine the list of genes that predict survival. Most of the genes in the good survival profile were associated with the four gene-expression signatures characteristic of germinal center B cells, proliferating cells, reactive stromal and immune cells in the lymph node, or major histocompatibility complex class II complex. A **genomic predictor of survival after chemotherapy in DLBCL** was constructed with 17 genes most strongly associated with good response to chemotherapy. This gene-based predictor and the international prognostic index were independent prognostic indicators. Identification of this genomic biomarker of outcome has important implications for the treatment of DLBCL patients: approximately 50% of the patients are placed in the favorable prognostic group, for whom anthracycline-based chemotherapy may be curative, while one-quarter of the patients are assigned to the poor prognosis group, for whom alternative therapies should be considered (133).

Independent of the subclassification of DLBCL, genomic biomarkers of outcome were pursued in a mixed population of DLBCL patients (134). Gene expression was applied to diagnostic tumor specimens from DLBCL patients who received standard cyclophosphamide, adriamycin, vincristine, and prednisone (CHOP)-based chemotherapy, and the gene expression signatures were analyzed by a supervised learning prediction method to identify cured *versus* refractory disease. The algorithm classified the patients into two categories with drastically different 5-year overall survival rates (70% vs. 12%). The outcome predictor marker included genes involved in regulation of responses to B-cell receptor signaling, critical serine/threonine phosphorylation pathways, and apoptosis. For example, three validated outcome predictor genes, *NOR1, PDE4B*, and *PKC*-β, regulate apoptotic responses to antigen-receptor engagement and, potentially, cytotoxic chemotherapy.

Thus the microarray-based gene expression profiling studies described above have provided rich sources of data that can be mined for associations with clinical outcome. Each of these studies has yielded a gene expression signature predictive of outcome. However, development of a clinically useful diagnostic tool requires that the composite gene signature be validated in independent studies utilizing separate populations of patients. Moreover, a clinical diagnostic test must be based on a more practical measurement platform that would not require extensive statistical expertise for data analysis. Toward development of a clinical diagnostic tool for outcome prediction in DLBCL, an RT-PCR-based assay was designed for a composite biomarker comprising six genes (135). A group of 36 genes predictive of outcome in DLBCL was selected for an initial screen. All the genes were derived from previously published microarray data sets and other data sources. Quantitative RT-PCR was used to measure the expression of these 36 genes as well as two internal control genes (PGK1 and GAPDH) in 66 specimens of DLBCL and in a reference sample. All patients were treated with anthracycline-based chemotherapy (CHOP or a CHOP-like regimen). A univariate analysis of the expression levels for the three selected genes was conducted, with overall survival as a dependent variable. The genes were then ranked according to their predictive power, measured as a univariate z score. A positive z score correlated with shorter overall survival, whereas a negative z score was associated with a longer survival. To facilitate routine measurement of the composite biomarker in a clinical environment, the number of genes selected as best predictors was minimized by setting a z parameter cutoff of ± 1.5. This resulted in a composite marker that included six genes: LMO2, BCL6, FN1, CCND2, SCYA3, and BCL2. This marker was used to rank the patient population tested based on their survival scores, thus dividing the patients into three groups according to the risk of death (low risk, score <0.063; medium risk, $0.063 \leq score < 0.093$; high risk, $score \geq 0.093$). The overall survival rates in the three groups were 65%, 49%, and 15% for the low-risk, medium-risk, and high-risk groups, respectively, illustrating the utility of the biomarker in predicting the increased survival in patients who respond to chemotherapy. When the model was applied to previously published microarray data sets (130, 134), it was able to predict survival, thus demonstrating the validity of the biomarker for DLCBL outcome prediction, regardless of the measurement method (RT-PCR or microarrays). If the clinical utility of this new biomarker is proven in prospective randomized clinical trials, it will enable routine stratification of DLBCL patients into groups for which different treatments will need to be used.

An important implication of the above-mentioned biomarker studies for drug discovery is that they established a process for identification of high-risk poor-outcome individuals (approximately one-third of all DLBCL patients), for whom novel therapeutic agents need to be designed. Thus one can envision that current clinical development programs for DLBCL may use the newly developed composite stratification markers to separately evaluate the efficacy of the therapeutic under development in the poor-outcome category of patients. Furthermore, new lymphoma drug discovery programs may benefit from early application of the predictive biomarker to isolate preclinical models that have the "poor-outcome" profile and therefore may better model the disease in the poorly served category of patients.

The biomarker discovery studies described above have all been conducted with clinical samples with a well-annotated history of disease progression and response to therapy. Microarray technology requires fresh-frozen samples conserved and processed by a uniform protocol. Such samples are difficult to obtain for most diseases, complicating biomarker discovery efforts. Most importantly, in drug discovery programs, it is always beneficial to obtain early correlates of drug sensitivity early in preclinical testing, that is, before the drug reaches the clinical development stage.

Because of the aforementioned factors, **cultured cell lines represent an attractive option for biomarker discovery**. Their utility has been proven by several recent biomarker studies. In particular, a panel of cancer cell lines used by the National Cancer Institute (NCI) in drug screening (NCI-60 panel) has recently been used to develop predictors of cytotoxic drug response (136). The cell lines in the panel were treated with docetaxel, and the most sensitive and most resistant lines were selected for further analysis. Genes were selected whose expression showed the strongest correlation with drug sensitivity, and Bayesian binary regression analysis was applied to develop a model that recognizes gene expression patterns associated with sensitivity to docetaxel. A gene expression signature comprised of 50 genes was then identified that classified cell lines on the basis of docetaxel sensitivity. Obviously, as the ultimate goal was to select patients for treatment, the signature developed in cell lines needed validation in patient samples. Previously published gene expression studies for breast cancer patients treated with docetaxel (137) were used to evaluate the predictive power of the in vitro docetaxel sensitivity predictor signature. The composite biomarker correctly predicted docetaxel response in 22 of 24 tumor samples analyzed, which corresponds to an overall accuracy of ~92%. Mann–Whitney U-test for statistical significance has revealed that the predictor signature can reliably distinguish between docetaxel responders and nonresponders, implying that gene expression biomarkers identified in cell lines may show utility in patients.

Many cancer types are treated with complex chemotherapeutic regiments comprised of multiple drugs with different mechanisms of action. As the study was taken to the next level of complexity, the individual in vitro signatures of drug sensitivity obtained in cell lines were evaluated for their capacity to predict response of patients to therapeutic regiments containing combinations of drugs. Specifically, the single-agent signatures for paclitaxel, 5-FU, adriamycin, and cyclophosphamide obtained in NCI-60 cell lines were applied to predict response in 51 patients to a chemotherapeutic regimen including all these drugs in a previously published breast cancer neoadjuvant study (138, 139). The predicted response based on the individual chemosensitivity signatures differentiated well between the responders ($n = 13$) and nonresponders ($n = 38$), with the exception of the prediction for 5-FU. The prediction of response to the complex neoadjuvant regimen based on a combined probability of sensitivity derived from the individual chemosensitivity predictions resulted in a statistically significant difference between the responders and nonresponders (P value < 0.0001). Thus the results of this groundbreaking biomarker discovery study establish the **value of gene expression biomarkers in stratification of patients for therapy with either single-agent or combination therapy**. The next step in moving the predictive biomarkers to the clinic would be setting up a clinical trial to evaluate the performance of a currently accepted combination of agents versus a biomarker-based selection of drugs (136).

In summary, combined efforts by biologists, clinicians, and mathematicians have enabled significant progress in the area of developing gene expression-based predictors of disease outcome and drug response. Substantial challenges remain, however, on the way to implementation of RNA-based biomarkers in drug development and clinical practice. One unsolved question is the relevance of individual genes comprising predictive expression signatures to the disease mechanisms. Despite initial sequencing of the human genome and ongoing efforts in functional annotation of human genes, our knowledge of the functional roles of most genes remains inadequate. This limits our ability to link the existing gene expression biomarkers to novel therapeutic candidates targeting specific intracellular pathways. Next, when a novel RNA biomarker has been successfully associated with drug sensitivity in a discovery setting, the subsequent step of clinical validation almost always represents a significant challenge. **A novel concept of clinical development** needs to be adopted, whereby genomics-based stratification would be a key consideration in the initial planning and trial design. In the absence of these new development strategies, clinical validation possibilities for many biomarker programs are limited to retrospective assessment in archived samples from previously run trials. This raises the issues of sample quality and variability in sample preparation protocols at different institutions. Clear realization of these issues is necessary to facilitate the advances of the RNA biomarker field to the next stage, where the emerging markers would transition to the clinic and thus enable genomics-based treatment selection for individual patients in clinical trials and, further, in routine clinical practice.

In conclusion, drug discovery organizations continue to face the challenge of matching therapeutic candidates with the right subpopulations of patients who are more likely to benefit from therapy. Patterns of genetic alterations define the cancer phenotypes and determine, among other features, the sensitivity of tumors to drugs. Composite gene expression biomarkers reflect the unique patterns of intracellular pathway deregulation in tumors, enabling the capture of the complexity of a tumor's phenotype. The biomarker discovery studies reviewed in this subchapter demonstrate that RNA-based biomarkers provide the opportunity to predict the disease outcome in an individual patient and, most importantly, match the right therapeutic with the right individual, thus optimizing the treatment response. The success of such studies for existing therapeutics justifies increased investments in biomarker research by pharmaceutical discovery organizations, with the intention of repeating this success for novel drug candidates entering the development pipeline. Multiple gene expression biomarker studies are now being run as part of drug discovery programs, with the ultimate goal of developing composite biomarkers predictive of response to the agents in development. Thus the rapid development of the gene expression biomarker field facilitates matching patients with the right drugs, warranting an optimistic view of personalized drug discovery in the decades to come.

3.4. CLINICAL VALIDATION OF GENOMIC BIOMARKERS

In early 2005, the FDA published a document titled "Guidance for Industry: Pharmacogenomic Data Submissions" (140), which outlines the framework for submission and review of genomics data and proposes new approaches to the interaction between sponsors and the agency (reviewed in (141)). Importantly, the document introduces the **concepts of exploratory and valid biomarkers** and defines these two terms. Valid biomarkers are further categorized into **known valid biomarkers and probable valid biomarkers**. Known valid biomarkers are defined as biomarkers that are measured in an analytical test system with well-established performance characteristics and for which there is widespread agreement in the medical or scientific community about the physiological, toxicological, pharmacological, or clinical significance of the results. Probable valid biomarkers are defined as biomarkers that are measured in an analytical test system with well-established performance characteristics and for which there is a scientific framework or body of evidence that appears to elucidate the physiological, toxicological, pharmacological, or clinical significance of the test results (141). The key difference between known valid and probable valid biomarkers is that the former have been validated independently by multiple laboratories, and the data related to their utility are available to the scientific community, whereas the utility of the latter is based on data generated by a single institution. Table 3.4 lists some known valid genomic biomarkers applied at the time when this book was written.

Exploratory biomarkers can be regarded as precursors of probable or known valid biomarkers (141). Their main role is to generate hypotheses based on preclinical studies and link existing preclinical knowledge on the drug targets with the variability in drug response. They can be utilized in compound selection and optimization as they provide useful information on the drug's potential efficacy and toxicity in preclinical model systems. Obviously, their transition to valid biomarkers would require that the sponsors accumulate a significant amount of clinical data. A process map has recently been developed by FDA scientists for validation of genomic safety biomarkers (141). The discovery stage may involve identification of biomarker candidates in a preclinical model system with a high-throughput microarray-based technology. Data obtained at this stage can be submitted to the FDA as a voluntary genomic data submission (VGDS) to obtain early feedback from the agency. The quality of data from this step will determine whether the next phase, method development, should be implemented. Method development would then involve conversion of a high-throughput discovery

Table 3.4 Known Valid Genomic Biomarkers [modified from (141)]

Genomic Biomarker	Context	Drug	Drugs with Similar Context
1. Patient stratification for drug efficacy			
Her2/neu overexpression	Overexpression of Her2/neu is the main criterion for selection of patients for therapy.	Herceptin	
C-KIT expression	Predicts response to the tyrosine kinase inhibitor imatinib mesylate (Gleevec) in gastrointestinal stromal tumors (GISTs).	Gleevec	
EGFR expression	Predicts clinical response to cetuximab. Patients enrolled in clinical studies for cetuximab were required to have immunohistochemical evidence of positive EGFR expression, as indicated by the DakoCytomation EGFR pharmDx™ test kit.	Cetuximab	
EGFR expression, alternative context	EGFR presence or absence	Erlotinib	Gefitinib
2. Patient stratification for drug toxicity			
*UGT1A1*28* allele	*UGT1A1* polymorphism predicts higher concentrations of the drug and an active metabolite and hence increased toxicity.	Irinotecan	
VKORC1 variants	Polymorphisms of vitamin K epoxide reductase complex subunit (*VKORC1*) predict increased toxicity of warfarin. Lower drug dose is recommended for patients with VKORC1.	Warfarin	
TPMT polymorphisms associated with low and intermediate TPMT enzyme activity	Nonfunctional alleles (such as *TPMT*2, TPMT*3A*, and *TPMT*3C*) yield TPMT deficiency or lower activity and result in increased risk of myelotoxicity. Testing of the TPMT genotype or the enzyme activity is recommended for azathioprine therapy.	Azathioprine	Thioguanine, mercaptopurine

(*continued overleaf*)

Table 3.4 Continued

Genomic Biomarker	Context	Drug	Drugs with Similar Context
UCD deficiency disorders	Urea cycle disorders associated with valproic acid. Before the initiation of valproate therapy, evaluation for UCD should be considered.	Valproic acid	Sodium phenylacetate and sodium benzoate
NAT variants	*N*-acetyltransferase polymorphisms associated with slow acetylation, leading to higher blood levels of the drug and hence increased toxicity.	Rifampin, isoniazid, and pyrazinamide	Isosorbide dinitrate and hydralazine hydrochloride
DPD deficiency	Dihydropyrimidine dehydrogenase deficiency is associated with increased risk of severe toxic reaction to 5-FU. Genotypic assays should be used to detect DPD variants associated with toxicity before initiation of therapy.	Capecitabine	Fluorouracil cream; fluorouracil topical solution and cream
CYP2C19 variants	*CYP2C19* variants (PM and EM) are associated with changes in drug exposure. Patients are monitored to determine whether it is necessary to adjust the dosage of drugs when taken concomitantly with omeprazole.	Omeprazole	Pantoprazole; voriconazole; esomeprazole; proguanil and atovaquone; nelfinavir; delavirdine; lansoprazole; amoxicillin; clarithromycin; lansoprazole; lansoprazole and naproxen

CYP2C9 variants	Patients carrying P450 2C9 variants associated with poor drug metabolism should be administered celecoxib with caution because reduced metabolic clearance may result in abnormally high plasma levels.	Celecoxib	Warfarin
CYP2C9 variants, alternative context	CYP2C9 metabolism and drug–drug interactions. Fenofibrate is a weak inhibitor of CYP2C19 and CYP2A6 and a mild to moderate inhibitor of CYP2C9 at therapeutic concentrations.	Fenofibrate	Bosentan; fluvastatin
CYP2D6 variants	CYP2D6 PM variants are associated with reduced enzyme activity. Fluoxetine and other agents metabolized by P450IID6 inhibit CYP2D6 and thus may make normal metabolizers resemble poor metabolizers. Therapy with medications that are predominantly metabolized by the P450IID6 system and that have a relatively narrow therapeutic index should be initiated at the low end of the dose range if a patient is receiving fluoxetine concurrently or has taken it in the previous 5 weeks.	Fluoxetine HCl	Fluoxetine HCl and olanzapine; atomoxetine; cevimeline hydrochloride; tolterodine; terbinafine; tramadol and acetamophen; clozapine; venlafaxine; aripiprazole; risperidone; metoprolol; propranolol; carvedilol; tiotropium bromide inhalation; propafenone; tamoxifen; thioridazine; timolol maleate; protriptyline HCl; vicoprofen-hydrocodone bitartrate and ibuprofen

platform into an analytical platform with well-defined performance characteristics that would reliably measure the biomarker candidates discovered (common biomarker performance characteristics are listed in Table 3.5). For example, quantitative RT-PCR assays can be developed to measure the expression of genes identified by microarray screening. If key characteristics of the assay such as sensitivity, specificity, and reproducibility are optimized, the sponsors may proceed to the next step, formulation and submission of a study protocol proposal. The key sections of the study protocol proposal have recently been outlined by FDA experts (141). Importantly, the proposal needs to contain a comprehensive description of the assay and a detailed outline of the statistical procedures used for biomarker testing, including the power and sample size considerations for sensitivity, specificity, and reproducibility and the choice of statistical methods. The protocol proposal is then reviewed by various functional groups within the FDA, including the genomics, pharmacology, and toxicology organizations. The process is expected to result in a written review and a face-to-face meeting and is likely to initiate a series of iterative changes followed by go/no-go decisions.

The next step in the process map is a dose ranging study, reflecting the fact that many exploratory biomarkers would belong to the toxicity biomarker category. Minimal evidence will be required to prove that the dose of the drug affects the marker as well as the current standard end points, such as histopathology changes. The dose ranging study will then be followed by a validation study to probe the context in which the biomarker may be considered valid. A rigorous correlative analysis will need to be performed to establish a tight association between the biomarker and the end point, such as histopathology or drug response. A go/no-go point can be made if the study results confirm the proposed mechanistic, diagnostic, or predictive context for genomic biomarker candidates. The validation study report is a comprehensive summary of the data obtained for the genomic biomarker, with a focus on the results illustrating the specificity and sensitivity of the biomarker relative to the currently accepted standard, such as

Table 3.5 Biomarker Performance Characteristics

Parameter	Definition
Sensitivity	The fraction of individuals with the disease that manifest positive biomarker test values (values above the predefined threshold)
Specificity	The fraction of individuals without the disease that manifest negative test values (values below the predefined threshold)
Accuracy	Concordance between the measured biomarker value and its true value
Precision	Concordance between independent measurements of the biomarker obtained under predefined testing conditions
Efficacy	The ability of the test to provide clinically significant results in a significant proportion of the patient population under study
Safety	The health benefits of the device outweigh any possible health risks.

histopathology. The report will be evaluated by the same regulatory team at the FDA that has reviewed the study protocol proposal. Criteria for the evaluation of study results will include the reproducibility, sensitivity, and specificity of the biomarker measurement. Again, the interactions between the sponsors and the agency are likely to be iterative, with multiple rounds of review. The regulatory review step will be completed when the FDA classifies the biomarker of interest as "probable valid."

Clinical validation of a genomic biomarker in prospectively designed clinical trials is the critical step in determining its utility. The main goals of the clinical validation study are (i) to define the sensitivity, specificity, and positive and negative predictive values and (ii) to prove clinical utility of the test. Several clinical trials designs have been proposed for validating genomic biomarkers (Fig. 3.4) (142). For genomic stratification markers, the trial must answer the question of whether the drug response rate is higher in the patient subpopulation positive for the marker relative to the response rate in the unselected patient population (Fig. 3.4A). This question can theoretically be answered by a single-arm clinical study as shown in Figure 3.4B, but this approach has a significant limitation. The treatment is administered only in the marker-positive subpopulation, which means that the response rates will need to be compared with historical data for the overall patient population. Additionally, the negative predictive value (NPV) of the marker is not assessed in this design, that is, the drug response rate in marker-negative patients is not measured. To calculate the required number of subjects for a single-arm clinical trial, the sensitivity and the positive predictive value (PPV) of the marker need to be estimated.

The advantage of a more complex randomized trial (Fig. 3.5C) is that it permits evaluation of the response rate in both selected and unselected patients and enables the assessment of the treatment specificity of the marker. To assess the utility of the marker in a setting similar to the phase III clinical trials for therapeutic agents, randomization may be applied to patients to receive treatment only if the marker is present or to receive therapy regardless of the status of the marker. This design permits immediate assessment of the clinical utility of the marker, because it enables a direct comparison of the clinical end point after treatment for the selected and unselected treatment arms. This design, similar to a phase III clinical trial for therapeutic agents, would require significant numbers of patients, but the trial can be stopped early if a prespecified level of predictivity is achieved early or if the marker clearly does not meet the utility criteria.

Finally, the study design depicted in Figure 3.5D may be used to determine whether the marker under study is specific to a certain treatment or is a general therapy response predictor in the disease considered (142). This design assesses the correlation between the presence of the marker and the response to two or more therapeutic agents. The marker is thus measured after enrollment, and the patient is assigned into one of the treatment groups (drug under study or current standard treatment). The primary objective for this design is to establish whether patients who carry the marker are significantly more likely to respond to the drug under study than to the standard regimen. A secondary objective is to show that

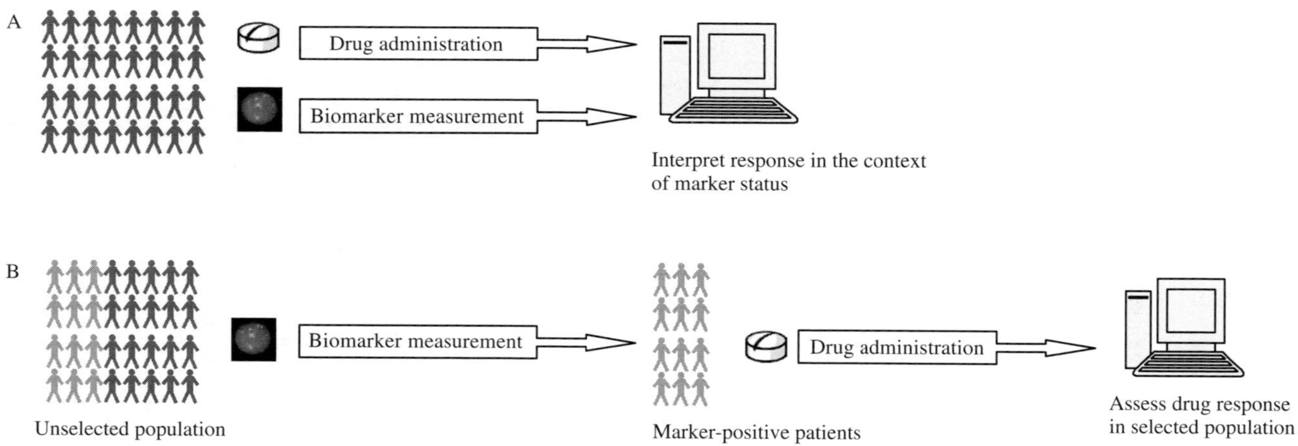

Figure 3.4 Designs of clinical trials aimed at assessing clinical utility of a predictive genomic biomarker. A) All patients are treated with the drug regardless of the marker status, and the drug response is analyzed together with the biomarker measurement results. B) The biomarker is detected upfront, and the treatment is only administered to patients carrying the marker. C) A randomized trial aimed at assessing the ability of the biomarker to improve the treatment outcome relative to the use of the drug in an unselected population. D) A trial design that compares the clinical benefit of a novel drug A and the current standard therapy B in a population selected with a biomarker. Modified from L. Pusztai and K. Hess (2004). Ann Oncol 15: 1731. See color insert.

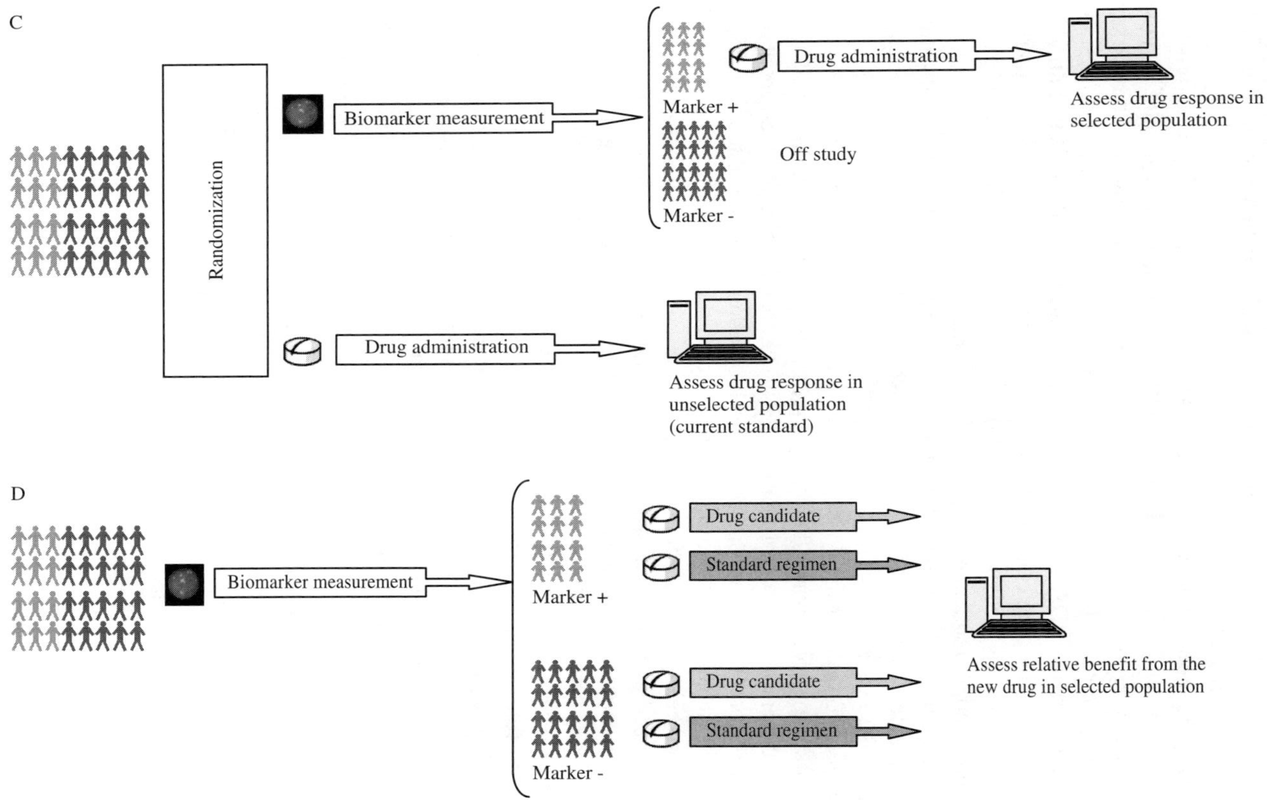

Figure 3.4 (Continued)

marker-negative patients will not benefit from the drug under study as much as the marker-positive patients do, and that for these patients the standard therapeutic regimen may be the preferred option. An early discontinuation option may be necessary to minimize the exposure of marker-negative patients to the drug if it is not likely to provide benefit.

Industry-wide acceptance of a biomarker is only likely after a large-scale cross-validation study that would allow a transition to the "known valid" status. The importance of this step has prompted the creation of a number of biomarker consortia, most notably the Critical Path Institute (www.c-path.org). It was established to support the FDA in its effort to implement the Critical Path Initiative (143) through developing collaborative projects, notably in the area of biomarker validation. Generally speaking, cross-validation of a genomic biomarker is a process of independent replication of the findings of the initial report by two or more separate companies. A full report containing the results of this cross-validation will be evaluated by the FDA to determine whether the biomarker may be categorized as "known valid."

It is noteworthy that the concept of validation should relate to the fitness of the biomarker to serve a particular purpose rather than blindly applying predefined criteria to different types of biomarkers (144). For example, for the Oncotype Recurrence Score™, a composite predictive biomarker of response to chemotherapy in breast cancer covered earlier in this chapter, the validation process should establish whether the use of the marker results in clinical benefit, but it does not necessarily need to be mechanistically understood or related to the disease mechanism (144). However, in the absence of mechanistic understanding, the marker cannot be considered valid according to the criteria proposed by the FDA (140). These criteria are important for biomarkers developed as surrogate end points, but are less relevant to individual components of composite genomic biomarkers (144). This is an important consideration for composite genomic biomarkers, such as gene expression signatures or copy number signatures predictive of drug response. While they can be useful in patient stratification, their validation does not imply that their components (individual genes in expression signatures or individual gene amplifications or deletions in copy number signatures) are related to the disease mechanism. In summary, the creation of the process map described above reflects the intention of the FDA to accelerate the acceptance of genomic biomarkers by the biomedical community through introducing a formal process whereby genomics is adopted by drug development organizations. This is rapidly becoming a critical task as initial research data on genomic biomarkers are translated into robust diagnostics that affect the discovery, development, and clinical application of therapeutic agents.

REFERENCES

1. Biomarkers and surrogate endpoints: preferred definitions and conceptual framework. *Clin Pharmacol Ther* 2001;69:89–95.

2. Vogel CL, Cobleigh MA, Tripathy D, Gutheil JC, Harris LN, Fehrenbacher L, Slamon DJ, Murphy M, Novotny WF, Burchmore M, Shak S, Stewart SJ, and Press M. Efficacy and safety of trastuzumab as a single agent in first-line treatment of HER2-overexpressing metastatic breast cancer. *J Clin Oncol* 2002;20:719–26.
3. Cobleigh MA, Vogel CL, Tripathy D, Robert NJ, Scholl S, Fehrenbacher L, Wolter JM, Paton V, Shak S, Lieberman G, and Slamon DJ. Multinational study of the efficacy and safety of humanized anti-HER2 monoclonal antibody in women who have HER2-overexpressing metastatic breast cancer that has progressed after chemotherapy for metastatic disease. *J Clin Oncol* 1999;17:2639–48.
4. Bosma PJ, Chowdhury JR, Bakker C, Gantla S, de Boer A, Oostra BA, Lindhout D, Tytgat GN, Jansen PL, Oude Elferink RP, et al. The genetic basis of the reduced expression of bilirubin UDP-glucuronosyltransferase 1 in Gilbert's syndrome. *N Engl J Med* 1995;333:1171–5.
5. Raijmakers MT, Jansen PL, Steegers EA, and Peters WH. Association of human liver bilirubin UDP-glucuronyltransferase activity with a polymorphism in the promoter region of the UGT1A1 gene. *J Hepatol* 2000;33:348–51.
6. Ando Y, Saka H, Ando M, Sawa T, Muro K, Ueoka H, Yokoyama A, Saitoh S, Shimokata K, and Hasegawa Y. Polymorphisms of UDP-glucuronosyltransferase gene and irinotecan toxicity: a pharmacogenetic analysis. *Cancer Res* 2000;60:6921–6.
7. Lander ES, Linton LM, Birren B, Nusbaum C, Zody MC, Baldwin J, Devon K, Dewar K, Doyle M, FitzHugh W, et al. Initial sequencing and analysis of the human genome. *Nature* 2001;409:860–921.
8. Venter JC, Adams MD, Myers EW, Li PW, Mural RJ, Sutton GG, Smith HO, Yandell M, Evans CA, Holt RA, et al. The sequence of the human genome. *Science* 2001;291:1304–51.
9. Rowley JD. The critical role of chromosome translocations in human leukemias. *Annu Rev Genet* 1998;32:495–519.
10. Bedi A. BCR-ABL gene rearrangement and expression of primitive hematopoietic progenitors in chronic myeloid leukemia. *Blood* 1993;81:2898–28902.
11. Druker BJ. Effects of a selective inhibitor of the Abl tyrosine kinase on the growth of Bcr-Abl positive cells. *Nat Med* 1996;2:561–6.
12. Druker BJ, Talpaz M, Resta DJ, Peng B, Buchdunger E, Ford JM, Lydon NB, Kantarjian H, Capdeville R, Ohno-Jones S, and Sawyers CL. Efficacy and safety of a specific inhibitor of the BCR-ABL tyrosine kinase in chronic myeloid leukemia. *N Engl J Med* 2001;344:1031–7.
13. Albertson DG, Collins C, McCormick F, and Gray JW. Chromosome aberrations in solid tumors. *Nat Genet* 2003;34:369–76.
14. Cheng KW, Lahad JP, Kuo WL, Lapuk A, Yamada K, Auersperg N, Liu J, Smith-McCune K, Lu KH, Fishman D, Gray JW, and Mills GB. The RAB25 small GTPase determines aggressiveness of ovarian and breast cancers. *Nat Med* 2004;10:1251–6.
15. Linn SC, West RB, Pollack JR, Zhu S, Hernandez-Boussard T, Nielsen TO, Rubin BP, Patel R, Goldblum JR, Siegmund D, Botstein D, Brown PO, Gilks CB, and van de Rijn M. Gene expression patterns and gene copy number changes in dermatofibrosarcoma protuberans. *Am J Pathol* 2003;163:2383–95.
16. Pollack JR, Sorlie T, Perou CM, Rees CA, Jeffrey SS, Lonning PE, Tibshirani R, Botstein D, Borresen-Dale AL, and Brown PO. Microarray analysis reveals a major direct role of DNA copy number alteration in the transcriptional program of human breast tumors. *Proc Natl Acad Sci USA* 2002;99:12963–8.
17. Heidenblad M, Lindgren D, Veltman JA, Jonson T, Mahlamaki EH, Gorunova L, Van Kessel AG, Schoenmakers EF, and Hoglund M. Microarray analyses reveal strong influence of DNA copy number alterations on the transcriptional patterns in pancreatic cancer: implications for the interpretation of genomic amplifications. *Oncogene* 2005;24:1794–801.

18. Levsky JM and Singer RH. Fluorescence in situ hybridization: past, present and future. *J Cell Sci* 2003;116:2833–8.
19. Hynes NE, and Lane HA. ERBB receptors and cancer: the complexity of targeted inhibitors. *Nat Rev Cancer* 2005;5:341–54.
20. Slamon DJ, Clark GM, Wong SG, Levin WJ, Ullrich A, and McGuire WL. Human breast cancer: correlation of relapse and survival with amplification of the HER-2/neu oncogene. *Science* 1987;235:177–82.
21. Carter P, Presta L, Gorman CM, Ridgway JB, Henner D, Wong WL, Rowland AM, Kotts C, Carver ME, and Shepard HM. Humanization of an anti-p185HER2 antibody for human cancer therapy. *Proc Natl Acad Sci USA* 1992;89:4285–9.
22. Harris M. Monoclonal antibodies as therapeutic agents for cancer. *Lancet Oncol* 2004;5:292–302.
23. Hudziak RM, Lewis GD, Winget M, Fendly BM, Shepard HM, and Ullrich A. p185HER2 monoclonal antibody has antiproliferative effects in vitro and sensitizes human breast tumor cells to tumor necrosis factor. *Mol Cell Biol* 1989;9:1165–72.
24. Slamon DJ, Leyland-Jones B, Shak S, Fuchs H, Paton V, Bajamonde A, Fleming T, Eiermann W, Wolter J, Pegram M, Baselga J, and Norton L. Use of chemotherapy plus a monoclonal antibody against HER2 for metastatic breast cancer that overexpresses HER2. *N Engl J Med* 2001;344:783–92.
25. Tedesco KL, Thor AD, Johnson DH, Shyr Y, Blum KA, Goldstein LJ, Gradishar WJ, Nicholson BP, Merkel DE, Murrey D, Edgerton S, and Sledge GW, Jr. Docetaxel combined with trastuzumab is an active regimen in HER-2 3+ overexpressing and fluorescent in situ hybridization-positive metastatic breast cancer: a multi-institutional phase II trial. *J Clin Oncol* 2004;22:1071–7.
26. Ross JS, Fletcher JA, Bloom KJ, Linette GP, Stec J, Clark E, Ayers M, Symmans WF, Pusztai L, and Hortobagyi GN. HER-2/neu testing in breast cancer. *Am J Clin Pathol* 2003;120 Suppl:S53–71.
27. Pauletti G, Dandekar S, Rong H, Ramos L, Peng H, Seshadri R, and Slamon DJ. Assessment of methods for tissue-based detection of the HER-2/neu alteration in human breast cancer: a direct comparison of fluorescence in situ hybridization and immunohistochemistry. *J Clin Oncol* 2000;18:3651–64.
28. Takano T, Ohe Y, Sakamoto H, Tsuta K, Matsuno Y, Tateishi U, Yamamoto S, Nokihara H, Yamamoto N, Sekine I, Kunitoh H, Shibata T, Sakiyama T, Yoshida T, and Tamura T. Epidermal growth factor receptor gene mutations and increased copy numbers predict Gefitinib sensitivity in patients with recurrent non-small-cell lung cancer. *J Clin Oncol* 2005;23:6829–37.
29. Cappuzzo F, Hirsch FR, Rossi E, Bartolini S, Ceresoli GL, Bemis L, Haney J, Witta S, Danenberg K, Domenichini I, Ludovini V, Magrini E, Gregorc V, Doglioni C, Sidoni A, Tonato M, Franklin WA, Crino L, Bunn PA, Jr., and Varella-Garcia M. Epidermal growth factor receptor gene and protein and Gefitinib sensitivity in non-small-cell lung cancer. *J Natl Cancer Inst* 2005;97:643–55.
30. Hirsch FR, Varella-Garcia M, Bunn PA, Jr, Di Maria MV, Veve R, Bremnes RM, Baron AE, Zeng C, and Franklin WA. Epidermal growth factor receptor in non-small-cell lung carcinomas: correlation between gene copy number and protein expression and impact on prognosis. *J Clin Oncol* 2003;21:3798–807.
31. Maris JM. The biologic basis for neuroblastoma heterogeneity and risk stratification. *Curr Opin Pediatr* 2005;17:7–13.
32. Mosse YP, Greshock J, Margolin A, Naylor T, Cole K, Khazi D, Hii G, Winter C, Shahzad S, Asziz MU, Biegel JA, Weber BL, and Maris JM. High-resolution detection and mapping of genomic DNA alterations in neuroblastoma. *Genes Chromosomes Cancer* 2005;43:390–403.

33. Wang Q, Diskin S, Rappaport E, Attiyeh E, Mosse Y, Shue D, Seiser E, Jagannathan J, Shusterman S, Bansal M, Khazi D, Winter C, Okawa E, Grant G, Cnaan A, Zhao H, Cheung NK, Gerald W, London W, Matthay KK, Brodeur GM, and Maris JM. Integrative genomics identifies distinct molecular classes of neuroblastoma and shows that multiple genes are targeted by regional alterations in DNA copy number. *Cancer Res* 2006;66:6050–62.
34. Paris PL, Andaya A, Fridlyand J, Jain AN, Weinberg V, Kowbel D, Brebner JH, Simko J, Watson JE, Volik S, Albertson DG, Pinkel D, Alers JC, van der Kwast TH, Vissers KJ, Schroder FH, Wildhagen MF, Febbo PG, Chinnaiyan AM, Pienta KJ, Carroll PR, Rubin MA, Collins C, and Van Dekken H. Whole genome scanning identifies genotypes associated with recurrence and metastasis in prostate tumors. *Hum Mol Genet* 2004;13:1303–13.
35. Weiss MM, Kuipers EJ, Postma C, Snijders AM, Siccama I, Pinkel D, Westerga J, Meuwissen SG, Albertson DG, and Meijer GA. Genomic profiling of gastric cancer predicts lymph node status and survival. *Oncogene* 2003;22:1872–9.
36. Rennstam K, Ahlstedt-Soini M, Baldetorp B, Bendahl PO, Borg A, Karhu R, Tanner M, Tirkkonen M, and Isola J. Patterns of chromosomal imbalances defines subgroups of breast cancer with distinct clinical features and prognosis. A study of 305 tumors by comparative genomic hybridization. *Cancer Res* 2003;63:8861–8.
37. Chin S-F, Wang Y, Thorne NP, Teschendorff AE, Pinder SE, Vias M, Naderi A, Roberts I, Barbosa-Morais NL, Garcia MJ, Iyer NG, Kranjac T, Robertson JFR, Aparicio S, Tavare S, Ellis I, Brenton JD, and Caldas C. Using array-comparative genomic hybridization to define molecular portraits of primary breast cancers. *Oncogene* 2007;26:1959–70.
38. Stange DE, Radlwimmer B, Schubert F, Traub F, Pich A, Toedt G, Mendrzyk F, Lehmann U, Eils R, Kreipe H, and Lichter P. High-resolution genomic profiling reveals association of chromosomal aberrations on 1q and 16p with histologic and genetic subgroups of invasive breast cancer. *Clin Cancer Res* 2006;12:345–52.
39. Han W, Han M-R, Kang J, Bae J-Y, Lee JH, Bae Y, Lee J, Shin H-J, Hwang K-T, Hwang S-E, Kim S-W, and Noh D-Y. Genomic alterations identified by array comparative genomic hybridization as prognostic markers in tamoxifen-treated estrogen receptor-positive breast cancer. *BMC Cancer* 2006;6:92.
40. Chibon F, Mariani O, Derre J, Mairal A, Coindre JM, Guillou L, Sastre X, Pedeutour F, and Aurias A. ASK1 (MAP3K5) as a potential therapeutic target in malignant fibrous histiocytomas with 12q14–q15 and 6q23 amplifications. *Genes Chromosomes Cancer* 2004;40:32–7.
41. Tonon G, Wong KK, Maulik G, Brennan C, Feng B, Zhang Y, Khatry DB, Protopopov A, You MJ, Aguirre AJ, Martin ES, Yang Z, Ji H, Chin L, and Depinho RA. High-resolution genomic profiles of human lung cancer. *Proc Natl Acad Sci USA* 2005;102:9625–30.
42. Ehlers JP, Worley L, Onken MD, and Harbour JW. DDEF1 is located in an amplified region of chromosome 8q and is overexpressed in uveal melanoma. *Clin Cancer Res* 2005;11:3609–13.
43. Bea S, Zettl A, Wright G, Salaverria I, Jehn P, Moreno V, Burek C, Ott G, Puig X, Yang L, et al. Diffuse large B-cell lymphoma subgroups have distinct genetic profiles that influence tumor biology and improve gene expression-based survival prediction. *Blood* 2005;106:3183–90.
44. Wreesmann VB, Sieczka EM, Socci ND, Hezel M, Belbin TJ, Childs G, Patel SG, Patel KN, Tallini G, Prystowsky M, Shaha AR, Kraus D, Shah JP, Rao PH, Ghossein R, and Singh B. Genome-wide profiling of papillary thyroid cancer identifies MUC1 as an independent prognostic marker. *Cancer Res* 2004;64:3780–9.
45. Smith LT, Mayerson J, Nowak NJ, Suster D, Mohammed N, Long S, Auer H, Jones S, McKeegan C, Young G, Bos G, Plass C, and Morrison C. 20q11.1 amplification in giant-cell tumor of bone: array CGH, FISH, and association with outcome. *Genes Chromosomes Cancer* 2006;45:957–66.

46. Schwaenen C, Nessling M, Wessendorf S, Salvi T, Wrobel G, Radlwimmer B, Kestler HA, Haslinger C, Stilgenbauer S, Dohner H, Bentz M, and Lichter P. Automated array-based genomic profiling in chronic lymphocytic leukemia: development of a clinical tool and discovery of recurrent genomic alterations. *Proc Natl Acad Sci USA* 2004;101:1039–44.
47. Lee JA and Lupski Jr. Genomic rearrangements and gene copy-number alterations as a cause of nervous system disorders. *Neuron* 2006;52:103–21.
48. Rovelet-Lecrux A, Hannequin D, Raux G, Le Meur N, Laquerriere A, Vital A, Dumanchin C, Feuillette S, Brice A, Vercelletto M, Dubas F, Frebourg T, and Campion D. APP locus duplication causes autosomal dominant early-onset Alzheimer disease with cerebral amyloid angiopathy. *Nat Genet* 2006;38:24–6.
49. Salehi A, Delcroix JD, Belichenko PV, Zhan K, Wu C, Valletta JS, Takimoto-Kimura R, Kleschevnikov AM, Sambamurti K, Chung PP, Xia W, Villar A, Campbell WA, Kulnane LS, Nixon RA, Lamb BT, Epstein CJ, Stokin GB, Goldstein LS, and Mobley WC. Increased App expression in a mouse model of Down's syndrome disrupts NGF transport and causes cholinergic neuron degeneration. *Neuron* 2006;51:29–42.
50. Olejniczak ET, Van Sant C, Anderson MG, Wang G, Tahir SK, Sauter G, Lesniewski R, and Semizarov D. Integrative genomic analysis of small-cell lung carcinoma reveals correlates of sensitivity to bcl-2 antagonists and uncovers novel chromosomal gains. *Mol Cancer Res* 2007;5:331–9.
51. FDA White Paper: "Drug-device co-development concept paper (draft)", available at http://www.fda.gov/cder/genomics/pharmacoconceptfn.pdf.
52. Nigro JM, Baker SJ, Preisinger AC, Jessup JM, Hostetter R, Cleary K, Bigner SH, Davidson N, Baylin S, Devilee P, et al. Mutations in the p53 gene occur in diverse human tumour types. *Nature* 1989;342:705–8.
53. Bodmer WF. Cancer genetics: colorectal cancer as a model. *J Hum Genet* 2006;51:391–6.
54. Tennis M, Krishnan S, Bonner M, Ambrosone CB, Vena JE, Moysich K, Swede H, McCann S, Hall P, Shields PG, and Freudenheim JL. p53 Mutation analysis in breast tumors by a DNA microarray method. *Cancer Epidemiol Biomarkers Prev* 2006;15:80–5.
55. Wikman FP, Lu ML, Thykjaer T, Olesen SH, Andersen LD, Cordon-Cardo C, and Orntoft TF. Evaluation of the performance of a p53 sequencing microarray chip using 140 previously sequenced bladder tumor samples. *Clin Chem* 2000;46:1555–61.
56. Cooper M, Li SQ, Bhardwaj T, Rohan T, and Kandel RA. Evaluation of oligonucleotide arrays for sequencing of the p53 gene in DNA from formalin-fixed, paraffin-embedded breast cancer specimens. *Clin Chem* 2004;50:500–8.
57. Ahrendt SA, Halachmi S, Chow JT, Wu L, Halachmi N, Yang SC, Wehage S, Jen J, and Sidransky D. Rapid p53 sequence analysis in primary lung cancer using an oligonucleotide probe array. *Proc Natl Acad Sci USA* 1999;96:7382–7.
58. Wen WH, Bernstein L, Lescallett J, Beazer-Barclay Y, Sullivan-Halley J, White M, and Press MF. Comparison of TP53 mutations identified by oligonucleotide microarray and conventional DNA sequence analysis. *Cancer Res* 2000;60:2716–22.
59. Bergh J, Norberg T, Sjogren S, Lindgren A, and Holmberg L. Complete sequencing of the p53 gene provides prognostic information in breast cancer patients, particularly in relation to adjuvant systemic therapy and radiotherapy. *Nat Med* 1995;1:1029–34.
60. Kovach JS, Hartmann A, Blaszyk H, Cunningham J, Schaid D, and Sommer SS. Mutation detection by highly sensitive methods indicates that p53 gene mutations in breast cancer can have important prognostic value. *Proc Natl Acad Sci USA* 1996;93:1093–6.
61. Baker SJ, Fearon ER, Nigro JM, Hamilton SR, Preisinger AC, Jessup JM, vanTuinen P, Ledbetter DH, Barker DF, Nakamura Y, White R, and Vogelstein B. Chromosome 17 deletions and p53 gene mutations in colorectal carcinomas. *Science* 1989;244:217–21.

62. Lin X and Howell SB. DNA mismatch repair and p53 function are major determinants of the rate of development of cisplatin resistance. *Mol Cancer Ther* 2006;5:1239–47.
63. Lowe SW, Bodis S, McClatchey A, Remington L, Ruley HE, Fisher DE, Housman DE, and Jacks T. p53 status and the efficacy of cancer therapy in vivo. *Science* 1994;266:807–10.
64. Lowe SW, Ruley HE, Jacks T, and Housman DE. p53-dependent apoptosis modulates the cytotoxicity of anticancer agents. *Cell* 1993;74:957–67.
65. Rodenhuis S, van de Wetering ML, Mooi WJ, Evers SG, Van Zandwijk N, and Bos JL. Mutational activation of the K-ras oncogene. A possible pathogenetic factor in adenocarcinoma of the lung. *N Engl J Med* 1987;317:929–35.
66. Sugio K, Ishida T, Yokoyama H, Inoue T, Sugimachi K, and Sasazuki T. ras gene mutations as a prognostic marker in adenocarcinoma of the human lung without lymph node metastasis. *Cancer Res* 1992;52:2903–6.
67. Rodenhuis S, Slebos RJ, Boot AJ, Evers SG, Mooi WJ, Wagenaar SS, Van Bodegom PC, and Bos JL. Incidence and possible clinical significance of K-ras oncogene activation in adenocarcinoma of the human lung. *Cancer Res* 1988;48:5738–41.
68. Bos JL. Ras oncogenes in human cancer: a review. *Cancer Res* 1989;49:4682–9.
69. Slebos RJ, Kibbelaar RE, Dalesio O, Kooistra A, Stam J, Meijer CJ, Wagenaar SS, Vanderschueren RG, Van Zandwijk N, Mooi WJ, et al. K-ras oncogene activation as a prognostic marker in adenocarcinoma of the lung. *N Engl J Med* 1990;323:561–5.
70. Graziano SL, Gamble GP, Newman NB, Abbott LZ, Rooney M, Mookherjee S, Lamb ML, Kohman LJ, and Poiesz BJ. Prognostic significance of K-ras codon 12 mutations in patients with resected Stage I and II non–small-cell lung cancer. *J Clin Oncol* 1999;17:668–75.
71. Rodenhuis S, Boerrigter L, Top B, Slebos RJ, Mooi WJ, van't Veer L, and Van Zandwijk N. Mutational activation of the K-ras oncogene and the effect of chemotherapy in advanced adenocarcinoma of the lung: a prospective study. *J Clin Oncol* 1997;15:285–91.
72. Pao W, Wang TY, Riely GJ, Miller VA, Pan Q, Ladanyi M, Zakowski MF, Heelan RT, Kris MG, and Varmus HE. KRAS mutations and primary resistance of lung adenocarcinomas to gefitinib or erlotinib. *PLoS Med* 2005;2:e17.
73. Gazdar AF, Shigematsu H, Herz J, and Minna JD. Mutations and addiction to EGFR: the Achilles "heal" of lung cancers? *Trends Mol Med* 2004;10:481–6.
74. Fukuoka M, Yano S, Giaccone G, Tamura T, Nakagawa K, Douillard JY, Nishiwaki Y, Vansteenkiste J, Kudoh S, Rischin D, Eek R, Horai T, Noda K, Takata I, Smit E, Averbuch S, Macleod A, Feyereislova A, Dong RP, and Baselga J. Multi-institutional randomized phase II trial of gefitinib for previously treated patients with advanced non-small-cell lung cancer (The IDEAL 1 Trial) [corrected]. *J Clin Oncol* 2003;21:2237–46.
75. Paez JG, Janne PA, Lee JC, Tracy S, Greulich H, Gabriel S, Herman P, Kaye FJ, Lindeman N, Boggon TJ, Naoki K, Sasaki H, Fujii Y, Eck MJ, Sellers WR, Johnson BE, and Meyerson M. EGFR mutations in lung cancer: correlation with clinical response to gefitinib therapy. *Science* 2004;304:1497–500.
76. Lynch TJ, Bell DW, Sordella R, Gurubhagavatula S, Okimoto RA, Brannigan BW, Harris PL, Haserlat SM, Supko JG, Haluska FG, Louis DN, Christiani DC, Settleman J, and Haber DA. Activating mutations in the epidermal growth factor receptor underlying responsiveness of non-small-cell lung cancer to gefitinib. *N Engl J Med* 2004;350:2129–39.
77. Pao W, Miller V, Zakowski M, Doherty J, Politi K, Sarkaria I, Singh B, Heelan R, Rusch V, Fulton L, Mardis E, Kupfer D, Wilson R, Kris M, and Varmus H. EGF receptor gene mutations are common in lung cancers from "never smokers" and are associated with sensitivity of tumors to gefitinib and erlotinib. *Proc Natl Acad Sci USA* 2004;101:13306–11.
78. Konopka JB, Watanabe SM, Singer JW, Collins SJ, and Witte ON. Cell lines and clinical isolates derived from Ph1-positive chronic myelogenous leukemia patients express c-abl proteins with a common structural alteration. *Proc Natl Acad Sci USA* 1985;82:1810–4.

79. Daley GQ, Van Etten RA, and Baltimore D. Induction of chronic myelogenous leukemia in mice by the P210bcr/abl gene of the Philadelphia chromosome. *Science* 1990;247:824–30.
80. Druker BJ, Sawyers CL, Kantarjian H, Resta DJ, Reese SF, Ford JM, Capdeville R, and Talpaz M. Activity of a specific inhibitor of the BCR-ABL tyrosine kinase in the blast crisis of chronic myeloid leukemia and acute lymphoblastic leukemia with the Philadelphia chromosome. *N Engl J Med* 2001;344:1038–42.
81. Druker BJ, Talpaz M, Resta DJ, Peng B, Buchdunger E, Ford JM, Lydon NB, Kantarjian H, Capdeville R, Ohno-Jones S, and Sawyers CL. Efficacy and safety of a specific inhibitor of the BCR-ABL tyrosine kinase in chronic myeloid leukemia. *N Engl J Med* 2001;344:1031–7.
82. Gorre ME, Mohammed M, Ellwood K, Hsu N, Paquette R, Rao PN, and Sawyers CL. Clinical resistance to STI-571 cancer therapy caused by BCR-ABL gene mutation or amplification. *Science* 2001;293:876–80.
83. Hirota S, Isozaki K, Moriyama Y, Hashimoto K, Nishida T, Ishiguro S, Kawano K, Hanada M, Kurata A, Takeda M, Muhammad Tunio G, Matsuzawa Y, Kanakura Y, Shinomura Y, and Kitamura Y. Gain-of-function mutations of c-kit in human gastrointestinal stromal tumors. *Science* 1998;279:577–80.
84. Demetri GD, von Mehren M, Blanke CD, Van den Abbeele AD, Eisenberg B, Roberts PJ, Heinrich MC, Tuveson DA, Singer S, Janicek M, Fletcher JA, Silverman SG, Silberman SL, Capdeville R, Kiese B, Peng B, Dimitrijevic S, Druker BJ, Corless C, Fletcher CD, and Joensuu H. Efficacy and safety of imatinib mesylate in advanced gastrointestinal stromal tumors. *N Engl J Med* 2002;347:472–80.
85. Chen LL, Trent JC, Wu EF, Fuller GN, Ramdas L, Zhang W, Raymond AK, Prieto VG, Oyedeji CO, Hunt KK, Pollock RE, Feig BW, Hayes KJ, Choi H, Macapinlac HA, Hittelman W, Velasco MA, Patel S, Burgess MA, Benjamin RS, and Frazier ML. A missense mutation in KIT kinase domain 1 correlates with imatinib resistance in gastrointestinal stromal tumors. *Cancer Res* 2004;64:5913–9.
86. Antonescu CR, Besmer P, Guo T, Arkun K, Hom G, Koryotowski B, Leversha MA, Jeffrey PD, Desantis D, Singer S, Brennan MF, Maki RG, and DeMatteo RP. Acquired resistance to imatinib in gastrointestinal stromal tumor occurs through secondary gene mutation. *Clin Cancer Res* 2005;11:4182–90.
87. Thomas RK, Baker AC, Debiasi RM, Winckler W, Laframboise T, Lin WM, Wang M, Feng W, Zander T, Macconaill LE, et al. High-throughput oncogene mutation profiling in human cancer. *Nat Genet* 2007;39:347–51.
88. Ting AH, McGarvey KM, and Baylin SB. The cancer epigenome—components and functional correlates. *Genes Dev* 2006;20:3215–31.
89. Feinberg AP and Tycko B. The history of cancer epigenetics. *Nat Rev Cancer* 2004;4:143–53.
90. Murrell A, Rakyan VK, and Beck S. From genome to epigenome. *Hum Mol Genet* 2005;14 Spec No 1:R3–R10.
91. Ushijima T. Detection and interpretation of altered methylation patterns in cancer cells. *Nat Rev Cancer* 2005;5:223–31.
92. Takai D and Jones PA. Comprehensive analysis of CpG islands in human chromosomes 21 and 22. *Proc Natl Acad Sci USA* 2002;99:3740–5.
93. Bird A. DNA methylation patterns and epigenetic memory. *Genes Dev* 2002;16:6–21.
94. Antequera F and Bird A. Number of CpG islands and genes in human and mouse. *Proc Natl Acad Sci USA* 1993;90:11995–9.
95. Richardson B. Impact of aging on DNA methylation. *Ageing Res Rev* 2003;2:245–61.
96. Egger G, Liang G, Aparicio A, and Jones PA. Epigenetics in human disease and prospects for epigenetic therapy. *Nature* 2004;429:457–63.
97. Baylin SB and Ohm JE. Epigenetic gene silencing in cancer—a mechanism for early oncogenic pathway addiction? *Nat Rev Cancer* 2006;6:107–16.

98. Jones PA and Baylin SB. The fundamental role of epigenetic events in cancer. *Nat Rev Genet* 2002;3:415–28.
99. Silber JR, Bobola MS, Ghatan S, Blank A, Kolstoe DD, and Berger MS. O6-methylguanine-DNA methyltransferase activity in adult gliomas: relation to patient and tumor characteristics. *Cancer Res* 1998;58:1068–73.
100. Esteller M, Garcia-Foncillas J, Andion E, Goodman SN, Hidalgo OF, Vanaclocha V, Baylin SB, and Herman JG. Inactivation of the DNA-repair gene MGMT and the clinical response of gliomas to alkylating agents. *N Engl J Med* 2000;343:1350–4.
101. Paz MF, Yaya-Tur R, Rojas-Marcos I, Reynes G, Pollan M, Aguirre-Cruz L, Garcia-Lopez JL, Piquer J, Safont MJ, Balana C, Sanchez-Cespedes M, Garcia-Villanueva M, Arribas L, and Esteller M. CpG island hypermethylation of the DNA repair enzyme methyltransferase predicts response to temozolomide in primary gliomas. *Clin Cancer Res* 2004;10:4933–8.
102. Hegi ME, Diserens AC, Godard S, Dietrich PY, Regli L, Ostermann S, Otten P, Van Melle G, de Tribolet N, and Stupp R. Clinical trial substantiates the predictive value of *O*-6-methylguanine-DNA methyltransferase promoter methylation in glioblastoma patients treated with temozolomide. *Clin Cancer Res* 2004;10:1871–4.
103. Esteller M, Hamilton SR, Burger PC, Baylin SB, and Herman JG. Inactivation of the DNA repair gene *O*-6-methylguanine-DNA methyltransferase by promoter hypermethylation is a common event in primary human neoplasia. *Cancer Res* 1999;59:793–7.
104. Esteller M, Gaidano G, Goodman SN, Zagonel V, Capello D, Botto B, Rossi D, Gloghini A, Vitolo U, Carbone A, Baylin SB, and Herman JG. Hypermethylation of the DNA repair gene *O*(6)-methylguanine DNA methyltransferase and survival of patients with diffuse large B-cell lymphoma. *J Natl Cancer Inst* 2002;94:26–32.
105. Toyota M, Sasaki Y, Satoh A, Ogi K, Kikuchi T, Suzuki H, Mita H, Tanaka N, Itoh F, Issa JP, Jair KW, Schuebel KE, Imai K, and Tokino T. Epigenetic inactivation of CHFR in human tumors. *Proc Natl Acad Sci USA* 2003;100:7818–23.
106. Satoh A, Toyota M, Itoh F, Sasaki Y, Suzuki H, Ogi K, Kikuchi T, Mita H, Yamashita T, Kojima T, Kusano M, Fujita M, Hosokawa M, Endo T, Tokino T, and Imai K. Epigenetic inactivation of CHFR and sensitivity to microtubule inhibitors in gastric cancer. *Cancer Res* 2003;63:8606–13.
107. Kantharidis P, El-Osta A, deSilva M, Wall DM, Hu XF, Slater A, Nadalin G, Parkin JD, and Zalcberg Jr. Altered methylation of the human MDR1 promoter is associated with acquired multidrug resistance. *Clin Cancer Res* 1997;3:2025–32.
108. Plumb JA, Strathdee G, Sludden J, Kaye SB, and Brown R. Reversal of drug resistance in human tumor xenografts by 2′-deoxy-5-azacytidine-induced demethylation of the hMLH1 gene promoter. *Cancer Res* 2000;60:6039–44.
109. Eifel P, Axelson JA, Costa J, Crowley J, Curran WJ, Jr., Deshler A, Fulton S, Hendricks CB, Kemeny M, Kornblith AB, Louis TA, Markman M, Mayer R, and Roter D. National Institutes of Health Consensus Development Conference Statement: adjuvant therapy for breast cancer, November 1–3, 2000. *J Natl Cancer Inst* 2001;93:979–89.
110. Tamoxifen for early breast cancer: an overview of the randomised trials. Early Breast Cancer Trialists' Collaborative Group. *Lancet* 1998;351:1451–67.
111. Polychemotherapy for early breast cancer: an overview of the randomised trials. Early Breast Cancer Trialists' Collaborative Group. *Lancet* 1998;352:930–42.
112. Effects of chemotherapy and hormonal therapy for early breast cancer on recurrence and 15-year survival: an overview of the randomised trials. *Lancet* 2005;365:1687–717.
113. Fisher B, Jeong JH, Bryant J, Anderson S, Dignam J, Fisher ER, and Wolmark N. Treatment of lymph-node-negative, oestrogen-receptor-positive breast cancer: long-term findings from National Surgical Adjuvant Breast and Bowel Project randomised clinical trials. *Lancet* 2004;364:858–68.

114. Perou CM, Sorlie T, Eisen MB, van de Rijn M, Jeffrey SS, Rees CA, Pollack JR, Ross DT, Johnsen H, Akslen LA, Fluge O, Pergamenschikov A, Williams C, Zhu SX, Lonning PE, Borresen-Dale AL, Brown PO, and Botstein D. Molecular portraits of human breast tumours. *Nature* 2000;406:747–52.

115. van 't Veer LJ, Dai H, van de Vijver MJ, He YD, Hart AA, Mao M, Peterse HL, van der Kooy K, Marton MJ, Witteveen AT, Schreiber GJ, Kerkhoven RM, Roberts C, Linsley PS, Bernards R, and Friend SH. Gene expression profiling predicts clinical outcome of breast cancer. *Nature* 2002;415:530–6.

116. van de Vijver MJ, He YD, van 't Veer LJ, Dai H, Hart AAM, Voskuil DW, Schreiber GJ, Peterse JL, Roberts C, Marton MJ, Parrish M, Atsma D, Witteveen A, Glas A, Delahaye L, van der Velde T, Bartelink H, Rodenhuis S, Rutgers ET, Friend SH, and Bernards R. A gene-expression signature as a predictor of survival in breast cancer. *N Engl J Med* 2002;347:1999–2009.

117. Sorlie T, Perou CM, Tibshirani R, Aas T, Geisler S, Johnsen H, Hastie T, Eisen MB, van de Rijn M, Jeffrey SS, Thorsen T, Quist H, Matese JC, Brown PO, Botstein D, Eystein Lonning P, and Borresen-Dale AL. Gene expression patterns of breast carcinomas distinguish tumor subclasses with clinical implications. *Proc Natl Acad Sci USA* 2001;98:10869–74.

118. Sorlie T, Tibshirani R, Parker J, Hastie T, Marron JS, Nobel A, Deng S, Johnsen H, Pesich R, Geisler S, Demeter J, Perou CM, Lonning PE, Brown PO, Borresen-Dale AL, and Botstein D. Repeated observation of breast tumor subtypes in independent gene expression data sets. *Proc Natl Acad Sci USA* 2003;100:8418–23.

119. Paik S, Shak S, Tang G, Kim C, Baker J, Cronin M, Baehner FL, Walker MG, Watson D, Park T, Hiller W, Fisher ER, Wickerham DL, Bryant J, and Wolmark N. A multigene assay to predict recurrence of tamoxifen-treated, node-negative breast cancer. *N Engl J Med* 2004;351:2817–26.

120. Cronin M, Pho M, Dutta D, Stephans JC, Shak S, Kiefer MC, Esteban JM, and Baker JB. Measurement of gene expression in archival paraffin-embedded tissues: development and performance of a 92-gene reverse transcriptase-polymerase chain reaction assay. *Am J Pathol* 2004;164:35–42.

121. Hornberger J, Cosler LE, and Lyman GH. Economic analysis of targeting chemotherapy using a 21-gene RT-PCR assay in lymph-node-negative, estrogen-receptor-positive, early-stage breast cancer. *Am J Manag Care* 2005;11:313–24.

122. Paik S, Tang G, Shak S, Kim C, Baker J, Kim W, Cronin M, Baehner FL, Watson D, Bryant J, Costantino JP, Geyer CE, Jr., Wickerham DL, and Wolmark N. Gene expression and benefit of chemotherapy in women with node-negative, estrogen receptor-positive breast cancer. *J Clin Oncol* 2006;24:3726–34.

123. Glas AM, Floore A, Delahaye LJ, Witteveen AT, Pover RC, Bakx N, Lahti-Domenici JS, Bruinsma TJ, Warmoes MO, Bernards R, Wessels LF, and Van't Veer LJ. Converting a breast cancer microarray signature into a high-throughput diagnostic test. *BMC Genomics* 2006; 7:278.

124. Nesbitt JC, Putnam JB, Jr., Walsh GL, Roth JA, and Mountain CF. Survival in early-stage non-small cell lung cancer. *Ann Thorac Surg* 1995;60:466–72.

125. Potti A, Mukherjee S, Petersen R, Dressman HK, Bild A, Koontz J, Kratzke R, Watson MA, Kelley M, Ginsburg GS, West M, Harpole DH, Jr., and Nevins Jr. A genomic strategy to refine prognosis in early-stage non-small-cell lung cancer. *N Engl J Med* 2006;355:570–80.

126. Chen HY, Yu SL, Chen CH, Chang GC, Chen CY, Yuan A, Cheng CL, Wang CH, Terng HJ, Kao SF, Chan WK, Li HN, Liu CC, Singh S, Chen WJ, Chen JJ, and Yang PC. A five-gene signature and clinical outcome in non-small-cell lung cancer. *N Engl J Med* 2007;356:11–20.

127. A clinical evaluation of the International Lymphoma Study Group classification of non-Hodgkin's lymphoma. The Non-Hodgkin's Lymphoma Classification Project. *Blood* 1997;89:3909–18.
128. Wright G, Tan B, Rosenwald A, Hurt EH, Wiestner A, and Staudt LM. A gene expression-based method to diagnose clinically distinct subgroups of diffuse large B cell lymphoma. *Proc Natl Acad Sci USA* 2003;100:9991–6.
129. Alizadeh AA, Eisen MB, Davis RE, Ma C, Lossos IS, Rosenwald A, Boldrick JC, Sabet H, Tran T, Yu X, et al. Distinct types of diffuse large B-cell lymphoma identified by gene expression profiling. *Nature* 2000;403:503–11.
130. Rosenwald A, Wright G, Chan WC, Connors JM, Campo E, Fisher RI, Gascoyne RD, Muller-Hermelink HK, Smeland EB, Giltnane JM, et al. The use of molecular profiling to predict survival after chemotherapy for diffuse large-B-cell lymphoma. *N Engl J Med* 2002;346:1937–47.
131. Savage KJ, Monti S, Kutok JL, Cattoretti G, Neuberg D, De Leval L, Kurtin P, Dal Cin P, Ladd C, Feuerhake F, Aguiar RC, Li S, Salles G, Berger F, Jing W, Pinkus GS, Habermann T, Dalla-Favera R, Harris NL, Aster JC, Golub TR, and Shipp MA. The molecular signature of mediastinal large B-cell lymphoma differs from that of other diffuse large B-cell lymphomas and shares features with classical Hodgkin lymphoma. *Blood* 2003;102:3871–9.
132. Rosenwald A, Wright G, Leroy K, Yu X, Gaulard P, Gascoyne RD, Chan WC, Zhao T, Haioun C, Greiner TC, et al. Molecular diagnosis of primary mediastinal B cell lymphoma identifies a clinically favorable subgroup of diffuse large B cell lymphoma related to Hodgkin lymphoma. *J Exp Med* 2003;198:851–62.
133. Staudt LM, and Dave S. The biology of human lymphoid malignancies revealed by gene expression profiling. *Adv Immunol* 2005;87:163–208.
134. Shipp MA, Ross KN, Tamayo P, Weng AP, Kutok JL, Aguiar RC, Gaasenbeek M, Angelo M, Reich M, Pinkus GS, Ray TS, Koval MA, Last KW, Norton A, Lister TA, Mesirov J, Neuberg DS, Lander ES, Aster JC, and Golub TR. Diffuse large B-cell lymphoma outcome prediction by gene-expression profiling and supervised machine learning. *Nat Med* 2002;8:68–74.
135. Lossos IS, Czerwinski DK, Alizadeh AA, Wechser MA, Tibshirani R, Botstein D, and Levy R. Prediction of survival in diffuse large-B-cell lymphoma based on the expression of six genes. *N Engl J Med* 2004;350:1828–37.
136. Potti A, Dressman HK, Bild A, Riedel RF, Chan G, Sayer R, Cragun J, Cottrill H, Kelley MJ, Petersen R, Harpole D, Marks J, Berchuck A, Ginsburg GS, Febbo P, Lancaster J, and Nevins Jr. Genomic signatures to guide the use of chemotherapeutics. *Nat Med* 2006;12:1294–300.
137. Chang JC, Wooten EC, Tsimelzon A, Hilsenbeck SG, Gutierrez MC, Elledge R, Mohsin S, Osborne CK, Chamness GC, Allred DC, and O' Connell P. Gene expression profiling for the prediction of therapeutic response to docetaxel in patients with breast cancer. *Lancet* 2003;362:362–9.
138. Rouzier R, Perou CM, Symmans WF, Ibrahim N, Cristofanilli M, Anderson K, Hess KR, Stec J, Ayers M, Wagner P, Morandi P, Fan C, Rabiul I, Ross JS, Hortobagyi GN, and Pusztai L. Breast cancer molecular subtypes respond differently to preoperative chemotherapy. *Clin Cancer Res* 2005;11:5678–85.
139. Rouzier R, Rajan R, Wagner P, Hess KR, Gold DL, Stec J, Ayers M, Ross JS, Zhang P, Buchholz TA, Kuerer H, Green M, Arun B, Hortobagyi GN, Symmans WF, and Pusztai L. Microtubule-associated protein tau: a marker of paclitaxel sensitivity in breast cancer. *Proc Natl Acad Sci USA* 2005;102:8315–20.
140. FDA Guidance for industry: Pharmacogenomics data submissions" available at http://www.fda.gov/cber/gdlns/pharmdtasub.pdf.

141. Goodsaid F and Frueh F. Process map proposal for the validation of genomic biomarkers. *Pharmacogenomics* 2006;7:773–82.
142. Pusztai L and Hess KR. Clinical trial design for microarray predictive marker discovery and assessment. *Ann Oncol* 2004;15:1731–7.
143. FDA White Paper: "Innovation or stagnation, challenge and opportunity on the critical path to new medical products", available at http://www.fda.gov/oc/initiatives/criticalpath/whitepaper.html.
144. Simon R. Validation of pharmacogenomic biomarker classifiers for treatment selection. *Cancer Biomark* 2006;2:89–96.

Chapter 4

Fundamental Principles of Toxicogenomics

4.1. INTRODUCTION

The term **toxicogenomics** has been interchangeably used to refer to the application of the gene expression profiling technologies or more broadly to the use of all "omics" technologies in the field of toxicology. While evaluation of the genome, proteome, and metabolome has definitely useful applications in toxicology, in this book we restrict the use of the term "toxicogenomics" to describe the utilization of transcriptomic data to detect, investigate, or characterize toxicological effects. The introduction of microarray technology in the 1990s was received with much enthusiasm and excitement by the toxicology community, because it was viewed as a unique, novel approach to generate large amounts of previously not accessible, relevant molecular data (1, 2). It was then proposed that microarray data could be used to extract sensitive indicators of toxicity and mechanistic clues that would markedly improve our understanding of adverse drug reactions and toxic effects. In fact, toxicogenomics was perceived by some as a technology that could revolutionize the practice of toxicology (3). This initial enthusiasm was manifested by the rapid adoption of toxicogenomics by the pharmaceutical industry, regulatory bodies, and academic institutions and was evidenced by the rapid growth of publications on the topic. The interest in rapidly moving the science into concrete applications has also triggered the creation of multiple consortia, initiatives, and workshops in an attempt to promote the maturation of this emerging field. These various activities have indeed helped to address technical issues, build the scientific base of the toxicology community, and establish guidelines for the submission of toxicogenomics data to journals, databases, and even regulatory agencies. Clearly, significant resources have already been invested in

Genomics in Drug Discovery and Development, by Dimitri Semizarov and Eric Blomme

the field, but one must admit that it is still unclear what the exact return will be on these investments. This is not surprising, since the impact of novel technologies on drug discovery and development is typically difficult to ascertain given the long cycle times of the pharmaceutical industry (4). Unrealistic expectations commonly accompany the introduction of novel technologies, which is also true for toxicogenomics. Indeed, the outcome of years of investigation and of active scientific exchanges has probably produced less than initially expected. Yet regular technical improvements, more focused studies, and several success stories are clear indications that toxicogenomics is one of the most promising advances in toxicology that have occurred in the past decade.

Most published toxicogenomics reports and presentations have so far consisted of retrospective evaluations of reference and tool compounds that induce well-characterized, general or specific toxicities. These types of studies were necessary to demonstrate the value of the technology and better define its most useful applications. However, they are not sufficient to determine whether and to what extent the technology can affect our ability to develop compounds with safer profiles and better-defined risks. A definite transition has occurred in the last few years, whereby several pharmaceutical companies have implemented toxicogenomics in their discovery and development programs. This will certainly result in insightful retrospective evaluations, and the experience from these pioneering companies will ultimately translate into better quantification of the value provided by toxicogenomics in the drug discovery and development process. At this point, it is too early to ascertain the ultimate impact of the technology, and given the lengthy drug development cycle, this impact may not become evident for several years. Given the lack of clear retrospective data and the immaturity of the field, implementing the technology to impact drug discovery and development remains a significant challenge in most institutions. In this chapter and in Chapters 5 and 6, one of our objectives is to provide practical advice to facilitate the use of toxicogenomics in drug discovery and development. Here, we introduce the basic principles of toxicogenomics with a particular focus on technical and practical aspects, as well as analysis methods and toxicogenomics databases. Specific applications of gene expression analysis in predictive, diagnostic, and mechanistic toxicology are covered in Chapters 5 and 6.

4.2. FUNDAMENTALS OF TOXICOGENOMICS

In the previous chapters, we have described how gene expression profiling methodologies enable interrogation of the transcriptome of tissues or cells at various time points. Molecular toxicologists have been particularly interested in applying gene expression microarrays to investigate the transcriptional response of tissues or cells following exposure to toxic or experimental compounds. These investigations have mostly used microarray platforms, since they generate a genome-wide view of gene regulation. However, it is noteworthy that data from microarray experiments frequently represent only an initial step in the formulation of hypotheses regarding mechanisms of toxicity and in the discovery of biomarkers of toxicity. Hence, studies using microarrays can lead, as will be illustrated on several occasions later, to the development of streamlined, specific assays that can routinely be used in the drug discovery and development process. These follow-up assays are typically developed with more cost-effective, less resource-intensive, and more user-friendly platforms with better throughput. Indeed, one may argue that the ability to transfer microarray-generated knowledge to more focused assays is an important requirement for toxicogenomics to realize its full potential in drug discovery and development. Therefore, when appropriate, we will address the feasibility and utility of transferring microarray-based findings to alternative technology platforms.

4.2.1. Principle of Toxicogenomics

Compounds can induce toxicity through a variety of mechanisms. Not all mechanisms of toxicity are fully understood, but years of research have elucidated some classic mechanisms of toxicity. Toxicants injuring cells through similar mechanisms are typically grouped into classes. Examples include such mechanisms as covalent binding of endogenous molecules, modulation of the activity of endogenous signaling molecules, or disruption of normal molecular functions. Several studies have demonstrated that members of a specific class of toxicants affect similar biological pathways leading to shared gene expression changes (Fig. 4.1). In other words, sets of genes differentially regulated with respect to vehicle controls (either up-regulation or down-regulation) can be used to

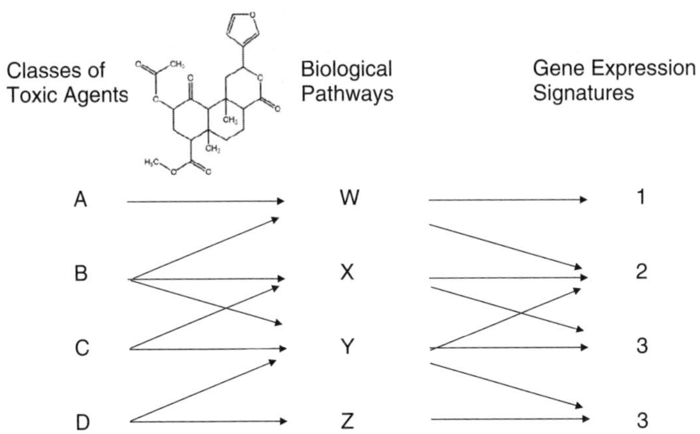

Figure 4.1 Principle of toxicogenomics. Members of specific classes of toxic agents cause changes in the cellular/tissue transcriptome and affect similar biological pathways as evidenced by shared gene expression changes. These shared gene expression changes represent signatures that can be used to predict, identify, or characterize toxic effects.

classify compounds according to mechanisms of toxicity, and in many situations these shared transcriptional effects can be spotted well before obvious phenotypic changes occur. These observations represent the foundation of toxicogenomics: Interrogating the transcriptional regulation of specific gene sets represents a precise and mechanism-based approach for toxicity detection. These gene sets can be used to monitor the presence or to predict the future occurrence of phenotypic changes of toxicological significance. They are referred to as **signatures of toxicity** and represent biomarkers that can be used for hazard identification and risk assessment. Chapters 2 and 3 introduce the concept of a gene expression signatures and discuss the principles of their identification and application in biomarker discovery, which are also pertinent to toxicology.

A particularly attractive aspect of toxicogenomics is the mechanistic clarity that this methodology offers when applied to toxicology. Traditional toxicology is a mostly observational discipline that uses only a limited number of end points with mechanistic value. In contrast, global evaluation of transcriptomic changes constitutes a rapid approach to detecting abnormalities in subcellular components or deregulation of biological pathways that can be used to formulate hypotheses on the mechanism of toxicity of a compound. This improved accuracy represents an opportunity for toxicologists to rapidly characterize mechanisms of toxicity. In Chapters 5 and 6, several examples are used to illustrate this aspect. It should, however, be emphasized that, albeit useful in generating hypotheses, microarray experiments typically require follow-up orthogonal experiments to confirm or refute these hypotheses.

4.2.2. Technical Reproducibility

Microarray technology has been evolving extremely rapidly, and technical improvements are constantly being introduced. Therefore, we will not comment extensively on the current status of platforms and tools for toxicogenomics, since this status would likely be rapidly obsolete. Rather, we focus on the essential questions and trends that are relevant to the implementation of gene expression profiling in toxicology.

To be useful as a tool for the evaluation of toxicity, a technology or assay must meet specific technical criteria. First, the assay must be technically repeatable and reproducible. Second, it must have an acceptable biological reproducibility, that is, the interindividual variability should be low enough to run the assay with a reasonable number of biological replicates (or in terms more familiar to toxicologists, with a reasonable group size). This aspect is covered in the next section. Finally, the end points generated must provide added value compared with the battery of end points currently available. Although obvious, this last point deserves reinforcement. There have been many claims or studies demonstrating that gene expression profiling can detect various toxic changes, already well-identifiable by several current, well-validated, and understood biomarkers. Such cases represent interesting academic exercises that confirm the robustness of the technology but are not practical to identify and understand toxicity with higher precision in drug discovery and development. This is further discussed in Chapters 5 and 6.

The questionable reproducibility and accuracy of the early commercial and custom microarray platforms have been rapidly corrected through technical improvement, and several studies have confirmed these advances for the current gene expression platforms (2, 5). In particular, the reliability and precision of microarray platforms in the context of biomarkers was recently confirmed by a consortium led by the U.S. Food and Drug Administration (FDA) (6). The MicroArray Quality Control (MAQC) project was initiated to rigorously address concerns of microarray data reliability. In the first phase of the project, expression data on four titration pools from two distinct reference RNA samples were generated at multiple test sites with a variety of microarray-based and alternative technology platforms. Results indicated intraplatform consistency across test sites, as well as a high level of interplatform concordance for genes identified as differentially expressed. A recent report also confirmed that high intralaboratory and interlaboratory reproducibility can be achieved with current commercial platforms, especially when strictly controlled standard operating procedures are used (7). This study conducted an interlaboratory comparison of gene expression measurements on the same microarray platform starting with identical input RNA. Measurement of a 70-gene breast cancer signature for breast cancer was highly robust and reproducible across three distinct laboratories.

Microarray results are influenced by a variety of variables, including the methods of array manufacturing, RNA extraction, probe labeling, hybridization conditions, and image acquisition and analysis. Rapid technological improvements have occurred in most of these areas, especially because of the advent of automation using robotics and better standardization of practices across the scientific community. Automation has tremendously impacted the repeatability and reproducibility of many laboratory techniques, including microarray technologies. In general, automated methods are associated with lower technical interexperiment variation than manual methods (8). This optimization has led to important procedural adjustments. For instance, it used to be common practice and highly desirable to verify specific gene expression changes observed with microarrays using independent methodologies. This verification step typically began with the same samples that were studied in the microarray experiment and employed a protocol designed to best answer the scientific question (9). For instance, if specific transcript levels were the parameter of interest, independent methods were selected to quantify mRNA levels. In contrast, if microarray data were used as an antecedent to identify changes in protein levels, techniques such as Western blots or immunohistochemistry would be deemed more relevant. Real-time reverse transcription-polymerase chain reaction (RT-PCR) was used most commonly, since this method is rapid, precise, sensitive, and relatively inexpensive and requires a minimal amount of input template material (9, 10). This so-called validation step was, however, time-consuming and resource-intensive. Past studies have shown with real-time RT-PCR that most microarray results were accurate, especially for highly regulated genes, and that differences between microarray- and PCR-generated data were mostly in the amplitude of the detected gene expression change (10–13). In other words, correlation between RT-PCR and microarray data is generally strong: The direction of change of expression levels (i.e., either up- or down-regulation) between two subsets of interest is accurately measured with most microarray platforms irrespective of the normalization procedure (13). However, it has been reported that fold changes in mRNA expression levels determined by RT-PCR are usually greater than those measured by microarrays (13).

In our experience, the current microarray platforms perform very well, and microarray data need only occasional verification with independent methodologies when used in toxicology. In fact, most toxicogenomics analysis approaches are focused on evaluating deregulation of pathways containing many genes, and therefore, in most instances, precision in quantification of a single gene only minimally affects the biological interpretation of the data. There are, however, instances where validation with an independent method is warranted or highly recommended. Such instances include, but are not limited to, the following situations where critical decisions will depend on these results.

- Results of a study are highly dependent on a single gene or a limited gene set that cannot be used to interrogate pathway deregulation.
- A multigene marker has been identified to monitor efficacy or safety in a program, and its practical use warrants the use of an alternative focused platform.
- Better quantification of expression levels is needed. This is the case when, for instance, a gene expression profiling study only detects a few significant gene expression changes and investigators want to confirm any effect on mRNA expression levels of gene products with a known toxicological role.
- The treatment-induced changes in gene expression levels are so pronounced that signals are saturated, and alternative methods are required to demonstrate dose response. An example of this situation is shown in Figure 4.2.

Beyond these situations, we typically do not consider RT-PCR validation a necessary component of a toxicogenomics experiment. The aforementioned improvement in reliability of microarray platforms clearly represents a major step toward the confident integration of this novel technology in the toxicology toolbox.

Several studies have also shown a high correlation between microarray-generated data and data obtained with lower-throughput gene expression platforms, such as Taqman™ Low Density Arrays (Applied Biosystems, Foster City, CA) or ArrayPlate (High Throughput Genomics, Inc., Tucson, AZ) (14, 15). Results from these studies indicate that microarray results can indeed be used to design smaller, more focused platforms for specific applications aimed at interrogating limited gene sets, leading to the possible introduction

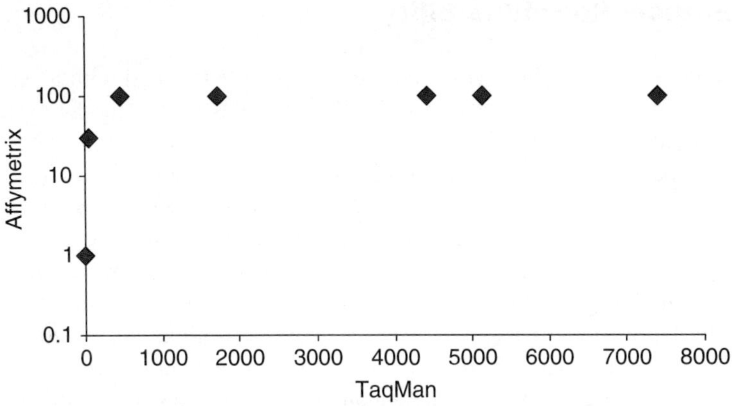

Figure 4.2 Correlation between differential expression levels of cytochrome P450, 1A1 (CYP1A1) measured with Affymetrix microarrays and RT-PCR (TaqMan). Data points represent the ratio between the CYP1A1 mRNA levels in the liver of rats treated with various compounds and the CYP1A1 mRNA levels in the liver of the corresponding vehicle-treated control rats. Note that the measurements from Affymetrix microarrays are saturated when CYP1A1 mRNA is highly induced (100-fold or more) over control liver levels.

of standardized, reproducible, and high-throughput diagnostic assays (16). Other studies have also demonstrated that current platforms have acceptable interplatform reproducibility for assessing toxic reactions (2, 17). In general, these investigations have found that the magnitude of any particular expression ratio may differ from one platform to the other, but that the overall directionality of the expression change for genes with above-background intensity of signal is well correlated across platforms (18). These differences in magnitude can be attributed in part to microarray configuration and platform-specific protocols, but the overall concordance between platforms is high enough to ensure consistency of biological interpretation, especially when gene expression changes are interpreted in the context of biological pathways. In other words, even if the exact lists of deregulated genes may differ between platforms, the biological themes represented by these genes are similar in essence. This is best illustrated by a study that used two distinct platforms to investigate the profiles induced by a prototypical cell cycle inhibitor (12). The authors observed good agreement between the two platforms and demonstrated excellent correlation with RT-PCR data; pathway analysis of the microarray data yielded themes consistent with cell cycle inhibition (12). Interplatform reproducibility is of particular importance in toxicogenomics, since, as will be discussed below, robust interpretation of input gene expression data is facilitated by comparisons with reference data that are often generated on different platforms. The evidence cited above also indicates the feasibility of correlating data from independent laboratories as long as experimental inconsistencies are addressed (6).

4.2.3. Biological Reproducibility

Biological studies, especially those conducted in vivo, are associated with a certain inherent level of interindividual variability that depends on the model used or the end point evaluated. Correct interpretation of observations or results, therefore, necessitates the use of experimental groups comprised of a sufficient number of biological replicates. This is especially true for toxicology studies, where the end points typically evaluated, such as serum chemistry or hematology panels, may show a high degree of variability. Guidelines are available that provide recommendations related to group sizes for toxicology studies of various durations; these are briefly addressed in Chapter 5. These recommendations are based on years of experience with toxicological evaluation of chemical agents using traditional end points and therefore may not be adequate for gene expression studies. Therefore, an understanding of the interindividual variability in gene expression data following exposure to toxicants is necessary to determine the appropriate number of animals needed to reliably interpret gene profiling data.

Gene expression data are definitely associated with some degree of interindividual variability inherent to any biological system, and this variability depends on the tissue evaluated. In general, gene expression profiles derived from homogeneous tissues such as liver or kidney show limited interindividual variability. In fact, in our experience, these expression profiles tend to be less variable than the functional or morphological end points typically used in toxicological assessment, such as serum chemistry, hematology, or histopathology. Only one published study to date has extensively evaluated interindividual differences in gene expression profiles in the liver. This study evaluated 24 vehicle-treated rats from two independent experiments (19). Not surprisingly, some interindividual variability was observed, but for the most part, differentially expressed genes were random and limited in number. This implies that the impact of individual animal variability on gene expression profiles can be properly controlled by using robust experimental designs, at least for the liver. While these observations cannot be generalized to compound treatments or to other, less homogeneous tissues, they demonstrate that liver-derived gene expression profiles have an acceptable biological reproducibility. We have experienced similar acceptable interindividual variability in other tissues, including kidney, spleen, heart, and bone marrow. All these tissues share the characteristic of being relatively homogeneous. For more heterogeneous tissues, the degree of biological variability is still unclear.

This limited variability for major target tissues of toxicity offers significant advantages, since it indicates that experimental group sizes do not need to be too large, thereby limiting overall animal use but also compound requirements to manageable levels. The latter is, as we will see below, of critical importance in the evaluation of compounds at early stages of drug discovery and probably represents one of the most attractive aspects of using gene expression profiling in discovery toxicology. In addition, this reasonable degree of variability indicates that useful reference data can be generated with a limited number of samples for every compound profiled, resulting in significant cost reduction.

4.2.4. Species Extrapolation

The ultimate objective of the toxicological evaluation of an experimental pharmaceutical compound is to identify potential hazards associated with compound exposure and assess their risks for humans. This evaluation is initially conducted in non-clinical studies with relatively standardized animal models and then continued during clinical trials via thorough monitoring of trial participants. There have been several well-publicized cases of human toxicities that were not predicted in animal studies or even clinical trials, suggesting that the current safety assessment paradigm is relatively imprecise. In fact, several groups have advocated the inadequacy of current preclinical models due to insufficient predictivity for human beings (3, 20). While improvements would clearly be helpful and would limit the exposure of humans to potentially toxic compounds, it is nonetheless worth noting that preclinical toxicology assessments are relatively efficient at detecting molecules toxic to humans. In fact, according to an industry report, 94% of human toxicities are detected in preclinical toxicological evaluations (21). And this number probably underrepresents the predictive value of animal studies, since large numbers of exploratory compounds are eliminated after toxicological testing in animals and are never pursued in the clinic. Nevertheless, any technology improving prediction of human adverse events in preclinical studies and clinical trials would strongly impact the productivity of pharmaceutical research and development (R&D) by enabling the termination of hazardous compounds at earlier stages and hopefully by limiting the occurrences of market withdrawal (22). These topics are further elaborated in Chapter 5, but it is important to underscore here that toxicogenomics is widely considered to be a technology with clear potential to improve the prediction of toxicological reactions in humans and to determine the human relevance of toxic effects identified in animal species.

Results from several studies have shown that, in general, toxicogenomics improves the robustness of cross-species extrapolation. For instance, felbamate is an antiepileptic drug that is remarkably non-toxic in preclinical species such as the rat, dog, or monkey, but produces devastating idiosyncratic toxicity in a small percentage of patients characterized by irreversible aplastic anemia and fulminant liver failure (23). The idiosyncratic reaction associated with felbamate treatment is thought to be the result of reactive metabolite formation, and this event may not be easily detected in animals by traditional approaches. A recent study explored the transcriptomic changes occurring in the liver of rats exposed to felbamate (24). As previously reported, felbamate did not induce overt hepatotoxicity in rats based on the evaluation of multiple functional or morphological parameters; however, compound-induced gene expression changes indicative of oxidative stress and reactive metabolite formation were identified with gene expression-based signatures previously established by these investigators. This particular example suggests that, at least in some situations, gene expression analysis is more sensitive than traditional functional end points in identifying toxicologically relevant cellular responses. This observation is not surprising, since responses to toxicants are typically evolutionarily conserved and consequently are expected to induce similar molecular responses, including transcriptomic changes. These molecular responses may not necessarily translate into overt morphological or functional responses consistent with tissue injury, but may indicate a disruption of tissue homeostasis that may have toxicological relevance in certain contexts. Through an analysis of gene expression changes and of biological pathways affected, one can rapidly and reliably identify the major cellular subsystems affected by compound exposure, better assess the relevance of a specific toxic change, and predict with increased accuracy how a specific species would react to a compound. This aspect is definitely an exciting attribute of toxicogenomics, but it is still premature to determine how practical it will be.

It is important to keep expectations realistic when dealing with novel technologies. Most importantly, all technologies have limitations that need to be clearly understood and addressed with the use of robust scientific designs. In the case of toxicogenomics, the lack of toxicologically relevant gene expression changes in a preclinical species cannot be construed as an indication of the absence of risks for humans. For instance, it is unlikely, and in theory impossible, that a toxicity induced by a reactive metabolite only produced in humans will be detected with standard animal studies. Therefore, toxicogenomics data need to be interpreted in the context of other concurrent preclinical evaluations that can better address situations unlikely to be uncovered by gene expression data. Hence, this human-specific reactive metabolite could be identified before human exposure in cross-species in vitro metabolism studies. If it is considered a potential hazard after an in silico evaluation or based on previous experience, then one could easily design an appropriate test system and take advantage of the sensitivity and mechanistic clarity of toxicogenomics to interrogate this potential hazard at a very early stage. Likewise, human-specific reactions may be better uncovered with human cell-based in vitro systems (25, 26). These concepts and approaches are further discussed in Chapters 5 and 6. Here, we would like to emphasize that, while a toxicogenomics analysis in the context of a particular toxicological study can improve the overall risk assessment to humans, there are instances where the technology will be unreliable.

4.3. ANALYSIS OF TOXICOGENOMICS DATA

The development of gene expression profiling platforms has reached a stage where most technical issues are well understood and controlled. Therefore, generating consistent and robust gene expression profiles does not represent a significant challenge for most organizations. Obviously, large core laboratories equipped with sophisticated automated machinery are likely to yield more consistent results, but even small-scale laboratories can now generate high-quality gene expression profiles for toxicogenomics analysis. In contrast, analysis of these high-content data represents the most complex step of the process, and extracting biological information from such information-rich data sets may represent a daunting challenge. This section provides an overview of the analytical tools available for the analysis of toxicogenomics data, which is probably the most critical part of the overall process. These tools are in general not unique to toxicogenomics and therefore are already covered to some degree in previous chapters. Hence, in this chapter we attempt to focus on the tools that are, in our experience, the most useful to address questions relevant to toxicology. Those readers who are interested in a more detailed analysis of current approaches to analysis of gene expression data are referred to an excellent review by Slonim (27).

4.3.1. Compound-Induced Gene Expression Changes

Preclinical toxicology studies are relatively standard in duration and experimental design, and typically contain a vehicle control group and multiple groups treated with increasing doses of a test article. The first step of a toxicogenomics analysis is to understand the transcriptional changes (i.e., the list of genes that are down- or up-regulated) that occur in the tissues or cells after treatment with a test article relative to the vehicle controls. With one-color microarray platforms, a certain degree of control exists, since ratios between different sample groups can be created after hybridization. In contrast, when two-color or two-channel microarray platforms are used, the hybridization design determines the ratios generated in the experiment and is therefore of critical importance.

Three common hybridization designs can be used in experiments using two-color microarrays. These designs are referred to as direct, reference, and loop (28, 29). In the direct design, the test article-treated samples are hybridized against their appropriate control samples. This allows for the identification of differentially expressed genes at a specific time. In the reference design, which is the most commonly used design within the biological community, all study samples are hybridized against a common reference sample; this approach is very useful for studies containing one control group for several treatment groups and is well suited for characterizing temporal relationships. Finally, the loop design consists of the sequential hybridization of all study samples against one another. It is seldom used, despite its potential higher precision and advantages for time course experiments (28, 30, 31).

Once gene expression profiles are generated, statistical procedures are required to identify the genes that are differentially regulated by treatment with a test article. These procedures can be found in a number of genomic analysis software packages that also often provide relevant information for each gene present on a specific array platform. Each of these commercial software packages has strengths and weaknesses, and their performance and features are constantly being improved, such that a discussion on the current status of these packages would be moot. For information purposes only, we provide in Table 4.1 a non-exhaustive list of commercially or publicly available software packages

Table 4.1 List of Commercially or Publicly Available Software Packages for Analysis of Gene Expression Experiments

Software	Attributes	URL
Spotfire	Commercial exploration, analysis, and visualization tools for various high-content data sets Links to public data sources (GeneCards, KEGG, SwissProt, NCBI GenBank, OMIM, and Entrez databases)	http://spotfire.tibco.com/products/
Database for Annotation, Visualization and Integrated Discovery (DAVID)	Public analysis tool from the U.S. National Institutes of Health Useful for pathway analysis with enriched biological themes, Gene Ontology (GO) terms, visualization of genes on BioCarta & KEGG pathway maps	http://david.abcc.ncifcrf.gov/
Rosetta Resolver® system	Commercial analysis tool and repository for gene expression data Built-in, standardized analysis methods, including technology-specific error models, ability to perform cross-species and/or cross-technology analyses (e.g., comparisons of Affymetrix vs. Agilent arrays, or optionally, qPCR versus microarrays)	http://www.rosettabio.com/products/resolver/default.htm
MatWorks	Commercial programming platform for analyzing and visualizing complex biological data and systems Allows for a broad range of analysis, simulation, algorithm development, and application deployment.	http://www.mathworks.com/products/bioinfo/
GoMiner™	Public tool for biological interpretation of gene expression data Useful for pathway analysis	http://discover.nci.nih.gov/gominer/
GeneSpring GX	Agilent visualization and analysis solution designed for use with gene expression data	http://www.chem.agilent.com/scripts/pds.asp?lpage = 27881

NetAffx™	Affymetrix tool to correlate GeneChip® array results with array design and annotation information (probe sequences and gene annotations) Useful for ortholog matching	http://www.affymetrix.com/analysis/index.affx
PANTHER	Commercial analysis tool for microarray data Useful for mapping genes to molecular function and biological processes and pathways	http://www.pantherdb.org/tools/genexAnalysis.jsp
Gene Ontology (GO)	Provides a controlled vocabulary to describe gene and gene product attributes in any organism. Useful for mapping genes to molecular function and biological processes and pathways	http://www.geneontology.org/
Ingenuity Pathway Analysis	Commercial pathway analysis tool Proprietary manually curated database Proprietary ontology Compound and toxicology contents	http://www.ingenuity.com/
GeneGo MetaCore™ and MetaDrug™	Software packages for functional analysis of experimental data, including gene expression data. Useful for pathway analysis Proprietary manually curated database Tools for data visualization, mapping and exchange, multiple networking algorithms and filters. Compound and toxicology contents	http://www.genego.com/metadrug.php
Ariadne Pathway Studio	Pathway analysis research and discovery tool Automated curated database powered by an internal text-to-knowledge technology (Medscan)	http://www.ariadnegenomics.com/

along with their major attributes. This list does not cover all commercial or public tools, and only tools for which we have sufficient familiarity with are listed. In addition, customized internal analysis tools are available in some organizations. In general, we believe that selection of a particular tool is a very personal choice. Hence, a user-friendly, average software or a well-understood program may be more appropriate for some investigators, while more elaborate analysis systems may be required by others. For novice groups, most genomic software providers offer trial periods, and this is probably the best way to select an analysis tool suitable for the organization. Furthermore, several publicly available web-based tools, such as DAVID from the National Institutes of Health (http://david.abcc.ncifcrf.gov/), are freely available to scientists worldwide.

Regardless of the tools used, a list of genes differentially regulated by a compound treatment can be generated with a level of significance assigned to them typically generated by ad hoc statistical methods designed specifically for microarray data. Standard statistical methods can help determine the significance of observed changes. These standard methods, such as the *t*-test or analysis of variance (ANOVA) *F* statistic, indicate the probability that a difference in the expression level of a specific gene occurred by chance. Although useful when dealing with a limited number of variables, these methods are not optimal for microarray experiments, where the variables number is in the thousands. Hence, with a *P* value at 0.01 and a microarray covering 10,000 genes, 100 genes would be identified as differentially expressed by chance, a number too high to correctly interpret data. This shortcoming led to the development of ad hoc statistical methods designed specifically for microarray data. For instance, significance analysis of microarrays (SAM) is a method that assigns a score to each gene on the basis of change in expression levels relative to the standard deviation of repeated measurements (32). For genes with scores greater than an adjustable threshold (delta), SAM uses permutations of the repeated measurements to estimate the false discovery rate (FDR), that is, the probability that a given gene identified as differentially expressed is a false positive. SAM is generally viewed as a robust and straightforward method that can be adapted to a broad range of experimental situations (32). More information can be found at the following URL: http://www-stat.stanford.edu/~tibs/SAM/. Similarly, the standard Bonferroni correction, by multiplying the uncorrected *P* value by the number of genes tested, restricts the number of genes falsely identified as differentially expressed (27). However, this method is thought to overcorrect because it assumes independence of the different tests, and using this correction often results in no statistically significant findings in experiments where changes in gene expression are clearly evident. Therefore, step-down methods are available to reduce the stringency of the analysis, but overall one must admit that a consensus on best statistical practices for toxicogenomics analysis remains elusive (33).

4.3.2. Visualization Tools

Microarray experiments typically lead to long lists of differentially expressed genes that are difficult for the human brain to comprehend, interpret, or interconnect. Several projection techniques can reduce the dimensionality of these multivariate data by embedding the variables and objects of the data in a visualizable space (34). These visualization tools can be used to generate a rapid high-level overview of the nature and extent of compound-induced gene expression changes (27). They are typically applied at an intermediate step of the toxicogenomics analysis. A variety of visualization tools are available in analysis software packages, but for the most part molecular toxicologists typically use principal component analysis (PCA) or unsupervised clustering tools (reviewed later in this subsection and in Chapter 1). These methods are unsupervised, because the dimension reduction is driven by the data rather than by any preset classification scheme or previous knowledge.

PCA is a statistical technique for reducing data dimensionality that is well-suited for analyzing and visualizing large data sets (34, 35). It is based on the concept that most of the data variation is driven by a small number of transformed variables. It projects the high-content data into a new space based on linear combinations of variables (principal components) that retain a large amount of the variation in the original data. These principal components account for much of the variance in the original data. With PCA, one can rapidly visualize and determine the dominant differences between several compound treatments or among several doses. An example of application of PCA to toxicogenomics data is shown in Figure 4.3. Examination of the principal components provides insight into and visualization of the genes that drive the most pronounced differences among experimental conditions.

Clustering methods organize multivariate data into groups with roughly similar patterns (27, 36). When applied to gene expression data, clustering identifies groups of genes showing similar expression patterns. These methods are ideal to visualize toxicogenomics data because they allow one to interrogate similarity and differences among treatments and doses, but also to identify genes with similar expression profiles and to group genes into functionally relevant classes. The literature on statistical clustering is large, and a multitude of choices is available for clustering methods. There is no objective guidance on which clustering methods are best suited for toxicogenomics, and it is well accepted that visualizing microarray data with more than one clustering technique can be quite useful in elucidating relevant biological patterns (36, 37).

Two techniques have been most frequently used in toxicogenomics studies. Hierarchical clustering produces a hierarchy of clusters not fixed in advance. As a starting point, variables form their own clusters, and the two clusters most closely related by some similarity metric (i.e., the two "nearest" clusters) are merged into

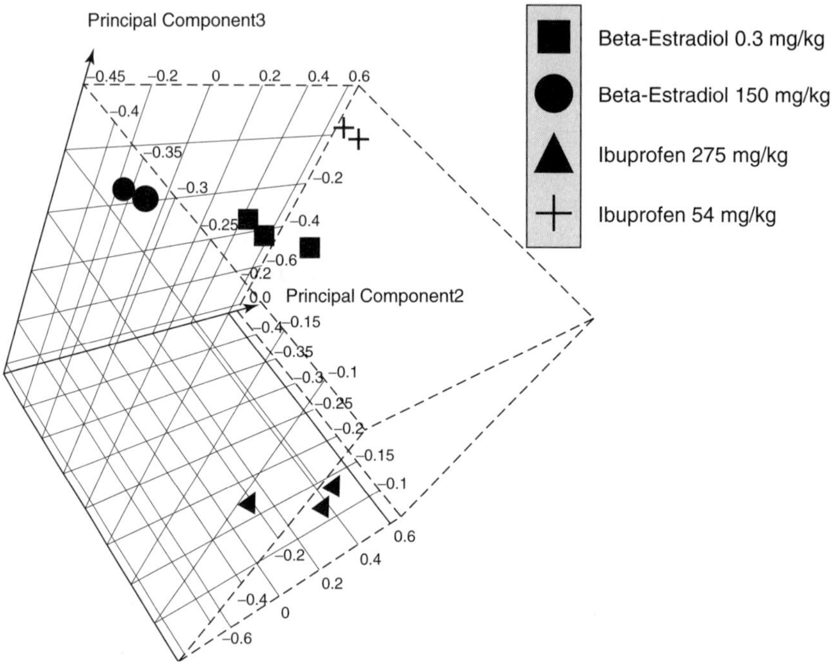

Figure 4.3 Principal component analysis (PCA) of liver gene expression profiles from rats treated daily for 3 days with either ibuprofen at 54 or 275 mg/kg/day or β-estradiol at 0.3 or 150 mg/kg/day. PCA is a useful visualization tool for toxicogenomics data sets. In this example, general treatment-related transcriptomic responses are easily visualized and can be semiquantified by comparing the variability observed within the various treatment groups. Examination of the principal components can also provide further insight into the identity of the genes that drive these pronounced differences.

a bigger cluster at each hierarchical level (27, 38, 39). The process of merging the two closest clusters is repeated until a single cluster remains. This process rearranges the data into a hierarchical tree structure that can be broken into the desired number of clusters by cutting across the tree at a particular height. The advantage of hierarchical clustering is that it does not require prespecification of the number of clusters and that tree structures are easy to visualize and interpret (39). The resulting tree or dendrogram permits a quantitative visualization of the closeness or dissimilarity between groups, and this comparison can be inspected at various heights. This is particularly useful when gene expression profiles for compound treatments are compared to a wide range of reference toxicogenomics data, a process that is sometimes referred to as a "guilt-by-association" approach. Finally, clustering facilitates the visualization of treatment-related transcriptomic changes that can direct the subsequent analysis. However, it is worth indicating that hierarchical clustering may not necessarily reflect the contained profiles,

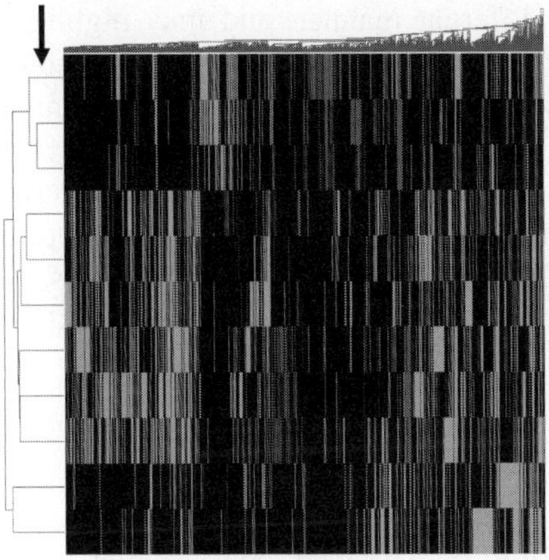

Figure 4.4 Heat map with hierarchical clustering illustrating the transcriptomic changes occurring in the liver of male rats treated with ibuprofen at 54 or 275 mg/kg/day or β-estradiol at 0.3 or 150 mg/kg/day. Since these profiles are similar to those used in Figure 4.3, the reader can compare the hierarchical clustering approach with the principle component analysis (PCA) method for visualization of compound-induced gene expression profiles. Genes shown in the horizontal axis include genes that were up- or down-regulated by at least twofold with a P value <0.01 (as determined with Rosetta Resolver software). Green and red shades indicate genes that are down-regulated and up-regulated, respectively in treated animals compared to their respective vehicle controls. Note the overall limited interindividual variability in gene expression profiles, indicating good biological reproducibility. By focusing on blocks of genes with consistent patterns of up- or down-regulation, the user can further identify the genes driving most of the response. The dendrogram on the left side (arrow) permits a quantitative visualization of the closeness or dissimilarity between experimental groups and between individuals from the same experimental group. See color insert.

as the higher in the tree one looks, the less related the genes within a cluster may be to each other (39). Figure 4.4 provides an example of the application of hierarchical clustering to toxicogenomics data.

K-means clustering is another method frequently used in molecular toxicology. It uses a centroid approach that requires that the user select the number of clusters either randomly or deliberately (38). The required preselection of a number of clusters is a difficult problem with this method and typically requires the use of several plausible k values. The algorithm then partitions the samples into the k clusters, optimizing some objective function (such as within-cluster similarity) by iteratively assigning samples to the nearest centroid's cluster and then adjusting the centroids to represent the new clusters' center points. The k-means clustering method is particularly useful for describing relationships between clusters.

The multitude of choices available can make it difficult to select the most appropiate high-level pattern visualization tools for a toxicogenomics evaluation. As stated by Slonim, there is no single best way to evaluate a clustering method and no single best clustering method for a data set (27). All these tools represent methods to visualize high-level data in a different manner, and may highlight diverse patterns in the data set. Therefore, they are best viewed as complementary approaches for analysis. For instance, PCA can be used to determine whether a test compound induces changes very divergent from other preselected reference compounds, but it is not well suited for detecting more subtle patterns within a treatment group. Conversely, hierarchical clustering is useful in detecting these subtle patterns but may overemphasize dissimilarities between samples, in contrast to PCA. Hence, it is recommended to employ multiple visualization methods in pattern discovery.

4.3.3. Class Prediction

As mentioned above, a fundamental principle of toxicogenomics is that toxic compounds injuring cells or tissues through similar mechanisms of toxicity will affect similar biological pathways, leading to shared gene expression changes. Therefore, it is possible to identify gene expression sets that will be consistently regulated by all members of the same class of toxicants, and these so-called gene expression signatures can be used as predictive or diagnostic biomarkers to classify compounds. In this section, we discuss the methodologies used to develop these gene expression-based models to classify compounds as toxic or non-toxic. Figure 4.5 illustrates the overall process of gene expression signature derivation. It is important to emphasize that these methodologies consist of sophisticated statistical methods, and therefore their successful implementation requires a strong biostatistical expertise in addition to proficiency in toxicology. Ideally, this should be accomplished by multidisciplinary teams of toxicologists, statisticians, and bioinformaticists.

After deciding on the end point and tissue of interest, the first step consists of selecting an appropriate training set composed of a sufficient number of gene expression profiles. These profiles need to be induced in the target tissue by a wide range of compounds known to cause (positive class) or not cause (negative class) the toxic end point of interest. Both the positive and negative classes need to cover as much chemical diversity as possible, as well as preferentially all mechanisms whereby the end point can occur. Ideally, these gene expression profiles should be generated on the platform that will be used on a routine basis. Because the training set is selected based on prior knowledge, the classification methodology is referred to as supervised. The selection of the training set is critical for the overall success. Failures to develop accurate models can often be explained by the limited size of training sets or by inappropriate classification of the gene expression profiles used in the training set. In our experience, classification of the gene expression profiles is not a trivial task because of the complexity of and the overlap between most toxic reactions. Therefore, it is highly recommended to involve a seasoned toxicologist or pathologist at this stage. For example, a common misconception is to consider as positive the gene expression profiles induced by a compound known to be associated with a particular toxic end

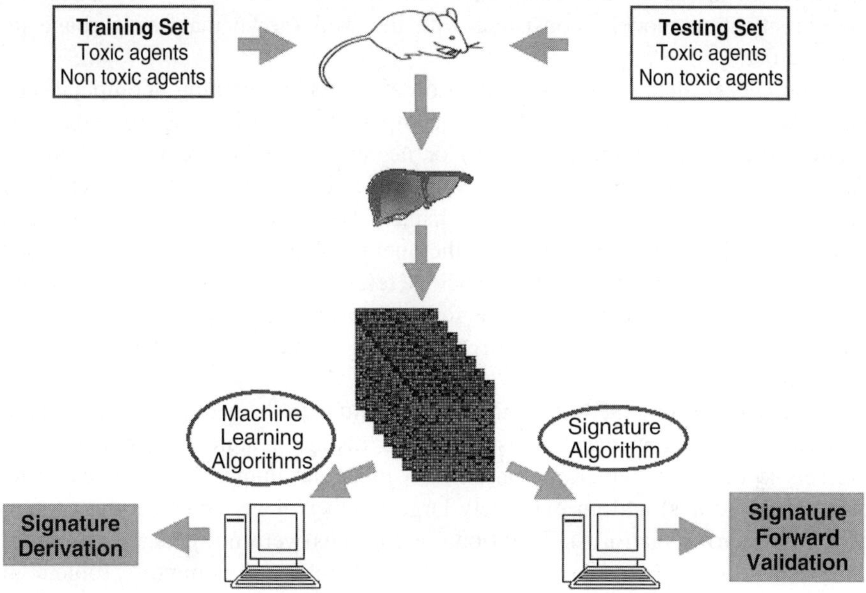

Figure 4.5 Derivation and validation of gene expression signatures for toxic end points. Gene expression signatures are derived with a training set composed of gene expression profiles induced by a wide range of structurally and pharmacologically diverse compounds that are toxic or non-toxic in the tissue studied (illustrated here as an example is the liver). This training set is analyzed with various machine learning algorithms (such as neural networks or support vector machines) that develop a model to classify compounds as toxic or non-toxic in the tissue of interest. This classification may be based on non-specific toxic changes (i.e., toxic vs. non-toxic) or may be refined to predict the occurrence or confirm the presence of specific toxic changes (for instance, bile duct hyperplasia or lipidosis for the liver). These predictive models are then evaluated in a forward validation step by using a testing set of expression profiles distinct from those of the training set. The objective of this validation step is to determine in an unbiased manner the prediction accuracy of the model, as well as its sensitivity and specificity.

point but administered at a dosage not associated with any toxic effect. The famous quote of Paracelsus (1493–1541) is particularly relevant here: "All things are poison and nothing (is) without poison. Solely the dose determines that a thing is not a poison." In other words, it is the combination of dose and compound and not the compound only that needs consideration when building a training set. Likewise, it is important to ensure that the compound at the dose chosen produces the anticipated toxic end point. This can be done, as will be discussed later, by anchoring the gene expression profiles to selected phenotypic end points that occur either concurrently or after longer exposure periods. Carefully built training sets considerably diminish the level of noise in the data, and ultimately improve the performance of the classification models, consistent with the well-known mantra "garbage in, garbage out."

The next step involves the use of machine-learning computational methods that are trained to classify compounds as belonging to either the positive (i.e., causing the end point) or negative class. There is an abundant list of methods for this purpose, and an in-depth discussion of each of them is beyond the scope of this chapter. Here, we review some general features of these methods and list the methodologies most commonly used by molecular toxicologists. The reader is referred to appropriate reviews and references available in the literature for additional information. Chapter 5 also provides specific examples of their application in the context of toxicological evaluations.

Machine-learning methodologies are not unique to the genomics field and are utilized extensively in various disciplines. A distinct feature of genomic training sets is that they typically consist of a moderate number of samples (gene expression profiles) and an extremely large number of variables (probe content of the microarray platform). Traditional algorithms were not designed to anticipate this "small n, high variable" feature. Therefore, a common problem in developing toxicological signatures is overfitting, which refers to situations in which machine-learning methodologies model the training sets too well, hindering generalization. Overfitting typically results in models that are excellent in classifying samples from the training sets but extremely poor in classifying samples naive to the training set.

Reducing the number of parameters by focusing on those relevant to the classification model is an approach to avoid overfitting. A wealth of dimension-reduction techniques is available. For instance, one common approach is to rank genes with respect to differences in expression levels between the positive and negative classes using parametric or non-parametric statistical tests, such as standard or permutation t- or F-tests, Wilcoxon statistics, or SAM (32, 40, 41). Using these tests, a smaller number of differentially expressed genes can be selected based on a specified level of significance and/or on the level of gene regulation (magnitude of the change in mRNA levels). Another approach entails the use of noise reduction methods, such as PCA (discussed above). Since the PCA method reduces data into rank-ordered components according to the amount of variance, one can indeed select the first n components as most informative to reduce the dimension of gene expression data and utilize only the most informative gene subsets (34). Wavelet transformation represents another common data compression technique. This mathematical function divides data into different frequency components and studies each component with a resolution matched to its scale (42). Wavelet transformation is useful, because it can efficiently compress and denoise microarray data by eliminating the less informative components of the data set.

Once the dimension of data is reduced to appropriate levels, various machine-learning methods can be used to develop the classification model. Each of these algorithms has unique strengths and weaknesses, and the selection of the appropriate machine learning methodology should be made based on the predetermined expectations for the signature. For instance, does the model need to distinguish between two classes or more? Some methods, such as support vector machines or linear discriminant analysis, are binary and therefore inadequate for generating multiclass models. In contrast, alternative algorithms like neural networks are amenable to separate multiple classes. Other questions are also relevant to selecting a machine-learning algorithm. How large should the multigene marker be? What is the desired sensitivity and specificity? Are the genes used in the model expected to confer some mechanistic indications of biological processes? Several machine-learning methods are typically referred to as black-box methods to indicate their inability to provide clues into biological processes (43). In Chapter 5, we discuss several examples of signatures of toxicity that have been developed with diverse methodologies.

Here we briefly describe the most commonly used machine-learning methods and their general features. We should first emphasize that there is a paucity of published studies that provide an unbiased comparison of the different methodologies available. There is, however, some consensus that simpler machine-learning methods usually outperform complex ones in situations where the number of variables is larger than the number of samples. Logistic regression and linear discriminant analysis use statistical inference to weigh the contributions of each gene expression value in sample prediction. These methods typically perform better when large data sets are available or when there is a clear distinction between the two classes of compounds (44). For example, our group was able to generate a robust classification model for peroxisome proliferators and aryl hydrocarbon receptor agonists with linear discriminant analysis and a training set relatively modest in size (14). Linear discriminant analysis was appropriate for that study, because these two classes of toxicants caused very distinctive gene expression changes.

Support vector machines (SVM) are a family of statistical machine-learning methods that are particularly suitable for genomics data analysis because of their robust performance with sparse and noisy data. SVMs have been shown to perform well in different areas of biological data analysis (45). They are well suited for large number of variables and consequently do not typically require prior reduction of data. In fact, there is evidence that SVM prediction improves when all data are used. In short, SVMs try to draw a demarcation in an n-dimensional gene expression space that most distantly separates two classes of the training set. If no demarcation is found, the samples are then mapped into a higher-dimensional space where a separator exists. SVMs are difficult to interpret but generally generate powerful models. Several studies have actually suggested that SVMs outperform other methodologies when used for gene expression-based class prediction (46, 47).

Computational neural networks (NNs) constitute an excellent alternative to generate multiclass models. They are analogous in concept to a biological nervous system, as they are composed of a number of interconnected processing elements called neurons or nodes tied together with weighted connections. Neural networks are capable of online learning, prediction, and decision-making by exploiting computational power to systematically search through a large number of possibilities (42). An iterative learning process begins by feeding input data into the network, which calculates the predicted output based on predetermined weights. Comparison of the predicted output and the targeted output leads to adjustment of the weights of the connections, and a new output is calculated. This process is reiterated until the network output closely matches the targeted output. Neural networks are considered excellent to address complex patterns and learn from new input data, and have gained increasing popularity for the classification of gene expression profiles in many disciplines. For instance, as illustrated in Chapter 5, our laboratory has used a NN algorithm to generate a predictive model of hepatoxicity in rats (48). Likewise, Piotrowski et al. used a combination of wavelets for noise reduction and data compression with NNs to generate a model to predict the mode of action of various toxic agents (42).

The choice and complexity of these various classification algorithms represent a challenge for molecular toxicologists. Similar to the visualization tools reviewed above, they should be viewed as complementary, and several methods should be evaluated when generating classification models. The ultimate criterion in selecting a classification methodology should be the prediction accuracy of the model. This accuracy is estimated during a validation step using a testing set of samples (expression profiles) distinct from those in the training set (48). This step is critical and the only approach to correctly estimate the performance of a classification model. In toxicology applications, the validation set should incorporate compounds with structural diversity compared to those included in the training set. It is also important to use compounds that adequately cover the chemical space used in an organization. Ultimately, these models are generated for use in the toxicological evaluation of novel exploratory compounds, and therefore they should classify these compounds as accurately as possible. In our laboratory, we regularly reevaluate the performance of the models built to ensure their optimal accuracy.

4.3.4. Network and Pathway Analysis

Although changes in the expression of individual genes may be useful for toxicological interpretation, it is usually more beneficial and informative to assess transcriptomic changes by examining regulated biological networks and pathways. Generally speaking, a **biological pathway** denotes a group of molecular entities acting and interacting together to drive a specific biological process (49). Biological pathways are built through accumulation of knowledge about molecular interactions that are available in various biological databases. They represent useful tools to explain and interpret experimental observations and generate improved hypotheses when dealing with information-rich data sets that measure the abundance of a large number of biological molecules.

Analysis based on biological pathways (typically referred to as **pathway analysis or functional enrichment**) increases the confidence that a change in expression of specific genes has biological implications and provides tremendous assistance in data interpretation in the context of toxicology. In general, pathway analysis also can identify more subtle changes in expression than the gene lists that result from univariate statistical analysis, because it allows for an evaluation of consistent changes in expression of groups of genes with related function (50). This is particularly exemplified by studies that compared results from similar samples across different laboratories and microarray platforms. In particular, the Hepatotoxicity Working Group of the International Life Sciences Institute (ILSI) Health and Environmental Sciences Institute (HESI) Technical Committee on Application of Genomics to Mechanism-Based Risk Assessment reported the results of their evaluation of the biological and gene expression responses in rats exposed to two reference compounds (clofibrate and methapyrilene) (51). RNA samples were distributed to 16 laboratories for analysis on six microarray platforms. Not surprisingly, discrepancies were seen in the expression levels of multiple individual genes when a gene-by-gene comparison was made. They were largely due to differences in the microarray platforms and approaches to data analysis. However, when the microarray data were considered in the context of biological pathways, a good overall agreement between the data sets was observed. In other words, analyses based on pathways are expedient for mitigating confounding factors and extracting the major biological themes in a gene expression data set.

Pathway analysis typically involves identification of biological pathways that contain an overrepresentation of genes differentially regulated in treated samples. The significance of such overrepresentation is that it points to pathway activation or inhibition. Detection of such an event requires the use of methodologies to map gene expression data into pertinent biological pathways based on their functional annotation and reported molecular interactions. There exist several commercial and public software packages for pathway analysis; some of them are listed in Table 4.1. A more general discussion of pathway analysis of microarray data can also be found in Chapter 2. Here, we briefly discuss several of these tools to illustrate their application in toxicogenomics. These tools were selected because of our familiarity with them and not necessarily because of the superiority of the products. Interested readers are encouraged to consult recent literature on the topic. In addition, it is noteworthy that some of the microarray databases (discussed later in this chapter) also offer a spectrum of analysis tools, including pathway analysis solutions. These analysis products rely on a robust knowledge base of molecular interactions, usually extracted from various public databases or peer-reviewed scientific articles.

The Ingenuity® pathway analysis system (offered by Ingenuity Systems, Mountain View, CA) is a Java web-based application that enables discovery, visualization, and exploration of molecular interaction networks in gene expression data. This program can be applied to identify biological mechanisms, affected pathways, and functions most befitting to the data sets evaluated.

Gene identifiers and the corresponding expression values are uploaded into the application, and the genes are mapped to a global molecular network developed based on the information contained in the Ingenuity® Pathway Knowledge Base. This pathway analysis program has been extensively used in toxicogenomics research. For instance, the Ingenuity® Pathways Knowledge Base has been utilized to query and interpret the transcriptional changes observed in bovine kidney epithelial MDBK cells after treatment with butyrate (52). The analysis successfully identified remarkable changes in genetic networks related to cell cycle, cell death, and DNA proliferation, providing a basis for a better understanding of the mechanisms by which butyrate affects kidney cells. Likewise, this pathway analysis methodology has identified a novel pharmacological mechanism for MK886 (a compound known to inhibit both 5-lipoxygenase-activating-protein or FLAP and peroxisome proliferator-activated receptor-α or PPAR-α) and the molecular basis of carcinogenic nitrosamine NNK [4-(methylnitrosamino)-1-(3-pyridyl)-1-butanone]-induced pulmonary neoplasia with relatively small-scale microarray studies (53, 54).

MetaCore™ and MetaDrug™ (GeneGo Inc., St. Joseph, MI) are two commercially available pathway analysis products that have also been applied to query and interpret toxicogenomics data (55). Other available tools include Pathway Assist™, PathArt™, or publicly available pathway databases such as GenMAPP, Biocarta, or the Kyoto Encyclopedia of Genes and Genomes (KEGG) pathway database (56, 57). All these products have unique underlying proprietary pathway knowledge bases that are built either manually or automatically with text mining tools or a combination of both. Again, there is no consensus as to which method is the best, and there is no published unbiased evaluation of the contents and algorithms of these databases. Consequently, some level of variability in pathway analysis results is expected depending on the tools used.

Most pathway analysis tools rely on existing functional annotation of genes and literature data, which may contain a degree of inaccuracy or can be species-specific. The annotations for some preclinical species may not be complete. In addition, abundant historical data in preclinical toxicology species are still scarce, and our experience is still insufficient for the meaningful interpretation of thousands of simultaneous gene expression changes, which may often appear disconnected. Only after a specific toxicological change has been observed consistently can its real toxicological significance be truly understood. The same concept holds true for gene expression changes. In a recent comprehensive study, over 300 liver microarray gene expression profiles for three different classes of compounds (genotoxic carcinogens, non-genotoxic carcinogens, and non-carcinogens) were clustered across 72 putative oncogenes (58). The three classes of compounds were strikingly interdispersed within the cluster, indicating that up-regulation of oncogene expression was not a surrogate marker for carcinogenesis, since both non-carcinogens and carcinogens upregulated the expression of these selected genes. This study clearly highlights the danger of focusing on deregulation of a single gene or a limited set of genes in a microarray experiment. Gene expression perturbations induced by toxicants typically reflect a large number of complex pharmacological, physiological, metabolic, and biochemical processes (48, 59). For proper interpretation, toxicologically relevant gene expression changes or deregulated biological pathways must be identified and segregated from those that are adaptative, beneficial, or unrelated to the development of toxicity. This requires the access to a significant wealth of toxicology reference data. These toxicology data repositories, including well-curated reference gene expression databases, are usually not accessible to the developers of most pathway analysis tools. Therefore, despite less attractive attributes, tools provided by commercial toxicogenomics reference providers may be more advantageous when evaluating gene expression data in the context of toxicology rather than pharmacology. In particular, the functional annotation of genes may be more relevant, and the biological pathways may be more "toxicology-centric" and adapted to the toxicological species of interest, such as the rat. New users are advised to evaluate as many pathway analysis software packages as possible before selecting the one that may best fit their specialized needs (56).

4.4. PRACTICAL AND LOGISTIC ASPECTS OF TOXICOGENOMICS

Gene expression profiling is an evolving technology with constant improvements in technical and analytical aspects. Technical issues have been aggressively addressed by various consortia or collaborations composed of leading academic, governmental, and industry centers (17, 60–65). In one of these pioneering collaborative efforts, a committee formed by the membership of the International Life Sciences Institute (ILSI) Health and Environmental Sciences Institute (HESI) addressed various issues, challenges, and opportunities afforded by toxicogenomics (66). The data generated by this collaborative scientific program were extremely beneficial to enhance our understanding of biological and technical variability associated with the technology. Progressive standardization of protocols has increased reproducibility across platforms and laboratories (61). While improvements will continue to occur, the current state of the technology is such that gene expression data are sufficiently robust to be used successfully in toxicological evaluation. In this chapter, technical aspects related to the instrumentation are not discussed in depth, since they are likely to be obsolete at the time of printing. Only practical aspects relevant to molecular toxicologists are covered, such as species considerations and experimental design. For other issues, the reader is referred to existing publications, as well as manufacturers' specifications and recommendations (9, 30, 62, 63, 65).

4.4.1. Species Considerations

The laboratory or Norway rat (*Rattus norvegicus*) is a commonly used experimental animal and definitely the favored species for toxicology testing in the pharmaceutical industry. A large fraction of in vivo studies contained in the preclinical evaluation package of a small molecule drug is conducted in rats. In particular, initial toxicological testing in discovery is typically performed in rats because of the reasonable compound requirement for this small species. This initial rat testing is exploratory in nature and is usually not designed to support regulatory submissions. Its objective is to rapidly determine potential toxic liabilities associated with novel compounds and make informed decisions on whether or not to pursue them further as possible development candidates. Additionally, in contrast to other species such as the dog or monkey, the rat genome is fairly well annotated, and historical rat gene expression data are quite abundant in the public and private domains. For these reasons, the rat represents the favored species for toxicogenomics studies and is the default species discussed in this chapter and in Chapter 5.

Microarray studies for toxicological evaluation have also been conducted in species other than the rat. The mouse (*Mus musculus*) genome is obviously very well studied given the extensive use of mice in biology, so it should come to no surprise that this species has also been commonly used in toxicogenomics research. In addition, several investigators have experimented with microarray platforms available for larger animal species, such as dogs and monkeys (67). For instance, several studies have successfully used oligonucleotide arrays to generate gene expression profiles in canine (*Canis familiaris*) models of osteoarthritis and heart diseases (2, 5, 68, 69). The information generated from these experiments confirmed previous data, but also was instrumental to formulate new hypotheses about the pathophysiology of these diseased conditions. However, several limitations have been recognized for these platforms. For instance, only a limited set of genes is represented on these first-generation canine arrays, precluding genome-wide analysis. Deficiencies in the annotation of the microarray platform definitely restrict the knowledge that can be acquired from these experiments compared to previous experiments conducted with rodent tissues, and may also lead to incorrect interpretation (68, 69). On the canine array used in these studies, only 7,340 of the 23,836 total probe sets are well annotated. Accordingly, some up- or down-regulated genes were obviously difficult to put in a pathophysiological context. However, the canine gene annotation database is constantly improving. A custom canine Affymetrix microarray has recently been designed that carries high-quality sequences taken from 48 tissues (70). Another Affymetrix-based canine array containing over 22,000 probe sets has been developed and used to evaluate the acute-phase response of dogs following lipopolysaccharide (LPS) exposure (71). This platform proved reliable in detecting expected gene expression changes and also identified novel mediators of the acute-phase response in dogs. Arrays of this type represent a tool with clear potential for mechanistic toxicology investigations in dogs and will likely lead to a better understanding of toxicity in this species. Furthermore, it may be useful in identifying toxicity biomarkers in dogs, a species often more relevant to humans than the rat in terms of toxicological responses.

Non-human primates, especially cynomolgus monkeys (*Macaca fascicularis*), are species important for toxicology. Whole genome sequence information is not available for non-human primates, and until recently, there were no commercial microarray platforms for gene expression studies in non-human primates. Genomic reagents for non-human primates have started to become available (31, 72). In addition, gene expression profiling can be conducted in these species, as according to most estimates, the primate mRNA sequences have a $>90\%$ homology to human sequences (73). Consequently, investigators have used microarray platforms designed for humans to conduct gene expression studies in primates (74, 75). Long oligonucleotide-based platforms appear more suitable than short oligonucleotide microarrays for this application, since they are less sensitive to single-base mismatches. Therefore, these platforms recognize many more genes from non-human primate RNA samples or have fewer "false negative" genes than short oligonucleotide microarray platforms (76). However, short oligonucleotide platforms can still provide reliable data for gene expression analysis, although the absolute gene expression measurements may be imprecise and the hybridization signals will be low (77). When a post hoc validation of microarray studies was performed with TaqMan® RT-PCR (Applied Biosystems, Foster City, CA), the PCR and array data agreed well for the majority of the comparisons made, confirming that cDNA- or oligonucleotide-based human platforms are valid alternatives for gene expression studies in non-human primate tissues (76–78). Only a few studies have used human gene expression platforms to investigate toxic mechanisms in monkey tissues. In general, these studies confirmed the acceptability and reliability of the technology. For instance, expression profiles have been generated from the liver of monkeys treated with ciprofibrate (79). Ciprofibrate is part of a well-characterized class of compounds called peroxisome proliferator-activated receptor α (PPARα) agonists. PPARα is a nuclear receptor that plays a central role in the hepatic uptake and β-oxidation of fatty acids. PPARα agonists are known to up-regulate the transcription of genes related to fatty acid metabolism and mitochondrial function. Using human Affymetrix microarrays, the authors confirmed that ciprofibrate treatment in monkeys is associated with up-regulation of genes belonging to the fatty acid metabolism and mitochondrial oxidative phosphorylation pathways, a finding consistent with the pharmacological activity of the compound.

While rats, mice, dogs, and cynomolgus monkeys represent the most common animal species used in pharmaceutical toxicology, several new animal species have been investigated in recent years and are gaining popularity for specific applications. For instance, zebra fish (*Danio rerio*) are of growing interest for various toxicology specialties, such as environmental toxicology, but also for discovery toxicology in the pharmaceutical industry (80). Besides the attractive features of this model, such as size, transparency, ease of breeding, and an excellent level of characterization, its genome sequence is known and can be analyzed and compared to that of other species' genomes (81). This makes it a useful model for toxicogenomics studies, although its genome is not yet fully annotated. There are currently multiple available commercial or academic zebra fish microarray platforms that represent a standardized toolset for transcriptional profiling (81). For instance, the hepatic transcriptional effects of arsenic exposure were studied in zebra fish with a 65-mer oligonucleotide platform (82). The transcriptome changes in the liver indicated that cellular and tissue injury stemmed from DNA and protein damage secondary to arsenic metabolism and oxidative stress. This study nicely demonstrated that in zebra fish transcriptomic changes analyzed in the context of biological networks can provide an understanding of the coordinated molecular response to toxicants. Nevertheless, there are also important limitations associated with the use of zebra fish in toxicogenomics. Most importantly, isolation of organ-specific RNA represents a challenge with this model (81).

4.4.2. Toxicogenomics Studies

4.4.2.1. Sample Considerations

Toxicogenomics studies can be conducted with either cell cultures or organs collected from animal studies. Cell cultures are very homogeneous, and gene expression changes induced by toxicants are typically repeatable in a laboratory and reliably reproduced in different laboratories under similar experimental conditions. Studies using tissues can be more challenging. Several organs, such as liver or heart, are relatively homogeneous, and gene expression profiles can be reproduced, as long as consistent tissue collection protocols are followed. Other tissues, such as brain, gastrointestinal tract, or testis, are more heterogeneous and complex, so that their gene expression profiles are less consistent. For instance, the brain is composed of a wide diversity of cell types (neurons, glia, etc.) with immense phenotypic and transcriptional diversity (67, 83, 84). In addition, depending on the region of the brain, marked differences in functions and transcriptomes are present between cells of the same type. This complexity represents a significant challenge for toxicogenomics studies, since heterogeneity in tissues with different proportions of various cell types may generate noise in gene expression measurements and thus significantly complicate their statistical analysis. Tissue heterogeneity may result in the identification of differentially expressed genes that may be unrelated to the cell type being studied or in the detection of toxicologically irrelevant genomic changes.

Technologies such as laser capture microdissection (LCM) can be used to obtain purified cell populations from heterogeneous tissues and to isolate RNA from specific cell types (85). This technology is discussed in detail in Chapter 2. The microdissection approach has been very successful for DNA analyses because of the high stability of DNA. However, transcriptomics evaluations with microdissected specimens are more difficult to conduct. First, RNA is much less chemically stable than DNA, and it is much more challenging to avoid degradation during dissection (86). Second, only minute amounts of RNA can be obtained with LCM and an RNA amplification step is required before microarray analysis (87, 88). Third, LCM and RNA amplification protocols are time- and labor-intensive and result in reduced capacity. Finally, gene expression changes in single-cell populations may not capture essential paracrine interactions between cells that may be critical to understand mechanisms of toxicity. With these limitations in mind, microdissection can nonetheless clearly help in generating useful profiles for biological interpretation. Indeed, the successive steps of the LCM procedure appear to have a minimal impact on gene expression profiles. A recent study compared gene expression profiles from laser microdissected and non-microdissected samples from rats treated for one month with clofibrate (89). Hematoxylin and eosin (H&E) staining and laser microdissection itself did not affect RNA quality, but the overall process of the LCM procedure impacted the samples, resulting in a bias in the gene expression profiles. Nonetheless, this bias did not impact data interpretation. In a different study, the protective mechanism of astaxanthin on renal glomerular cells was evaluated with Affymetrix GeneChip arrays and glomerular cells isolated by LCM (90). This approach enabled identification of the pathways regulated after astaxanthin treatment of *db/db* mice.

Consistent tissue collection protocols are critical for the correct interpretation of toxicogenomics data, because inadequate collection procedures can lead to the identification of gene expression changes related to the collection procedure rather than to the toxicity being evaluated. Even in the case of rather homogeneous organs like heart or kidney, sampling differences may impact gene expression profiles. For instance, the impact of sampling differences on gene expression profiles in the liver has been evaluated (91). Using liver from rats treated with toxic and non-toxic doses of acetaminophen, the authors evaluated the transcriptomic response in the left and median hepatic lobes. Their results showed clear differences in the number and nature of differentially expressed genes for both doses of acetaminophen. Likewise, relatively homogeneous tissues like kidney are composed of several compartments that are clearly different in structure, function, and transcriptomic profile. Toxicants may induce changes in only specific compartments or in a cellular subpopulation. Failure to cover all compartments of a tissue would limit one's ability to detect toxicant-induced gene expression changes. This has prompted our laboratory to establish collection protocols for all tissues that are routinely evaluated. Thus we collect kidneys in a manner such that appropriate and consistent proportions of cortex, medulla, and papilla are included for RNA extraction. Similarly, the median liver lobe is collected in its entirety for gene expression studies.

Vehicle treatment and time of collection may also introduce some variability in gene expression analysis. Many tissues are known to have so-called circadian clocks that regulate transcriptional rhythms thought to be important for the daily timing of physiological processes (92). Several studies have evaluated the extent of circadian gene regulation in various tissues. For instance, up to 10% of the genes expressed in mouse liver and heart have been found to be subject to circadian regulation (93). Another study investigated the relative levels of gene expression in rat liver and kidney as a function of the time of day and of the feeding regime and found several key metabolic pathways to be regulated (94). These results clearly indicate the importance of having an appropriate control group to filter out any gene regulation unrelated to compound exposure.

High-quality RNA samples are critical to obtain reliable gene expression profiles (60). For instance, it has been demonstrated by using samples of degraded RNA that up to 75% of differences in gene expression changes can be attributed to differences in RNA integrity between samples (95). Another study also confirmed the detrimental impact of RNA degradation on data quality, although data resulting in valid findings may still be generated (96). Since RNA is very susceptible to nuclease digestion, tissues intended for gene expression analysis must be collected immediately after sacrifice and flash frozen in liquid nitrogen or preserved in an appropriate RNA stabilization solution. Afterwards, they can be stored for prolonged periods of time at -80°C without significant RNA degradation. Inappropriate collection procedures or storage conditions will result in RNA degradation, as revealed by an overall poor RNA quality after RNA extraction procedures.

Most common RNA extraction procedures are adequate to isolate high-quality RNA from freshly collected samples (60). Nevertheless, even if appropriate collection and extraction methods are used, it is recommended to always evaluate RNA quality after extraction to justify the costly step of hybridization to microarrays and to ensure that interpretation of the microarray data is feasible. The traditional gold standard method to assess RNA quality is based on determining the ratio between the 28S and 18S ribosomal RNAs. Intact RNA has a 28S-to-18S RNA band ratio of 1.8. However, this method is not precise enough to detect subtle changes in RNA integrity (97, 98). RNA quality can also be assessed by microcapillary electrophoresis with instruments such as the Bioanalyzer from Agilent Technologies (Palo Alto, CA). This instrument uses a specially designed software program to assess RNA quality based on electrophoretic tracings and calculates an RNA integrity number (RIN) that ranges from 10 (RNA intact) to 1 (RNA completely degraded) (99). Another freely available classifier, the Degradometer, calculates a degradation factor and the true 28S-to-18S ratio based on the peak heights (95). Both software systems provide precise and reliable data on RNA integrity (98).

4.4.2.2. Experimental Design in Toxicogenomics Studies

The design of in vivo and in vitro toxicogenomics studies is determined by the toxicological issues to be addressed. Specific designs are discussed when the different applications of toxicogenomics are reviewed in Chapter 5. However, it is useful at this point to briefly cover several general considerations.

Duration of Dosing. In studies designed to investigate the mechanism of a toxicity, time-course experiments may be very powerful to capture the initial molecular events involved in the development of the toxic effect. Since gene expression changes are transient, determining the most appropriate time points and their frequency represents a critical aspect of the experimental design. The molecular events that may be the most pertinent to comprehend a mechanism of toxicity may occur well before obvious phenotypic changes can be detected with morphological or functional end points. Therefore, evaluating gene expression profiles during or before the development of a toxic phenotype is usually more insightful than doing so when the change has already occurred and is fully established. In addition, when evaluating mechanisms of toxicity, it is important to consider the systemic and tissue kinetic profiles of exposure to the toxic agent. This is best illustrated by a study of acetaminophen toxicity in mice. Mice were administered non-toxic and toxic doses of acetaminophen, and hepatic RNA samples were obtained 4 and 24 hours after the treatment and hybridized to microarrays (100). Hepatic toxicity was observed at both 4- and 24-hour time points at the toxic dose of acetaminophen. Time course expression profiles were analyzed and demonstrated that the most informative transcriptomic activity occurs approximately 4 hours after acetaminophen administration. In contrast, down-regulation of these genes was observed after 24 hours, and this coincided with the development of overt toxicity, but also with the decrease in systemic exposure to the compound. These data illustrate how frequent sampling but also evaluation of expression profiles during maximal systemic or tissue exposure to test articles can facilitate the delineation of the molecular mechanisms of toxicity.

In the case of a study used to rank-order several potential lead molecules for candidate selection, the study needs to be designed with a consideration for compound availability, the type of reference database, and the performance characteristics of the predictive signatures of interest. In other words, it is important to design the study based on the tools available for interpretation. For instance, in our laboratory, a database of hepatic gene expression profiles has been generated based on 3-day repeat-dose studies in rats. Consequently, we typically assess our compounds for hepatotoxicity in 3- or 5-day repeat-dose studies. In contrast, other companies have integrated a toxicogenomics component in their 2- to 4-week rat repeat-dose toxicology studies (101). The objective is usually to be proactive in case of unexpected toxicological changes in the studies or to enhance knowledge and expertise in gene expression analysis. In addition, these types of practices are useful to promote a better acceptance of these new technologies in a company.

Compound Dosage. In a mechanistic in vivo study, the dosage of the test article should be selected based on the best chance to consistently reproduce the toxicological change of interest without incorporating an unacceptable level of noise. For instance, dosages that are too high may be associated with moribundity and changes in body weight or food consumption. These effects may obscure gene expression changes related to a specific toxic mechanism with those related to secondary effects, such as general stress, decreased food intake or water consumption, and secondary or tertiary toxic mechanims. In studies designed to predict toxic effects, the method used to develop the predictive signatures should dictate the dosages used. Validation of predictive signatures addresses to some extent their predictive power and accuracy for various dosing paradigms. For instance, in our laboratory, our in vivo gene expression databases and signatures have been developed based on a low and a high dose. The low dose corresponds to an estimated pharmacological dose (a dose resulting in an exposure similar to that achieved at a dose causing a 50% response rate or ED_{50} in the most appropriate animal model), while the high dose corresponds to a maximal tolerated dose (MTD) for a 3- to 5-day study (defined by the highest dose before rats exhibit clinical signs of toxicity, significant lack of body weight gain, or a significant decrease in food consumption). Because these studies are typically conducted early in a project, for the vast majority of cases only limited information is available for selecting optimal dosages. Consequently, dose setting can be a challenging task and requires extensive communication among all project stakeholders. An adequate dose selection is, however, crucial for proper decision-making and, in our experience, is central to a successful prediction of toxicological changes.

Samples. When conducting an in vivo study, all major tissues and organs, including blood, should be collected for concurrent histopathological and clinical pathological examination. Hematology and serum chemistry panels are relatively inexpensive to generate compared to the current cost of microarrays. They can be performed quickly in a discovery setting, and provide critical information that assists in the correct interpretation of gene expression data. In particular, they provide information related to changes in the homeostasis of the tissue being analyzed and to the overall health status of the animal. Similarly, clinical observations are an important component of the interpretation of toxicogenomics data. For instance, if clinical observations and analysis of the clinical pathology and histopathology changes suggest that an animal is moribund, gene expression changes may be more reflective of the overall poor condition of the animal than of a specific mechanism of toxicity. Therefore, in certain situations, gene expression profiling studies may not be warranted.

In our laboratory, for predictive studies, we typically select and prioritize the tissues to be evaluated with microarrays based on a prior histopathological and clinical pathological evaluation. Such an assessment is extremely useful in determining which critical questions need to be addressed with toxicogenomics. For instance, if treatment with a test article results in clear evidence of a dose-limiting hepatotoxicity following histopathological evaluation or as indicated by marked elevations of serum transaminases, a toxicogenomics analysis of liver samples would not be necessary to terminate with confidence the compound from further development. However, if an understanding of the mode of action of this hepatotoxicity can be beneficial in selecting an alternative series or for screening purposes, a microarray analysis of the liver would be warranted, since it would result in molecular data pertaining to the mechanism of toxicity.

4.5. TOXICOGENOMICS REFERENCE DATABASES

4.5.1. Utility of Reference Databases in Toxicogenomics

The concept of databases is not novel to toxicology. In general, most toxicological end points can only be precisely interpreted in the context of reference or historical reference data. Genome-wide transcriptomic changes are especially challenging to interpret without access to reference data. After exposure to a compound, genes differentially expressed in a tissue can be rapidly identified and biological pathways can be interrogated with relative ease by using appropriate software tools. But without proper historical databases and experience, the toxicological significance of these data is difficult to ascertain and may be either ignored or overinterpreted. Compound-induced gene expression changes reflect multiple complex, interacting molecular processes, that may or may not be related to a toxic reaction (58, 59). Simply stated, not all gene expression changes induced by toxicants are toxicologically significant. In addition, changes in a transcript level may not be very specific and could easily be misinterpreted. The induction of cytochrome P450 1A1 (CYP1A1) represents a good illustration of this situation. CYP1A1 is a member of a family of xenobiotic metabolizing enzymes involved in detoxification of polycyclic aromatic compounds and in their bioactivation to putative carcinogens (102). CYP1A1 induction is a hallmark of aryl hydrocarbon receptor (AhR) pathway activation and is involved in the pathogenesis of the toxicity caused by 2,3,7,8-tetrachlorodibenzo-*p*-dioxin (103, 104). Therefore, without understanding the types of compounds that can cause increased CYP1A1 mRNA levels, one may easily argue that up-regulation of CYP1A1 mRNA levels represents a strong signal of activation of the AhR pathway and of dioxin-like toxicity. We and others have evaluated repositories of gene expression profiles from rat livers to show that large numbers of non-AhR activating compounds significantly up-regulate CYP1A1 expression levels at toxic doses (105). Perhaps more importantly, many of these compounds are FDA-approved drugs from multiple therapeutic classes that have been used safely in humans for years and therefore are very unlikely to be associated with any kind of dioxin-like toxicity. Therefore, it is clear that up-regulation of CYP1A1 mRNA levels is not a specific marker of dioxin-like toxicity or any other toxicity. This does not mean that CYP1A1 induction is an irrelevant change that should be disregarded. In Chapter 5, we actually present a case where CYP1A1 levels are clearly associated with a toxic change. As with any other markers, gene expression changes should not be interpreted in isolation: Their toxicological significance should critically be interrogated with appropriate and robust reference data.

Reference data are also critical when one formulates mechanistic hypotheses regarding the mode of action of a particular toxicant. Indeed, toxicologically relevant gene expression changes must be differentiated from those that are adaptive, beneficial, or unrelated to the development of the toxicological change. This differentiation can, in some situations, be achieved by using a limited number of reference profiles and, since gene expression changes are typically transient, a limited number of end points. However, the contextual information from large, established reference databases unarguably represents the optimal tool to properly interpret gene expression data. Large repositories of data offer the ability to associate unique gene expression changes or specific perturbations of biological pathways with unique or various classes of compounds or with specific toxicological mechanisms (106). In addition, large numbers of compounds, tissues, corroborative toxicological and pathological changes, and gene expression data in these reference databases allows one to strengthen statistical inferences.

Finally, repositories of gene expression profiles are irreplaceable when predictive gene expression-based models are generated. Statistical algorithms used for classification of compounds, such as support vector machines or neural networks, can only be developed by using a robust training set that covers a wide range of pharmacological activities, mechanisms of toxicity, and chemical structures. Furthermore, proper validation of these models requires access to a sufficiently large testing set composed of gene expression profiles distinct from those used in the training set, such that a good approximation of the performance characteristics of the model can be generated.

In summary, any toxicogenomics effort would greatly benefit from access to a wealth of carefully selected reference data. While for individual projects a focused small database of expression profiles can be successfully used, covering the large diversity of pharmacology, toxicology, and chemistry routinely encountered in drug discovery and development with a database represents a considerable investment of resources that very few institutions can achieve. This situation prompted the creation of several public and private toxicogenomics databases. Before reviewing these databases, we first discuss various general issues associated with the design and development of toxicogenomics databases.

4.5.2. Design and Development of Toxicogenomics Reference Databases

To be useful for toxicogenomics, a database needs to contain gene expression profiles induced in appropriate tissues and species by a wide variety of reference compounds (pharmaceutical agents, prototypical toxicants, control compounds, etc.) at multiple exposure levels and time points (40107–110). The reference compounds profiled in the database should reflect a variety of toxic mechanisms and sufficient chemical diversity, and should represent distinct structure-activity relationships (111). The nature of the toxicity and transcriptomic changes induced by compounds is largely dependent on the exposure achieved and the duration of the exposure. Therefore, it is critical to understand the dose response and incorporate into the database multiple doses resulting in pharmacological and toxicological exposures, as this allows one to differentiate gene expression changes related to pharmacological activity from the toxicologically relevant transcriptomic effects. It is also important to monitor the associated clinical signs and pathological changes to differentiate nonspecific transcriptomic changes related to moribundity from toxicity-specific gene regulation. Furthermore, monitoring compound-induced transcriptomic responses over a time period may be necessary to identify gene expression changes linked to a time-dependent toxic response and increases the chances of observing a true toxic effect. Therefore, evaluating several time points can be extremely valuable.

In our experience, transcriptomic effects occuring following acute or short exposures to compounds, albeit interesting, are associated with significant interindividual variability and may not be as informative as the changes induced once a steady-state tissue exposure has been achieved. Therefore, we try to avoid evaluating gene expression changes following a single exposure to compounds and favor a short repeat-dosing paradigm. Consequently, our critical reference data sets consist of profiles induced following daily exposure for 3–5 days.

The number of biological replicates required for each time point or dose is an important consideration in the design of a reference database. Biological replicates are critical to control for biological and technical variability associated with gene expression profiles. There are very few reports evaluating what an optimal group size should be for toxicogenomics studies in rats. Several reports on group size analysis have advocated using group sizes in the 10–14 replicate magnitude, a number that would make toxicogenomics too costly to be practical (112). Recent statistical reports, however, have shown that even small group sizes can produce a large amount of useful information, and that new methods to select only adequately powered probe sets may improve subsequent analysis (113, 114). In our experience, the interindividual variability of gene expression changes in rats is relatively minimal when sufficiently long dosing periods (i.e., more than a single dose and a single day) are used. Consequently, the use of three animals per group and per time point has generally been adequate to generate meaningful gene expression profiles.

It is advantageous to include a minimum of complementary technical and biological information into a toxicogenomics database. As stated previously, gene expression data can only be correctly understood in the context of the experimental conditions used in a study. This prompted the publication in 2001 by the Microarray Gene Expression Database group (MGED; http://www.mged.org/) of the **"Minimum Information About a Microarray Experiment" (MIAME) guidelines** for the reporting and publication of microarray experiments (115). These guidelines describe the minimum information required to ensure that microarray data can be easily interpreted and repeated. They do not specify the format in which the information should be provided, but only its content. They have become a standard for the submission of gene expression data to scientific journals and are used as a reference by some public and commercial databases. In addition, these recommendations have facilitated the development of MIAME-compatible microarray analysis software packages. Obviously, it is impossible to capture all the possible factors that may affect gene expression profiles; however, a minimum of standardization is an appropriate start to exchange, mine, and compare data. In fact, experience with databases for other types of biological data demonstrates that structured and consistent annotation and standardization are critical to building useful reference repositories.

Experimental inconsistencies can complicate the use of reference data and therefore should be minimized. In particular, the use of different platforms to generate the reference profiles can limit the value of some experimental data sets. Significant improvements have been achieved in the ability to extrapolate data from one platform to another and in establishing more consistent gene nomenclature; however, the use of different platforms still increases variability (65). Even when the same platform is used, gene expression data may differ across laboratories because of differences in protocols or instrumentation (29). These issues have been addressed by several major multi-institutional efforts involving academic, industrial, and governmental laboratories with the objectives of identifying, validating, and implementing standards that could be used for gene expression analysis (6, 64,116–119). Confounding experimental factors obviously still exist that may limit the comparison of gene expression data from one laboratory to another, and these should be carefully considered when selecting a database.

As we mentioned above, most toxicogenomics work in the pharmaceutical industry has been conducted with rats, since this species represents the favored small preclinical model to conduct in vivo toxicological studies. Ideally, one should compare gene expression profiles generated in the same species or even in the same strain. Practically, this may not always be feasible because of a lack of reference data in some species. Therefore, certain circumstances require extrapolation across species by genome mapping. Despite the good annotation of the genomes of the major preclinical species currently used in toxicology, the mapping of orthologous genes is not always precise and requires significant additional efforts (108). In addition, extrapolation across species assumes that all species react similarly to a specific toxicant, an assumption that may not always be correct for various reasons. For instance, different rat strains may respond differently to some toxic agents, and it is unclear how this divergence in phenotypic or functional response may translate at the transcriptomic level. Available data suggest that extrapolation across different rat strains may be possible and that after exposure to toxicants comparable transcriptomic responses are observed across different rat strains (120). Likewise, a predictive model of hepatotoxicity built on transcript profiles from one rat strain has successfully been used to classify gene expression profiles from another strain (45). However, some studies have demonstrated variation in pharmacological or toxicological responses correlating with differences in gene expression (121–123).

The optimal reference profiles should be derived from similar tissues or cells. We have already discussed several particularities and limitations associated with certain organs for toxicogenomics. In addition, it is important to recognize that routinely conducting toxicogenomics analyses of some tissues adds little value, and therefore the need for reference gene expression profiles in these tissues is rather limited. Hence, routine toxicogenomics profiling of tissues like stomach or thyroid gland, to take extreme examples, is unlikely to be conducted in an R&D organization. It is difficult to trace situations in which traditional preclinical toxicological evaluations fail to detect at an early stage target organ toxicity, since these types of preclinical safety data are typically not publicly available (124). Nonetheless, published data indicate that while detecting toxicity with traditional approaches is quite reliable in the cardiovascular, hematopoietic, and gastrointestinal systems, early detection of toxicity in the hepatic or urinary systems is less common (4). Therefore, in addition to their favorable tissue characteristics, liver and kidney represent tissues of choice for the use of toxicogenomics as a predictive tool. This is further supported by the frequency of these tissues as target organs of toxicity. Tissues like heart or spleen are quite homogeneous and amenable to large-scale gene expression profiling, and therefore also represent attractive options for toxicogenomics studies. Finally, as discussed in Chapter 5, while significant challenges still exist for the successful generation of gene expression profiles in this matrix, whole blood represents an extremely attractive tissue that could tremendously impact the future of toxicogenomics as a routine tool in toxicology. Very few reference data are available for this tissue in either public or reference databases, but given the potential advantages of what has also been called hemogenomics (i.e., genomic profiling of blood), it is likely that such reference data will sooner or later be available.

4.5.3. Existing Toxicogenomics Databases

There are multiple public, commercial, and proprietary compendia of gene expression profiles that can be used for the analysis of gene expression data sets in the context of toxicology (29, 125). Not all of these repositories have been specifically built for toxicology, but they still represent a useful source of toxicogenomics reference data. Several databases have been developed recently for toxicology and may be more appropriate for toxicogenomics, but may not always contain all necessary data. For instance, the National Center for Toxicogenomics, affiliated with the National Institute of Environmental Health Sciences, has created a reference knowledge database (Chemical Effects in Biological Systems or CEBS) that can be used to decipher mechanisms of action using any 'omics data set, but it is unclear what will be available in this database (126–128). Likewise, companies such as GeneLogic (Gaithersburg, MD) and Iconix (Mountain View, CA) have created large toxicogenomics reference databases that contain gene expression profiles induced by a variety of prototypical compounds, as well as corroborating toxicological and pathological end points (44, 111, 129).

Public repositories of gene expression profiles offer considerable scientific benefit to the toxicology community, but it is important to realize that there are many technical and logistic issues that challenge their effective use (125, 130). Profiles may have been generated with various, sometimes obsolete, gene expression platforms and protocols. Study designs may differ in terms of time points or doses used, and important accompanying data, such as clinical pathology or histopathology findings, may not be available, hindering the interpretation of gene expression changes. In contrast, commercial databases do not have these inconsistencies, are more complete, and offer the advantage of being more standardized. However, highly standardized and comprehensive databases require an enormous investment to develop and are generally provided in a proprietary model at a significant cost, such that they may not be affordable to most users (106).

To provide access to these proprietary data on as-needed basis, commercial providers are exploring innovative approaches, and these arrangements may ultimately benefit the toxicology community and promote the wider acceptance of toxicogenomics in the pharmaceutical industry. For instance, Affymetrix and Iconix partnered to offer the ToxFX™ Analysis Suite (http://www.affymetrix.com/products/application/toxicogenomics.affx). This product provides an automated toxicogenomics analysis that interrogates more than 50 toxicological outcomes, signals the genes that are likely to be most important in relation to those problems, and reports genes and biological pathways that are most likely to play a role in any predicted toxicity. Similarly, Iconix Biosciences (http://www.iconixpharm.com/) recently released its proprietary database DrugMatrix® online, making its content and customized tools available to a much broader group of investigators.

There are too many databases available to avoid significant omission and to justify an extensive coverage of all of them in this chapter. Therefore, a short description of selected potentially useful databases is provided here. These databases were selected based on our level of familiarity with their features, and not necessarily based on superiority. Table 4.2 also summarizes the major features and lists web sites for the databases discussed. From our experience, it is useful to evaluate several databases to confirm that the desired profiles and corroborating data sets are available, that required and useful analysis tools are offered, and that there is sufficient annotation and experimental similarity for the reference profiles. It is noteworthy that data from various sources can be combined for a meta-analysis, whereby data from a particular database may be analyzed with tools from a different source.

4.5.3.1. Chemical Effects in Biological Systems (CEBS)

The CEBS toxicogenomics data repository has been designed to promote a systems biology approach to understanding the biological effects of toxic agents (131). It has been developed by the National Center for Toxicogenomics (National Institute of Environmental Health Sciences) as a public toxicogenomics knowledge repository that combines data sets from transcriptomics, proteomics, metabonomics, and conventional toxicology with metabolic, toxicological pathway, and gene regulatory network information relevant to environmental toxicology and human disease. It is designed to enable users to perform integrative mining of conventional toxicology and 'omics data. CEBS data sets are fully documented in the experimental protocol and are searchable by compound, structure, toxicity or pathology end point, gene, gene family, single nucleotide polymorphism

Table 4.2 Public Databases for Toxicogenomics

Database	Attributes	URL
Gene Expression Omnibus (GEO)	World largest public repository Adherence to MIAME guidelines Toxicology data available Exploration, analysis and visualization tools	http://www.ncbi.nlm.nih.gov/projects/geo/
ArrayExpress	Large public repository Adherence to MIAME guidelines Toxicology data available in Tox-MIAMExpress Expression data from normal human and mouse tissues	http://www.ebi.ac.uk/arrayexpress/
Chemical Effects in Biological Systems (CEBS)	Evolving public toxicogenomics repository from the National Institute of Environmental Sciences (NIEHS) National Center for Toxicogenomics (NCT) Designed to house data from complex studies having multiple data steams (genetic, proteomic, metabonomics data) Exploration and analysis tools	http://cebs.niehs.nih.gov/
Environment, Drugs, and Gene Expression (EDGE)	Public toxicogenomics repository Standardized experimental conditions including standardized microarray platform Mostly focused on mouse liver microarray data Useful bioinformatics tools (clustering, BLAST searching, rank analysis, classification tools)	http://edge.oncology.wisc.edu/
Symatlas	Product of the Genomics Institute of the Novartis Research Foundation Expression data from a large panel of normal human and mouse tissues or cell culture models	http://symatlas.gnf.org/SymAtlas/
DbZach System	Toxicogenomics database allowing data mining and full knowledge-based understanding of toxicological mechanisms Contains correlating clinical chemistry parameters and histopathological data.	http://dbzach.fst.msu.edu/

(SNP), pathway, and network as a function of dose, time, and the phenotype of the target tissue (127). CEBS contains gene expression profiles from mouse and rat studies, along with excellent descriptions of studies and annotation of accompanying data, such as clinical pathology or histopathology (132). However, its contents are still too limited for optimal use, and it is unclear how extensively and rapidly this database will grow.

4.5.3.2. ArrayTrack®

ArrayTrack® has been developed by the National Center for Toxicological Research (NCTR) of the U.S. Food and Drug Administration (FDA) and provides an integrated solution for managing, analyzing, and interpreting microarray gene expression data (133, 134). ArrayTrack® is not a public repository of gene expression data but a useful alternative tool to analyze gene expression data. ArrayTrack® is publicly available online, and the prospective user can request a local installation version by contacting the NCTR. ArrayTrack® is MIAME-compliant and stores microarray data and associated experimental parameters associated with toxicogenomics studies. ArrayTrack® also provides useful statistical and visualization tools, as well as a rich collection of functional information about genes, proteins, and pathways for biological interpretation. The key feature of ArrayTrack® is a direct link between the analysis output and the functional information to facilitate the interpretation of results. With ArrayTrack®, users can easily select a statistical method to be applied to a stored microarray data set to generate a list of differentially expressed genes. The resulting gene list can be directly linked to pathways and gene ontology for functional analysis. The core facilities of the NCTR produce large volumes of genomic, proteomic, and metabonomic data with standardized experimental procedures. ArrayTrack® fully integrates these high-content data sets with results represented in public repositories as well as with conventional in vitro and in vivo toxicology data. The system enables data curation in accordance with standard ontology and provides a collection of robust tools for data analysis and knowledge mining. ArrayTrack® is logically constructed of three linked components: (i) a library that mirrors critical data in public databases; (ii) a database (MicroarrayDB) that stores MIAME-compliant microarray experiment information; and (iii) tools that operate on experimental and public data for knowledge discovery. With ArrayTrack®, one can select an analysis method and apply it to chosen microarray data stored in the MicroarrayDB; the analysis results can then be linked directly to gene information in the library.

4.5.3.3. Gene Expression Omnibus

The National Center for Biotechnology Information (NCBI) Gene Expression Omnibus (GEO) repository is a freely accessible archive of high-throughput molecular abundance data, which predominantly contains microarray-generated gene expression profiles (135). This database has a flexible design that can handle diverse formats of both unprocessed and processed data in a MIAME-compatible infrastructure that promotes fully annotated submissions. GEO stores an enormous amount of individual gene expression data derived from over 100 organisms, submitted by over 1500 laboratories, and addresses a wide range of biological phenomena. GEO was launched in 2000 before the MIAME guidelines were proposed, but incremental improvements to database design and curation standards have been applied since then. All data submitted to GEO are now syntactically validated for correct document structure, organization, and provision of basic elements; each submission is inspected by curators for content integrity (136). These steps make this database a unique and rich source of public gene expression data. GEO also provides several user-friendly web-based interfaces and applications to enable effective exploration, query, and visualization of gene expression data at the level of individual genes or entire studies.

4.5.3.4. ArrayExpress

ArrayExpress is a public database supported by the European Bioinformatics Institute (EBI) and designed for high-throughput functional genomics data (137). This public resource for microarray data has two major goals: to serve as an archive providing access to microarray data supporting publications and to build a knowledge base of gene expression profiles (130). ArrayExpress consists of two tightly integrated databases: the ArrayExpress Repository, which is a MIAME-supportive public archive of microarray data, and the ArrayExpress Data Warehouse, which is a database of gene expression profiles selected from the repository and consistently reannotated and optimized for queries. Archived experiments can be queried by experimental attributes, such as keywords, species, array platform, authors, journals, or accession numbers. Gene expression profiles can be queried by gene names and properties, such as Gene Ontology terms, and gene expression profiles can be visualized with provided tools, such as Expression Profiler (EP), a microarray data mining, analysis, and visualization tool. ArrayExpress contains gene expression and other microarray data from over 50,000 hybridizations and over 1,500,000 individual expression profiles comprising more than 1200 studies and covering 70 different species (137). Most data are related to peer-reviewed publications.

4.5.3.5. DbZach

dbZach is a modular relational database with associated data insertion, retrieval, and mining tools that manages traditional toxicology and complementary toxicogenomics data to facilitate comprehensive data integration, analysis, and sharing (138). This database complies with the MIAME standards and represents a toxicogenomics data management system with analysis tools (139).

4.5.3.6. ToxExpress®

ToxExpress® is a commercial product and service offered by GeneLogic Inc (Gaithersburg, MD). This commercial module consists of a large repository of gene expression profiles derived from human and animal tissues or cells exposed to well-characterized pharmaceutical and chemical agents. The most useful gene expression profiles are derived from preclinical toxicology studies evaluated across various time points and doses and are complemented by concurrent clinical pathology information. The database is comprised of gene expression profiles for over 13,000 individual vehicle- and drug-treated animal tissue samples and related in-life toxicology data. Tools provided by the system also include several statistical packages for data and pathway analysis.

4.5.3.7. DrugMatrix®

DrugMatrix® is a comprehensive toxicogenomics database provided by Iconix Biosciences (Mountain View, CA), now part of Entelos (Foster City, CA). It is relevant here to disclose that Abbott has acquired access to this commercial database, and therefore we are particularly familiar with this product. This database contains gene expression profiles from tissues collected from short-term repeat-dose rat studies with a wide variety of marketed and withdrawn drugs, toxicants, and reference standards (111). Test compounds were administered at a fully efficacious dose and a maximum tolerated dose (MTD), which represents the upper limit of tolerability based on 5-day daily-dosing range-finding studies (106). Over 600 compounds have been profiled in various tissues, including liver, kidney, heart, bone marrow, thigh skeletal muscle, spleen, and intestine. These expression profiles were generated by using standardized protocols with the CodeLink RU1 rat microarray and the Affymetrix RG230$_2$.0 rat GeneChip array and are complemented with highly standardized data points of serum chemistry, hematology, histopathology, or organ weights (106). Furthermore, DrugMatrix® contains a wealth of other useful information, such as compound information (chemical structure, pharmacology data, adverse event reports), literature information, and in vitro pharmacology data (key binding and enzyme assay data across a total of 130 different molecular targets). The database is also complemented with various useful analysis tools that enable searching for related reference gene expression profiles, interpreting gene expression changes in the context of proprietary, toxicologically relevant biological pathways, and identifying toxic signals with a proprietary battery of Drug Signatures (106).

4.6. CONCLUSION

In this chapter, we have attempted to provide a comprehensive review of the basic principles of toxicogenomics. Given the exploding number of publications related to gene expression analysis in general, and to toxicogenomics in particular, it is not surprising that this chapter will contain inevitable omissions when the book is published. We apologize in advance to our colleagues whose work was not included in this chapter. In addition, because toxicogenomics is evolving at an accelerated pace, we have tried to focus on major principles and critical points, and have avoided distracting the reader with technical details that are likely to become irrelevant in the near future. Likewise, analysis tools and gene expression repositories are constantly being developed, adjusted, and improved, such that any scientist involved in this field is required to keep abreast of new developments through the literature and scientific meetings. While the concepts reviewed here have been simplified for the general reader, it should be emphasized that the complexity of toxicogenomics is best managed by teams of individuals with complementary expertise. In contrast to traditional toxicology, toxicogenomics requires a broader and slightly different expertise, often only achieved by multidisciplinary teams composed of toxicologists, molecular biologists, bioinformaticians, and biostatisticians. The formation of productive teams composed of people with strikingly different scientific backgrounds, diverse expertise, and sometimes conflicting interests can be a significant challenge that should not be underestimated. In the next two chapters, we build on the general principles discussed here and explain how toxicogenomics can be applied in the context of drug discovery and development.

REFERENCES

1. Schena M, Shalon D, Davis RW, Brown PO. Quantitative monitoring of gene expression patterns with a complementary DNA microarray. *Science* 1995;270:467–470.
2. Gao Z, Xu H, DiSilvestre D, Halperin VL, Tunin R, Tian Y, Yu W, Winslow RL, Tomaselli GF. Transcriptomic profiling of the canine tachycardia-induced heart failure model: global comparison to human and murine heart failure. *J Mol Cell Cardiol* 2006;40:76–86.
3. Lovett RA. Toxicogenomics. Toxicologists brace for genomics revolution. *Science* 2000;289:536–537.
4. Greaves P, Williams A, Eve M. First dose of potential new medicines to humans: how animals help. *Nat Rev Drug Discov* 2004;3:226–236.
5. Oyama MA, Chittur S. Genomic expression patterns of cardiac tissues from dogs with dilated cardiomyopathy. *Am J Vet Res* 2005;66:1140–1155.
6. Shi L, Reid LH, Jones WD, Shippy R, Warrington JA, Baker SC, Collins PJ, de Longueville F, Kawasaki ES, Lee KY, Luo Y, Sun YA, Willey JC, Setterquist RA, Fischer GM, Tong W, Dragan YP, Dix DJ, Frueh FW, Goodsaid FM et al. The MicroArray Quality Control (MAQC) project shows inter- and intraplatform reproducibility of gene expression measurements. *Nat Biotechnol* 2006;24:1151–1161.
7. Ach RA, Floore A, Curry B, Lazar V, Glas AM, Pover R, Tsalenko A, Ripoche H, Cardoso F, d'Assignies MS, Bruhn L, Van't Veer LJ. Robust interlaboratory reproducibility of a gene expression signature measurement consistent with the needs of a new generation of diagnostic tools. *BMC Genomics* 2007;8:148.
8. Raymond F, Metairon S, Borner R, Hofmann M, Kussmann M. Automated target preparation for microarray-based gene expression analysis. *Anal Chem* 2006;78:6299–6305.

9. Chuaqui RF, Bonner RF, Best CJ, Gillespie JW, Flaig MJ, Hewitt SM, Phillips JL, Krizman DB, Tangrea MA, Ahram M, Linehan WM, Knezevic V, Emmert-Buck MR. Post-analysis follow-up and validation of microarray experiments. *Nat Genet* 2002;32 Suppl:509–514.
10. Rajeevan MS, Vernon SD, Taysavang N, Unger ER. Validation of array-based gene expression profiles by real-time (kinetic) RT-PCR. *J Mol Diagn* 2001;3:26–31.
11. Rajeevan MS, Ranamukhaarachchi DG, Vernon SD, Unger ER. Use of real-time quantitative PCR to validate the results of cDNA array and differential display PCR technologies. *Methods* 2001;25:443–451.
12. Bosotti R, Locatelli G, Healy S, Scacheri E, Sartori L, Mercurio C, Calogero R, Isacchi A. Cross platform microarray analysis for robust identification of differentially expressed genes. *BMC Bioinformatics* 2007;8 Suppl 1:S5.
13. Dallas PB, Gottardo NG, Firth MJ, Beesley AH, Hoffmann K, Terry PA, Freitas JR, Boag JM, Cummings AJ, Kees UR. Gene expression levels assessed by oligonucleotide microarray analysis and quantitative real-time RT-PCR—how well do they correlate? *BMC Genomics* 2005;6:59.
14. Yang Y, Abel SJ, Ciurlionis R, Waring JF. Development of a toxicogenomics in vitro assay for the efficient characterization of compounds. *Pharmacogenomics* 2006;7:177–186.
15. Sawada H, Taniguchi K, Takami K. Improved toxicogenomic screening for drug-induced phospholipidosis using a multiplexed quantitative gene expression ArrayPlate assay. *Toxicol* In Vitro 2006;20:1506–1513.
16. Abruzzo LV, Lee KY, Fuller A, Silverman A, Keating MJ, Medeiros LJ, Coombes KR. Validation of oligonucleotide microarray data using microfluidic low-density arrays: a new statistical method to normalize real-time RT-PCR data. *Biotechniques* 2005;38:785–792.
17. Irizarry RA, Warren D, Spencer F, Kim IF, Biswal S, Frank BC, Gabrielson E, Garcia JG, Geoghegan J, Germino G, Griffin C, Hilmer SC, Hoffman E, Jedlicka AE, Kawasaki E, Martinez-Murillo F, Morsberger L, Lee H, Petersen D, Quackenbush J, et al. Multiple-laboratory comparison of microarray platforms. *Nat Methods* 2005;2:345–350.
18. Petersen D, Chandramouli GV, Geoghegan J, Hilburn J, Paarlberg J, Kim CH, Munroe D, Gangi L, Han J, Puri R, Staudt L, Weinstein J, Barrett JC, Green J, Kawasaki ES. Three microarray platforms: an analysis of their concordance in profiling gene expression. *BMC Genomics* 2005;6:63.
19. Boorman GA, Irwin RD, Vallant MK, Gerken DK, Lobenhofer EK, Hejtmancik MR, Hurban P, Brys AM, Travlos GS, Parker JS, Portier CJ. Variation in the hepatic gene expression in individual male Fischer rats. *Toxicol Pathol* 2005;33:102–110.
20. Knight A. Systematic reviews of animal experiments demonstrate poor human clinical and toxicological utility. *Altern Lab Anim* 2007;35:641–659.
21. Olson H, Betton G, Robinson D, Thomas K, Monro A, Kolaja G, Lilly P, Sanders J, Sipes G, Bracken W, Dorato M, Van Deun K, Smith P, Berger B, Heller A. Concordance of the toxicity of pharmaceuticals in humans and in animals. *Regul Toxicol Pharmacol* 2000;32:56–67.
22. Kaplowitz N. Idiosyncratic drug hepatotoxicity. *Nat Rev Drug Discov* 2005;4:489–499.
23. Dieckhaus CM, Thompson CD, Roller SG, Macdonald TL. Mechanisms of idiosyncratic drug reactions: the case of felbamate. *Chem Biol Interact* 2002;142:99–117.
24. Leone AM, Kao LM, McMillian MK, Nie AY, Parker JB, Kelley MF, Usuki E, Parkinson A, Lord PG, Johnson MD. Evaluation of felbamate and other antiepileptic drug toxicity potential based on hepatic protein covalent binding and gene expression. *Chem Res Toxicol* 2007;20:600–608.
25. Liguori MJ, Anderson LM, Bukofzer S, McKim J, Pregenzer JF, Retief J, Spear BB, Waring JF. Microarray analysis in human hepatocytes suggests a mechanism for hepatotoxicity induced by trovafloxacin. *Hepatology* 2005;41:177–186.

26. Liguori MJ, Blomme EA, Waring JF. Trovafloxacin-induced gene expression changes in liver-derived in vitro systems: comparison of primary human hepatocytes to HepG2 cells. *Drug Metab Dispos* 2007;36:223–33.
27. Slonim DK. From patterns to pathways: gene expression data analysis comes of age. *Nat Genet* 2002;32 Suppl:502–508.
28. Vinciotti V, Khanin R, D'Alimonte D, Liu X, Cattini N, Hotchkiss G, Bucca G, de Jesus O, Rasaiyaah J, Smith CP, Kellam P, Wit E. An experimental evaluation of a loop versus a reference design for two-channel microarrays. *Bioinformatics* 2005;21:492–501.
29. Hayes KR, Bradfield CA. Advances in toxicogenomics. *Chem Res Toxicol* 2005;18:403–414.
30. Churchill GA. Fundamentals of experimental design for cDNA microarrays. *Nat Genet* 2002;32 Suppl:490–495.
31. Wallace JC, Korth MJ, Paeper B, Proll SC, Thomas MJ, Magness CL, Iadonato SP, Nelson C, Katze MG. High-density rhesus macaque oligonucleotide microarray design using early-stage rhesus genome sequence information and human genome annotations. *BMC Genomics* 2007;8:28.
32. Tusher VG, Tibshirani R, Chu G. Significance analysis of microarrays applied to the ionizing radiation response. *Proc Natl Acad Sci USA* 2001;98:5116–5121.
33. Lin DY. An efficient Monte Carlo approach to assessing statistical significance in genomic studies. *Bioinformatics* 2005;21:781–787.
34. Landgrebe J, Wurst W, Welzl G. Permutation-validated principal components analysis of microarray data. *Genome Biol* 2002;3:RESEARCH0019.
35. Raychaudhuri S, Stuart JM, Altman RB. Principal components analysis to summarize microarray experiments: application to sporulation time series. *Pac Symp Biocomput* 2000; 455–466.
36. Azuaje F. Clustering-based approaches to discovering and visualising microarray data patterns. *Brief Bioinform* 2003;4:31–42.
37. Wu LF, Hughes TR, Davierwala AP, Robinson MD, Stoughton R, Altschuler SJ. Large-scale prediction of *Saccharomyces cerevisiae* gene function using overlapping transcriptional clusters. *Nat Genet* 2002;31:255–265.
38. Datta S, Datta S. Comparisons and validation of statistical clustering techniques for microarray gene expression data. *Bioinformatics* 2003;19:459–466.
39. Sherlock G. Analysis of large-scale gene expression data. *Curr Opin Immunol* 2000;12: 201–205.
40. Bushel PR, Hamadeh HK, Bennett L, Green J, Ableson A, Misener S, Afshari CA, Paules RS. Computational selection of distinct class- and subclass-specific gene expression signatures. *J Biomed Inform* 2002;35:160–170.
41. Golub TR, Slonim DK, Tamayo P, Huard C, Gaasenbeek M, Mesirov JP, Coller H, Loh ML, Downing JR, Caligiuri MA, Bloomfield CD, Lander ES. Molecular classification of cancer: class discovery and class prediction by gene expression monitoring. *Science* 1999; 286:531–537.
42. Piotrowski PL, Sumpter BG, Malling HV, Wassom JS, Lu PY, Brothers RA, Sega GA, Martin SA, Parang M. A toxicity evaluation and predictive system based on neural networks and wavelets. *J Chem Inf Model* 2007;47:676–685.
43. Krafczyk S, Tietze S, Swoboda W, Valkovic P, Brandt T. Artificial neural network: a new diagnostic posturographic tool for disorders of stance. *Clin Neurophysiol* 2006;117:1692–1698.
44. Castle AL, Carver MP, Mendrick DL. Toxicogenomics: a new revolution in drug safety. *Drug Discov Today* 2002;7:728–736.
45. Steiner G, Suter L, Boess F, Gasser R, de Vera MC, Albertini S, Ruepp S. Discriminating different classes of toxicants by transcript profiling. *Environ Health Perspect* 2004;112:1236–1248.

46. Brown MP, Grundy WN, Lin D, Cristianini N, Sugnet CW, Furey TS, Ares M, Jr., Haussler D. Knowledge-based analysis of microarray gene expression data by using support vector machines. *Proc Natl Acad Sci USA* 2000;97:262–267.
47. Furey TS, Cristianini N, Duffy N, Bednarski DW, Schummer M, Haussler D. Support vector machine classification and validation of cancer tissue samples using microarray expression data. *Bioinformatics* 2000;16:906–914.
48. Yang Y, Blomme EA, Waring JF. Toxicogenomics in drug discovery: from preclinical studies to clinical trials. *Chem Biol Interact* 2004;150:71–85.
49. Sivachenko AY, Yuryev A. Pathway analysis software as a tool for drug target selection, prioritization and validation of drug mechanism. *Expert Opin Ther Targets* 2007;11:411–421.
50. Curtis RK, Oresic M, Vidal-Puig A. Pathways to the analysis of microarray data. *Trends Biotechnol* 2005;23:429–435.
51. Ulrich RG, Rockett JC, Gibson GG, Pettit SD. Overview of an interlaboratory collaboration on evaluating the effects of model hepatotoxicants on hepatic gene expression. *Environ Health Perspect* 2004;112:423–427.
52. Li CJ, Li RW, Wang YH, Elsasser TH. Pathway analysis identifies perturbation of genetic networks induced by butyrate in a bovine kidney epithelial cell line. *Funct Integr Genomics* 2007;7:193–205.
53. Abdel-Aziz HO, Takasaki I, Tabuchi Y, Nomoto K, Murai Y, Tsuneyama K, Takano Y. High-density oligonucleotide microarrays and functional network analysis reveal extended lung carcinogenesis pathway maps and multiple interacting genes in NNK [4-(methylnitrosamino)-1-(3-pyridyle)-1-butanone] induced CD1 mouse lung tumor. *J Cancer Res Clin Oncol* 2007;133:107–115.
54. Mayburd AL, Martlinez A, Sackett D, Liu H, Shih J, Tauler J, Avis I, Mulshine JL. Ingenuity network-assisted transcription profiling: Identification of a new pharmacologic mechanism for MK886. *Clin Cancer Res* 2006;12:1820–1827.
55. Ekins S, Bugrim A, Brovold L, Kirillov E, Nikolsky Y, Rakhmatulin E, Sorokina S, Ryabov A, Serebryiskaya T, Melnikov A, Metz J, Nikolskaya T. Algorithms for network analysis in systems-ADME/Tox using the MetaCore and MetaDrug platforms. *Xenobiotica* 2006;36:877–901.
56. Ekins S. Systems-ADME/Tox: resources and network approaches. *J Pharmacol Toxicol Methods* 2006;53:38–66.
57. Kanehisa M, Goto S, Kawashima S, Okuno Y, Hattori M. The KEGG resource for deciphering the genome. *Nucleic Acids Res* 2004;32:277–80.
58. Fielden MR, Pearson C, Brennan R, Kolaja KL. Preclinical drug safety analysis by chemogenomic profiling in the liver. *Am J Pharmacogenomics* 2005;5:161–171.
59. Guerreiro N, Staedtler F, Grenet O, Kehren J, Chibout SD. Toxicogenomics in drug development. *Toxicol Pathol* 2003;31:471–479.
60. Fuscoe JC, Branham WS, Melvin CD, Desai VG, Moland CL, Han T, Shi L, Tong W, Scully AT, Delongchamp RR. Technical issues involved in obtaining reliable data from microarray experiments. *Regul Res Perspect* 2006;6:1–22.
61. Bammler T, Beyer RP, Bhattacharya S, Boorman GA, Boyles A, Bradford BU, Bumgarner RE, Bushel PR, Chaturvedi K, Choi D, Cunningham ML, Deng S, Dressman HK, Fannin RD, Farin FM, Freedman JH, Fry RC, Harper A, Humble MC, Hurban P, et al. Standardizing global gene expression analysis between laboratories and across platforms. *Nat Methods* 2005;2:351–356.
62. Shi L, Tong W, Goodsaid F, Frueh FW, Fang H, Han T, Fuscoe JC, Casciano DA. QA/QC: challenges and pitfalls facing the microarray community and regulatory agencies. *Expert Rev Mol Diagn* 2004;4:761–777.
63. Shi L, Tong W, Su Z, Han T, Han J, Puri RK, Fang H, Frueh FW, Goodsaid FM, Guo L, Branham WS, Chen JJ, Xu ZA, Harris SC, Hong H, Xie Q, Perkins RG, Fuscoe JC. Microarray

scanner calibration curves: characteristics and implications. *BMC Bioinformatics* 2005;6 Suppl 2:S11.

64. Waring JF, Ulrich RG, Flint N, Morfitt D, Kalkuhl A, Staedtler F, Lawton M, Beekman JM, Suter L. Interlaboratory evaluation of rat hepatic gene expression changes induced by methapyrilene. *Environ Health Perspect* 2004;112:439–448.
65. Mattes WB. Annotation and cross-indexing of array elements on multiple platforms. *Environ Health Perspect* 2004;112:506–510.
66. Pennie W, Pettit SD, Lord PG. Toxicogenomics in risk assessment: an overview of an HESI collaborative research program. *Environ Health Perspect* 2004;112:417–419.
67. Sugai T, Kawamura M, Iritani S, Araki K, Makifuchi T, Imai C, Nakamura R, Kakita A, Takahashi H, Nawa H. Prefrontal abnormality of schizophrenia revealed by DNA microarray: impact on glial and neurotrophic gene expression. *Ann NY Acad Sci* 2004;1025:84–91.
68. Ojaimi C, Qanud K, Hintze TH, Recchia FA. Altered expression of a limited number of genes contributes to cardiac decompensation during chronic ventricular tachypacing in dogs. *Physiol Genomics* 2007;29:76–83.
69. Burton-Wurster N, Mateescu RG, Todhunter RJ, Clements KM, Sun Q, Scarpino V, Lust G. Genes in canine articular cartilage that respond to mechanical injury: gene expression studies with Affymetrix canine GeneChip. *J Hered* 2005;96:821–828.
70. Holzwarth JA, Middleton RP, Roberts M, Mansourian R, Raymond F, Hannah SS. The development of a high-density canine microarray. *J Hered* 2005;96:817–820.
71. Higgins MA, Berridge BR, Mills BJ, Schultze AE, Gao H, Searfoss GH, Baker TK, Ryan TP. Gene expression analysis of the acute phase response using a canine microarray. *Toxicol Sci* 2003;74:470–484.
72. Datson NA, Morsink MC, Atanasova S, Armstrong VW, Zischler H, Schlumbohm C, Dutilh BE, Huynen MA, Waegele B, Ruepp A, de Kloet ER, Fuchs E. Development of the first marmoset-specific DNA microarray (EUMAMA): a new genetic tool for large-scale expression profiling in a non-human primate. *BMC Genomics* 2007;8:190.
73. Hacia JG. Genome of the apes. *Trends Genet* 2001;17:637–645.
74. Marvanova M, Menager J, Bezard E, Bontrop RE, Pradier L, Wong G. Microarray analysis of nonhuman primates: validation of experimental models in neurological disorders. *FASEB J* 2003;17:929–931.
75. Huff JL, Hansen LM, Solnick JV. Gastric transcription profile of *Helicobacter pylori* infection in the rhesus macaque. *Infect Immun* 2004;72:5216–5226.
76. Walker SJ, Wang Y, Grant KA, Chan F, Hellmann GM. Long versus short oligonucleotide microarrays for the study of gene expression in nonhuman primates. *J Neurosci Methods* 2006;152:179–189.
77. Ji W, Zhou W, Gregg K, Yu N, Davis S, Davis S. A method for cross-species gene expression analysis with high-density oligonucleotide arrays. *Nucleic Acids Res* 2004;32:e93.
78. Dillman JF, III, Phillips CS. Comparison of non-human primate and human whole blood tissue gene expression profiles. *Toxicol Sci* 2005;87:306–314.
79. Cariello NF, Romach EH, Colton HM, Ni H, Yoon L, Falls JG, Casey W, Creech D, Anderson SP, Benavides GR, Hoivik DJ, Brown R, Miller RT. Gene expression profiling of the PPAR-alpha agonist ciprofibrate in the cynomolgus monkey liver. *Toxicol Sci* 2005;88:250–264.
80. Rubinstein AL. Zebrafish assays for drug toxicity screening. *Expert Opin Drug Metab Toxicol* 2006;2:231–240.
81. Alestrom P, Holter JL, Nourizadeh-Lillabadi R. Zebrafish in functional genomics and aquatic biomedicine. *Trends Biotechnol* 2006;24:15–21.

82. Lam SH, Winata CL, Tong Y, Korzh S, Lim WS, Korzh V, Spitsbergen J, Mathavan S, Miller LD, Liu ET, Gong Z. Transcriptome kinetics of arsenic-induced adaptive response in zebrafish liver. *Physiol Genomics* 2006;27:351–361.
83. Mirnics K, Pevsner J. Progress in the use of microarray technology to study the neurobiology of disease. *Nat Neurosci* 2004;7:434–439.
84. Galvin JE, Ginsberg SD. Expression profiling and pharmacotherapeutic development in the central nervous system. *Alzheimer Dis Assoc Disord* 2004;18:264–269.
85. Caretti E, Devarajan K, Coudry R, Ross E, Clapper ML, Cooper HS, Bellacosa A. Comparison of RNA amplification methods and chip platforms for microarray analysis of samples processed by laser capture microdissection. *J Cell Biochem* 2007;103:556–563.
86. Fink L, Bohle RM. Laser microdissection and RNA analysis. *Methods Mol Biol* 2005;293:167–185.
87. Schindler H, Wiese A, Auer J, Burtscher H. cRNA target preparation for microarrays: comparison of gene expression profiles generated with different amplification procedures. *Anal Biochem* 2005;344:92–101.
88. Luzzi VI, Holtschlag V, Watson MA. Gene expression profiling of primary tumor cell populations using laser capture microdissection, RNA transcript amplification, and GeneChip microarrays. *Methods Mol Biol* 2005;293:187–207.
89. Michel C, Desdouets C, Sacre-Salem B, Gautier JC, Roberts R, Boitier E. Liver gene expression profiles of rats treated with clofibric acid: comparison of whole liver and laser capture microdissected liver. *Am J Pathol* 2003;163:2191–2199.
90. Naito Y, Uchiyama K, Mizushima K, Kuroda M, Akagiri S, Takagi T, Handa O, Kokura S, Yoshida N, Ichikawa H, Takahashi J, Yoshikawa T. Microarray profiling of gene expression patterns in glomerular cells of astaxanthin-treated diabetic mice: a nutrigenomic approach. *Int J Mol Med* 2006;18:685–695.
91. Irwin RD, Parker JS, Lobenhofer EK, Burka LT, Blackshear PE, Vallant MK, Lebetkin EH, Gerken DF, Boorman GA. Transcriptional profiling of the left and median liver lobes of male f344/n rats following exposure to acetaminophen. *Toxicol Pathol* 2005;33:111–117.
92. Panda S, Antoch MP, Miller BH, Su AI, Schook AB, Straume M, Schultz PG, Kay SA, Takahashi JS, Hogenesch JB. Coordinated transcription of key pathways in the mouse by the circadian clock. *Cell* 2002;109:307–320.
93. Storch KF, Lipan O, Leykin I, Viswanathan N, Davis FC, Wong WH, Weitz CJ. Extensive and divergent circadian gene expression in liver and heart. *Nature* 2002;417:78–83.
94. Kita Y, Shiozawa M, Jin W, Majewski RR, Besharse JC, Greene AS, Jacob HJ. Implications of circadian gene expression in kidney, liver and the effects of fasting on pharmacogenomic studies. *Pharmacogenetics* 2002;12:55–65.
95. Auer H, Lyianarachchi S, Newsom D, Klisovic MI, Marcucci G, Kornacker K. Chipping away at the chip bias: RNA degradation in microarray analysis. *Nat Genet* 2003;35:292–293.
96. Schoor O, Weinschenk T, Hennenlotter J, Corvin S, Stenzl A, Rammensee HG, Stevanovic S. Moderate degradation does not preclude microarray analysis of small amounts of RNA. *Biotechniques* 2003;35:1192–1201.
97. Copois V, Bibeau F, Bascoul-Mollevi C, Salvetat N, Chalbos P, Bareil C, Candeil L, Fraslon C, Conseiller E, Granci V, Maziere P, Kramar A, Ychou M, Pau B, Martineau P, Molina F, Del Rio M. Impact of RNA degradation on gene expression profiles: assessment of different methods to reliably determine RNA quality. *J Biotechnol* 2007;127:549–559.
98. Imbeaud S, Graudens E, Boulanger V, Barlet X, Zaborski P, Eveno E, Mueller O, Schroeder A, Auffray C. Towards standardization of RNA quality assessment using user-independent classifiers of microcapillary electrophoresis traces. *Nucleic Acids Res* 2005;33:e56.

99. Schroeder A, Mueller O, Stocker S, Salowsky R, Leiber M, Gassmann M, Lightfoot S, Menzel W, Granzow M, Ragg T. The RIN: an RNA integrity number for assigning integrity values to RNA measurements. *BMC Mol Biol* 2006;7:3.
100. Williams DP, Garcia-Allan C, Hanton G, LeNet JL, Provost JP, Brain P, Walsh R, Johnston GI, Smith DA, Park BK. Time course toxicogenomic profiles in CD-1 mice after nontoxic and nonlethal hepatotoxic paracetamol administration. *Chem Res Toxicol* 2004;17:1551–1561.
101. Foster WR, Chen SJ, He A, Truong A, Bhaskaran V, Nelson DM, Dambach DM, Lehman-McKeeman LD, Car BD. A retrospective analysis of toxicogenomics in the safety assessment of drug candidates. *Toxicol Pathol* 2007;35:621–635.
102. Whitlock JP, Jr. Induction of cytochrome P4501A1. *Annu Rev Pharmacol Toxicol* 1999;39:103–125.
103. Mimura J, Fujii-Kuriyama Y. Functional role of AhR in the expression of toxic effects by TCDD. *Biochim Biophys Acta* 2003;1619:263–268.
104. Nebert DW, Dalton TP, Okey AB, Gonzalez FJ. Role of aryl hydrocarbon receptor-mediated induction of the CYP1 enzymes in environmental toxicity and cancer. *J Biol Chem* 2004;279:23847–23850.
105. Hu W, Sorrentino C, Denison MS, Kolaja K, Fielden MR. Induction of cyp1a1 is a non-specific biomarker of aryl hydrocarbon receptor activation: results of large scale screening of pharmaceuticals and toxicants in vivo and in vitro. *Mol Pharmacol* 2007;71:1475–1486.
106. Ganter B, Snyder RD, Halbert DN, Lee MD. Toxicogenomics in drug discovery and development: mechanistic analysis of compound/class-dependent effects using the DrugMatrix database. *Pharmacogenomics* 2006;7:1025–1044.
107. Hamadeh HK, Bushel PR, Jayadev S, Martin K, DiSorbo O, Sieber S, Bennett L, Tennant R, Stoll R, Barrett JC, Blanchard K, Paules RS, Afshari CA. Gene expression analysis reveals chemical-specific profiles. *Toxicol Sci* 2002;67:219–231.
108. Thomas RS, Rank DR, Penn SG, Zastrow GM, Hayes KR, Pande K, Glover E, Silander T, Craven MW, Reddy JK, Jovanovich SB, Bradfield CA. Identification of toxicologically predictive gene sets using cDNA microarrays. *Mol Pharmacol* 2001;60:1189–1194.
109. Waring JF, Jolly RA, Ciurlionis R, Lum PY, Praestgaard JT, Morfitt DC, Buratto B, Roberts C, Schadt E, Ulrich RG. Clustering of hepatotoxins based on mechanism of toxicity using gene expression profiles. *Toxicol Appl Pharmacol* 2001;175:28–42.
110. Waring JF, Cavet G, Jolly RA, McDowell J, Dai H, Ciurlionis R, Zhang C, Stoughton R, Lum P, Ferguson A, Roberts CJ, Ulrich RG. Development of a DNA microarray for toxicology based on hepatotoxin-regulated sequences. *EHP Toxicogenomics* 2003;111:53–60.
111. Ganter B, Tugendreich S, Pearson CI, Ayanoglu E, Baumhueter S, Bostian KA, Brady L, Browne LJ, Calvin JT, Day GJ, Breckenridge N, Dunlea S, Eynon BP, Furness LM, Ferng J, Fielden MR, Fujimoto SY, Gong L, Hu C, Idury R, et al. Development of a large-scale chemogenomics database to improve drug candidate selection and to understand mechanisms of chemical toxicity and action. *J Biotechnol* 2005;119:219–244.
112. Jeffery IB, Higgins DG, Culhane AC. Comparison and evaluation of methods for generating differentially expressed gene lists from microarray data. *BMC Bioinformatics* 2006;7:359.
113. Page GP, Edwards JW, Gadbury GL, Yelisetti P, Wang J, Trivedi P, Allison DB. The Power-Atlas: a power and sample size atlas for microarray experimental design and research. *BMC Bioinformatics* 2006;7:84.
114. Seo J, Hoffman EP. Probe set algorithms: is there a rational best bet? *BMC Bioinformatics* 2006;7:395.
115. Brazma A, Hingamp P, Quackenbush J, Sherlock G, Spellman P, Stoeckert C, Aach J, Ansorge W, Ball CA, Causton HC, Gaasterland T, Glenisson P, Holstege FC, Kim IF, Markowitz V, Matese JC, Parkinson H, Robinson A, Sarkans U, Schulze-Kremer S, et al. Minimum

information about a microarray experiment (MIAME)-toward standards for microarray data. *Nat Genet* 2001;29:365–371.

116. Chu TM, Deng S, Wolfinger R, Paules RS, Hamadeh HK. Cross-site comparison of gene expression data reveals high similarity. *Environ Health Perspect* 2004;112:449–455.
117. Shi L, Tong W, Fang H, Scherf U, Han J, Puri RK, Frueh FW, Goodsaid FM, Guo L, Su Z, Han T, Fuscoe JC, Xu ZA, Patterson TA, Hong H, Xie Q, Perkins RG, Chen JJ, Casciano DA. Cross-platform comparability of microarray technology: intra-platform consistency and appropriate data analysis procedures are essential. *BMC Bioinformatics* 2005;6 Suppl 2:S12.
118. Waring JF, Ulrich RG, Flint N, Morfitt D, Kalkuhl A, Staedtler F, Lawton M, Beekman JM, Suter L. Interlaboratory evaluation of rat hepatic gene expression changes induced by methapyrilene. *Environ Health Perspect* 2004;112:439–448.
119. Chu TM, Deng S, Wolfinger R, Paules RS, Hamadeh HK. Cross-site comparison of gene expression data reveals high similarity. *Environ Health Perspect* 2004;112:449–455.
120. Luhe A, Suter L, Ruepp S, Singer T, Weiser T, Albertini S. Toxicogenomics in the pharmaceutical industry: hollow promises or real benefit? *Mutat Res* 2005;575:102–115.
121. Herradon G, Ezquerra L, Nguyen T, Wang C, Siso A, Franklin B, Dilorenzo L, Rossenfeld J, Alguacil LF, Silos-Santiago I. Changes in BDNF gene expression correlate with rat strain differences in neuropathic pain. *Neurosci Lett* 2007;420:273–276.
122. Salas E, Alonso E, Sevillano J, Herradon G, Bocos C, Morales L, Ramos MP, Alguacil LF. Morphine differentially regulates hsp90beta expression in the nucleus accumbens of Lewis and Fischer 344 rats. *Brain Res Bull* 2007;73:325–329.
123. Kulkarni SG, Harris AJ, Casciano DA, Mehendale HM. Differential protooncogene expression in Sprague Dawley and Fischer 344 rats during 1,2-dichlorobenzene-induced hepatocellular regeneration. *Toxicology* 1999;139:119–127.
124. Stevens JL. Future of toxicology—mechanisms of toxicity and drug safety: where do we go from here? *Chem Res Toxicol* 2006;19:1393–1401.
125. Mattes WB, Pettit SD, Sansone SA, Bushel PR, Waters MD. Database development in toxicogenomics: issues and efforts. *Environ Health Perspect* 2004;112:495–505.
126. Waters MD, Fostel JM. Toxicogenomics and systems toxicology: aims and prospects. *Nat Rev Genet* 2004;5:936–948.
127. Xirasagar S, Gustafson SF, Huang CC, Pan Q, Fostel J, Boyer P, Merrick BA, Tomer KB, Chan DD, Yost KJ, III, Choi D, Xiao N, Stasiewicz S, Bushel P, Waters MD. Chemical effects in biological systems (CEBS) object model for toxicology data, SysTox-OM: design and application. *Bioinformatics* 2006;22:874–882.
128. Waters MD, Fostel JM. Toxicogenomics and systems toxicology: aims and prospects. *Nat Rev Genet* 2004;5:936–948.
129. Ganter B, Tugendreich S, Pearson CI, Ayanoglu E, Baumhueter S, Bostian KA, Brady L, Browne LJ, Calvin JT, Day GJ, Breckenridge N, Dunlea S, Eynon BP, Furness LM, Ferng J, Fielden MR, Fujimoto SY, Gong L, Hu C, Idury R et al. Development of a large-scale chemogenomics database to improve drug candidate selection and to understand mechanisms of chemical toxicity and action. *J Biotechnol* 2005;119:219–244.
130. Brazma A, Parkinson H, Sarkans U, Shojatalab M, Vilo J, Abeygunawardena N, Holloway E, Kapushesky M, Kemmeren P, Lara GG, Oezcimen A, Rocca-Serra P, Sansone SA. ArrayExpress—a public repository for microarray gene expression data at the EBI. *Nucleic Acids Res* 2003;31:68–71.
131. Waters M, Boorman G, Bushel P, Cunningham M, Irwin R, Merrick A, Olden K, Paules R, Selkirk J, Stasiewicz S, Weis B, Van Houten B, Walker N, Tennant R. Systems toxicology and the Chemical Effects in Biological Systems (CEBS) knowledge base. *EHP Toxicogenomics* 2003;111:15–28.

132. Fostel J, Choi D, Zwickl C, Morrison N, Rashid A, Hasan A, Bao W, Richard A, Tong W, Bushel PR, Brown R, Bruno M, Cunningham ML, Dix D, Eastin W, Frade C, Garcia A, Heinloth A, Irwin R, Madenspacher J et al. Chemical effects in biological systems—data dictionary (CEBS-DD): a compendium of terms for the capture and integration of biological study design description, conventional phenotypes, and 'omics data. *Toxicol Sci* 2005;88:585–601.

133. Tong W, Cao X, Harris S, Sun H, Fang H, Fuscoe J, Harris A, Hong H, Xie Q, Perkins R, Shi L, Casciano D. ArrayTrack—supporting toxicogenomic research at the U.S. Food and Drug Administration National Center for Toxicological Research. *Environ Health Perspect* 2003;111:1819–1826.

134. Tong W, Harris S, Cao X, Fang H, Shi L, Sun H, Fuscoe J, Harris A, Hong H, Xie Q, Perkins R, Casciano D. Development of public toxicogenomics software for microarray data management and analysis. *Mutat Res* 2004;549:241–253.

135. Barrett T, Edgar R. Gene expression omnibus: microarray data storage, submission, retrieval, and analysis. *Methods Enzymol* 2006;411:352–369.

136. Edgar R, Barrett T. NCBI GEO standards and services for microarray data. *Nat Biotechnol* 2006;24:1471–1472.

137. Parkinson H, Kapushesky M, Shojatalab M, Abeygunawardena N, Coulson R, Farne A, Holloway E, Kolesnykov N, Lilja P, Lukk M, Mani R, Rayner T, Sharma A, William E, Sarkans U, Brazma A. ArrayExpress—a public database of microarray experiments and gene expression profiles. *Nucleic Acids Res* 2007;35:D747–D750.

138. Burgoon LD, Boutros PC, Dere E, Zacharewski TR. dbZach: A MIAME-compliant toxicogenomic supportive relational database. *Toxicol Sci* 2006;90:558–568.

139. Burgoon LD, Zacharewski TR. dbZach toxicogenomic information management system. *Pharmacogenomics* 2007;8:287–291.

Chapter 5

Toxicogenomics: Applications to In Vivo Toxicology

5.1. THE VALUE OF TOXICOGENOMICS IN DRUG DISCOVERY AND DEVELOPMENT

In Chapter 4, we provide a detailed review of the principles and practical aspects of toxicogenomics. The current chapter focuses on the applications of this technology to in vivo toxicology studies, while Chapter 6 covers the use of toxicogenomics in in vitro systems. The emphasis is on studies using rats, since this species is the most widely used for toxicogenomics research, but we will make references to and use examples from other species when appropriate. Effectively integrating new technologies in a research and development (R&D) organization and measuring their impact on the productivity and success rate represent a significant challenge, which stems from the complexity and long cycle times of drug discovery and development. To address this challenge, we will use as many published or personal examples as possible that can best illustrate when and where toxicogenomics provides added value to drug discovery and development.

Before discussing specific applications of toxicogenomics, an attempt will be made to review the objectives and contents of a typical toxicology program during the development of small-molecule therapeutic agents. Since some readers may not be familiar with toxicology, we start this chapter with an overview of the practice of toxicology in the pharmaceutical industry. We should point out that this chapter mostly covers the use of toxicogenomics in programs focused on developing small molecules rather than biologics, since the former are usually associated with more toxicology issues. However, it is acknowledged that toxicology can also be encountered with biologic agents, and, when appropriate,

Genomics in Drug Discovery and Development, by Dimitri Semizarov and Eric Blomme

examples of the use of toxicogenomics in the field of biologic agents will be provided. Furthermore, toxicogenomics is also relevant to industries other than the pharmaceutical industry, and gene expression profiling has been used successfully to address toxicity triggered by agents other than pharmaceuticals. Therefore, although this book is focused on the use of genomics technologies in drug discovery and development, examples from the field of environmental, pesticide, chemical or cosmetic toxicology are used as illustrations, when appropriate.

The value that toxicogenomics can generate in drug discovery and development is best illustrated when put in the appropriate context. The discovery and development of new chemical entities (NCEs) is a long, difficult, complex and risky endeavor that results in many more failures than successes. Despite growing investments in R&D, the approval rate of novelty medicines has been steadily decreasing (1, 2). The best approach to address this productivity problem is to understand the major causes of failure observed in drug discovery and development. A better understanding of the problems can identify areas that need improvements, but also processes or technologies that are unnecessary and should consequently be eliminated because of their high cost and poor predictive value.

The most striking aspect of drug discovery and development is the high failure rate of experimental molecules at the late, costly stages of development. This has promoted a "fail early, fail often" approach advocating early, rapid, and multidirectional characterization of compounds, so that compounds unlikely to succeed can be terminated as early as possible. For that purpose, robust technologies have been developed to characterize important pharmacological, physicochemical, and biological properties of compounds. This is best illustrated with the improvements in our ability to predict the pharmacokinetic properties of compounds. In the late 1980s, poor pharmacokinetic properties were the main reason for the termination of compounds in the clinic (around 40%), while today they account for only around 10% of clinical failures (2, 3). This significant improvement in the attrition rate due to pharmacokinetic properties can partly be explained by the use of novel preclinical tools to predict with better accuracy the pharmacokinetic properties of experimental compounds (4). In contrast, toxicity still represents the most significant cause of attrition at all stages of drug development (5). In part, this reflects the complexity and diversity of toxic reactions, but also a lack of innovation occurring in the field of toxicology. Indeed, there have been limited advances in developing more sensitive and precise methods to detect and characterize toxic changes. Consequently, toxicology is applied mostly at advanced stages of drug discovery and still relies almost exclusively on animal studies identical to those used decades ago. These traditional animal studies are lengthy, costly, and not always predictive of the human situation, and they require significant amounts of compound. Moreover, the current battery of biomarkers of toxicity, albeit well understood and validated, has in general low sensitivity. Toxicogenomics represents a novel technology that has the potential to improve this situation by identifying likely failures related to toxicology at earlier stages (6, 7).

5.2. BASIC PRINCIPLES OF TOXICOLOGY IN DRUG DISCOVERY AND DEVELOPMENT

Toxicogenomics, defined in this book as the use of gene expression profiling in toxicology, should not be viewed as a specialized discipline that can be used in a vacuum. Initial claims suggested that toxicogenomics would change the way toxicological evaluations are performed and that it might even eliminate the need to conduct animal studies (8). In our opinion, toxicogenomics will definitely influence the discipline of toxicology, but it should be considered as an additional, complementary tool for the toxicologist. Consequently, an understanding of the general contents of a toxicology program for the development of small-molecule therapeutic agents is pertinent to fully gauging the value that genomics technologies can bring to preclinical toxicology and risk assessment. As indicated above, this chapter is mostly focused on the development of small molecules; however, whenever appropriate, comments are also provided regarding the use of toxicogenomics for the preclinical safety assessment of biological therapeutic agents. The current section provides a brief overview of the preclinical or non-clinical toxicological evaluation of experimental compounds. This overview is aimed at non-toxicologists and should not be viewed as a comprehensive guide to conducting non-clinical studies of drug candidates. Additional information can be found in the references provided or in regulatory guidance documents, as well as the excellent book chapter by Larson (9).

5.2.1. Preclinical Safety Assessment

Once a development candidate is selected, its preclinical safety assessment needs to be conducted. The main objectives of a non-clinical toxicological evaluation are to ensure that clinical trials are conducted with safe, sufficiently understood compounds and that only drugs shown to be safe are approved for commercial use. For these reasons, a minimal amount of toxicological characterization is required before compounds can be tested in clinical trials and for approval of novel drugs. In the pharmaceutical industry, toxicology testing of compounds follows a fairly well-defined process, whereby toxicity is evaluated with a somewhat harmonized battery of tests. These tests are based on years of experience in evaluating a wide range of therapeutic agents and are selected to maximize the likelihood of detecting toxic effects that are frequent and serious enough to terminate a compound or at least warrant further testing in clinical trials. It is well recognized that these studies do not detect all potential adverse events that may occur in the clinic. In particular, there are well-publicized cases of toxic reactions occurring in humans that have not been predicted in animal studies (10). Nevertheless, it is important to reemphasize that preclinical toxicology studies are relatively effective at detecting molecules toxic to humans.

Only limited published studies have thoroughly evaluated without bias the utility of preclinical studies in predicting toxic effects in humans (11). The largest published retrospective evaluation analyzed data on 150 drugs that caused adverse events or toxicity in humans (12). According to this industry report, 94% of human toxicities are detected preclinically (12). Moreover, this number is likely to underrepresent the predictive value of animal studies, because a vast majority of exploratory compounds that show toxicity in animal studies are rapidly eliminated and are never pursued in the clinic. For these cases, there is a complete lack of detailed published data (11).

Preclinical toxicology studies are typically conducted according to the Good Laboratory Practice (GLP) principles, which basically refer to specific rules set by the regulatory authorities to ensure the consistency and reliability of results. FDA rules for GLP can be found in the Code of Federal Regulations (21CFR58). The details of regulatory expectations for toxicology studies can be found in the guidance documents provided by the FDA (http://www.fda.gov/cder/guidance/) or the guidelines from the International Conference on Harmonization (ICH; http://www.ich.org/). The ICH is a joint regulatory-industry undertaking for international harmonization of the regulatory guidance documents across Europe, Japan, and the United States. The ICH was established in April 1990 and represents an ongoing project aimed at reaching scientific consensus on various regulatory issues. The objective is to normalize the various activities and their design during the development of compounds that are registered in multiple regions in order to reduce the redundancy and cost of approval in different regions. The ICH covers several topics, including safety.

For the sake of simplicity, the preclinical safety assessment process can be divided into five major components that are briefly described below.

5.2.1.1. Genetic Toxicology

The objective of genetic toxicology is to identify the potential genetic toxicity (also called genotoxicity) of compounds, since genotoxic compounds, or compounds that directly or indirectly damage DNA the have the potential to be human carcinogens. The tests required to determine the genotoxicity of a compound are delineated in the Guidance on Specific Aspects of Regulatory Genotoxicity Tests for Pharmaceuticals (ICH Topic S2A; http://www.ich.org/LOB/media/MEDIA493.pdf) and Genotoxicity: A Standard Battery for Genotoxicity Testing of Pharmaceuticals (ICH Topic S2B; http://www.ich.org/LOB/media/MEDIA494.pdf), which discuss methods for conducting the assays and the required battery of tests, respectively. These tests rely on detection of DNA damage (or its immediate outcome such as chromosome breakage or point mutations. Appropriate in vitro tests include the Ames test (bacterial mutation test) in *Salmonella sp*. and *Escherichia coli*, the mouse lymphoma thymidine kinase gene mutation assay (mammalian cell mutation test), or a chromosome aberration test performed in mammalian cell lines, such as CHO or V79 cells or human peripheral blood lymphocytes (13). These in vitro tests are highly sensitive to DNA-damaging effects and detect nearly all agents likely to be genetically hazardous. However, while the bacterial mutation assay has relatively high specificity, the in vitro mammalian cell assays have low specificity such that many compounds that are positive in vitro for genotoxic effects are weakly active, or even inactive in animals (13, 14).

In vivo tests include the bone marrow or peripheral blood micronucleus test in rodents, the rat liver unscheduled DNA repair induction (UDS) assay, and the analysis of chromosomal aberrations in rodent bone marrow cells. Only one of these tests is required by regulatory agencies; typically, the bone marrow micronucleus assay is the assay of choice. These in vivo assays are especially important for compounds for which high in vivo exposure is expected, for compounds with positive in vitro results, or for compounds with structure similar to that of known mutagens or clastogens.

ICH S2A discusses the interchangeability of these different test systems and the interpretation of test results, while ICH S2B specifies that before a first-in-human (FIH) trial, a bacterial gene mutation assay and an assay for chromosomal damage in mammalian cells in vitro are required. In practice, most sponsors also perform the assay for chromosomal aberrations in rodent bone marrow before FIH trials.

5.2.1.2. *Single-Dose Toxicity*

The objective of single-dose or acute toxicity studies is to determine the maximum tolerated dose (MTD) and/or approximate lethal dose (ALD) of a single dose of a test compound. Therefore, these studies must incorporate doses that will cause enough toxic changes to be considered adverse and above what could be considered tolerable for the animal. For regulatory purposes, this evaluation is usually done in two rodent species (typically rat and mouse). However, single-dose range-finder studies are also frequently done in a non-rodent species (most commonly in dogs or sometimes in a non-human primate), and these data are typically sufficient for regulatory purposes. ICH Topic M3 (http://www.ich.org/LOB/media/MEDIA506.pdf) indicates that a single-dose toxicity evaluation must be conducted in two mammalian species but refers to local guidances for more details regarding the study design and the end points to evaluate.

5.2.1.3. *Repeat-Dose Toxicity*

The evaluation of compounds in repeat-dose toxicity studies represents the bulk of a non-clinical toxicology program. It consists of a series of studies of progressively increasing duration (from 2 weeks up to 9 months) conducted in both sexes and in two species, a rodent (typically the rat) and a non-rodent (most generally, the dog or sometimes a non-human primate). Various guidelines are useful to understand the duration needed (ICH Topic M3 and ICH Topic S4; http://www.ich.org/LOB/media/MEDIA497.pdf), timing for completion (ICH Topic M3), contents, and end points to evaluate (Redbook, a set of guidelines by the Center for Food Safety and Applied Nutrition for safety assessment of food ingredients; http://www.cfsan.fda.gov/~redbook/red-toca.html). Studies conducted to support regulatory filings [Investigational New Drug Applications

(IND), New Drug Applications (NDA), Biologic License Applications (BLA)] must be conducted under GLP regulations. The objectives of these evaluations are to:

- Identify the target organs of toxicity
- Define the dose response, onset, and duration of the toxicities identified
- Determine the threshold doses and systemic exposures at which the toxicities occur, thereby allowing estimations of safety margins and the maximum recommended safe starting dose (MRSD) in the clinic
- Evaluate the reversibility of the toxicities
- Determine potential surrogate markers to monitor these toxicities in humans during clinical trials

To accomplish these goals, repeat-dose toxicity studies need to be designed so that toxicity can be correctly characterized in terms of dose, reversibility, and incidence. Typically, these studies include at least three dose groups plus a concurrent vehicle control group. A group of animals is exposed to doses not associated with any adverse effects (identified as the No Adverse Effect Level, or NOAEL), while another group is exposed to levels sufficient to induce toxic changes in target organs but not associated with toxic changes severe enough to compromise interpretation of results (15). A dose between the NOAEL and the toxic dose is also useful to characterize dose response. The NOAEL is used to define the MRSD, the dose initially used for dose escalation studies in the clinic. Therefore, achieving a NOAEL with sufficient safety margins is critical and requires a careful design of the studies, so that sufficiently high doses can be used in the clinic to test the clinical hypothesis (16).

The NOAEL is a professional opinion based on the design of the study, therapeutic indication(s) of the drug, expected pharmacology, and spectrum of off-target effects (i.e., type and severity of the toxicities identified) (15). As pointed out previously, determining a NOAEL can be quite complicated, and gene expression profiling data have the potential to enable a more reliable establishment of the NOAEL, especially in early short-term studies, where changes may be subtle and where harmful properties may be difficult to ascertain (15).

5.2.1.4. Reproductive Toxicity

Evaluation of the potential harmful effects of experimental compounds on the reproductive function requires rather complex preclinical studies. Reproductive toxicity refers to the adverse effects of molecules on any aspect of the reproductive cycle, including impairment of reproductive function and toxic changes (malformations, death) in the embryo (17). Special studies must be conducted to evaluate the effects of experimental agents on the highly complex reproductive cycle (17). A considerable amount of guidance is available in ICH Topic S5 (http://www.ich.org/LOB/media/MEDIA498.pdf). This guidance provides an overview of the combination of studies for most pharmaceutical compounds. The

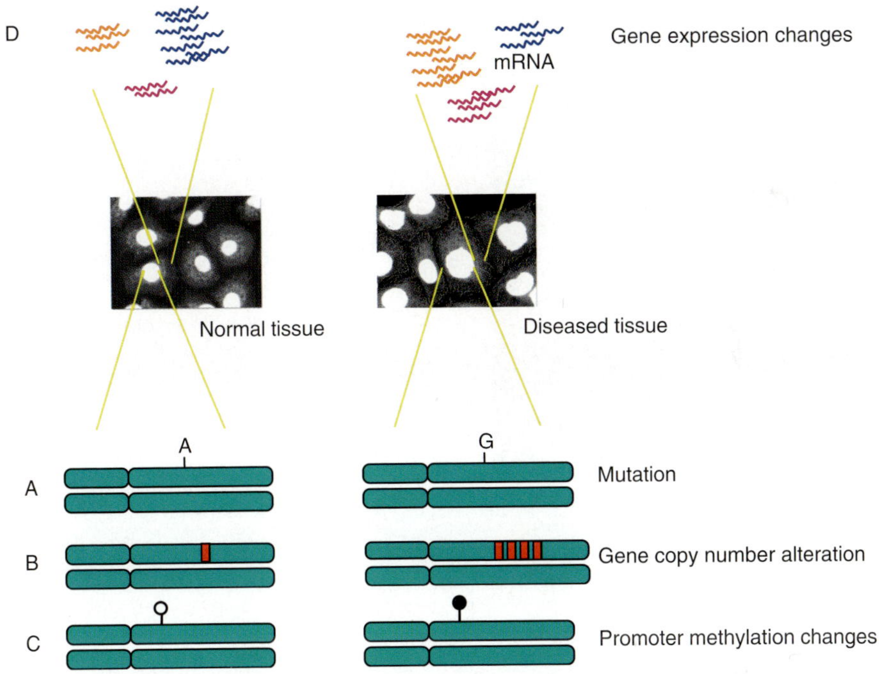

Figure 1.1 Genomic alterations found in diseased tissue. Common alterations at the DNA level include single-point mutations (A), gene copy number alterations (B), and epigenetic changes, such as abnormal promoter methylation (C). Single-point mutations represent insertions, substitutions, or deletions of individual base pairs in DNA. Copy number changes (gains or losses) may affect individual genes but may also involve large regions, such as entire chromosomal arms or whole chromosomes. One or both copies of a locus may be lost, resulting in a heterozygous or homozygous deletion, respectively. Copy number gains may vary in amplitude from one extra copy to dozens of additional copies. The amplified DNA sequences may either be incorporated into the mother chromosome or organized as extrachromosomal material. DNA methylation normally occurs at cytosine residues that are followed by a guanine (CpG islands). Methylation of CpG islands in the promoter regions of genes causes gene silencing. All these alterations at the DNA level may result to altered gene expression (D), thus affecting the phenotype of the cell.

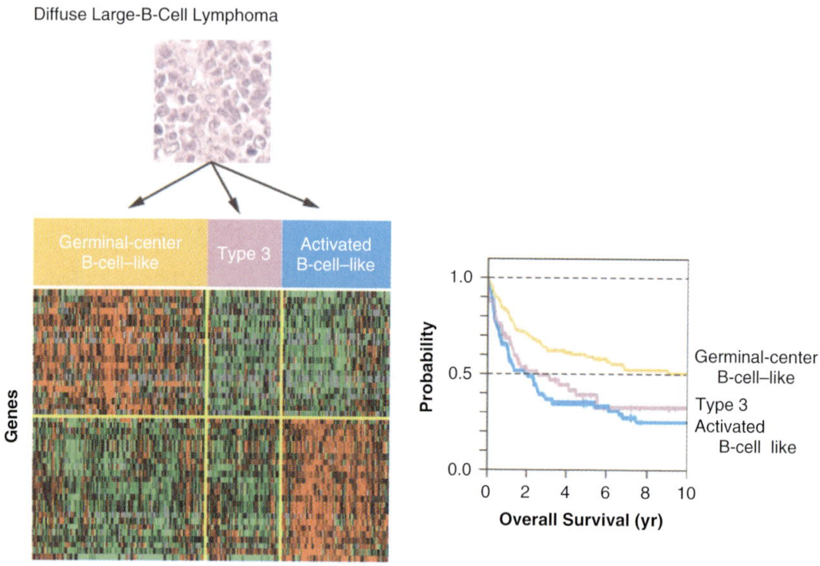

Figure 1.2 Utility of gene expression profiling in the identification of clinically relevant disease subtypes. See text for full caption.

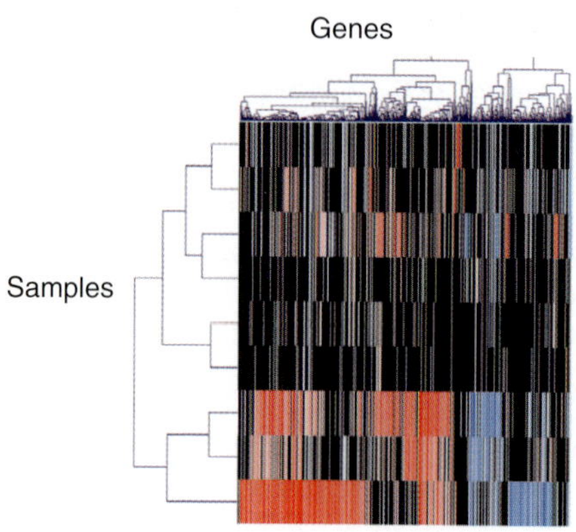

Figure 2.5 An example of a heat map obtained by hierarchical two-dimensional clustering of nine samples. See text for full caption.

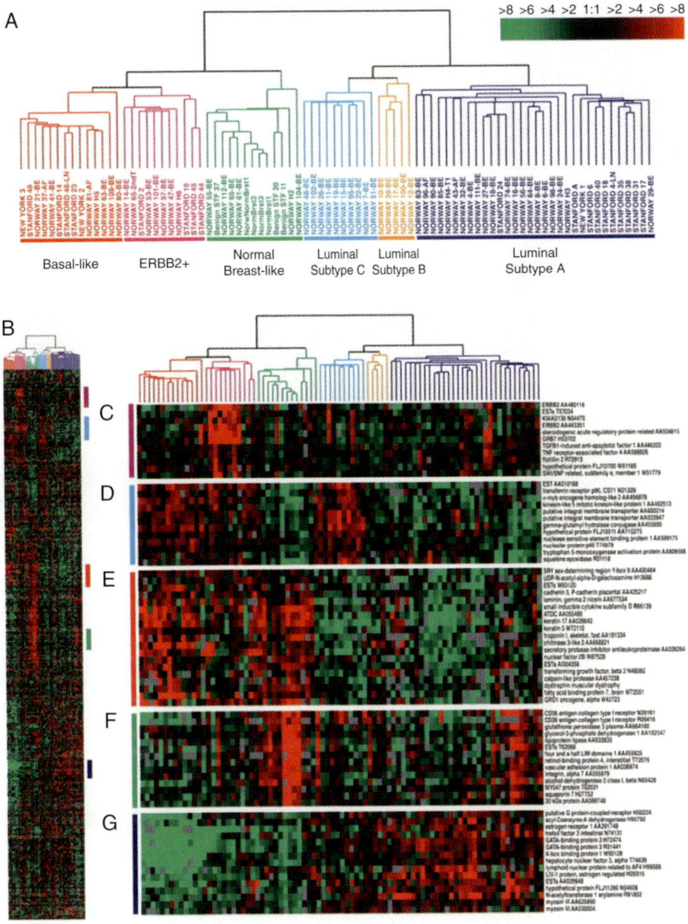

Figure 2.6 Gene expression patterns of 85 samples representing 78 breast carcinomas, three benign tumors, and four normal tissues, analyzed by hierarchical clustering. *A*) The tumor specimens were divided into five (or six) subtypes based on differences in gene expression. The cluster dendrogram showing the six subtypes of tumors are colored as luminal subtype A, dark blue; luminal subtype B, yellow; luminal subtype C, light blue; normal breast-like, green; basal-like, red; and ERBB2+, pink. *B*) The full cluster diagram scaled down. The colored bars on the right represent the inserts presented in *C–G*. *C*) *ERBB2* amplicon cluster. *D*) Novel unknown cluster. *E*) Basal epithelial cell-enriched cluster. *F*) Normal breast-like cluster. *G*) Luminal epithelial gene cluster containing *ER*. Figure is reproduced with permission from Sorlie et al. (2001), *Proc Natl Acad Sci USA* 98: 10869–10874 (ref. 40).

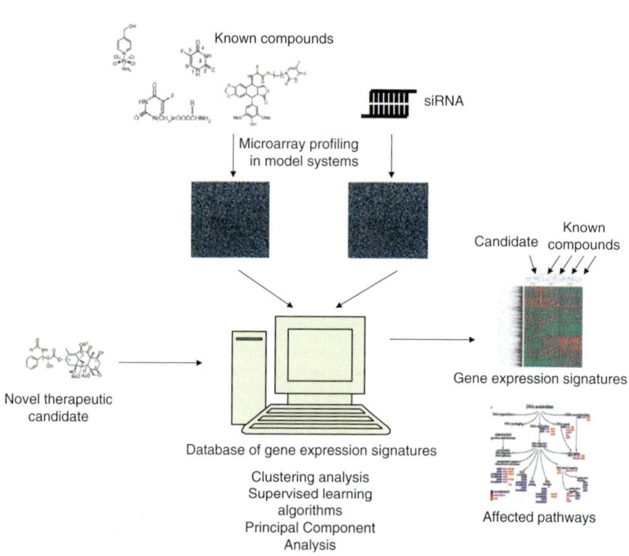

Figure 2.7 Creation of a genomics database for compound selection and optimization. Gene expression signatures of target inhibition and elimination are obtained with known inhibitors and siRNA, respectively, for multiple targets. Novel compounds synthesized to inhibit the target are profiled with the same microarray and their gene expression signatures are utilized to identify the affected pathways and targets.

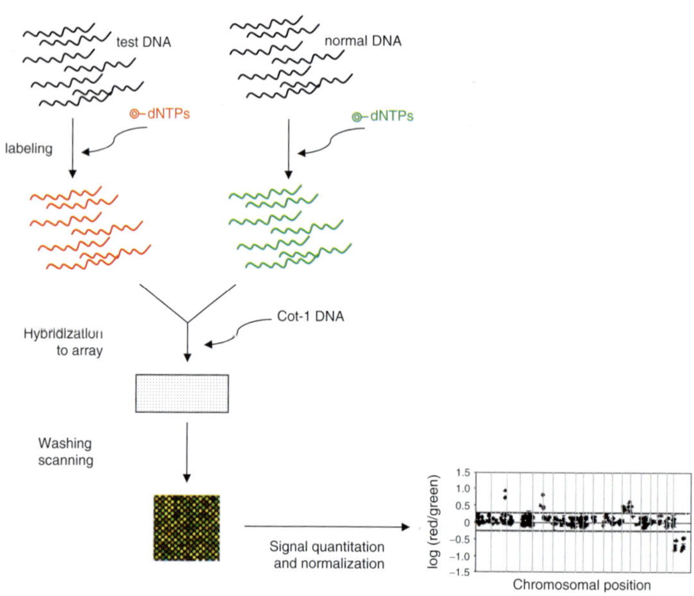

Figure 2.9 Two-color procedure for comparative genomic hybridization (CGH). The test gDNA and the reference normal gDNA are labeled with two different fluorophores. The gDNA samples are mixed and hybridized to a CGH array. Cot-1 DNA is added to eliminate the signal from repetitive sequences. After hybridization, the array is washed and scanned to generate signal intensities for all regions of interest.

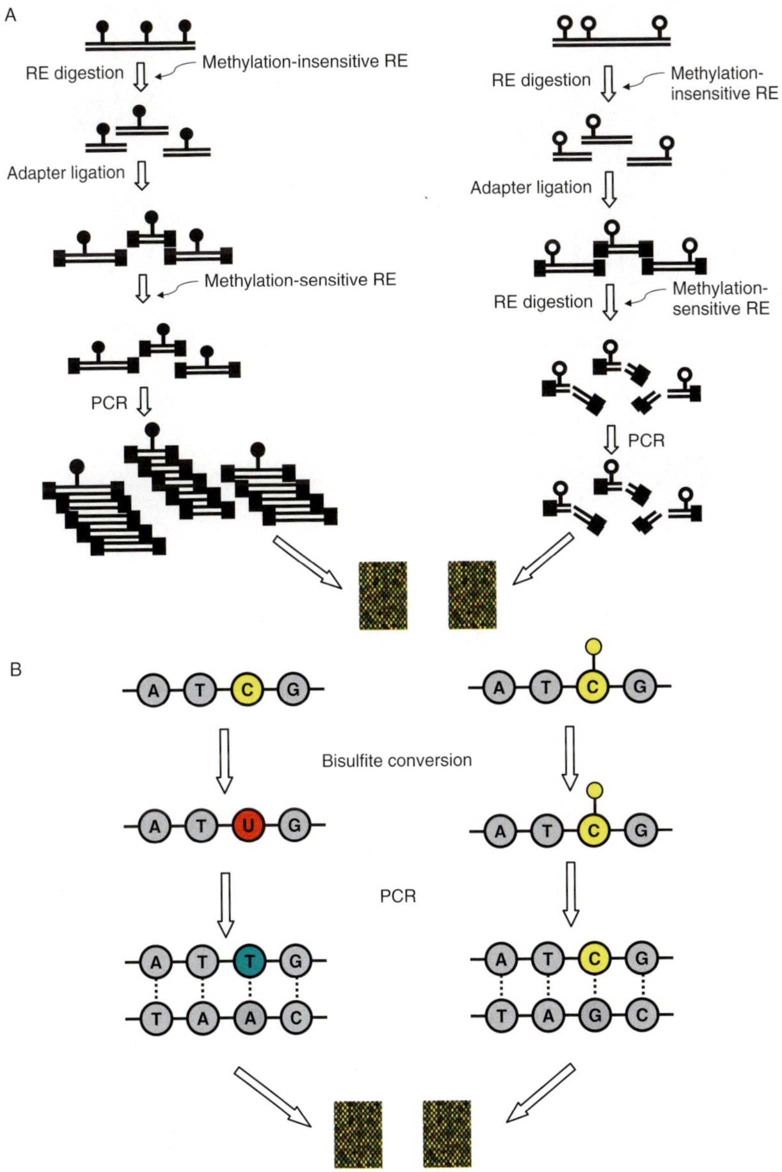

Figure 2.13 A) Microarray methodology for methylation profiling based on the use of methylation-sensitive restriction enzymes. See text for full caption.

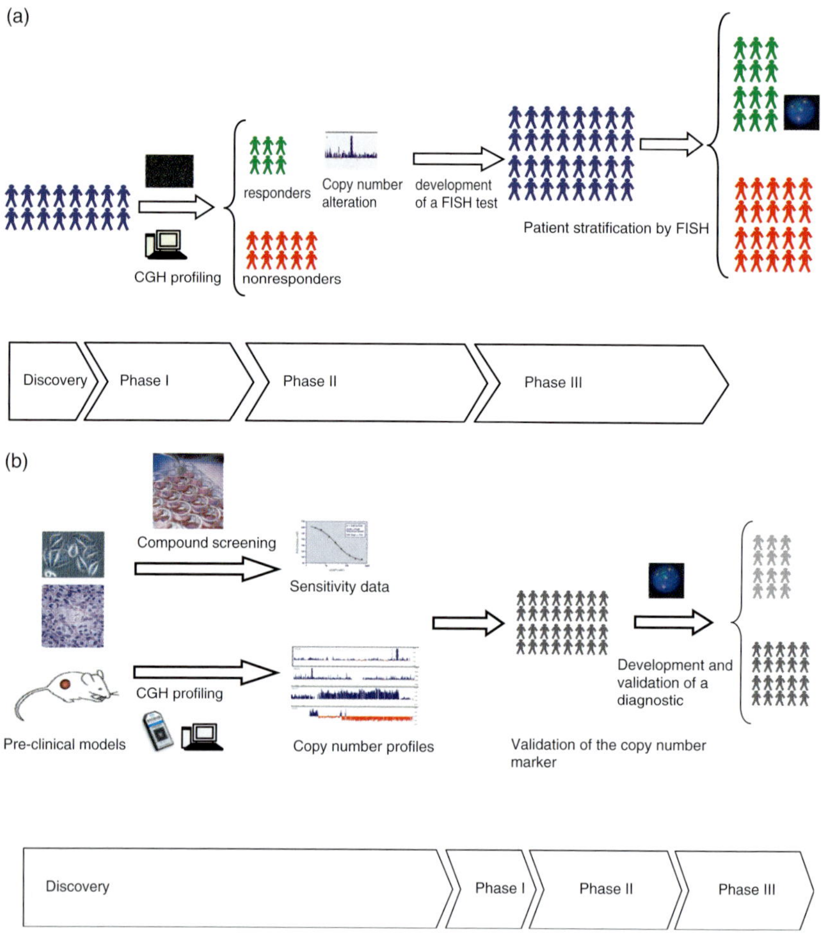

Figure 3.1 Applications of gene copy number profiling in the discovery of patient stratification biomarkers. A) Approach based on CGH profiling of patients enrolled in clinical trials. Tumor samples are profiled by CGH to identify gene copy number abnormalities associated with drug response. FISH probes for the marker regions are then designed and validated in a larger patient population (phase III trials or additional studies designed to validate the diagnostic). B) Proactive approach based on early identification of drug sensitivity markers in preclinical model systems. Cell lines and xenografts used to screen the drug candidate are profiled to determine gene copy number abnormalities associated with drug sensitivity. FISH probes are then developed for the sensitivity marker regions and applied to analyze the marker in the patients enrolled in phase II and phase III clinical trials.

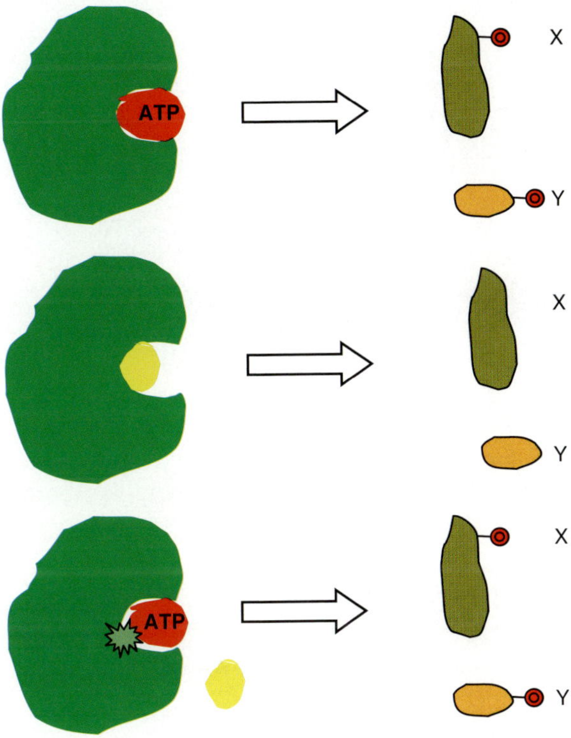

Figure 3.2 Possible effects of mutations in the drug target on the drug efficacy. In this example, the drug target is a protein kinase, and the drug is a small-molecule protein kinase inhibitor, which phosphorylates substrates X and Y. The wild-type enzyme is efficiently inhibited by the drug, resulting in abrogation of substrate phosphorylation. The enzyme that carries a mutation does not bind the drug as efficiently, as the phosphorylation continues to occur in the presence of the drug.

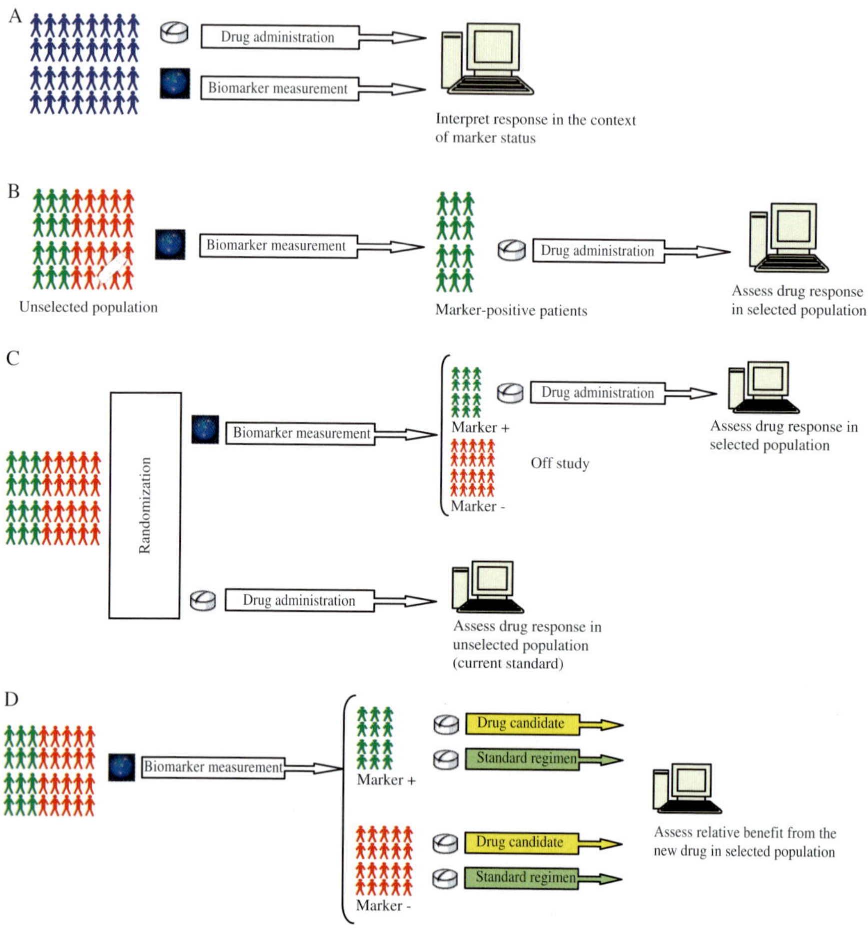

Figure 3.4 Designs of clinical trials aimed at assessing clinical utility of a predictive genomic biomarker. A) All patients are treated with the drug regardless of the marker status, and the drug response is analyzed together with the biomarker measurement results. B) The biomarker is detected upfront, and the treatment is only administered to patients carrying the marker. C) A randomized trial aimed at assessing the ability of the biomarker to improve the treatment outcome relative to the use of the drug in an unselected population. D) A trial design that compares the clinical benefit of a novel drug A and the current standard therapy B in a population selected with a biomarker. Modified from L. Pusztai and K. Hess (2004). Ann Oncol 15: 1731.

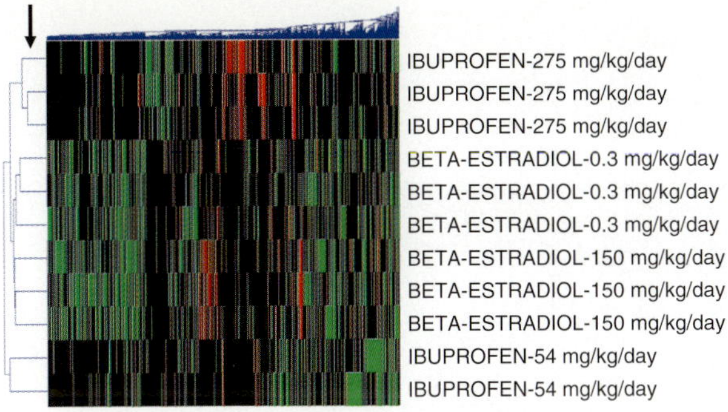

Figure 4.4 Heat map with hierarchical clustering illustrating the transcriptomic changes occurring in the liver of male rats treated with ibuprofen at 54 or 275 mg/kg/day or β-estradiol at 0.3 or 150 mg/kg/day. Since these profiles are similar to those used in Figure 4.3, the reader can compare the hierarchical clustering approach with the principle component analysis (PCA) method for visualization of compound-induced gene expression profiles. Genes shown in the horizontal axis include genes that were up- or down-regulated by at least twofold with a P value < 0.01 (as determined with Rosetta Resolver software). Green and red shades indicate genes that are down-regulated and up-regulated, respectively in treated animals compared to their respective vehicle controls. Note the overall limited interindividual variability in gene expression profiles, indicating good biological reproducibility. By focusing on blocks of genes with consistent patterns of up- or down-regulation, the user can further identify the genes driving most of the response. The dendrogram on the left side (arrow) permits a quantitative visualization of the closeness or dissimilarity between experimental groups and between individuals from the same experimental group.

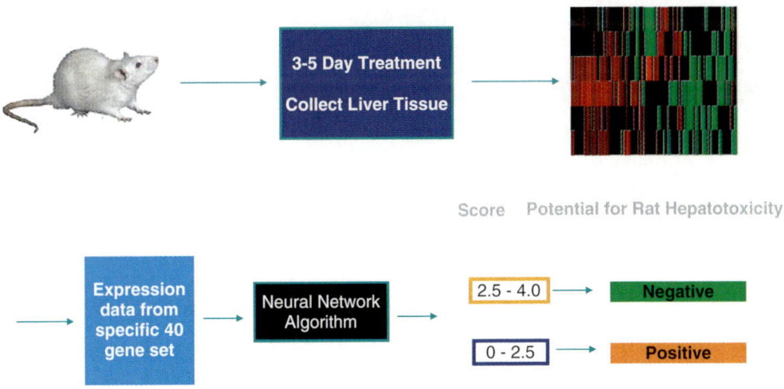

Figure 5.1 Predictive genomics assay for rat hepatotoxicity. This assay was developed with an internal rat liver gene expression database and an artificial neural network algorithm. Using microarray-generated gene expression profiles from male Sprague Dawley rats treated for 3–5 days with a variety of paradigm compounds, the neural network algorithm classifies the compounds based on their potential to cause hepatotoxicity in rats on exposure of longer duration (2 weeks of daily dosing or longer). This assay is based on a preselected 40-gene set, and its output is a score ranging from 0 to 4. A low score indicates a high probability that the test article will induce hepatotoxicity in rats in repeat-dose studies of longer duration at similar exposure levels. A cutoff point of 2.5 was selected based on a small validation set to distinguish negative (i.e., nonhepatotoxic) from positive (i.e., hepatotoxic) compounds. A longer forward validation using 52 compounds (9 hepatotoxic compounds and 43 nonhepatotoxic compounds) from our own chemical space demonstrated that this predictive assay had a 96% accuracy with a 98% specificity and a 89% sensitivity.

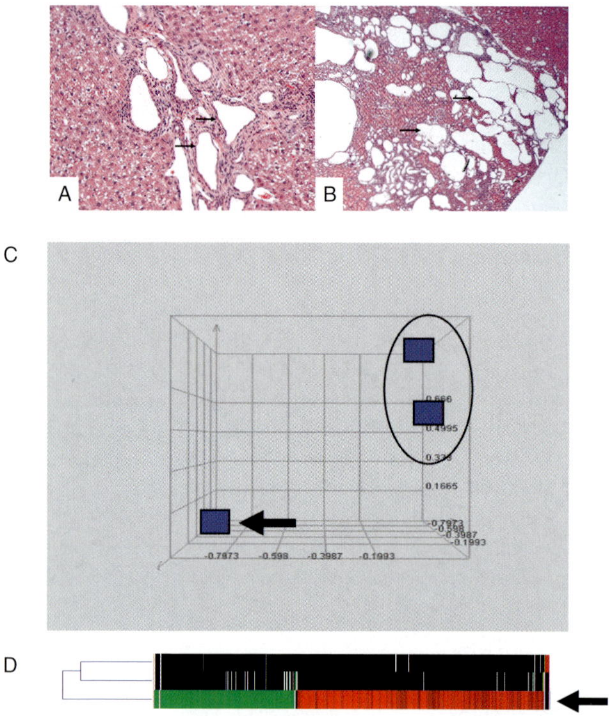

Figure 5.2 Example of a spontaneous change in the liver (A; 200 × magnification) and kidney (B; 20 × magnification) of a male Sprague-Dawley rat and its impact on tissue gene expression profiling. In this specific example, the rat suffered from a congenital, genetic condition called polycystic kidney disease, leading to the presence of cystic bile ducts in the liver (A; arrows) and tubules in the kidneys (B; arrows). These cystic ducts and tubules are associated with a variety of degenerative changes, such as interstitial fibrosis or chronic inflammation. Not surprisingly, a transcriptomic analysis of these tissues indicated marked changes compared to controls, which could suggest toxic changes. Illustrated here is a principal component analysis (C) and an agglomerative cluster analysis (D) of the gene expression changes observed in the kidney of this specific rat (thick arrows) and two other rats from the same treatment group. As expected, the rat with congenital polycystic kidney disease is very different at the transcriptomic level from the other two rats from the same treatment group, and many differentially expressed genes can be identified in this rat. Without concurrent histological evaluation, these transcriptomic changes would likely have been incorrectly interpreted as indicative of toxicity.

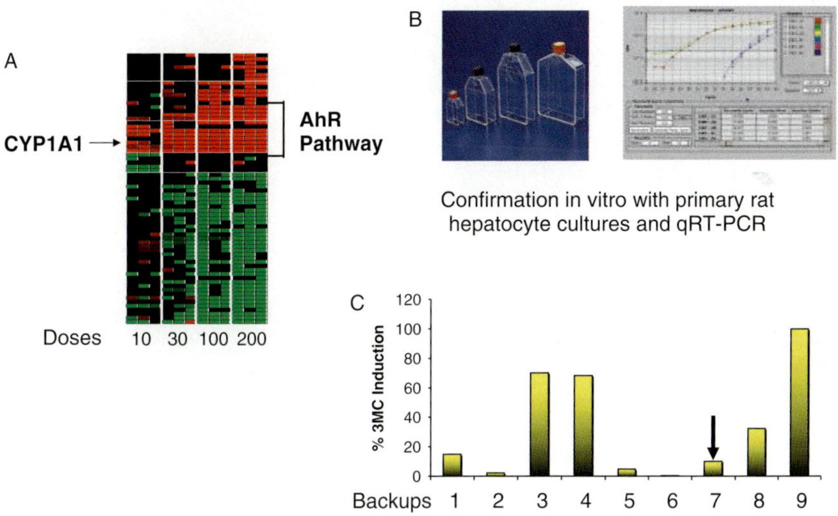

Figure 5.5 Use of toxicogenomics to elucidate metabolism issues. In this example, three male rats per group were treated for 5 days with an experimental compound at 10, 30, 100, and 200 mg/kg/day. Treatment with the compound was associated with dose-dependent increases in liver weight and histological evidence of mild centrolobular hepatocellular hypertrophy at 100 and 200 mg/kg/day. On transcriptomic analysis of the liver summarized on the heat map (A), this compound was shown to significantly induce CYP1A1 mRNA levels (arrow), as well as the aryl hydrocarbon receptor (AhR) pathway. This CYP1A1 induction was associated with a significant decrease in exposure over repeat dosing since the compound was also a substrate of CYP1A1 (autoinduction phenomenon). This mRNA induction of CYP1A1 was also detected in vitro with primary rat hepatocytes (B) and was shown to be relevant to humans with primary human hepatocytes. Backup compounds from the same series were then screened in vitro to rapidly identify a suitable backup not associated with this potential liability. In the bar graph shown in C, results from 9 compounds are shown (results are expressed as % of induction compared to 3-methyl-cholanthrene (3-MC), the positive control compound in this study. Backup 7 (arrow) was selected based on additional consideration and was shown to not be associated with autoinduction in a follow-up 5-day rat toxicity study.

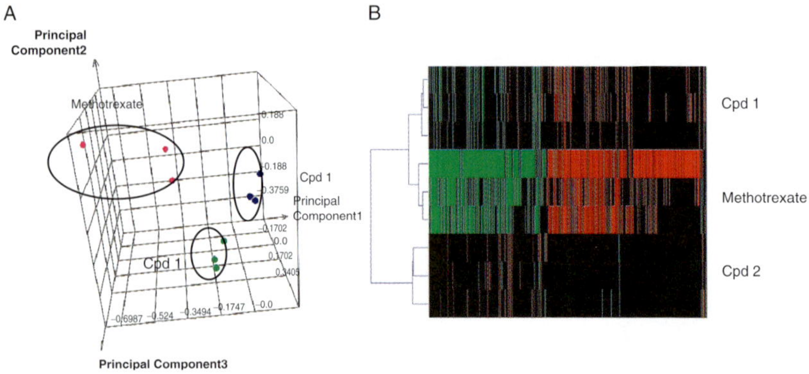

Figure 5.6 Gene expression changes in jejunal epithelial scrapings following treatment with experimental compounds. In this study, male rats were treated daily and orally for 5 days with methotrexate (a compound known to induce small intestinal injury), compound 1 (an internal experimental compound associated with limited epithelial changes in the small intestine), or compound 2 (an internal compound with no known intestinal effect). Jejunal epithelial scrapings, rather than samples of the whole jejunum, were collected and used for transcriptomic analysis, to limit the evaluation to the jejunal epithelium. Results are visualized here with principal component analysis (PCA; A) or a heat map with hierarchical clustering (B). Note that animals tightly cluster per treatment group, and that consistent changes in gene expression can be detected. In the heatmap, green indicates down-regulation, while red indicates up-regulation. In this particular example, using these differentially expressed genes, we could demonstrate that both compound 1 and methotrexate induced pathways consistent with epithelial proliferation and repair, a effect considered secondary to a primary toxic insult.

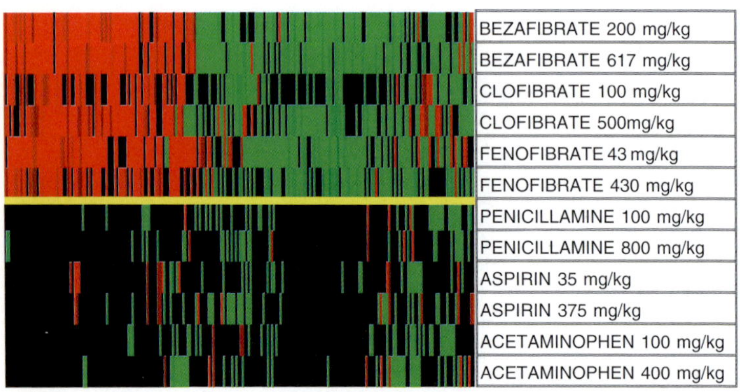

Figure 5.7 Heat map from an agglomerative hierarchical clustering analysis illustrating gene expression changes in the liver of rats following treatment with prototypical peroxisome proliferators. Male rats were treated orally and daily with various fibrates (bezafibrate, clofibrate, fenofibrate) at 2 doses for 5 days (the low dose representing a multiple of the therapeutic dose and the high dose representing a maximum tolerated dose). Fibrates are prototypical peroxisome proliferators. In addition, rats were treated with several non-peroxisome proliferators (penicillamine, aspirin, acetaminophen) using a similar dose selection. Shown here are the genes that were regulated at a P value < 0.01 and with at least a two-fold change. Green indicates down-regulation, while red indicates up-regulation. Each treatment group corresponds to three animals that were pooled in silico. Note that the three peroxisome proliferators altered the expression levels of large numbers of genes. These differentially expressed genes can be used to easily identify test agents that act as peroxisome proliferators.

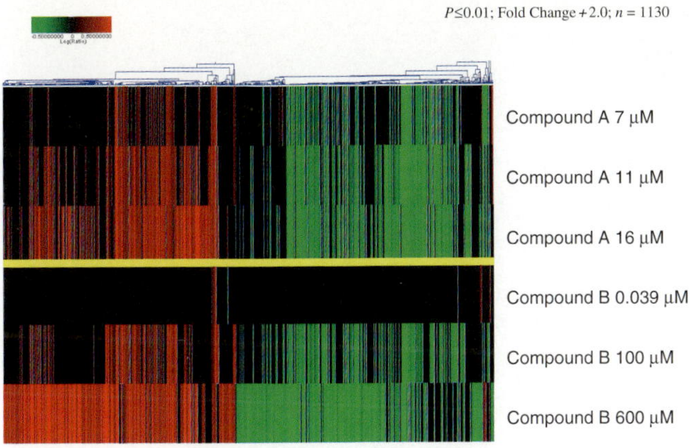

Figure 6.2 Heat map illustrating the transcriptomic effect on primary rat hepatocytes by treatment with two experimental compounds (compounds A and B). In this experiment, primary rat hepatocytes were treated with increasing doses of two experimental compounds. For compound A, 11 μM represents the concentration causing approximately 20% cell death after a 24-hour exposure, while the cytotoxic concentration for compound B was estimated to be around 200 μM. The lowest concentrations used for both compounds are 10-fold multiples of efficacious in vitro concentrations. Note that at concentrations lower than cytotoxic doses, limited gene expression changes are detected. In our experience, robust and reproducible gene expression profiles can only be obtained at concentrations high enough to cause some detectable cytotoxic changes in cells. In our laboratory, compounds are characterized in primary rat hepatocytes at concentrations sufficient to cause death of 20% of cells after a 24-hour exposure. Genes shown ($n = 1130$) are genes that were up- or downregulated by at least twofold with a P value < 0.01. Green indicates down-regulation, while red indicates up-regulation.

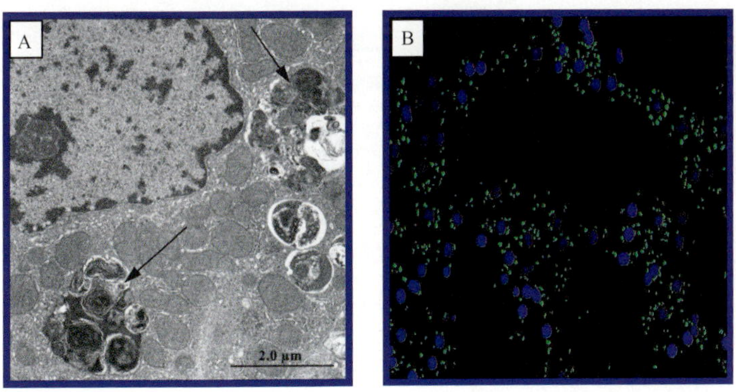

Figure 6.3 Drug-induced phospholipidosis. A. Electron microscopic evaluation of the liver of a rat treated for 2 weeks with an experimental compound inducing hepatic phospholipidosis. Phospholipidosis is the excessive cytoplasmic accumulation of phospholipids, a normal cellular component. Histologically, it is characterized by various levels of cytoplasmic vacuolation in a wide range of tissues. This vacuolation is due to membranous lamellar inclusions called lamellar bodies, detectable by electron microscopy evaluation (arrow). B. Detection of phospholipidosis with fluorescent microscopy. Screening in vitro approaches for phospholipidosis use fluorescent dyes or fluorescence-labeled phospholipids in cultures of hepatocytes or HepG2 cells. Illustrated here are primary rat hepatocytes exposed for 24 hours to amiodarone, a drug known to induce phospholipidosis in rats. A fluorescent probe (BODOPY-C_{12}-HPC) was used to detect the cytoplasmic lamellar bodies (green granules). (Courtesy of Abbott Department of Exploratory and Investigative Technologies.)

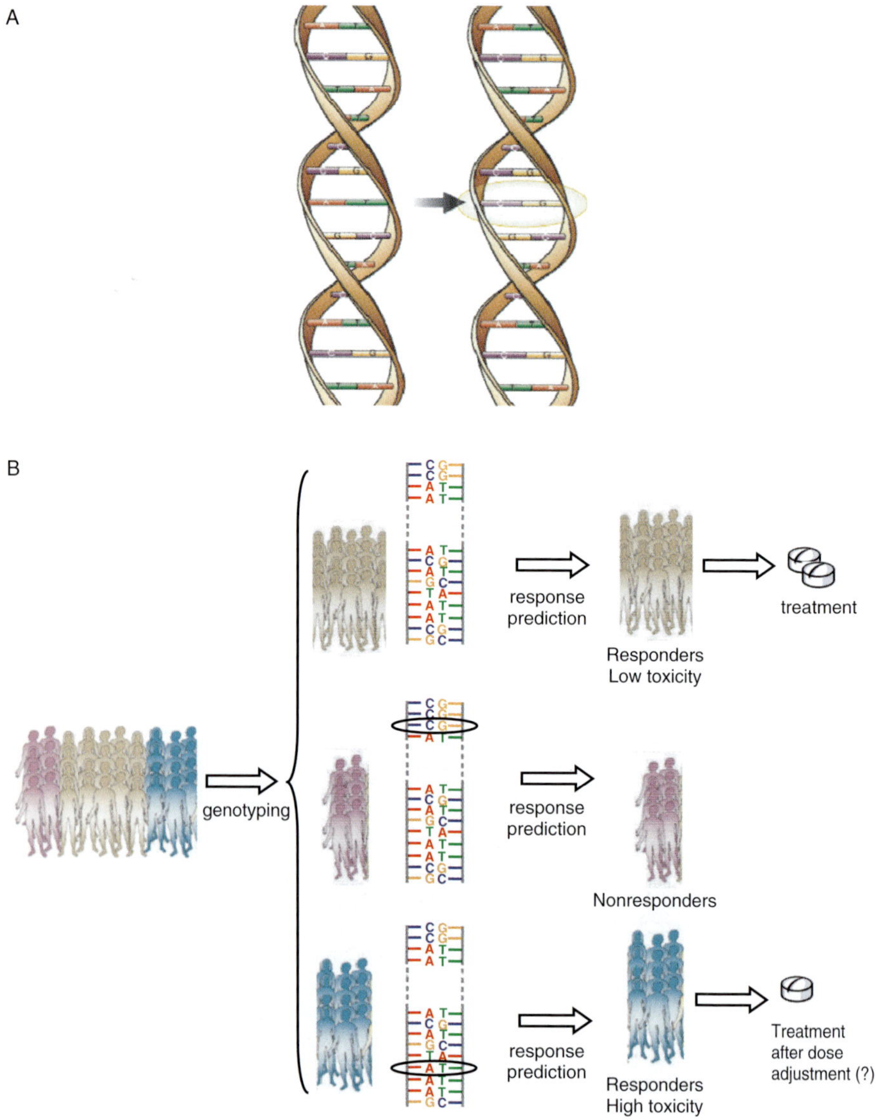

Figure 7.1 A) A single-base substitution in the DNA sequence (e.g., A:T → G:C) is the most common form of polymorphism in the human genome. B) If associated with drug efficacy or toxicity, SNPs can be used to predict drug response and select patients for therapy.

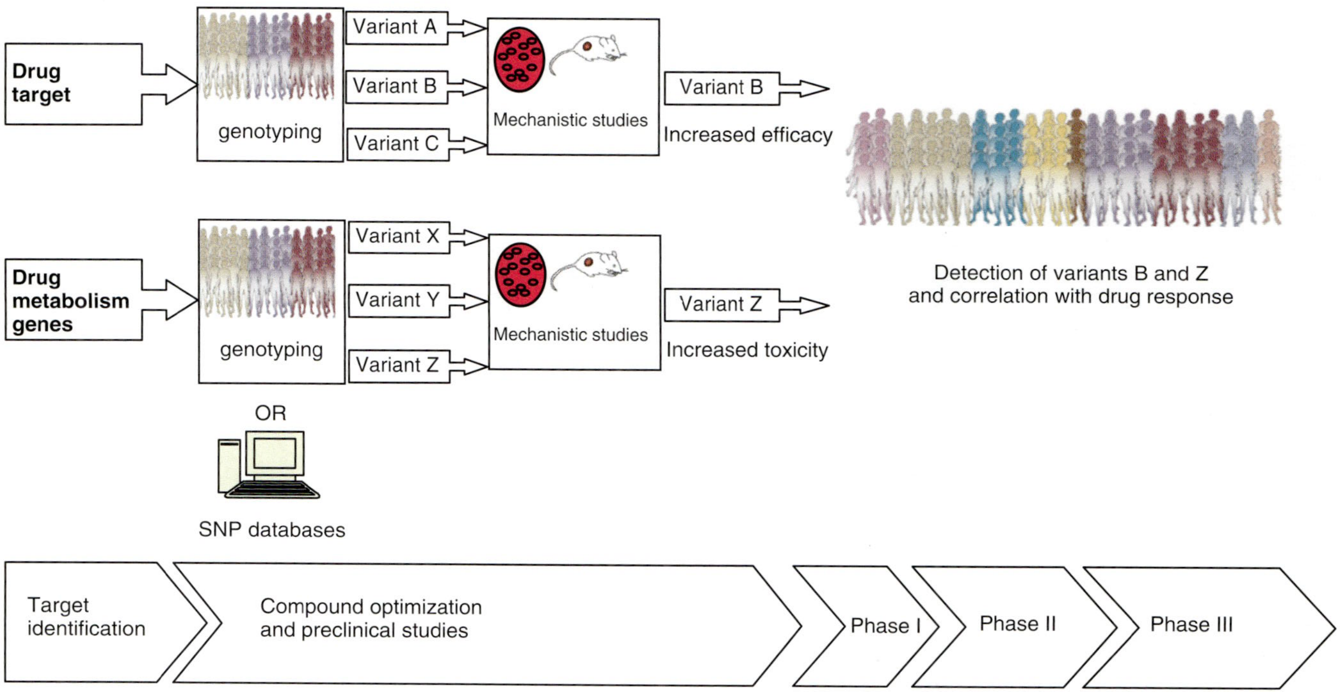

Figure 7.2 Candidate gene approach to pharmacogenetics studies in drug discovery. Drug target and known genes involved in drug metabolism are genotyped in an ethnically diverse population to identify existing variants. Alternatively, human SNP databases can be used. Once the spectrum of variation is defined for the candidate genes, possible associations with drug activity are explored in model systems, such as cell lines or animal models. The polymorphisms that affect the drug's mechanism in model systems are then studied in humans as clinical trials are initiated.

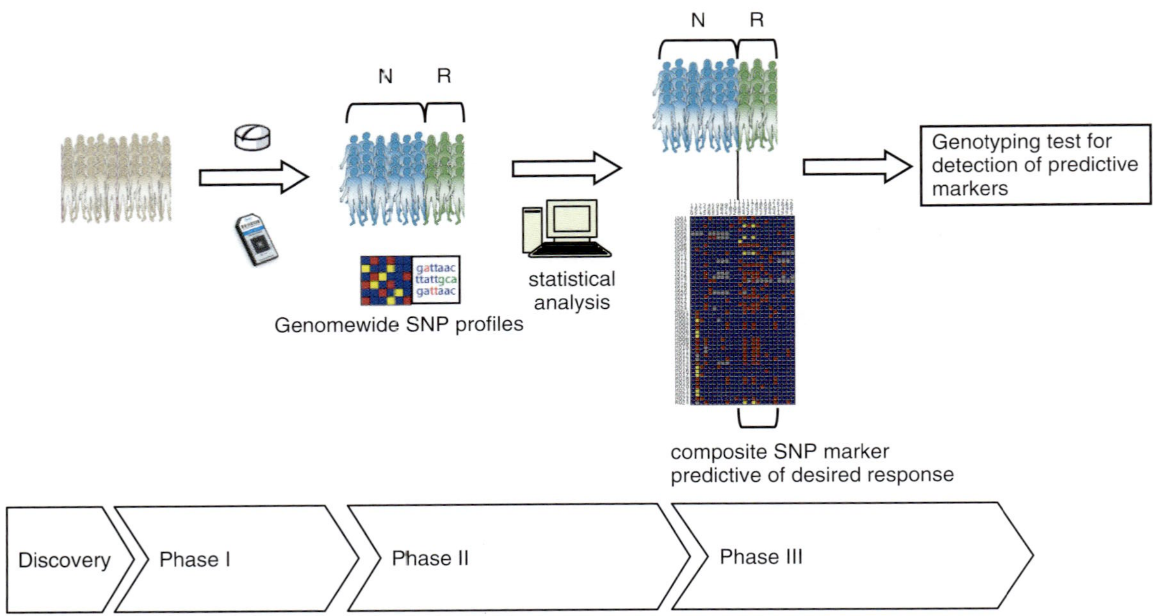

Figure 7.3 A) Genome-Wide approach to identification of polymorphic markers of drug response. Patients enrolled in the clinical trial are genotyped with a high-throughput method, such as microarray-based genotyping. After the response data have been collected, the genome-wide SNP profiles are correlated with the drug response to identify SNP profiles associated with the desired outcome (good efficacy or low toxicity). Once composite markers predictive of response are identified, a genotyping test may be designed to interrogate these markers in an independent cohort of patients.

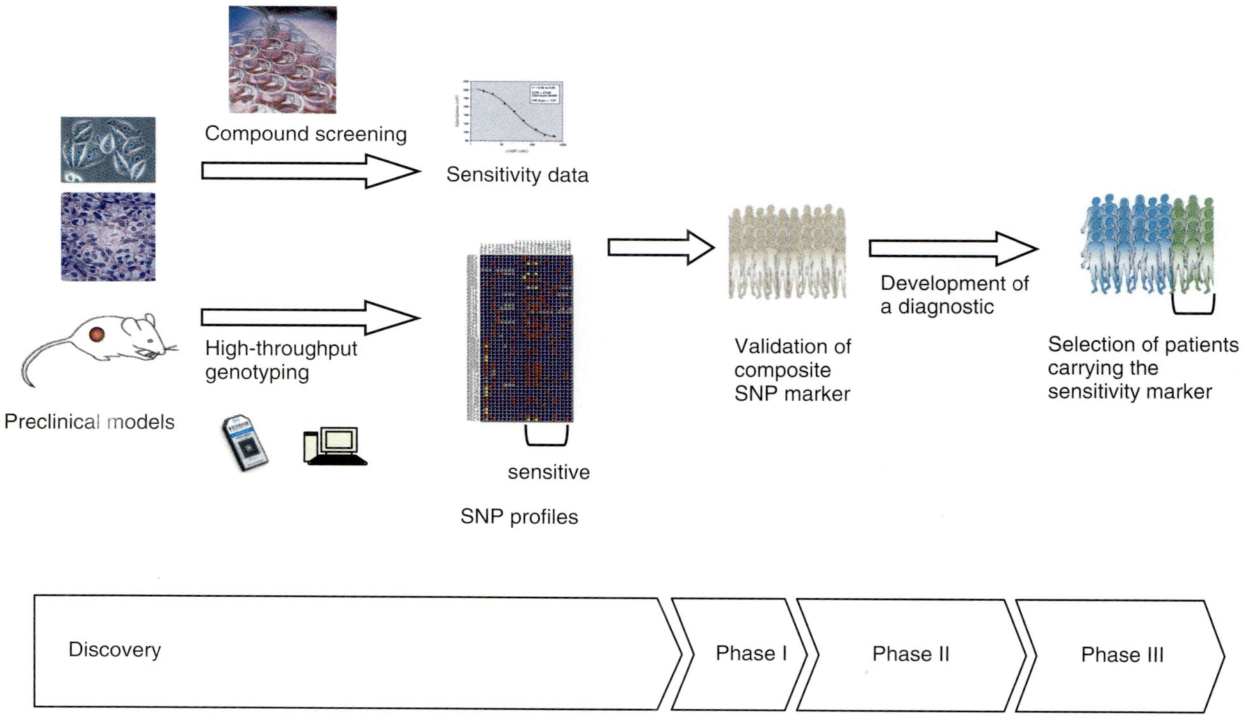

Figure 7.3 B) Early application of the genome-wide approach in drug discovery. Preclinical drug screening is coupled with high-throughput genotyping of the preclinical models to identify polymorphisms associated with sensitivity to the drug. Once composite SNP markers are identified in preclinical models, they can be explored in clinical trials, and if correlation with response is confirmed, a diagnostic genotyping test can be developed.

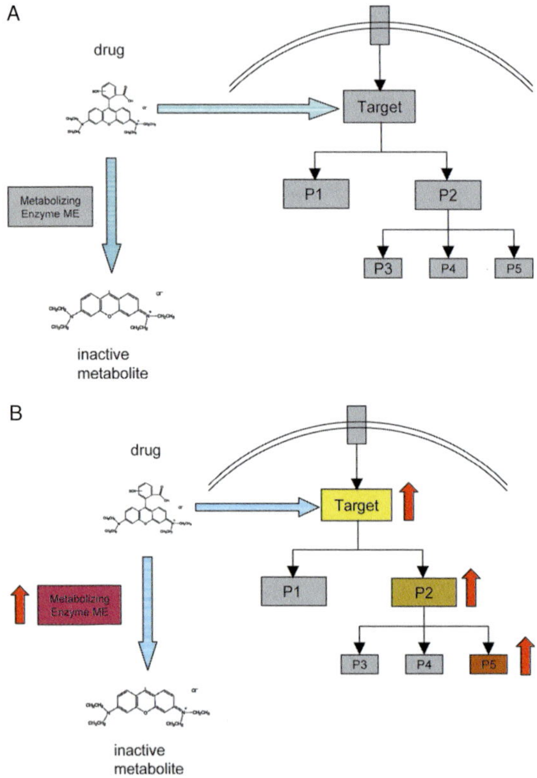

Figure 7.4 The pathway controlling a drug's mechanism may be affected by several polymorphisms. A) The target signals through proteins P1 and P2. Protein P2 induces P3, P4, and P5, which are implicated in the disease phenotype. The drug is converted into an inactive metabolite by a metabolizing enzyme ME. The dose of the drug is optimized for the most frequent genotype or the "average" individual. B) Polymorphic variants exist for the target, as well as SNPs in proteins P2, P5, and ME. Each of the polymorphisms causes an increase in expression or activity of the respective protein. When present individually in a patient, none of these SNPs significantly affects the efficacy of the drug, because the corresponding changes in expression are small. However, when all these SNPs are present in the same individual, their combined effect on the drug efficacy is significant. The higher concentrations of the target and its downstream modulators as well as the higher activity of the drug metabolizing enzyme contribute to the lower efficacy of the drug. Therefore, patients with this combination of variants may not respond to the drug.

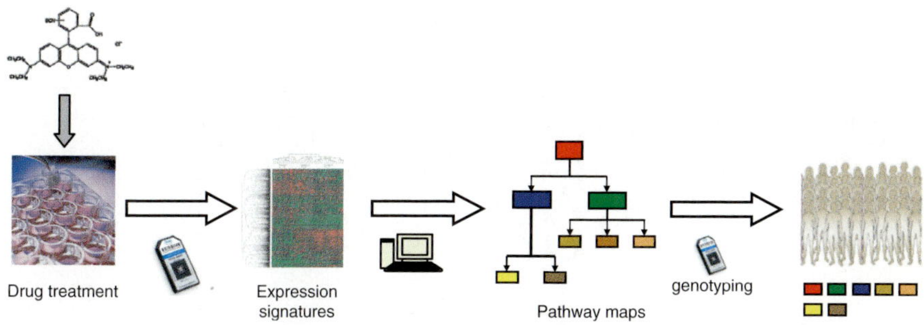

Figure 7.5 Use of gene expression microarrays to facilitate the pathway approach in pharmacogenetic studies. To elucidate the pathway involved in the drug's mechanism, cells are treated with the drug at different time points and profiled with expression microarrays. The resulting gene expression signatures are mapped to intracellular pathways. Genes involved in the activated pathways are then scanned for polymorphisms. The polymorphisms identified are considered candidate markers for drug efficacy because they may affect the drug's mechanism.

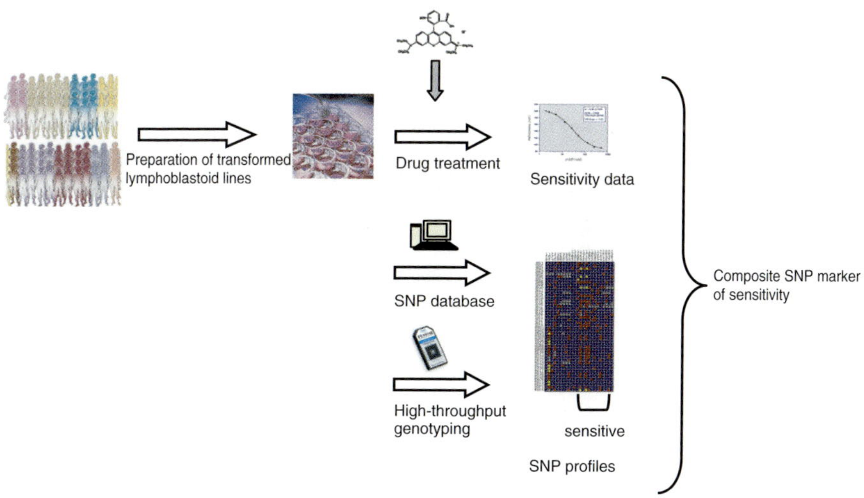

Figure 7.6 Use of cultured cell lines in pharmacogenetic studies in oncology to identify markers of sensitivity to anticancer drugs. Transformed lymphoblastoid cell lines are prepared from an ethnically diverse panel of individuals and cultured in vitro. The sensitivity of cell lines to drugs is determined with a cytotoxicity assay, and the genotypes of cells are obtained by high-throughput genotyping.

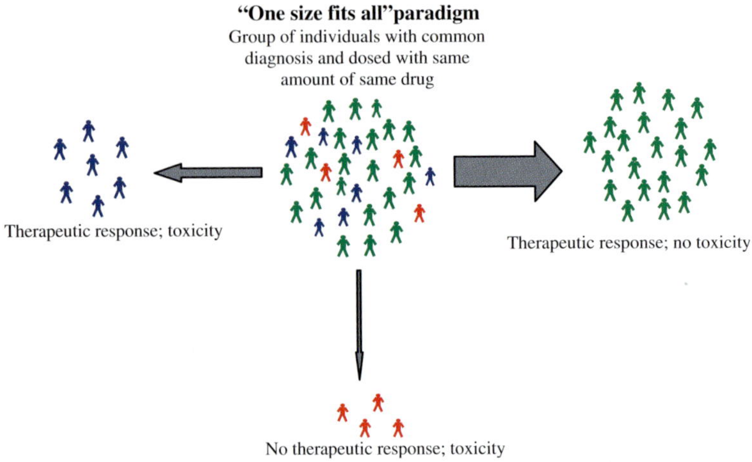

Figure 8.1 The goal of pharmacogenetics is to predict which individuals will respond well, will not respond, or will be at risk for toxicity, based on genetic variability; a goal of personalized medicine is to then use pharmacogenetic information to tailor drug therapy to each individual.

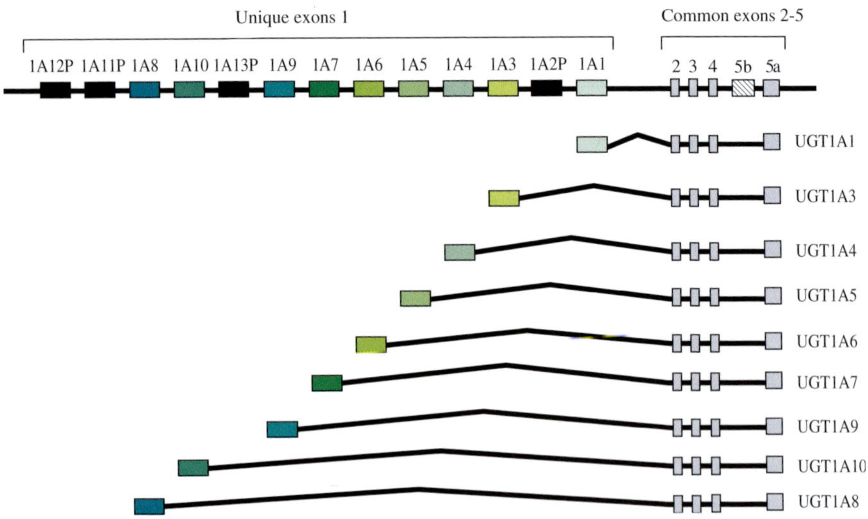

Figure 8.3 Schematic representation of the *UGT1A* locus and transcripts. The genomic structure includes 13 first exons, and four common exons (2–5). The black boxes represent pseudogenes; the hatched box represents the newly identified exon 5b. UGT1A transcripts generated from the 13 first exons and common exons 2, 3, 4, and 5a are shown.

fertility and early embryonic development study is typically conducted in rats to evaluate the effects of compounds in males and females from before mating to implantation. The pre- and post-natal development studies involve both rats and rabbits. First-generation dams are dosed from the implantation stage through the weaning of the first generation, and observations are continued through to sexual maturity of the first generation with assessment of the behavior and reproductive function of selected first-generation males and females. Finally, embryo-fetal development studies are conducted in rats and rabbits to evaluate visceral and skeletal abnormalities in offsprings. In addition, the repeat-dose toxicity studies covered above assess the effects of compounds on male and female reproductive organs.

5.2.1.5. Carcinogenicity

Evaluation of carcinogenicity in animals is conducted to assess the risk that experimental compounds may induce cancer in humans. In contrast to genetic toxicology, this assessment involves chronic studies that may also detect compounds that can induce tumors without a direct damaging effect on DNA, the so-called non-genotoxic carcinogens. Non-genotoxic carcinogens have been postulated to act through several tumorigenic mechanisms, namely, by increasing cell proliferation or survival, interfering with intercellular communication, causing endocrine modifications, inducing immunosuppression, or stimulating the formation of reactive oxygen species (18).

Carcinogenicity is typically assessed in 2-year studies conducted in both mice and rats. Such studies typically require a significant expenditure of both capital and active pharmaceutical ingredient (API) and utilize large numbers of animals (typically over 50/sex/group). As such, these studies are usually not required before FIH and are conducted relatively late in the development process, usually long after clinical trials have begun (19). ICH Topic S1A (http://www.ich.org/LOB/media/MEDIA489.pdf) defines the need for carcinogenicity studies of pharmaceutical agents. In short, these tests are required before registration for new pharmaceutical agents intended for chronic or intermittent use over 6 months in duration. Carcinogenicity studies may also be required for drugs administered for less than 6 months if factors are present that are suggestive of potential cancer risk in humans. In contrast, for compounds known to be genotoxic, no long-term carcinogenicity bioassays are typically required, since these compounds are presumed to be carcinogenic. Likewise, for compounds that are not systemically bioavailable (for instance, compounds topically applied) or compounds that will be prescribed to a target population with short life expectancy (such as cancer patients), these studies may also not be required.

Because of questions related to the value and relevance of these bioassays, alternative approaches have been explored. In particular, it has been proposed to limit the 2-year bioassay to one species, the rat, since data generated in mouse bioassays do not appear to contribute to human carcinogenic risk assessment (20). Also, transgenic mouse models have the potential to increase the sensitivity of the bioassay and thus decrease the duration of the study to 6 months (21). ICH Topic S1B (http://www.ich.org/LOB/media/MEDIA490.pdf) provides guidance regarding the selection of the rodent species (the rat by default) for the standard carcinogenicity assay and the use of alternative complementary carcinogenicity models (such as transgenic mouse models) and mechanistic studies to interpret a positive response and its relevance to humans. ICH Topic S1C (http://www.ich.org/LOB/media/MEDIA491.pdf) provides guidance on the selection of the appropriate doses for these bioassays. These doses are typically selected after 3-month dose-ranging subchronic studies in both males and females.

5.2.2. Discovery Toxicology

In the past, only limited preclinical toxicity evaluation was performed before compounds were selected for development. The current paradigm followed by most pharmaceutical R&D organizations is to increase the effort toward understanding and detecting toxicological liabilities at earlier stages and to increase the amount of toxicology-related information on compounds before the candidate selection step. This paradigm shift resulted from the realization that toxicity represents the single most common cause of attrition during development, even before the initiation of clinical trials (5). In most cases, this change has been implemented by running smaller and shorter versions of the studies discussed above before the candidate selection step. These discovery toxicology studies are not designed to generate a robust assessment of the potential toxicological liabilities of compounds for humans. Rather, their objective is to define at an early stage development-limiting toxicities of compounds with studies not conducted under the GLP. For instance, most companies now routinely screen discovery compounds for genetic toxicology with miniaturized, non-GLP versions of the Ames test or the micronucleus assay. Since these tests are often associated with a binary outcome (positive or negative), they are extremely useful for decision making (5). In addition, since there is often no acceptable safety margin for genotoxic compounds in many therapeutic indications, any positive signal can be used to terminate compounds with undesirable properties at an early stage. Likewise, several discovery in vivo and ex vivo assays are now routinely used to identify compounds with cardiovascular liabilities, while small rodent pilot studies are conducted to generate early toxicity signals (5, 22).

Since there are no common guidelines for discovery toxicology, each organization has its own testing strategy, which is often influenced by a past institutional experience or by preferences and beliefs of the leaders of an organization. It would be beyond the scope of this chapter to cover in depth all the available tests and studies used in the industry. Several important applications that can benefit from toxicogenomics are covered in the reminder of the chapter when predictive toxicology is discussed. For those readers interested in learning more, Kramer et al. and Stevens have recently published personal perspectives on this topic (5, 16). Usually new emerging technologies are evaluated and initially applied in discovery rather than in development. They can markedly influence the way toxicology is interrogated in discovery. Currently, toxicogenomics is mostly applied in discovery, but it is important to indicate that this technology can also be used in development, and as it matures the spectrum of its applications will likely widen to include clinical development.

5.3. TOXICOGENOMICS IN PREDICTIVE TOXICOLOGY

Here we use the term **predictive toxicology** to refer to the use of short-term, typically non-GLP tests or assays that can predict toxic changes that would occur in long-term studies. A perfect example of a predictive toxicology application that is discussed in this chapter is the use of various short-term models to predict the outcome of a 2-year carcinogenicity study. In addition, the term "*predictive toxicology*" is often used to refer to studies that attempt to extrapolate toxic reactions from preclinical species to humans (23–25). However, since the latter application requires a good mechanistic understanding of toxicity, we discuss this application under **mechanistic toxicology**.

Assays based on gene expression profiling represent only a fraction of the spectrum of tests used to predict toxic effects. Furthermore, toxicogenomics assays may utilize cell-based or in vivo systems. In this section, we limit our discussion to in vivo toxicogenomics approaches, while Chapter 6 is completely dedicated to applications of toxicogenomics to in vitro models.

Large-scale gene expression profiling represents an extremely sensitive approach to detecting genes and signaling pathways that are deregulated (either activated or inhibited) in cells or tissues after exposure to compounds. In fact, several studies have demonstrated that specific, toxicologically relevant transcriptional changes frequently occur before the development of the morphological and functional changes that are typically used to detect toxicity with clinical observations, hispathological examination, or clinical pathology measurements (7, 23, 26, 27). In our experience as well, for the vast majority of toxic changes in well-studied tissues such as liver, kidney, spleen, or heart, gene expression changes indicative of a specific toxic reaction can be detected earlier than with the traditional toxicological end points. Exceptions obviously exist, since certain toxic reactions may not require transcriptional changes to occur. For instance, compounds may interfere with specific ion channels or receptors in the central nervous system (CNS) or in the heart and lead to functional changes, such as CNS hyperactivity (convulsions, tremors, etc.) or ventricular fibrillation. Ultimately, if these toxic reactions do not lead to death, they may cause secondary transcriptional changes; however, these gene expression changes would not occur before clinical signs are observed. Likewise, it is still unclear whether toxicogenomics is sensitive enough to detect toxic changes that do not occur before extended periods of exposure to compounds, as is the case for toxicity identified in subchronic or chronic studies. To address that topic, we discuss below in this chapter the use of gene expression profiling to predict carcinogenicity in the liver of rats. Nevertheless, it has been our experience that certain changes occurring after extended periods of dosing (4 weeks or more) are quite challenging to detect by gene expression profiling in short-term studies (several days). This can probably be explained by the lack of experience in the toxicology community with that particular experimental design and by the unavailability of appropriate reference data. However, this phenomenon also reflects the complexity of toxic reactions, and it is likely that some of these chronic effects are not necessarily associated with early, specific transcriptional changes.

Despite this caveat, toxicogenomics represents an attractive novel approach to identifying toxic changes earlier in the drug discovery process, and this clearly can result in significant improvements in the productivity of drug discovery (7, 23, 24). In discovery toxicology, the objective of the initial in vivo evaluations is to detect development- or dose-limiting toxic changes with good accuracy and to understand the safety liabilities of a chemical series as early as possible. At this stage, compound availability represents a critical limiting factor in conducting robust in vivo studies. Typically, non-GLP rat exploratory toxicology studies have been conducted with males and females, using 5 animals/sex/group for 5–7 days of daily dosing. Using so many animals is usually considered necessary, since at this stage, toxic changes may be subtle and a sufficient number of animals is required to properly interpret the data generated. In fact, detecting toxicity that may occur after 2–4 weeks of dosing is far from reliable with this type of study design. These studies require at least 5–10 g of compounds and therefore cannot be conducted until late in the process. In addition, data generation and interpretation are labor-intensive and lengthy. More sensitive and predictive technologies have the potential to detect toxic changes with a smaller number of animals and shorter periods of dosing. In our experience, incorporating toxicogenomics as an end point in these exploratory rat toxicology studies improves the detection rate of toxic changes, and can reduce the number of rats used to three per group. Furthermore, because of the increased sensitivity provided by this additional end point, studies can be limited to males only, thereby further decreasing compound requirement.

Typically, our discovery exploratory toxicology studies in rats are conducted with as little as 1–2 g of compounds, an amount that can be generated by medicinal chemists at the bench. Because of the lower compound requirement, the studies can be conducted 2–6 months earlier than traditional rat exploratory studies. While an earlier toxicological evaluation represents an obvious improvement in terms of cycle times, it also leads to indirect process changes with profound implications. First, discovery teams are encouraged to evaluate multiple series in parallel, thereby increasing their probability to discover a good development candidate. Second, discovery teams are more prone to evaluate several potential candidates in parallel, rather than limit their focus to a single one. Therefore, should their initial candidate fail, the time to bring a backup into development can be significantly shortened. Last, but not least, the higher sensitivity of toxicogenomics increases the probability of detecting undesirable changes that may later contribute to termination. Therefore, toxicogenomics can facilitate the selection of molecules with an optimized toxicological profile, thereby ultimately reducing late-stage attrition. This decreased late-stage attrition can and should be retrospectively quantified, since it allows an organization to evaluate its efficiency and to potentially refine its processes or methods on an as-needed basis to maximize productivity.

Application of toxicogenomics in predictive toxicology is based on the development of predictive gene expression signatures. In Chapter 4, we discuss the general principles behind the generation of predictive gene expression signatures. In this section, specific examples are provided in the context of drug discovery and development. It is important to realize that despite a significant investment in toxicogenomics by pharmaceutical R&D organizations, relatively few gene expression signatures have been published so far. In addition, it is still unclear whether these tools are routinely used in decision-making in various organizations. Finally, gene expression signatures should not be viewed as the only tools for predicting toxic outcomes. Rather, they complement the battery of conventional analytical methods, such as hematology, serum chemistry, and histopathology. With these considerations in mind, we now discuss in more detail several aspects of predictive toxicogenomics, using hepatotoxicity, nephrotoxicity, and carcinogenicity as examples. We will then conclude the section by providing an overview of the use of predictive toxicogenomics in our organization.

5.3.1. Prediction of Hepatotoxicity

5.3.1.1. Hepatotoxicity: an Important Toxicology Problem in Drug Discovery and Development

The liver is a common, if not the most common, target organ for toxicants (28). The vulnerability of the liver to toxic substances stems from its critical synthetic, metabolic, and detoxifying functions as well as from its high exposure to compounds (28). Indeed, the liver is a highly vascular multifunctional organ intimately involved with multiple critical functions, including the detoxification and metabolism of a variety of endogenous and exogenous compounds. Nearly all orally dosed compounds pass through the liver before entering the systemic circulation, resulting in particularly high tissue exposures. Drug-induced hepatotoxicity is a major cause of failure in drug development and a frequent cause of adverse drug reactions (29–31). For instance, between 1992 to 2002, 43%, 25%, and 35% of drugs undergoing phase I, II, and III studies respectively, have been terminated because of the development of toxicity, a significant part because of hepatic injury (32). Likewise, between 1992 and 2002, 27% of market withdrawals in the United States and the European Union markets were due to hepatotoxicity. The most prevalent etiology for a "black box warning" is hepatotoxicity (32, 33). In preclinical toxicology studies, the failure rate due to hepatotoxicity is more difficult to determine because of the lack of detailed published reports. However, toxic changes in the liver are considered to be among the most common findings in preclinical toxicology studies (28).

In preclinical in vivo studies, hepatic toxicity is traditionally identified through the integrated evaluation of clinical pathology parameters and histopathological findings. Serum alanine aminotransferase (ALT) is a critical parameter for the identification of drug-induced hepatic injury in both preclinical and clinical studies. Therefore, this enzyme and another transaminase enzyme (aspartate aminotransferase or AST) are generally used to detect hepatocellular injury in rats, dogs, and monkeys (34). Because of the limited value of ALT activity in assessing hepatotoxicity in pigs, measurement of serum sorbitol dehydrogenase (SDH) activity is recommended in this species. In addition, glutamate dehydrogenase (GLDH) has been proposed as a sensitive and specific biomarker of acute hepatic injury in the rat (35). Additional parameters that are usually determined in preclinical animal studies and used for the assessment of hepatobiliary injury include serum alkaline phosphatase (ALP), γ-glutamyl transferase (GGT), and bilirubin. All these clinical pathology parameters can be determined at different time points during a typical study, but in rats this is typically done only at the end of the study. To determine reversibility of potential changes, a treatment-free recovery period can be added for a subset of study animals. At the end of the treatment period or the recovery period the animals are necropsied, and a number of organs and/or tissues are preserved for histopathological examinations. For histopathological evaluation of the liver, samples are collected from different liver lobes and histological sections are prepared from different sites. Such integrated evaluation of various serum biomarkers and histopathology is a robust approach for the detection of hepatotoxicity, but changes in these parameters may occur only after prolonged periods of dosing and may be subtle and challenging to interpret. Therefore, additional methods are needed for reliable early detection of hepatoxicity.

Because of the high incidence of hepatotoxicity, most of the early efforts in toxicogenomics have focused on the liver. Additionally, the liver is a rather homogeneous tissue, as opposed to tissues such as intestines or brain. The liver is composed mostly of hepatocytes (which represent approximately 80% of the total liver volume) that share similar biochemical functions, translating into relatively uniform and reproducible gene expression profiles (36). This homogeneity facilitates identification and interpretation of gene expression changes. For these reasons, a wealth of published and proprietary gene expression information is available for the liver, and there is substantial evidence that hepatotoxicity can be predicted with toxicogenomics methods with a relatively high accuracy.

5.3.1.2. Predictive Genomic Models of Hepatotoxicity

Like many other groups, our laboratory initially focused on the liver as a proof-of-principle tissue. Our primary objective was to establish whether gene expression profiles could be used to predict hepatotoxicity in rats. The effort was started by building a proprietary compendium of microarray-generated liver gene expression profiles derived from 3-day rat toxicology studies using over 50 hepatotoxicants and non-hepatotoxicants. Each compound was administered to three male Sprague-Dawley rats/group at a high dose (a dose expected to induce hepatotoxicity after at least 1 week of treatment) and at a lower, non-hepatotoxic dose. Accordingly, since the studies were of 3-day duration, treatment with most hepatotoxic compounds was not associated with significant changes in serum chemistry parameters (such as hepatic transaminases) or with histopathological findings suggestive of hepatotoxicity. Despite the lack of obvious hepatic changes, we were nonetheless able to select with an ANOVA analysis a group of genes that differentiated the expression profiles induced by the hepatotoxic compounds from those caused by the non-hepatotoxic compounds. We then used an artificial neural network algorithm coupled with principal component analysis (PCA) for dimensionality reduction and generated a quantitative model to classify compounds. This model classifies compounds according to a composite score that indicates the probability that the test compound would induce hepatotoxicity in rats on continued dosing (Fig. 5.1).

When the model was developed, we conducted a forward validation step using an independent set of gene expression profiles induced by hepatotoxic and non-hepatotoxic compounds that were not part of the training set. The neural network algorithm classified the testing sets of compounds with 89% accuracy. This excellent predictive performance promoted the use of this predictive classifier in the context of exploratory 3- to 5-day repeat-dose rat studies for various internal projects. However, because of the limited size and diversity of the testing set, it was still unclear whether this estimated accuracy would hold when proprietary compounds are evaluated.

Therefore, we conducted a reevaluation of the performance of this predictive model once sufficient internal data became available. Assessment of the predictive power for our internal compounds would ensure that our model adequately covers the changes induced by compounds derived from our internal chemical space. For 52 compounds, data were available from short-term (3–5 days) and long-term (2 weeks) rat studies. Using the gene expression profiles induced by these compounds in the rat liver after short-term dosing, we estimated the predictive accuracy of our predictive model by correlating the prediction with the histopathological and serum chemistry changes observed in the long-term studies. The model correctly classified eight compounds as hepatotoxic and 42 compounds as non-hepatotoxic. Only two compounds were incorrectly classified, indicating a sensitivity of 89%, a specificity of 98%, and a predictive accuracy of 96%. In addition, we interrogated the diagnostic accuracy of the model and demonstrated that it represents an excellent diagnostic tool that could be used as a complement to the current, well-established biomarkers of hepatoxicity. In fact, the model can correctly classify compounds in most situations in which toxic changes are already present based on evaluation of serum chemistry or histopathological changes.

Similar and alternative approaches have been used by other groups with comparable results. For instance, Dai et al. have used the same repository of gene expression profiles as us in an attempt to build a quantitative model to predict compound-induced liver injury (37). In this study, the authors preselected 238 genes that were significantly different between the hepatotoxic class and the non-hepatotoxic class as candidate markers for their classifier. These genes were selected based on their error-weighted log ratio. The number of dimensions was further reduced with wavelets, and a classifier was generated with a neural network algorithm. The expected error and generalizability of their classifier was then interrogated with a validation data set, which had not been used to establish the model. Based on this validation data set, the sensitivity, specificity and accuracy of the model were determined to be approximately 91%, 88%, and 89%, respectively.

Ruepp et al. have used support vector machines (SVMs) to classify experimental compounds according to several well-established hepatoxic end points (38). They created a database of gene expression profiles from the liver of

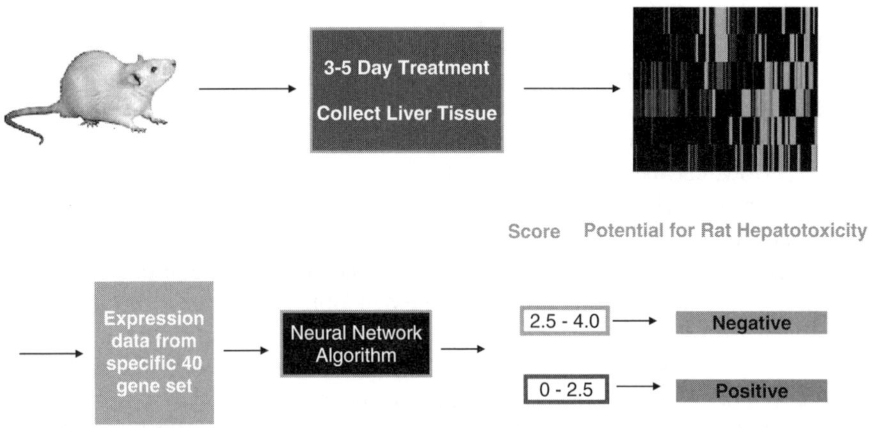

Figure 5.1 Predictive genomics assay for rat hepatotoxicity. This assay was developed with an internal rat liver gene expression database and an artificial neural network algorithm. Using microarray-generated gene expression profiles from male Sprague Dawley rats treated for 3–5 days with a variety of paradigm compounds, the neural network algorithm classifies the compounds based on their potential to cause hepatotoxicity in rats on exposure of longer duration (2 weeks of daily dosing or longer). This assay is based on a preselected 40-gene set, and its output is a score ranging from 0 to 4. A low score indicates a high probability that the test article will induce hepatotoxicity in rats in repeat-dose studies of longer duration at similar exposure levels. A cutoff point of 2.5 was selected based on a small validation set to distinguish negative (i.e., non-hepatotoxic) from positive (i.e., hepatotoxic) compounds. A longer forward validation using 52 compounds (9 hepatotoxic compounds and 43 non-hepatotoxic compounds) from our own chemical space demonstrated that this predictive assay had a 96% accuracy with a 98% specificity and a 89% sensitivity. See color insert.

male Wistar rats treated with various model compounds. Histopathological findings and serum chemistry parameters were used to anchor the gene expression profiles to selected end points (steatosis, cholestasis, direct acting, peroxisomal proliferation, or non-toxic/control). The resulting predictive model successfully discriminated two experimental compounds based on their hepatotoxic liability (steatosis vs. non-hepatotoxic) and recognized a hepatotoxic liability in the absence of pathological findings. The result of this study confirms our observations that toxicologically relevant gene expression changes frequently occur earlier than alterations detected with conventional methods, such as serum chemistry or histopathology.

Predictive gene expression signatures can also be developed for very specific hepatic toxic end points, as long as a robust training set can be built. For instance, bile duct hyperplasia is a histopathological finding that is often difficult to detect at early stages, since it typically occurs after prolonged periods of treatment and is not always associated with parallel changes in serum chemistry parameters. In collaboration with Iconix Biosciences (Mountain View, CA), we have derived a multigene biomarker for bile duct hyperplasia using liver transcript profiles from the commercial DrugMatrix® toxicogenomic database (39, 40). Our objective was to predict, within 1–5 days of dosing, the occurrence of bile duct hyperplasia in rats that would only be detectable histopathologically after weeks of continued dosing. We used gene expression profiles induced by several compounds known to induce bile duct hyperplasia, as well as negative control compounds. None of the compounds used in the training set induced bile duct hyperplasia after 5 days of dosing, as evidenced by a thorough histopathological evaluation. A signature was derived with a sparse linear programming (SPLP) algorithm, as described by Fielden et al. (41). The performance characteristics (sensitivity and specificity) of the predictive signature were then estimated with gene expression profiles from the liver of male Sprague-Dawley rats treated for 1 or 5 days with eight model compounds. Based on this forward validation data set, the signature had an accuracy of 88% with a sensitivity of 90% and a specificity of 83%. Further testing of the signature with a wide variety of our internal exploratory compounds confirmed its robust performance. It should be noted, however, that the signature was more accurate when used with gene expression profiles derived from 3- or 5-day studies than with those from 1-day studies. As we have already mentioned in Chapter 4, single-dose studies are typically associated with more interindividual variability, potential adaptive changes, and poorly characterized tissue kinetics. These factors contribute to the high level of noise in 1-day hepatic gene expression profiles. We therefore highly recommend reducing the noise in these experiments as much as possible by prolonging the dosing, so that steady-state tissue concentrations can be reached.

5.3.1.3. Additional Toxicogenomics Approaches to Predict Hepatotoxicity

Signatures and predictive models represent an advantageous and efficient approach to analyze toxicogenomics data, since they can usually be interpreted in a binary mode (i.e., presence or absence of specific toxic changes). While these tools have a definite value, their performance must be more clearly understood if one wants to use them in decision making. Clearly, no predictive models will be 100% accurate, and in a discovery setting they certainly do not need to be. Complementary analytical procedures can nevertheless be used to further confirm the hepatotoxicity potential of compounds, and increase one's confidence in the predictive ability of the signatures. In addition, predictive models can only be built if sufficient reference data are available to develop, train, and test them. Frequently, a sufficient amount of reference data is not available, and the resources required to develop these models may not be accessible in small to midsize R&D organizations. The lack of predictive signatures or models should not, however, preclude investigators from trying to predict hepatotoxicity with toxicogenomics.

The experience with toxicogenomics at Bristol-Myers Squibb has recently been published (27). The data reported confirm that significant transcriptional changes can be observed before histopathological alterations and, therefore, that one can use transcriptional changes to predict morphological changes in the liver. In this retrospective evaluation, several target tissues, including liver, were profiled at dose levels and times known to precede a previously identified toxicity. Of particular interest is the fact that dose groups with significant transcriptomic effects in various tissues (liver, brain, retina, vagina, testis, and stomach) were more likely to develop pathology even if they were not initially accompanied by histopathological changes. Based on an analysis of studies evaluating 33 compounds, marked global transcriptional changes (>4% of total transcripts present on the microarray at a P value <0.01) were shown to be a robust biomarker for toxic doses.

Another approach to analyzing toxicogenomics data is to use unsupervised procedures such as clustering or PCA to evaluate similarities that may exist with profiles from various reference treatments. This approach is often referred to as "guilt by association," since it assumes that closely similar transcriptomic profiles are likely to reflect similar toxic outcomes in the liver. This approach can only be applied if a sufficient number of hepatic reference profiles induced by various toxicants and non-toxicants are available; however, this number may be smaller than that required for development of predictive signatures. For instance, to evaluate the predictability and accuracy of gene expression profiles in toxicology, our laboratory treated male Sprague-Dawley rats with 15 different prototypical hepatotoxic compounds (allyl alcohol, amiodarone, Aroclor 1254, arsenic, carbamazepine, carbon tetrachloride, diethylnitrosamine, dimethylformamide, diquat, etoposide, indomethacin, methapyrilene, methotrexate, monocrotaline, and 3-methylcholanthrene) (42). These agents are known to cause a variety of hepatocellular injuries including necrosis, DNA damage, cirrhosis, hypertrophy, and hepatic tumors. When the gene expression profiles were clustered and correlated with the histopathology and serum chemistry findings, a strong association was observed. In addition, specific genes were identified whose regulation correlated strongly with the effects on clinical chemistry parameters. Overall, these results demonstrated that clustering can identify potential hepatotoxic treatments. Obviously, in this approach, it is impossible to understand the accuracy of the prediction, and there is no clear threshold separating a toxic from a non-toxic response. Consequently, clustering should be considered a signal-generation method, rather than a tool for decision making. It can be applied in several ways depending on the circumstances. If multiple compounds are available, clustering results can be used to prioritize compounds, assuming that compounds clearly more similar to reference toxicants in terms of transcriptomic activity are less desirable. In contrast, in situations where only one or two compounds are tested, this approach would simply suggest that close monitoring of liver function and morphology may be warranted in studies of longer duration.

5.3.2. Prediction of Nephrotoxicity

5.3.2.1. Kidney as a Target Organ of Toxicity

The kidney is one of the major organs routinely evaluated in toxicology studies. It is vulnerable to damage by toxic compounds because of its high blood flow, its role in xenobiotic metabolism and clearance, and its basic function of concentrating urine (43). The frequency at which renal toxicity occurs relative to other drug-induced toxicities has been reported to be anywhere between 2% and 20% (44). There are various nephrotoxic compounds that affect different subcellular compartments, such as glomeruli, proximal tubules, distal tubules, and papilla. Renal toxicity can be difficult to detect not only in preclinical studies, but also in the clinic. The most widely used serum biomarkers include serum creatinine and blood urea nitrogen (BUN) (45). These biomarkers that measure renal function are very insensitive, and increases in serum concentrations occur only with substantial functional deficit. In other words, BUN and serum creatinine are late indicators of renal damage and do not detect early histopathological injury.

Proteinuria and albuminuria can also be used as a measure of glomerular or tubular damage, and may be more sensitive than BUN or serum creatinine. Additional biomarkers have been proposed, such as the lysosomal enzyme *N*-acetyl glucosaminidase (NAG) and the brush border enzyme γ-glutamyl transpeptidase (GGT) (44). However, these two urinary markers display significant intra- and interindividual variability, precluding their use in toxicology. Histopathology represents an additional, robust end point to detect renal injury in toxicology studies, but is difficult to translate to the clinic.

Since the current serum or urinary biomarkers lack sufficient sensitivity and provide little to no information regarding the segment of the nephron affected or the mechanism of toxicity, there has been a significant effort in the industry to discover and validate better biomarkers of renal damage. Genomics technologies present a unique opportunity to develop new methods for predicting toxicity in the kidney.

5.3.2.2. Predictive Genomic Models of Nephrotoxicity

Generally speaking, approaches similar to those used in the liver may also be followed to develop predictive models of nephrotoxicity. As mentioned in Chapter 4, consistent sample collection is critical to generating reproducible gene expression profiles. This is especially true for the kidney, since this tissue is composed of multiple compartments that may or may not be affected by the compound. In our laboratory, we routinely collect the left kidney for gene expression profiling while the right kidney is used for histopathological examination. Evaluating one kidney histopathologically is sufficient to capture most compound-induced changes, since the vast majority of them affect the right and left kidneys to comparable levels. Before freezing, the left kidney is sliced sagittally in half, and one half is used to generate gene expression profiles. This collection procedure ensures that consistent portions of cortex, medulla, and papilla are included in the sample.

In the last few years, several predictive models of nephrotoxicity have been published. For instance, scientists at Iconix Biosciences took advantage of their comprehensive compendium of gene expression profiles (the DrugMatrix® database) to build a predictive gene expression signature of renal tubular toxicity for rats (39, 41). The value of their study stems from the large number of expression profiles included in both the training and testing sets (64 nephrotoxic and 21 non-nephrotoxic compounds). This large number of compounds in the training set increases the probability that different mechanisms of toxicity and a variety of chemical structures would be covered. The gene expression profiles were derived from the kidneys of male Sprague-Dawley rats treated for 5 days, a time point at which no obvious toxic changes were evident, as shown by the lack of histopathological observations and the absence of changes in serum chemistry parameters (i.e., serum creatinine and BUN). However, the authors confirmed the presence of renal tubular changes for their positive class of compounds by conducting 4-week studies. Based on the testing set of expression profiles, the signature correctly predicted the future presence or absence of renal tubular injury in 76% of the compound treatments, a performance far better than that of the current well-accepted serum biomarkers, creatinine and BUN.

Evidence that transcriptomic data have the potential to improve the prediction of renal toxic changes has also been published by others. For instance, a study evaluated time- and dose-dependent gene expression changes associated with proximal tubular injury in the rat (46). Male Sprague-Dawley rats were treated with various prototypical nephrotoxic compounds to generate renal gene expression profiles 1, 3, and 7 days after initiation of dosing. The profiles clustered based on the severity and type of pathology in individual animals. Further, the expression changes were indicative of tubular toxicity, showing hallmarks of tubular degeneration/regeneration and necrosis. Finally, with a support vector machine (SVM)-based approach and a training set of 120 gene expression profiles, a predictive classifier was developed that was able to predict with a 100% selectivity and a 82% sensitivity the type of pathology in a testing set composed of 28 gene expression profiles.

Given the excellent performance of our predictive model of rat hepatotoxicity, our laboratory was also interested in generating a predictive model of nephrotoxicity based on the compendium of renal gene expression profiles in the DrugMatrix® database. Our approach was similar to that used to build our hepatotoxicity model discussed in the previous section. Briefly, after careful curation of the reference data, we identified 173 gene expression profiles that met our inclusion criteria. These profiles covered 62 compounds (33 compounds toxic to the kidney and 29 compounds non-toxic) dosed for 3 or 5 days in male rats. Discriminative probe sets between these classes of compounds were selected by ANOVA analysis, and a quantitative model was developed with an artificial neural network algorithm. In a forward validation procedure, this predictive nephrotoxicity model correctly classified 10 of 11 compounds (6 nephrotoxic and 5 non-nephrotoxic). We are now routinely applying this assay in the context of our exploratory, non-GLP toxicology rat studies. Our experience with the liver model facilitates the use of this predictive model for decision-making. However, it is clear that a more robust and extensive forward validation will be required in the future in the form of a retrospective evaluation of the performance of the model.

5.3.3. Prediction of In Vivo Carcinogenicity

5.3.3.1. Value Created by Toxicogenomics in the Assessment of Carcinogenicity

For many years, the carcinogenic potential of experimental therapeutic agents has been evaluated with a combination of in vitro assays and in vivo rodent models, as previously discussed. In vitro genotoxicity assays are discussed in Chapter 6, and here we limit our discussion to the potential applications of genomics technologies to predict with better accuracy the outcomes of 2-year chronic bioassays. There is no doubt that the current standard protocols for the assessment of potential carcinogenicity in rodents have greatly advanced the fields of toxicology and risk assessment (47). They have been facilitated by improved experimental control of variables and the significant amount of experience accumulated with the common laboratory strains of rats and mice (47). However, these studies have important limitations that can be partly addressed with genomics technologies.

The most obvious limitation is related to the timing of these studies. Rodent bioassays are long and costly (2 years of in-life studies plus several additional months for histopathological assessment, data interpretation, and report generation). In addition, since additional preliminary studies are required to design these studies and because of the large amount of active pharmaceutical ingredient (API) needed, they cannot be performed until late in a program. Alternative methods using genetically engineered mouse models have been proposed to shorten the dosing period (21). However, given the lack of recent literature reports on these models, it does not appear that their current use is extensive. Toxicogenomics applied in a predictive mode using shorter dosing paradigms would also allow for a profound reduction in the duration of dosing, thereby reducing the amount of API required and enabling an earlier assessment of the compound. The objective of a predictive genomics model for carcinogenicity would not necessarily be to replace the current regulatory carcinogenicity studies. Rather, such predictive model would allow sponsors to interrogate at an earlier stage the likelihood that their molecules have an in vivo carcinogenicity potential. This additional piece of information would result in more informed decisions on compounds at earlier stages or in the design of proactive investigative efforts to examine the relevance of this risk to humans. Ultimately, these models would also decrease the likelihood of exposing humans to potentially hazardous compounds.

The end point of the rodent bioassays consists essentially of macroscopically and microscopically detectable tumors, which may also spontaneously occur in naive, aging animals and whose incidence varies tremendously among animals, thereby requiring a significant number of animals to obtain a sufficient statistical power (48). Some of these tumors occur in unexpected locations and are characterized by unusual histology without correlates to human neoplasia, making extrapolation of risks to humans a challenging, if not obscure exercise (47). This has led to a long, ongoing debate regarding the relevance of some of the rodent responses to predicting human risk (10). Beyond issues related to the choices of species, strains, doses, and the route of exposure or time course of exposure, the traditional rodent lifetime bioassays seldom include investigations of mechanisms of action or of differences in metabolism (47). This black box approach to risk assessment could clearly be improved with the introduction of molecular mechanistic end points, such as those generated by genomics investigations. Although the main objective of genomics models is to predict the risk of carcinogenicity, it is noteworthy that such models may also provide useful insights into the mechanism of action of a compound. This additional level of information offers the opportunity to address at a very early stage questions related to the relevance of rodent responses to humans.

5.3.3.2. Predictive Genomic Models of Carcinogenicity

Since the current protocols for carcinogenicity assessment have been in place for over three decades, a large body of data has been accumulated on a variety of chemicals. This historical data set represents a unique resource for the development and validation of predictive models. Using a variety of standard rodent carcinogens, several groups have taken advantage of these reference data sets to evaluate the feasibility of predicting a carcinogenic response with studies ranging from several days to several weeks.

Kramer et al. were the first group to report findings of interest (19). In their approach, these investigators used a 5-day repeat-dose toxicity study in male Sprague-Dawley rats in an attempt to predict hepatic carcinogenicity. More specifically, their objective was to use gene expression profiling to identify candidate molecular markers that may predict hepatic carcinogenicity induced by either non-genotoxic or genotoxic compounds (19). Although there are likely to be numerous mechanisms by which hepatic carcinogenesis is mediated in rodents, the authors hypothesized that a small number of critical genes could be used to detect genotoxic and non-genotoxic carcinogens in the early events of carcinogenesis, since broad mechanisms, such as oxidative stress and disruption in the balance between proliferation and apoptosis, were likely being shared by most carcinogens. The study included three dose levels of a number of well-characterized compounds, including five non-genotoxic carcinogens (bemitradine, clofibrate, doxylamine, methapyrilene, and phenobarbital), one genotoxic carcinogen (2-acetylaminofluorene), one carcinogen that may act via genotoxicity (tamoxifen), a mitogen (isoniazid), and a non-carcinogenic hepatotoxicant (4-acetylaminofluorene). Mining of the hepatic gene expression profiles generated on a cDNA microarray platform resulted in the identification of several candidate molecular markers, including transforming growth factor-β-stimulated clone 22 (*TSC-22*) and NAD(P)H cytochrome P450 oxidoreductase (*CYP-R*), which were down-regulated and up-regulated, respectively.

The expression levels of these two candidate markers correlated with an estimated carcinogenic potential of each compound and dose level. Obviously, it is expected that mining large numbers of genes for a relatively small set of experiments would generate correlations that may occur by chance. However, the role and function of the two identified markers suggest a mechanistic role in hepatic carcinogenesis. For instance, CYP-R catalyzes the transfer of electrons from NADPH to heme oxygenase, cytochrome b5 and a variety of cytochromes P450. As correctly stated by the authors, up-regulation of this gene may simply represent a surrogate marker for cytochrome P450 induction, since many non-genotoxic carcinogens are also cytochrome inducers (49). Alternatively, *CYP-R* induction may reflect a role for oxidative stress in rodent hepatocarcinogenesis. An independent study designed to confirm these results, however, could not demonstrate that CYP-R mRNA up-regulation was a reliable acute predictive marker of hepatic carcinogenesis (50).

TSC-22 was first isolated as a TGF-β-inducible gene in a mouse osteoblast cell line (51). TSC-22 is the only member of a novel family of transcription factors that contains a transcriptional repressor domain and does not have a DNA-binding domain but displays transcriptional repressor activity after complexing with a binding partner, THG-1 (52, 53). TSC-22 is postulated to be a tumor suppressor, since it has been shown to be down-regulated in several tumors or cancer cell lines and to be induced by anticancer agents (54, 55). To determine whether TSC-22 mRNA down-regulation can reliably predict carcinogenesis, a separate group investigated its modulation during a long-term study with clofibric acid in a different strain of rat (the Fisher 344 rat) using a classic initiation-promotion regimen (50). Briefly, male F344 rats were given a single non-necrogenic injection of diethylnitrosamine and fed a diet with or without clofibric acid for up to 20 months. Gene expression measurements revealed that TSC-22 mRNA was strongly down-regulated early in the study. This down-regulation was sustained throughout the study but disappeared on withdrawal of the test article. Consistent with these results, it was demonstrated in a separate study that TSC-22 mRNA is down-regulated by a factor of two to three in the mouse liver after a 2-week exposure to non-genotoxic carcinogens, such as oxazepam and Wy-14,643, as well as in spontaneous liver tumors and in B6C3F1 mouse liver tumors that were induced by several chemicals in 2-year carcinogenicity studies (53, 56). In addition, the same authors used a small interference RNA (siRNA) against TSC-22 to demonstrate that it down-regulates growth arrest and DNA damage-inducible 45β (GADD45b) expression, suggesting a mechanism by which TSC-22 mRNA down-regulation may alter the balance between cell death and survival (53).

These corroborating findings by three independent groups suggest that TSC-22 mRNA levels may represent an early marker to identify non-genotoxic carcinogens. However, more data will be required to better understand its real predictive value. For instance, the lack of regulation of TSC-22 mRNA levels following exposure to several non-carcinogens was confirmed (53). However, no down-regulation was detected after treatment for 2 weeks with other carcinogens, such as *o*-nitrotoluene or methyleugenol, indicating that this marker is unlikely to have a universal predictive value (53). In fact, it appears that no single marker would suffice to detect at an early stage all hepatic carcinogens, given the range of mechanisms potentially involved.

Other studies have used gene expression profiling to investigate the transcriptomic responses following exposure to hepatic carcinogens. In particular, an attempt was made to determine whether known genotoxic carcinogens would induce a common set of genes belonging to defined biological pathways and whether these genes could be used as a predictive signature for hepatic genotoxic carcinogens (57). To achieve this goal, potent carcinogens [dimethylnitrosamine, 2-nitrofluorene, aflatoxin B1, and 4-(methylnitroamino)-1-(3-pyridyl)-1-butanone or NNK] were dosed daily for 14 days in Wistar rats. Gene expression analysis of livers indicated that biological pathways, such as DNA damage response, specific detoxification response, or cell proliferation and survival, were primarily impacted. These shared pathway effects are consistent with the expected early events of tumorigenesis, suggesting that interrogation of these pathways could be predictive of future tumor development.

Encouraged by these positive preliminary results, the investigators extended their analysis with a subsequent study that included non-genotoxic carcinogens (methapyrilene, diethylstilbestrol, Wy-14643, piperonylbutoxide) (58). Their objectives were to determine whether carcinogens at doses known to induce liver tumors in the 2-year rat bioassay deregulate characteristic sets of genes in a short-term in vivo study with male Wistar rats and whether these deregulated genes represent defined biological pathways. Non-genotoxic carcinogens impacted cellular pathways distinct from those affected by the genotoxic carcinogens (including oxidative DNA or protein damage, cell cycle progression, tissue regeneration), and demonstrated pathway effects that were consistent with compound-specific mechanisms. These results confirmed the notion that while neither a single gene nor a single pathway will be sufficient to discriminate between the two classes of carcinogens, the use of combinations of pathway-associated gene expression profiles represents a feasible approach to predicting the genotoxic or non-genotoxic carcinogenic potential of a compound in short-term rat studies.

These pioneering studies represented critical steps in demonstrating the potential of genomics to predict carcinogenicity. However, because they relied on a limited set of reference profiles and lacked a sufficiently robust set of gene expression profiles induced by non-carcinogens, it is impossible to estimate the predictive value of the markers identified. In addition, these studies did not take advantage of sophisticated algorithms to define robust multigene markers that could be used in a quantitative or semiquantitative assessment. As previously discussed, to generate an accurate predictive gene expression signature, a sufficient repository of gene expression profiles from a larger variety of carcinogens and non-carcinogens administered at different doses and different time points would be required. Ultimately, this can only be effectively accomplished through a collaborative effort in the scientific community. Nevertheless, recent independent reports have used this approach, which could result in an in silico refinement and validation of the proposed predictive signatures of carcinogenicity.

With gene expression profiles from the livers of rats treated for up to 14 days with a variety of genotoxic and non-genotoxic carcinogens, as well as non-carcinogens, classifiers were generated with a support vector machine (SVM) algorithm (59). These classifiers were then tested with a set of independent gene expression profiles. Results indicated an accuracy of up to 88%, depending on the marker gene set evaluated. Likewise, with hepatic gene expression profiles from rats treated for 5 days with one hundred structurally and mechanistically diverse non-genotoxic hepatocarcinogens and non-hepatocarcinogens, a signature of hepatic carcinogenicity was derived (60). An independent validation of the signature on 47 test chemicals indicated assay sensitivity and specificity of 86% and 81%, respectively (60). Interestingly, this study also evaluated in parallel several proposed short-term in vivo indicators of hepatic carcinogenicity (such as liver weight, histological evidence of hepatocellular hypertrophy or hepatic necrosis, serum ALT activity, induction of cytochrome P450 genes, and repression of TSC-22 or α2-macroglobulin mRNA). The gene expression signature was shown to be more accurate than the other indicators measured. In addition, as indicated by the authors, the genomic analysis offers the additional benefit of providing data to better understand the modes of action of the test compounds, since the gene expression profiles can be compared to the reference profiles induced by various test chemicals for which mechanistic information may be available.

Finally, a third group took advantage of a large internal database of hepatic gene expression profiles from male rats treated for 1 day with more than one hundred compounds to build gene expression signatures for non-genotoxic hepatic carcinogens (61). A gene selection algorithm yielded six genes (nuclear transport factor 2, *NUTF2*; progesterone receptor membrane component 1, *Pgrmc1*;

liver uridine diphosphate glucuronyltransferase, phenobarbital-inducible form, *UDPGTr2*; metallothionein 1A, *MT1A*; suppressor of lin-12 homolog, *Sel1h*; and methionine adenosyltransferase 1 alpha, *Mat1a*), which identified non-genotoxic carcinogens with an 88.5% prediction accuracy, as estimated by cross-validation. This six-gene signature also predicted non-genotoxic carcinogens with an 84% accuracy when samples were hybridized to a different microarray platform. Analysis of these data also confirmed that the differentially expressed genes mostly belong to biologically and cancer-relevant pathways.

5.3.4. Gene Expression-Based Biomarkers in Other Tissues and the Promise of Hemogenomics

While the vast majority of published reports on predictive approaches using genomics have focused on the liver and to a smaller extent on the kidney, conceptually, similar approaches can be applied to other tissues. In our laboratory, for instance, we have generated signatures or predictive models for such tissues as spleen and heart, and others have reported useful results in skeletal muscle (27). However, for several tissues, predictive gene expression-based models are of limited benefit for the toxicologist or are less likely to perform at acceptable levels of accuracy. First, predictive signatures for tissues that are seldom the target of development-limiting toxicity are of no or very limited utility, and hence the cost of generating these genomic biomarkers can hardly be justified. Second, as already alluded to in Chapter 4, certain tissues present an inherent structural complexity that challenges the development of reproducible gene expression profiles. We already discussed the heterogeneity of the brain and gastrointestinal tracts as examples, but several other tissues present this problem as well. Third, for some tissues, robust toxicity biomarkers already exist. An example is bone marrow, in which toxicity can easily be monitored with standard hematology panels. In these instances, the incremental value of genomics biomarkers is so low that one might question the rationale of such an investment. This does not suggest that genomic investigations in these tissues do not provide any value, but the value of genomics experiments here would be mostly in generating a mechanistic understanding of toxic events, rather than in predicting them.

In our opinion, blood represents a tissue with great potential for predictive toxicogenomics for several reasons. While genomics signatures represent an exciting scientific advancement, their use has so far been limited to preclinical samples, since human tissues are available only in rare circumstances. In contrast, blood is easily accessible for repeated measurements without the need for invasive collection procedures. Multiple studies have already demonstrated the feasibility of measuring specific transcripts in human blood samples and using them as molecular markers for evaluation of disease risk or diagnosis of pathological or toxicological conditions, or as predictors of drug response (62–64). Furthermore, blood could serve as a surrogate for other tissues to monitor perturbations throughout the organism (65, 66). It is believed that white blood cells can monitor the state of the various components of an organism, and that this monitoring activity is reflected in their transcriptome. Using blood in predictive toxicogenomics could therefore result in major advantages for toxicologists. First, blood-based predictive genomic signatures would limit the generation of expression profiles to one, rather than multiple, tissues since it is potentially feasible to monitor the state of various tissues by studying blood. Second, it would become possible to bridge preclinical studies with clinical trials. One of the main reasons for compound termination in preclinical development is the presence of toxic changes that cannot be appropriately monitored in the clinic because of the lack of appropriate biomarkers. Blood-based genomics signatures could represent a new source of valid biomarkers to monitor specific toxic events in clinical trials. Several studies suggest that this can indeed be achieved, since various insults or pathological conditions in rats and humans have been shown to induce characteristic gene expression changes in blood (66–72). Likewise, peripheral blood has been shown to be an appropriate matrix for molecular markers of diseases such as Parkinson disease or Huntington disease (73, 74). Finally, a recent study has shown that diagnostic molecular profiles of radiation exposure can be generated in mouse blood and transferred to humans (75).

Before genomics signatures of toxicity can be generated with blood, several technical and scientific challenges must be addressed (66). Unlike homogeneous tissues such as liver, blood is composed of a heterogeneous cell population that is subject to changes related or unrelated to the toxicity. For instance, decreases in white blood cell counts usually driven by lymphopenia are frequent observations in toxicology studies. These variations in cell population could make it difficult to determine whether gene expression profiles reflect alterations in cell population, changes in the transcriptome of a specific cell subpopulation, or simply intraindividual sample variation (76). In addition, it may be challenging to define the cell type that is driving the observed gene expression alterations. Gene expression profiling of blood is also complicated by the presence of globin transcripts that make up as much as 70% of whole blood mRNA. Their abundance leads to significant technical issues exemplified by decreased percentages of "present" calls on microarrays and by an increased variability between replicates (77). This technical difficulty can be resolved by using one of the existing globin reduction protocols or white blood cell isolation procedures (such as buffy coats), but these additional steps add a non-negligible level of complexity and, in our experience, may increase the interassay variability (66, 77). Finally, it is well recognized that the field of blood genomics, also referred to as hemogenomics, is relatively immature and that substantial efforts in standardization of practical collection procedures are still required (66).

Currently, no experimental data exist to support the hypothesis that blood-derived gene expression profiles can be used to predict or diagnose toxic events occurring in other parts of the organism. We believe that blood-derived gene expression signatures may be useful in detecting toxic changes in other tissues. The critical question, however, is whether these signatures will provide any additional value compared to the battery of blood-based biomarkers currently available to the toxicologist. Several groups are starting to address the question of preclinical and even clinical utility, but very limited data have been publicly released so far. In our laboratory, one of the current objectives is to build a small repository of blood-derived gene expression profiles. Blood samples have been obtained from rats treated with carefully selected compounds causing known and well-characterized toxic changes in a specific tissue. Reference gene expression profiles will be used to evaluate the feasibility of blood-derived genomics signatures. Our current efforts are devoted to establishing the necessary protocols to reach an acceptable level of technical reproducibility and repeatability. Once established, these protocols will be used to determine whether biological variability in blood is sufficiently low to generate useful signatures. If an adequate level of biological reproducibility is demonstrated, multigene classifiers will be derived for specific toxic changes. We will first interrogate end points that are more straightforward to detect and then will follow up with more challenging biomarkers.

Thus it is not yet clear whether hemogenomics has any practical use in toxicology, but it is important to reemphasize that blood-based genomics signatures may be useful in scanning for the presence of toxic changes occurring anywhere in an organism. Therefore, they may represent a useful complement to the current serum chemistry and hematology biomarkers that are routinely used in toxicology. Signals from these signatures could then be further investigated by evaluating the transcriptome of the tissues identified as targets of toxicity either to confirm the presence of toxicity or to investigate its mechanism.

5.3.5. Integration of Toxicogenomics in Discovery Toxicology

In the last decade, the value of early in vivo toxicology studies, mostly using rodents and in particular rats, has been well recognized. These non-GLP exploratory studies are now commonly conducted in several pharmaceutical R&D organizations (5). The objectives of these studies are to detect at an early stage potential toxicology liabilities associated with compounds or series of compounds and to generate dose-ranging information necessary for proper design of regulated toxicology studies. In other words, data from these early in vivo toxicology studies should not be viewed as a part of a risk assessment scheme, but mostly as useful pieces of information to make internal decisions and, if possible, to shorten the time to the initiation of the regulated GLP toxicology studies.

Since there are no guidelines for designing such studies, various strategies have been used. As we have already indicated, these in vivo testing approaches are often influenced by a past institutional experience or by the preferences and beliefs of the leaders of an organization. In addition, the capabilities and available expertise, as well as various logistic aspects, dictate to some extent the appropriate course of action for a specific organization. Therefore, in this section, we only discuss the strategy currently followed by our organization. We explain how toxicogenomics has been integrated to increase the value of early exploratory rat toxicology studies. Obviously, such an approach may not be appropriate for every organization, and with the advent of new technologies or the introduction of additional biomarkers, it may require modifications. However, we believe that it represents a good illustration of how one may be able to take advantage of the predictive value of toxicogenomics to impact the productivity of a discovery organization.

We typically evaluate compounds with desirable properties as early as possible in short-term exploratory toxicology studies using male Sprague-Dawley rats. Before these studies are conducted, sufficient characterization (pharmacology and pharmacokinetic studies and in vitro genotoxicity screens) has been performed to ensure that the compounds have a good chance of being viable. Appropriate dosages are selected based on the information available for the compound. Dosage selection represents probably the most challenging part of the study design, and experience in conducting such studies is definitely a prerequisite. It is beyond the scope of this chapter to review all alternatives used to select optimal doses. It suffices to say that the objective is to have at least one dosage group in which dose-limiting toxicity can be identified and characterized. Exceptions may exist, for instance, when saturation of absorption for orally dosed compounds occurs. However, the primary objective of the study is to generate data sufficient to make decisions on the future fate of a compound, so appropriate data must be generated. These studies are typically of 5-day duration, but can be of shorter or longer duration in some cases. It is important to point out that there is no industry consensus on what represents the optimal duration for these studies, with proponents for shorter studies and proponents for longer (up to 2 week long) studies. Obviously, longer studies are more likely to detect toxic effects; however, they also require more resources, in particular more active pharmaceutical ingredient (API). In our experience, 5-day duration represents the best compromise, since it is most often sufficient to reach steady-state tissue levels and since it can address such issues as tissue accumulation, while requiring a reasonable amount of API. As mentioned before, API is often the reason why studies cannot be conducted at early stages, and therefore compound requirement must be a major criterion when evaluating technologies to characterize compounds in discovery. Our internal data suggest that, at least for our chemical space, the vast majority of dose- and development-limiting toxic effects can be detected within 5 days of daily dosing, particularly when toxicogenomics data are invoked.

The incremental value of including females in these early studies is also a subject of current debate. There is definitely a non-negligible risk in conducting studies with males only, since one may argue that differences in sensitivity between sexes or female-specific toxic effects may exist. This is an issue that can be addressed to some extent with toxicogenomics. By providing mechanistic information on the toxicity observed, gene expression profiling may help limit the risk of using only one sex, especially when evaluating gene expression profiles in the context of a robust database of profiles. While certain toxic effects may appear sex-specific based on phenotypic evaluation, they usually trigger unique pathways in the other sex that may not be obvious at the phenotypic level but can be interrogated at the molecular level. We currently do not have sufficient data to corroborate this, but we have observed pathway effects consistent with this assumption. Thus one may be able to use gene expression profiles in males to predict the occurrence of female-specific toxic effects in non-sexual organs. These signals may be used to justify a rapid follow-up study using female rats. It is also not infrequent to note differences in systemic exposure to compounds between male and female rats. Assuming that most toxic changes are driven by exposure and not by sex-specific responses, these situations can be easily addressed before initiating longer-term studies by conducting pharmacokinetic studies in females at selected doses to confirm that exposures in females are similar to those seen in males. If differences in exposures are observed, then a better understanding of how differently female rats may respond on a toxicological level may be warranted.

We have already indicated that gene expression profiles after exposure to compounds show relatively low interindividual variability, in contrast to what may be seen with conventional hematology and serum chemistry parameters. In our opinion, this represents a very useful feature of toxicogenomics, since it allows one to reduce the number of rats used in studies. This reduction has important implications both in terms of API requirement and in terms of animal welfare issues. In our studies, we have found that using three male rats per dosage group is often sufficient, and that the incremental value of adding more animals is usually not justified.

Obviously, the predictive aspect of toxicogenomics represents its most interesting feature in the context of these studies. However, toxicogenomics has important limitations. As an example, a case of spontaneous pathology and its impact on tissue gene expression profiles is illustrated in Figure 5.2. Spontaneous changes are not uncommon in rats and may be difficult to identify with gene expression profiling in isolation. Because of years of experience reviewing rat tissues, the pathologist is probably best positioned to classify these changes as spontaneous. The example used in Figure 5.2 exemplifies the reasons why it is inappropriate to implement toxicogenomics in a vacuum. The value of toxicogenomics is realized when used as a complement to, rather than as a replacement for, the current robust, well-validated end-points routinely used by toxicologists. Figure 5.3 summarizes how we have implemented toxicogenomics in our testing strategy.

5.4. TOXICOGENOMICS IN MECHANISTIC TOXICOLOGY

Toxic changes are commonly identified in preclinical studies. In fact, the objective of preclinical toxicology studies is to define the toxicological profiles of experimental compounds, and consequently, these studies in most instances do reveal toxic changes, especially at the high dose level. These toxic changes may

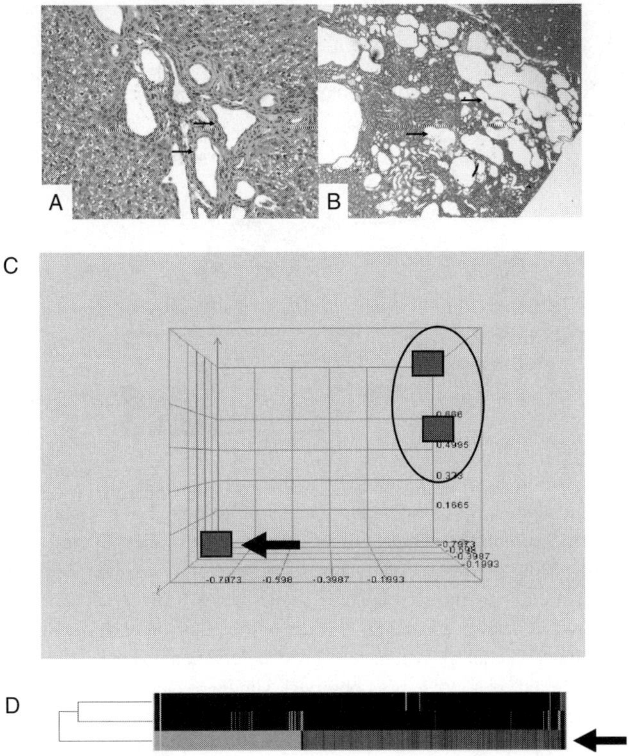

Figure 5.2 Example of a spontaneous change in the liver (A; 200 × magnification) and kidney (B; 20 × magnification) of a male Sprague-Dawley rat and its impact on tissue gene expression profiling. In this specific example, the rat suffered from a congenital, genetic condition called polycystic kidney disease, leading to the presence of cystic bile ducts in the liver (A; arrows) and tubules in the kidneys (B; arrows). These cystic ducts and tubules are associated with a variety of degenerative changes, such as interstitial fibrosis or chronic inflammation. Not surprisingly, a transcriptomic analysis of these tissues indicated marked changes compared to controls, which could suggest toxic changes. Illustrated here is a principal component analysis (C) and an agglomerative cluster analysis (D) of the gene expression changes observed in the kidney of this specific rat (thick arrows) and two other rats from the same treatment group. As expected, the rat with congenital polycystic kidney disease is very different at the transcriptomic level from the other two rats from the same treatment group, and many differentially expressed genes can be identified in this rat. Without concurrent histological evaluation, these transcriptomic changes would likely have been incorrectly interpreted as indicative of toxicity. See color insert.

prompt the decision to discontinue the compound or help to determine a No Adverse Effect Level (NOAEL) to ensure that sufficient safety margins exist to proceed. Various toxic changes are common and mechanistically well understood, so that toxicologists can adequately assess risk or terminate a molecule. When several alternative backup compounds are available, it may make more sense to

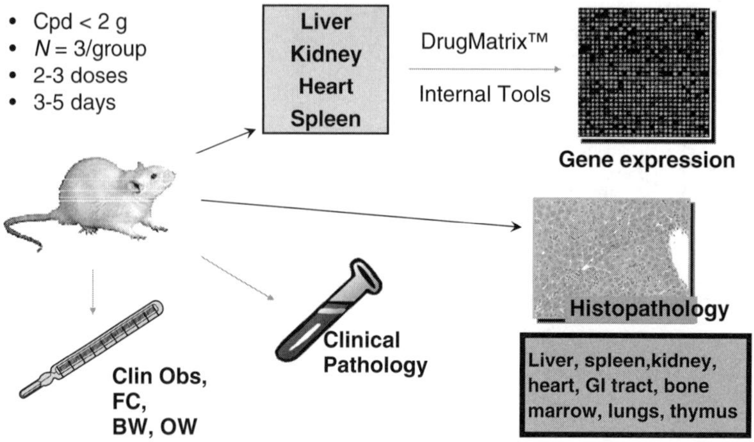

Figure 5.3 Integration of toxicogenomics in the exploratory toxicology paradigm. Toxicogenomics represents a useful supplement to the current battery of toxicological end points and allows for a more refined characterization of compounds in short-term rat studies. In these 3- to 5-day exploratory studies, we typically use three male rats per dose level and two doses with the objective of reaching a toxic dose at the high dose level. These studies can typically be conducted with less than 2 g of compound, a considerable advantage in discovery. Evaluation of the primary end points (in-life observations, clinical pathology parameters, gross observations) allows for rapid decisions at the end of the dosing period. If no toxicity is observed with these end points, histopathological evaluation of a limited tissue set (liver, spleen, kidney, heart, gastrointestinal tract, bone marrow, lungs, thymus) is performed. If no toxic changes are identified by histopathology, the absence of toxicity is confirmed with a battery of gene expression signatures for several tissues, such as liver, kidney, heart, or spleen. In addition, gene expression profiles are generated when there is a need to understand a mechanism of toxicity (MOT).

rapidly advance a backup compound than to investigate a development-limiting toxic change associated with the original molecule. Thus, on a business level, not all toxic changes are worth investigating. There are, however, many instances where the mechanistic understanding of a toxic change provides significant additional value (Box 5-1). For instance, the relevance to humans of a specific toxic effect detected in animals may not be obvious. Also, a toxic effect may be detected very late in a program, for example, during the chronic or lifetime animal studies. In that situation, given the amount of resources already invested in a compound, it may make a lot of business sense to investigate the toxicity further at the molecular level. This may help rescue the compound if mechanistic studies demonstrate the irrelevance of the toxicity to humans or if robust biomarkers are identified to monitor these changes in humans. Mechanistic studies may also assist in the selection of backup compounds in a program plagued by a specific development-limiting toxic change. Indeed, understanding the mechanism of a specific toxicity facilitates the development of appropriate counterscreens to identify backup compounds unlikely to induce the same toxicity. Finally, some toxic changes seen only at very high exposures may still be development-limiting when no robust biomarkers exist to monitor them in the clinic. This situation is a frequent cause of termination of otherwise very attractive compounds. By interrogating the transcriptional effects induced by a compound, one may be able to identify biomarkers that can subsequently be validated and used in clinical studies to monitor toxicity.

BOX 5-1 *Value of Mechanistic Toxicological Investigations*

- Understanding of the relevance to humans of a preclinical toxicological finding
- Improved risk assessment and better extrapolation of preclinical findings to the clinic
- Identification of specific biomarkers to monitor a toxic reaction preclinically or in the clinic
- Rational selection of counterscreens for the efficient prioritization of backup compounds unlikely to induce a specific toxic change
- Differentiation of on-target from off-target effects: better prioritization of targets in the discovery portfolio

Microarray-based gene expression profiling is a reliable and efficient technology to rapidly generate large numbers of molecular data for a particular toxicity. However, it would be misleading to suggest that genomics data are typically sufficient to develop a good understanding of the molecular mechanism of toxicity. Instead, such data should be analyzed to formulate mechanistic hypotheses that can be confirmed or refuted in follow-up functional experiments. There are instances, however, when gene expression profiles can be used with sufficient confidence to make decisions.

This section discusses the use of gene expression profiling to understand toxic mechanisms in the context of drug discovery and development. Since the introduction of gene expression microarrays, a significant number of reports have been published that describe the use of gene expression profiling to investigate mechanisms of toxicity. These publications originate from academic, industrial, and regulatory laboratories and, although scientifically interesting, may not always be relevant to a toxicologist working in pharmaceutical R&D. Here, we restrict our discussion to practical approaches to addressing toxicity issues and illustrate them by using selected examples from the literature or from our personal work. We apologize in advance to our colleagues if their work is not included in the following pages of this chapter.

5.4.1. Toxicogenomics to Investigate Mechanisms of Hepatoxicity

Years of investigational studies have led to a good understanding of various mechanisms of hepatoxicity. Not surprisingly, it is primarily in the liver that the value of toxicogenomics in deciphering mechanisms of toxicity has been demonstrated. In general, this has allowed various investigators to confirm that, in most instances, gene expression profiles were extremely useful in deciphering mechanisms of changes observed in rats or mice. Hepatic hypertrophy is probably the best example to illustrate the use of gene expression profiling in elucidating the molecular mechanism of a phenotypic change.

Liver enlargement, detected through increased liver weight or by histopathological evaluation, is a relatively frequent observation in rodents treated with xenobiotic agents. There are several well-characterized mechanisms of drug-induced hepatomegaly, such as increases in smooth endoplasmic reticulum content, increases of cytochrome P450 monooxygenase (CYP) activity, peroxisome proliferation, hypertrophy of mitochondria, or enhanced hepatocellular proliferation (78). These changes may be considered adaptive and not necessarily of toxicological significance, or may be rodent-specific and therefore not relevant to humans (79, 80). Nevertheless, they can also be associated with evidence of liver injury, as evidenced by increased serum ALT activity levels, hepatocellular apoptosis, and necrosis, as can be seen with enzyme induction responses (81, 82). In addition, compounds causing liver hypertrophy in rats may be non-genotoxic carcinogens, that is, agents that can induce liver tumors in the lifetime rodent carcinogenicity studies without causing direct chemical DNA damage (i.e., these compounds would not be detected as genotoxic agents with the genotoxicity assays). While it is well understood that findings in rodent carcinogenicity studies may not always be relevant to humans, a positive test result triggers appropriate experiments to confirm the mechanism of the toxicity and determine its relevance to humans (19). Thus an early understanding of the mechanism behind a case of rodent hepatic enlargement represents a definite advantage, since it helps in interpreting and positioning liver findings.

Hamadeh et al. were among the first to test the applicability of gene expression profiling to distinguishing between different classes of hepatic enzyme inducers, compounds frequently associated with liver enlargement in rats (83). In this study, rats were treated with three classic peroxisome proliferators (clofibrate, Wy-14,643, and gemfibrozil) and an enzyme inducer, phenobarbital. Treatment of rats with phenobarbital leads to increased liver weights and hepatocellular hypertrophy due to induction of various drug-metabolizing enzymes, including cytochrome P450 2B (CYP2B), 2C, and 3A (84). As expected, rats treated for a single day with either test article showed no histopathological changes, whereas after 2 weeks of treatment hepatocellular hypertrophy was confirmed histopathologically in all animals. Gene expression profiles were generated from the livers of the rats treated for a single day. The three prototypical peroxisome proliferators induced gene expression patterns distinct from those caused by phenobarbital.

In addition, the authors identified a set of 22 genes to accurately classify blinded liver hypertrophy-causing compounds as inducers of peroxisome proliferation or phenobarbital-like compounds. In this pioneering study, the authors addressed relatively straightforward mechanisms associated with hepatic enlargement. However, their approach represents a good illustration of how specific mechanisms of toxicity associated with a similar phenotypic outcome can be distinguished by gene expression profiling. In particular, it illustrates how extremely useful collecting gene expression profiles at multiple time points during mechanistic studies can be, since the pathogenesis of a change may be more apparent from the gene expression profiles before obvious phenotypic changes occur.

In our laboratory, we have routinely been using gene expression profiling to study various hepatotoxic findings. For instance, a few years ago, our laboratory reported an investigational study to define the mechanisms of hepatotoxicity associated with an experimental thienopyridine inhibitor of NF-κB-mediated expression of cellular adhesion molecules (81). This compound, known as A-277249, induced hepatic changes in rats in a 2-week repeat-dose toxicity study. These changes were characterized by hepatic enlargement, elevations of serum transaminases (ALT and AST), alkaline phosphatase, and GGT, and histopathological changes (hepatocellular hypertrophy and hyperplasia). It was assumed that mechanistically relevant transcriptomic changes may occur earlier than the morphological change. Therefore, a 3-day repeat-dose rat toxicity study was repeated for this compound at doses known to be associated with the phenotypic changes. Gene expression profiles of the livers were generated with microarrays. The compound was shown to induce extensive changes in gene expression. Using our internal repository of gene expression profiles induced in rat liver by prototypical hepatotoxicants, we applied agglomerative hierarchical clustering to demonstrate that A-277249 induced transcriptomic changes similar to those caused by well-characterized activators of the aryl hydrocarbon nuclear receptor (AhR). This "guilt-by-association" analysis suggested that the hepatic changes induced by A-277249 were, at least in part, mediated by AhR either directly or through NF-κB (81).

AhR is a nuclear receptor that mediates responses to various toxicants, such as halogenated aromatic toxicants (85). In this particular example, the toxicogenomics evaluation strongly indicated that this chemical class may not be optimal and that an AhR-mediated mechanism was likely the cause of the hepatotoxicity. One should, however, be cautious not to overinterpret gene expression changes, because they reflect a large number of complex, interacting molecular processes that may or may not be related to a toxic reaction (86). Simply stated, not all gene expression changes induced by toxicants are toxicologically relevant. Activation of AhR represents a good ilustration of the danger associated with overinterpretation. Induction of cytochrome P450 1A1 (CYP1A1) is a hallmark of AhR pathway activation and is involved in the pathogenesis of the toxicity caused by 2,3,7,8-tetrachlorodibenzo-*p*-dioxin (87, 88). CYP1A1 is a member of a family of xenobiotic metabolizing enzymes involved in detoxification of polycyclic aromatic compounds; it also contributes to the generation of mutagenic and reactive metabolites (89). Therefore, one may interpret these data to suggest that up-regulation of CYP1A1 mRNA may be used as evidence of AhR pathway activation and of dioxin-like toxicity. Such an oversimplification would be erroneous. Indeed, using large databases of reference gene expression profiles from rat livers, we and others have demonstrated that numerous non-AhR-activating compounds significantly up-regulate CYP1A1 expression in rat liver at toxic doses (90). Furthermore, a significant portion of these compounds consist of FDA-approved drugs from multiple classes that do not represent risk for humans (90).

Nevertheless, CYP1A1 induction may be relevant to some hepatic and pharmacokinetic changes. Recently, we evaluated an experimental compound in a 5-day exploratory rat toxicology study. This compound moderately increased liver weight and induced centrolobular hepatocellular hypertrophy at exposures close to the efficacious ones. In addition, the plasma exposures after 5 days of daily dosing (as calculated by the area under the curve or AUC for a 24-h period) were decreased by as much as 50%. To investigate the mechanisms of these phenomena, we conducted a toxicogenomics analysis of the liver. Treatment with the compound was found to be associated with a significant up-regulation of CYP1A1 mRNA, which was confirmed by RT-PCR. Furthermore, we observed increased CYP1A1 activity in isolated hepatic microsomes and elevated CYP1A1 expression by immunohistochemistry (Fig. 5.4). An in vitro metabolism screen showed that the test article was metabolized preferentially by CYP1A1, suggesting that pharmacokinetic changes could be explained by hepatic autoinduction. Overall, these results indicated that the phenotypic and pharmacokinetic changes observed in rats were likely related to induction of CYP1A1. This understanding prompted us to establish a strategy for selecting backup compounds from the same chemical series by screening compounds in vitro with primary rat hepatocytes and evaluating CYP1A1 induction by quantitative reverse transcription-PCR. This strategy resulted in the rapid selection of compounds not associated with

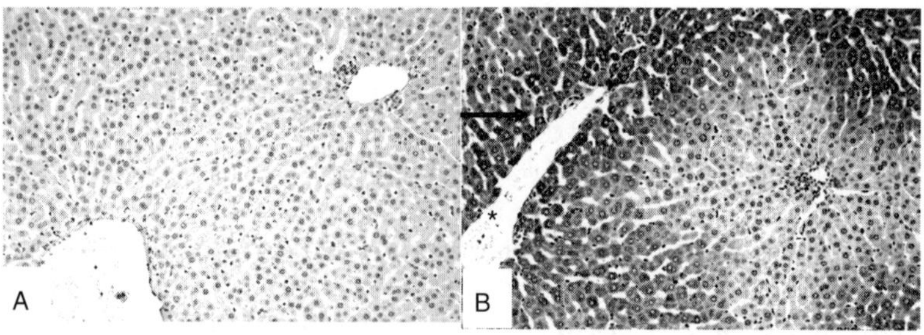

Figure 5.4 Immunohistochemical detection of cytochrome P450 1A1 (CYP1A1) in rat liver. Shown here are sections of liver from a vehicle-treated rat (A) and from a rat treated for 5 days with a CYP1A1 inducer (B) These sections were processed for immunohistochemical detection of CYP1A1. A dark cytoplasmic stain indicates the presence of CYP1A1. Note that CYP1A1 cannot be detected in the vehicle-treated rat, in contrast to the marked induction observed in the test article-treated rat (arrow). * = central vein. Magnification 200 ×.

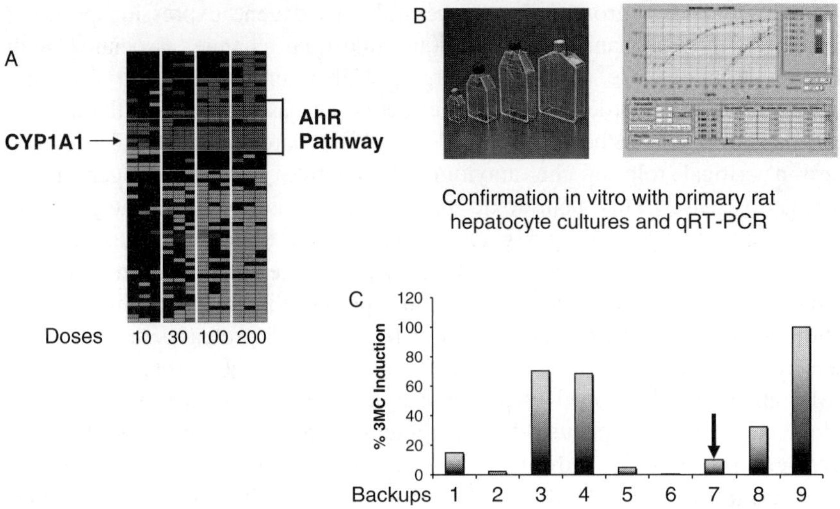

Figure 5.5 Use of toxicogenomics to elucidate metabolism issues. In this example, three male rats per group were treated for 5 days with an experimental compound at 10, 30, 100, and 200 mg/kg/day. Treatment with the compound was associated with dose-dependent increases in liver weight and histological evidence of mild centrolobular hepatocellular hypertrophy at 100 and 200 mg/kg/day. On transcriptomic analysis of the liver summarized on the heat map (A), this compound was shown to significantly induce CYP1A1 mRNA levels (arrow), as well as the aryl hydrocarbon receptor (AhR) pathway. This CYP1A1 induction was associated with a significant decrease in exposure over repeat dosing since the compound was also a substrate of CYP1A1 (autoinduction phenomenon). This mRNA induction of CYP1A1 was also detected in vitro with primary rat hepatocytes (B) and was shown to be relevant to humans with primary human hepatocytes. Backup compounds from the same series were then screened in vitro to rapidly identify a suitable backup not associated with this potential liability. In the bar graph shown in C, results from 9 compounds are shown (results are expressed as % of induction compared to 3-methyl-cholanthrene (3-MC), the positive control compound in this study. Backup 7 (arrow) was selected based on additional consideration and was shown to not be associated with autoinduction in a follow-up 5-day rat toxicity study. See color insert.

CYP1A1 induction, phenotypic changes in the liver of rats, and autoinduction (Fig. 5.5).

5.4.2. Intestinal Toxicity and Notch Signaling

Large-scale gene expression profiling may theoretically be used to investigate the mechanism of action of compounds in virtually any tissue. However, the complexity of some tissues may hinder the use of this method. Since the gastrointestinal tract represents a good example of a heterogeneous tissue, this section discusses the application of genomics technologies to understand the mechanism of gastrointestinal toxicity.

Two independent groups have successfully used gene expression profiling to elucidate the mechanism of a unique intestinal change associated with inhibition of a putative therapeutic target. Alzheimer disease is a disabling neurodegenerative disorder for which effective disease-modifying therapeutic treatments have not yet been developed (91, 92). There is substantial evidence toward a critical role of the amyloid-beta peptide in the pathogenesis of Alzheimer disease. This neurotoxic peptide abnormally accumulates in the brain of patients and self-aggregates to form toxic oligomers, causing neurodegeneration. Amyloid-beta is produced from a precursor protein (amyloid precursor protein, or APP) by two sequential proteolytic cleavages that involve β- and γ-secretases (91, 93). Hence, these two aspartyl proteases are attractive targets for pharmacological intervention, and functional γ-secretase inhibitors (FGSIs) are being developed as potential agents for the therapeutic control of amyloid buildup in Alzheimer disease. Potent FGSIs have been identified by several groups (94, 95). However, these compounds are usually not specific to γ-secretase cleavage of APP, and they also inhibit the processing of other γ-secretase substrates, such as Notch and a subset of cell surface receptors and proteins involved in embryonic development, hematopoiesis, cell adhesion, and cell-cell contacts (96). In particular, FGSIs can block the cleavage of the cell fate regulator Notch-1, which plays an important role in the differentiation of the immune system and the gastrointestinal tract (94, 95, 97).

During the preclinical evaluation of experimental FGSIs from various chemical series, a unique gastrointestinal change was observed in rats. This change was characterized by an increase in gastrointestinal weight, distended stomach and small and large intestines, and a mucoid enteropathy related to goblet cell hyperplasia (97, 98). This metaplastic change in the ileum of rats was suggestive of disruption in Notch signaling. In a first investigation, the authors conducted a microarray analysis of the ileum from FGSI-treated rats (98). The analysis of the differential expression responses supported a perturbation in Notch signaling perturbation. These gene expression studies also identified that the gene encoding the serine protease adipsin was significantly up-regulated after treatment with FGSIs. The biological significance of this interesting finding was confirmed by demonstrating elevated levels of the adipsin protein in the gastrointestinal contents and feces of FGSI-treated rats, as well as an increased number of ileal crypt enterocytes expressing adipsin by immunohistochemistry. These data suggested that adipsin could be exploited as a specific, sensitive, and non-invasive biomarker of FGSI-induced gastrointestinal toxicity.

In a second study, an independent group of investigators evaluated the same phenomenon with a reverse approach. Their objective was to evaluate whether FGSIs known to modulate Notch processing alter differentiation in tissues, whose architecture is governed by Notch signaling (97). Han Wistar rats were dosed for up to 5 days with experimental FGSIs from three chemical series, which differed in their in vitro inhibitory activity against Notch processing. The potent inhibitors of Notch processing caused a dose-dependent goblet cell metaplasia in the intestinal tract, in contrast to the weak inhibitors. Gene expression profiling of the duodenal samples from animals with gastrointestinal changes revealed significant time-dependent deregulation of expression of various genes, including that coding for adipsin, as well as several others encoding various panendocrine, hormonal, differentiation, and transcription factors. Of particular interest, Rath1, a transcriptional activator of gut secretory lineage differentiation, was shown to be up-regulated. It was also shown that both Rath1 and adipsin proteins were increased in quantity in feces, confirming their potential value as non-invasive biomarkers of intestinal goblet metaplasia.

Other published studies have addressed the application of genomics to the intestinal tract. For instance, recent studies has evaluated global gene expression profiles induced by several agents [curcumin and tunicamycin or (−)-epigallocatechin-3-gallate] in the small intestine of C57BL/6J mice (99–101). These experiments were useful as they improved our understanding of the biological effects of these agents. However, in the context of toxicological evaluations in pharmaceutical R&D, one may argue that these types of investigations do not represent the highest priority. In our experience, toxicogenomics studies generate the most value when addressing specific toxicology issues, rather than when they are conducted to generally increase the understanding of mechanisms of action for various compounds.

In the past few years, we have established that evaluating gene expression changes in the intestines of treated rats can be extremely valuable in investigating several findings. Specifically, we have on several occasions evaluated the level of expression of various CYP enzymes in the intestines to elucidate the mechanism behind reduced oral bioavailability of compounds after multiple dosing. In addition to autoinduction due to increased CYP activity in the liver, the intestinal metabolism can play a significant role in these cases. For instance, it has been shown that enhanced intestinal metabolism significantly contributes to the decrease in oral bioavailability of rifampicin in rats following prolonged exposure (102). While the preferred method to quantify induction of CYP enzymes is based on microsomal activity, quantification of mRNA levels by RT-PCR represents a rapid and accurate procedure to detect induction (103).

In addition to CYP enzyme induction assays, we have conducted several microarray analyses of the small and large intestines to further our understanding of the gastrointestinal effects of various novel formulations or to explain unexpected findings such as epithelial hyperplasia or intestinal degeneration and atrophy. In contrast to the studies mentioned above, we did not use RNA isolated from the entire tissue or from all components of the intestine. Indeed, it would be erroneous to consider the intestine as a single tissue and to neglect the strong regional differences present among the various parts of the gastrointestinal tract. In addition, the intestinal tract is composed of multiple compartments, such as the mucosa, submucosa, and muscular layers, complicating the interpretation of differential gene expression of a single compartment. To facilitate the analysis and to improve reproducibility, our protocol involved scraping of intestinal epithelial cells from consistent regions of the small intestines before RNA isolation. This approach ensures that gene expression is interrogated only in the

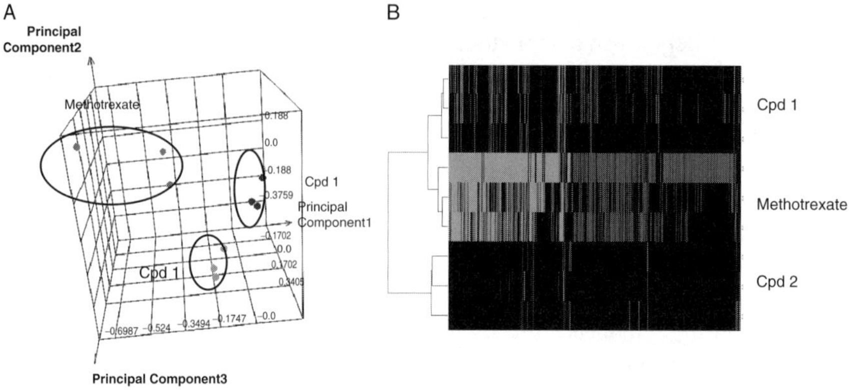

Figure 5.6 Gene expression changes in jejunal epithelial scrapings following treatment with experimental compounds. In this study, male rats were treated daily and orally for 5 days with methotrexate (a compound known to induce small intestinal injury), compound 1 (an internal experimental compound associated with limited epithelial changes in the small intestine), or compound 2 (an internal compound with no known intestinal effect). Jejunal epithelial scrapings, rather than samples of the whole jejunum, were collected and used for transcriptomic analysis, to limit the evaluation to the jejunal epithelium. Results are visualized here with principal component analysis (PCA; A) or a heat map with hierarchical clustering (B). Note that animals tightly cluster per treatment group, and that consistent changes in gene expression can be detected. In the heatmap, green indicates down-regulation, while red indicates up-regulation. In this particular example, using these differentially expressed genes, we could demonstrate that both compound 1 and methotrexate induced pathways consistent with epithelial proliferation and repair, a effect considered secondary to a primary toxic insult. See color insert.

epithelial layer, the compartment most often impacted by toxicological changes. In our experience, this protocol provided a very high degree of reproducibility with good repeatability. This procedure is illustrated in Figure 5.6. The large intestine is less amenable to mucosal scrapings, and for this organ using RNA extracted from specific sections may be more appropriate.

5.4.3. Cardiac Toxicity

For experimental therapeutic agents the liver is probably the most common target organ of toxicity, but cardiac toxicity is sufficiently prevalent in drug discovery and development to warrant interest by toxicologists. In development, pharmaceutical agents have been withdrawn from the market because of cardiovascular toxicity, while development candidates have been discontinued because of unanticipated cardiac findings in chronic toxicology studies. In discovery, it may be challenging to identify backup molecules to replace a lead compound that triggers development-limiting cardiovascular toxicity. Several preclinical systems have been developed to monitor functional deficits of the cardiovascular system. They have been shown to be quite useful in ensuring the safety of patients in clinical trials or in prescribing agents with potential cardiac effects (22, 25, 104). In contrast, no robust biomarkers exist for detecting and monitoring the progression of structural damage or altered cellular homeostasis in the myocardium (105). However, there is a significant interest in avoiding compounds that affect myocardiocyte homeostasis or cause structural damage, since these changes would likely be development-limiting when only low safety margins can be defined preclinically. In our experience, conventional biomarkers of cardiac injury, such as serum creatine kinase (CK), lactate dehydrogenase (LDH), aspartate aminotransferase (AST), glutamate dehydrogenase (GLD), and aldolase, are of limited diagnostic value in the evaluation of cardiac injury.

Recent data have shown that cardiac troponins represent more sensitive and specific biomarkers for detecting myocardial injury in the rat (105, 106). For instance, a recently published study using a isoproterenol model of cardiac injury in Hanover Wistar rats clearly demonstrated a close association between the histopathological assessment of myodegeneration and the serum levels of both cardiac troponins I and T (106). Maximal cardiac troponin levels preceded the maximal severity of the histopathological lesion observed. Generally, the troponins demonstrated superiority over the aforementioned conventional biomarkers of cardiac toxicity. However, while serum levels of cardiac troponins appear to provide better value than other markers, they do not precede the development of cardiac structural damage as identified histologically. Also, their half-life is relatively short, resulting in a short diagnostic window. For these reasons, troponins are unlikely to be sufficiently robust as biomarkers of cardiac toxicity.

Histological changes induced by cardiac toxicants are relatively limited in nature. In their acute state, they are usually characterized by myocardiocyte degeneration, apoptosis, or necrosis associated with secondary inflammatory infiltrates. At subacute to chronic stages, cardiac toxicity may be associated with chronic active inflammation and fibrosis. At these advanced stages of cardiac injury, where changes are non-specific and can be classified as "end-stage," it is difficult to formulate hypotheses regarding possible toxic mechanisms. Specialized techniques may be used to better define and potentially understand the mechanisms of toxicity. For instance, ultrastructural examination has been conducted to reveal that mitochondria likely represent a primary target organelle of toxicity (107). However, morphological changes observed by histopathology or electron microscopy rarely point at the exact mechanism of cardiac toxicity. For this reason, gene expression profiling represents an ideal approach to interrogation of molecular mechanisms of toxicity in the heart. However, just as in the case of the intestinal tract (considered in the previous subsection), it would be erroneous to regard the heart as a homogeneous tissue and to collect samples at random. In our experience, a consistent sample collection protocol is one of the most critical aspects of a successful toxicogenomics evaluation in the heart. The right and left ventricles differ transcriptionally, and are probably even more remote when compared to auricles. These differences have prompted us to establish a consistent collection protocol, whereby the heart is sectioned in half so that each section contains approximately similar portions of left and right ventricles and auricles. One half is used for gene expression analysis, while the other half is preserved in formalin to generate histopathological slides. The main disadvantage of this protocol is that it generates artifacts in the histology specimens. However, these artifacts can easily be recognized and disregarded by an experienced pathologist.

Doxorubicin has been frequently used to interrogate cardiac toxicity. Doxorubicin is an anthracycline antibiotic with broad-spectrum chemotherapeutic activity in hematologic malignancies and solid tumors (108). The use of this agent is, however, limited by its cardiotoxicity that may occur months to years after treatment (109). The toxicity also occurs in animals, providing excellent preclinical models to understand the mechanism of cardiotoxicity of the anthracycline antibiotics. Doxorubicin-induced chronic cardiotoxicity is always associated with dilated cardiomyopathy and histological lesions varying from myofibrillar loss to cytoplasmic vacuolization (mostly caused by dilation of the sarcoplasmic reticulum). This morphological pattern is observed in both humans and laboratory animals and reflects the action of metabolites formed after uptake and biotransformation of doxorubicin inside cardiomyocytes (110). Several studies have proposed a mechanism of cardiotoxicity whereby reactive oxygen species (ROS) are formed via redox cycling of the semiquinone radical mediated by a number of oxidoreductases endowed with one-electron Q-reductase activity (such as microsomal NAD[P]H-cytochrome P450 reductases, nuclear cytochrome b_5 reductase(s), or mitochondrial NADH dehydrogenase) (110–112). Formation of ROS results in damage to cellular macromolecules, such as DNA, proteins, or lipids, as well as disturbances in mitochondrial function (111, 112).

The cardiotoxicity of anthracycline compounds is difficult to detect with serum chemistry and light microscopy after a single dose in the rat (26, 108, 113). However, experimental evidence indicates that the damage caused by anthracyclines on cardiomyocytes is immediate after each injection and that the functional efficiency of the myocardium can be affected long before the morphological alterations become detectable (113). Not surprisingly, shortly after rats are treated with doxorubicin, large numbers of differentially expressed genes can be detected in the heart, and these early trancriptomic effects are useful to elucidate the mechanisms of acute toxicity associated with anthracycline compounds. We have used doxorubicin on several occasions as a positive control compound during investigative studies designed to elucidate the mechanism of cardiotoxicity associated with various experimental compounds (26). In these studies, we have consistently demonstrated that an acute treatment with doxorubicin deregulates several biological pathways, including pathways related to mitochondrial function and calcium regulation, consistent with its known mechanism of toxicity (26). These findings suggest that toxicogenomics represents a convenient tool for mechanistic investigations of drug-induced cardiotoxicity. Interestingly, these transcriptomic changes indicative of mitochondrial dysfunction occur well before the decrease in ATP production, which can be shown with isolated mitochondria from the heart of doxorubicin-treated rats (114, 115). These findings also confirm previous observations that transcriptomic effects relevant to the mechanism of toxicity can occur well before the actual organelle dysfunction is detected, and thus they suggest that the timing of the transcriptomic evaluation is critical. This phenomenon also enables effective selection of exploratory compounds without cardiac toxicity liabilities. This point can be illustrated by a recent study published by our discovery organization (116).

Our team was interested in developing inhibitors of acetyl-coA carboxylase (ACC) as potential therapeutic agents for the treatment of hyperlipidemia. The ACC enzymes catalyze the carboxylation of acetyl-coenzyme A (CoA) to form malonyl-CoA, which is used by fatty acid synthase for de novo lipogenesis (117). They also act as potent allosteric inhibitors of carnitine palmitoyltransferase 1 (CPT1), a mitochondrial membrane protein that shuttles long-chain fatty acyl-CoAs into the mitochondria, where they are oxidized (118). In rodents and in humans, there are two isoforms of ACC (ACC1 and ACC2) encoded by distinct genes (119, 120). ACC2 is the favored therapeutic target of the two enzymes, since ACC1-knockout mice are embryonically lethal, while mice lacking ACC2 show a favorable metabolic profile, as evidenced by resistance to high-fat diet-induced obesity and decreased lipid contents in both liver and adipose tissue (121, 122).

Evaluation of some of our early series of ACC2 inhibitors revealed neurological and cardiovascular liabilities in preclinical models, precluding further evaluation of this chemotype. These observations prompted us to develop counterscreens to select compounds without neurological and cardiovascular toxicity. In particular for the cardiotoxicity, we were interested in studying the structure-toxicity relationship to confirm that this liability was associated with the presence in the molecules of a specific alkyne linker (116). We evaluated these compounds in our favored toxicology model, the rat, so that it would be possible to take advantage of our repository of reference expression profiles. A study was conducted with male Sprague-Dawley rats administered daily for 3 days a toxic dose of an inhibitor representative of the problematic series. In addition, we used an inactive enantiomer of this inhibitor as a control to demonstrate that the toxicity was not related to an on-target effect, and doxorubicin as a positive control, since we knew that doxorubicin should induce relevant gene expression changes after 3 days of treatment.

As expected, no evidence of cardiotoxicity was seen by histopathological and serum chemistry evaluation after this short-term dosing period for any of the compounds tested. The gene expression profiles induced by both enantiomers in the heart were very similar to those induced by doxorubicin, confirming their cardiotoxic liability and also suggesting similar mechanisms of toxicity (116). In addition, the expression profiles induced by the enantiomers had strong correlations with a number of reference expression profiles in the DrugMatrix® commercial database (Iconix Biosciences, Mountain View, CA), especially with those induced by cardiotoxicants or cardiotonic agents, such as cyclosporine A, haloperidol, or norepinephrine, again suggesting cardiotoxic liability. Finally, when differentially expressed genes were analyzed in the context of biological pathways, the mitochondrial oxidative phosphorylation pathway was found to be impacted the most, implyng an effect on mitochondrial function, as has also been shown for doxorubicin and other cardiac toxicants. Overall, this straightforward investigation confirmed that the cardiotoxic potential of this series of compounds was structurally based and was not due to the inhibition of the target. This allowed us to rapidly select safer chemotypes for this program by replacing the alkyne linker with alternative linker groups. The cardiovascular profile of these novel structures was improved drastically, as evidenced by the lack of the toxicity-associated transcriptomic effects in the heart of treated rats.

5.4.4. Testicular Toxicity

The male and female gonads (ovaries and testes, respectively) are highly specialized tissues that produce mature gametes through an elaborate process of germ cell proliferation and differentiation, which is modulated by a complex network of endocrine and paracrine inputs (123). Because of their similarity in structure, regulation, and output, it is not uncommon for toxic agents to affect both tissues. To illustrate how gene expression profiling can be used to interrogate mechanisms of gonadal toxicity, we focus here on the testis, but similar approaches are likely to be applicable to the ovary.

Toxicology of the male reproductive system is receiving increased interest fueled by a growing number of reports on falling sperm counts and rising reproductive disorders in the human population (124, 125). There is a wide variety of possible mechanisms and manifestations of toxicity, but despite widely cited evidence that exposures to various environmental toxicants can lead to male infertility, little is really understood about the mechanisms involved (126). The testis is a known target of toxicity for a wide range of therapeutic and particularly environmental agents (123). Morphologically, some potent toxic agents can trigger widespread cell death in the testis of experimental animals, an event that can be detected by microscopic evaluation after even single doses or very short exposures. For instance, this phenomenon has been observed for a variety of chemotherapeutic agents developed for oncology indications. In these situations, testicular toxicity is expected based on the mechanism of action of the compounds. The chemotherapeutic agent cisplatin is a heavy metal coordination compound that produces DNA cross-links, leading to its therapeutic benefits but also to toxic effects, which include among others negative effects on spermatogenesis (127, 128). In contrast, initial effects on testicular function of agents developed for other therapeutic indications may not be evident, as subtle disturbances in spermatogenesis may not be detected at early stages, unless more sophisticated techniques, such as spermatogenic or tubular staging, are used (124). For these compounds, changes may not even be present at the morphological level until longer-term in vivo studies (4 weeks or longer) are conducted. In other words, potentially development-limiting testicular changes may be unexpectedly uncovered quite late in a program, and without a proper understanding of the toxic mechanism it may be challenging to determine whether the toxicity is of relevance to humans, decide what backup compound to move forward, or ascertain whether the testicular changes are due to the inhibition of the target or to an off-target effect.

Currently, no robust and sensitive biomarker exists for predicting testicular toxicity in preclinical studies or clinical trials. For this reason, testicular toxicity can be development-limiting, since changes identified in animals may not be monitored in humans. Semen analysis to evaluate sperm count, motility, and abnormality can be useful in detecting testicular dysfunction, but is not routinely conducted during early preclinical safety evaluation of compounds (129, 130). Likewise, serum follicle-stimulating hormone (FSH) can be a useful indicator of testicular dysfunction, but is generally not sensitive enough to detect early toxic changes in preclinical studies or in clinical trials (131, 132). Inhibin B has been proposed as a potentially useful biomarker of testicular toxicity. Inhibin B is a glycoprotein produced predominantly by Sertoli cells. Blood inhibin B levels appear to reflect the degree of testicular function, with high levels observed in normal, fertile males and lower levels in individuals with testicular damage (133). In fact, epidemiological and clinical studies in humans have shown that when evaluated in combination with serum FSH, inhibin B levels have predictive value for detecting male infertility (134–136). This has prompted toxicologists to propose inhibin B as a potential biomarker of testicular toxicity for use in rodent toxicity studies, but also to monitor testicular function in clinical trials. An international consortium under the auspices of the International Life Science Institute (ILSI) Health and Environmental Sciences Institute (HESI) has been evaluating the suitability and limitations of inhibin B to detect modest testicular dysfunction in rats, but the data communicated so far have not confirmed that this marker would be of significant utility (133).

Gene expression profiling is a promising approach to discovery of new sensitive markers of testicular injury and to furthering our understanding of the mechanisms of testicular toxicity. This is obviously of particular interest to the pharmaceutical industry, since it would help terminate compounds toxic to the testis at an earlier stage. In addition, this would be extremely relevant to the field of environmental toxicology, since a significant number of pesticides and other environmental toxicants are associated with known or suspected testicular effects.

Compared to toxicogenomics studies in the liver or kidney, there have been relatively few published reports using gene expression profiling to investigate mechanisms of testicular toxicity. Because of the interest of environmental toxicologists in the male reproductive tract, the majority of reports have focused on environmental toxicants or pesticides. However, these examples provide some useful insights for the evaluation of therapeutic agents. In general, mechanistic studies of testicular toxicants are quite challenging because of the complexity of the testis as a tissue and because of the elaborate mechanisms regulating testicular function. The testis is an extremely heterogeneous tissue composed of several different cell types with unique and critical functional properties. These cell types display a complex pattern of paracrine and endocrine interactions and cellular interdependence. In the testis, the four main target cells for toxic agents are the Leydig cells, the Sertoli cells, the peritubular cells, and the germ cells. Effects on each of these cellular targets for chemical-induced disruption of spermatogenesis are associated with characteristic initial morphological lesions of the testis. Ultimately, all cell types can be indirectly affected at the functional and morphological levels, complicating the determination of the primary target of toxicity through morphological evaluation.

Because of the critical role of endocrine-paracrine interactions and cellular interdependence, in vitro models are of limited use in investigating mechanisms of testicular toxicity. Existing cell lines of Sertoli or Leydig cell origin do not retain major functional properties, and primary cultures of testicular cells are labor-intensive and technically challenging and can only be used for short-term experiments (137). For instance, Leydig cell cultures are frequently used to interrogate specific mechanisms of toxicity. Cultures are prepared with Percoll gradient centifugation methods, but are usually not pure (138). Additionally, such preparations only yield a limited number of cells that need to be used within a short time period. These primary cultures are, however, very appropriate to confirm or refute hypotheses. For instance, adult Leydig cells have been used to further understand the mechanism by which Aroclor 1254, a polychlorinated biphenyl, disrupts gonadal function (139). It was demonstrated by RT-PCR that treatment with Aroclor 1254 down-regulates the transcripts for various enzymes involved in steroidogenesis, including steroidogenic acute-regulatory (StAR) protein, cytochrome P450 side-chain cleavage enzyme, and 3β- and 17β-hydroxysteroid dehydrogenase (HSD). These results, combined with a demonstration of decreased basal and luteinizing hormone (LH)-stimulated testosterone and estradiol production, confirmed that Leydig cells represent a primary target cell for Aroclor 1254 and that the mechanism of toxicity is partly related to steroidogenesis.

We have, on several occasions, stressed the importance of consistent collection procedures for toxicogenomics analysis. The testis being a paired organ, the vast majority of toxic or even spontaneous changes occur bilaterally. Therefore, in our laboratory, one testis is routinely collected for histopathological evaluation, while the other one is flash frozen for gene expression analysis. This ensures that spontaneous changes unrelated to treatment with a test article can be discovered before analysis and discarded so as not to compromise the analysis. It also allows us to anchor significant gene expression changes to morphological changes.

Several toxicogenomics studies investigating the molecular basis of testicular toxicity have been published (126, 140–144). For instance, the transcriptomic effects of exposure of mice to bromochloroacetic acid, a known testicular toxicant, were detected with a custom nylon DNA array. Differentially expressed genes included genes with known functions in fertility, such as Hsp70-2 and SP22, as well as genes encoding proteins involved in cell communication, adhesion, and signaling, suggesting that bromochloroacetic acid may disrupt cellular interactions between Sertoli cells and spermatids (126, 145). Likewise, several studies have used gene expression profiling to understand the mechanisms of testicular toxicity of phthalate compounds, including di-(*n*-butyl)phthalate (DBP)

(144, 146). Phthalates have been used as additives in industrial products since the 1930s and are ubiquitous environmental contaminants to which the general population is exposed through consumer products, diet, or medical treatments (147). An association between some phthalates and testicular toxicity has been demonstrated, with some evidence linking phthalate exposure to the decline of human male fertility, especially in developed countries (147). DBP is a plasticizer widely used in products such as food wraps, blood bags, or cosmetics, but it is also a male reproductive toxicant with demonstrated antiandrogenic effect (148). These effects on the male reproductive system and the fact that DBP metabolites are detectable in human urine raise sufficient concerns to justify a comprehensive study of the toxic mechanisms as a way to improve risk assessment (149). The molecular mechanism of this antiandrogenic effect was further studied by transcriptomic analysis of fetal rat testes exposed in utero to DBP, which identified the down-regulation of several steroidogenic enzymes (144). In agreement with these results, DBP has been shown to impair cholesterol uptake and transport and steroidogenesis in the fetal rat testis, while regulating a significant number of steriodogenesis- and spermatogenesis-related genes in immature male Sprague-Dawley rats (146, 150). Similar approaches have been used to investigate the mechanisms of toxicity of various other testicular toxicants, such as the triazole fungicides (143). In general, these studies have generated robust data that were successfully used to formulate plausible hypotheses regarding mechanisms of toxicity. However, in most instances, these hypotheses need to be tested with specifically designed orthogonal experiments.

In pharmaceutical discovery, gene expression profiling would be potentially valuable as a tool to detect testicular toxicants at early stages in order to avoid compound termination at advanced stages. A recent study addressed the feasibility of this predictive approach (151). Using four prototypical testicular toxicants (2,5-hexanedione, a Sertoli cell toxicant; ethylene glycol monomethyl ether or EGME, a germ cell toxicant; cyclophosphamide, a spermatogonia toxicant; and sulfasalazine), this study evaluated whether gene expression profiling could facilitate the identification of compounds with testicular toxicity liabilities. Testicular gene expression profiles were generated 6 hours after treatment of rats. Only one agent induced testicular histopathological changes at this time point. Several differentially expressed genes were nonetheless detected whose functions were consistent with the known activity of the test toxicants. These findings are of interest as they confirm that mechanistically relevant gene modulation in the reproductive system may occur before the development of morphological changes. However, it is still unclear whether predictive models can be generated for the testis. Moreover, one may actually question the preclinical value of such models, especially given the resources required for their development and validation.

Our laboratory has also attempted to determine the utility of microarray analyses in mechanistic investigations of testicular toxicity. Our initial focus was on characterizing the transcriptomic effects of prototypical toxicants, for which sufficient mechanistic information was available. In a pilot study, we treated male rats for 1 or 4 days with selected testicular toxicants. These reference compounds were selected to cover several mechanisms of toxicity and injury to different target cells. As expected, such short-term treatments did not result in significant testicular histopathological changes, although all toxicants used in the study were known to be potent testicular toxicants. The 1-day time point was associated with large interindividual variability for each test compound, precluding us from generating robust and consistent expression profiles. In addition, very few differentially expressed genes appeared to have mechanistic relevance at this time point. This is consistent with our observations for other tissues. In general, single-dose experiments are associated with significant interindividual variability and therefore are inappropriate for mechanistic studies. In contrast, the 4-day time point was associated with robust and reproducible gene expression profiles. For most toxicants, the number of differentially expressed genes was quite small compared to what we typically observe in other tissues, such as liver or kidney. However, these transcript changes were consistent with known mechanisms of action or were extremely useful to formulate novel hypotheses.

For instance, two of the compounds used were the halogenated acetic acids dibromoacetic acid (DBAA) and dichloroacetic acid (DCAA). Halogenated acetic acids are major disinfection by-products of water chlorination and ozonation (152). Daily oral treatment of male rats with DBAA and DCAA at high doses (250 and 500 mg/kg/day, respectively) induced specific early morphological changes in the testis, characterized by failed spermiation or failure of release by Sertoli cells of mature step 19 spermatids (153, 154). Disorganization, distortion, and degeneration of late spermatids are observed at later time points (152). While these morphological changes strongly suggest that the Sertoli cell is the target cell of DBAA and DCAA, our toxicogenomics evaluation using whole testes indicated that both compounds induced a consistent down-regulation of cytochrome P450c17α (CYP17) (155). CYP17 is expressed in the testis by Leydig cells and represents a key enzyme for the production of gonadal testosterone, which catalyzes the conversion of progesterone to hydroxyprogesterone and andronestenedione (156). Consequently, decreased CYP17 expression and activity levels of this enzyme result in accumulation of progesterone and decreased testosterone production by Leydig cells. In other words, these gene expression data suggest that halogenated acetic acids induce testicular toxicity partly through an effect on testicular testosterone production, and that Leydig cells represent an important cellular target. Since these transcript changes were potentially central to the mechanism of toxicity, we used real-time RT-PCR to confirm that CYP17 mRNA was down-regulated by treatment with DBAA and DCAA. Moreover, we demonstrated the biological implications of this transcript change by showing evidence that testicular testosterone concentrations were significantly decreased in the testis of rats treated for 4 days with DBAA or DCAA. Orthogonal in vitro studies using primary cultures of rat Leydig cells confirmed that both halogenated acetic acids directly reduce CYP17 mRNA and activity levels, leading to a significant reduction in human chorionic gonadotrophin (hCG)-stimulated testosterone production.

This investigation is of interest for several reasons. First, it demonstrates that transcriptomics-based approaches are feasible and useful in studying the toxic mechanisms in the testis. Despite minimal transcriptomic changes, microarray platforms can readily identify differentially expressed genes with mechanistic relevance to the toxicity, as long as exposures are sufficiently long. Second, it reemphasizes the importance of analyzing gene expression data with an open mind, rather than using them only to confirm earlier hypotheses or known mechanisms of toxicity. Because gene expression analysis allows for a global evaluation of changes at specific time points, novel controversial mechanisms may be uncovered. Once generated, these novel hypotheses should be interrogated with appropriately designed studies and a different analytical tool. Third, this investigation illustrates the value of evaluating drug-induced transcriptomic changes in a whole tissue. In previous chapters, we indicated that techniques such as laser-capture microdissection (LCM) can be utilized to measure gene expression changes in selected cell populations. However, by isolating the presumed target cells with these dissection procedures, one may overlook transcriptional changes in other cell populations that may be just as important to elucidate the mechanism of toxicity. In the case of halogenated acids, one might easily have speculated based on morphology that Sertoli cells are the primary cellular target and therefore might have focused on this cell type only. This approach would have missed the potential critical role that decreased CYP17 expression and the resulting lowered testosterone production play in the pathogenesis of DBAA-induced testicular toxicity.

5.5. TOXICOGENOMICS AND TARGET-RELATED TOXICITY

Rapid advances in our understanding of the molecular basis of various diseases have resulted in an unprecedented number of novel potential targets for therapeutic intervention. As we discuss in previous chapters, new therapeutic targets can be identified through various approaches, with genomics technologies rapidly becoming a key contributor to target discovery. New targets need to be validated to demonstrate that modulating their activity results in a therapeutic benefit. In general terms, target validation may follow a three-step process (157). First, sufficient data should be accumulated to show that there is a good chance that a given target has causal relevance to the disease process. Second, experimental evidence in available cell- or animal-based experimental models should be generated to confirm that modulation of target activity results in the expected outcome. Third, the impact of intervention via the target must be clinically interrogated and demonstrated. Obviously, the third component cannot be achieved before compounds are tested in the clinic. There is, however, unquestionable value in achieving the first two steps (which may be termed preclinical validation), as this enables rational selection of targets most likely to succeed in the clinic.

This traditional target validation procedure typically does not include an evaluation of the effects of modulating the target activity on normal cellular, organ, or body function. In other words, the potential safety liabilities associated with a particular target are seldom interrogated at an early stage. Not surprisingly, a substantial number of so-called attractive targets do not result in any novel therapeutic agents, partly because of their safety liabilities. One may therefore argue that developing a good understanding of the potential toxicity associated with a target should also constitute a critical phase of target drugability assessment, and target validation in general. This is especially true for most genomics-derived targets, since they frequently play critical roles in normal cellular function, and consequently modulation of their activity can lead to what is often referred to as target-related toxicity, mechanism-based toxicity, or on-target toxicity.

There are several ways to leverage gene expression profiling to predict potential target-related toxicity. Ideally, this evaluation should be conducted at the target identification or validation stages, since this information could be useful in prioritizing the portfolio of novel targets. Obviously, it is unlikely that data generated at this stage will be completely conclusive, but in a prioritization exercise they may help to focus resources on programs most likely to succeed when considered in the context of anticipated therapeutic benefits and commercial opportunities. In addition, these early investigations may yield useful information regarding likely target organs or other toxicity issues that may prove extremely useful in designing an optimal testing strategy for a program. For instance, these data may trigger the incorporation of specific studies or assays to address potential safety liabilities as early as possible in an effort to further validate or terminate a target.

5.5.1. Target Expression in Normal Tissues

One approach to studying potential on-target toxicity is to evaluate the expression of targets in normal tissues from both preclinical species and humans. This can be done systematically by developing quantitative expression tissue maps (23, 158, 159). For instance, DT-diaphorase (also known as NAD(P)H:quinone oxidoreductase 1 or NQO1) is highly expressed in many tumors compared to normal tissues and is a critical bioactivator of mitomycin C, an antitumor antibiotic of the quinone family (160). For these reasons, DT-diaphorase may represent a potentially selective target for oncology. To address the lack of information about the cell-specific expression of DT-diaphorase, a study was conducted to map the distribution of this enzyme in normal human tissues by immunohistochemistry (161). DT-diaphorase was found to be highly expressed in glomerular podocytes, suggesting a potential mechanistic link with the renal toxicity observed with mitomycin C, in which injury to the glomerular filtration apparatus appears to be the primary damage (162). In addition, these results indicated a potential renal toxicity liability for compounds metabolized by DT-diaphorase or for DT-diaphorase inhibitors.

Conceptually, such tissue maps could be used to identify tissues that are more likely to be affected by toxic changes, based on the assumption that expression of a particular target suggests a role in tissue function or homeostasis. In addition, these data could reveal differences in expression between species, suggesting species-specific reactions to compounds. The first step in generating tissue maps requires the availability of complete tissue repositories from the relevant preclinical species, as well as from human beings. Target expression can be then be evaluated at the level of mRNA or protein expression or enzymatic activity (in the case of an enzyme) or by using assays such as receptor binding assays. Evaluation of mRNA levels is technically the easiest way to generate expression data. In particular, with microarrays, it is possible to profile the tissues of interest, and thus simultaneously generate data for all novel targets. However, mRNA levels do not necessarily correlate well with protein or activity levels, and secondary analytical methods, such as Western blots or activity assays, are highly recommended to confirm or complement these mRNA data. Furthermore, this approach measures mRNA concentrations in the whole tissue, not reflecting the mRNA levels in small tissue compartments or cell types, where the target of interest may play a critical role. Thus complex tissues should ideally be studied with techniques that enable a better spatial understanding of expression, such as in situ hybridization or immunohistochemistry.

Expression profiles of various tissues represent useful information, but by themselves they can only identify potential target tissues of toxicity. They are not sufficient to predict whether target-related toxicity will occur, and if so, what therapeutic margins may exist for a particular target. Indeed, novel targets are frequently identified based on overexpression in diseased samples, with the assumption that the target may play a more important role in the diseased tissues than in normal tissues. In other words, the difference in expression levels may result in safety margins sufficient enough for development, and determining this safety window based on expression levels of a target is probably not feasible, especially at an early stage. Hence, it would be incorrect to conclude that expression in normal tissues would translate into non-drugability or on-target toxicity. Experimental modulation of the target still represents the most reliable approach to study potential on-target toxicity.

5.5.2. Target Modulation

As we mentioned earlier, target validation involves modulation of target activity to demonstrate experimentally that intervention through this target will likely result in the expected therapeutic outcome. This is typically conducted in cell-based systems and/or animal models. Cell-based systems are extremely useful in developing an early understanding and confirmation of the function of a target when programs are initiated. They are, however, less suitable for addressing target-related toxicity, especially because, in our experience, they reveal little about potential therapeutic indexes. In this section, we focus almost exclusively on animal models.

Modulation of a target may be achieved by several approaches, as described below.

5.5.2.1. Genetically Modified Animals

Genetically modified animals, mostly mouse models, are widely used models for elucidating the molecular mechanisms of various diseases. They represent convenient tools for target identification and validation, since there appears to be a good correlation between the knockout phenotype and the anticipated drug efficacy (163). These models may be extremely valuable in confirming the functional significance of a target for a specific disease process, but they also have some important drawbacks as tools for target validation and for studying target-related toxicity. Indeed, for knockout mouse models, the lack of a gene product throughout development might affect organogenesis and cause compensatory up-regulation of other genes (164). On one hand, these mice may not be viable or may have development-related defects that may have little bearing on various therapeutic interventions. On the other hand, the compensatory activity of other related pathways may mask effects that could occur after pharmacological intervention on a target. For these reasons, conditional knockout strategies to functionally silence a gene on demand may be more appropriate for validating a target before specific antagonists become available (164, 165). In addition, the process of generating these mice is labor-intensive and time-consuming. There are significant hurdles to the use of these animal models in drug discovery, ranging from proprietary and commercial rights to logistical issues, that make it difficult to obtain them early in a program (166). Furthermore, unpredicted phenotypic alterations are not uncommon in these models, and some of those may be strain-specific and not always relevant to the effects of target modulation by compounds (167). Finally, mice are generally poor predictors of toxic changes relevant to humans. Thus murine knockout models may not represent the most efficient approach to early identification of target-related toxicity.

5.5.2.2. Tool Compounds

The term **tool compounds** is used to refer to molecules that can modulate the activity of the target in the desired direction (either activation or inhibition). They are typically small molecules but may also be biological agents, such as recombinant endogenous ligands or monoclonal antibodies. Early in a program, these compounds are unlikely to have good drug-like properties, and may not have been well characterized in terms of selectivity and pharmacology. However, such compounds may still be used, as long as sufficient systemic exposure can be achieved for a short repeat-dose study. In our experience, discovery scientists favor the use of mice for in vivo studies, especially since compound requirements are lower in this species. However, we strongly recommend using rats as the testing species for this early characterization, since they are generally better predictors of human toxicities and are the reference small animal species for the regulatory toxicology studies.

Because tool compounds are typically not well characterized, they may induce toxic changes that are related to unrecognized off-target activities rather than the primary pharmacology. In other words, because of the uncertainty associated with the pharmacology of these compounds, effects observed in vivo might not relate to the target itself (164). Therefore, it is critical to include appropriate controls in these experiments. To control for toxicity related to a chemical class or off-target effects, several tool compounds from different chemical classes should ideally be used. Inactive compounds with close structural similarities (such as inactive enantiomers, for instance) represent excellent controls for toxic changes related to a particular chemistry. The use of these types of controls is illustrated in the section on cardiotoxicity with our investigation of experimental ACC2 inhibitors (116). As a reminder, in that study an inactive enantiomer of the test compound was used to confirm that the cardiac toxicity was related to the chemistry, rather than to inhibition of the target.

In our laboratory, the favored approach to target assessment is to conduct 3-day repeat-dose studies in male rats using carefully selected tool compounds. Doses, regimens (i.e., once or twice daily), and routes of administration are selected to achieve efficacious exposures (as determined by concurrent studies in relevant preclinical efficacy models), as well as higher exposures, so that target-related safety margins can be assessed. Early in a program, it may be challenging to attain sufficiently high exposures with certain compounds because of their poor physicochemical and biological properties. Therefore, the actual therapeutic safety windows may remain unclear. To evaluate results from these studies, we complement the traditional end points collected during toxicology assessment (clinical observation, body weight changes, food consumption, clinical pathology, necropsy and histology findings) with a toxicogenomics evaluation of various tissues. Tissues for gene expression profiling are selected and prioritized based on the biology of the target (generated from the expression tissue maps and the literature) and on the evaluation of clinical pathology and histopathology changes. For instance, if there is evidence of liver injury based on elevated serum ALT activity, we would evaluate the liver at the transcriptomic level. Gene expression changes are interpreted in the context of available reference databases and our battery of tools, including the gene expression signatures already discussed in this chapter and in Chapter 4. For these studies, gene expression analysis may be applied to confirm that the target is modulated and to evaluate whether the toxic changes observed are due to target inhibition or an off-target effect.

5.5.2.3. Gene Silencing

The term **gene silencing** implies inhibition of gene function by such tools as antisense oligonucleotides, ribozymes, or siRNA (163, 168, 169). The mRNA is a relatively easy target for nucleic acid-based gene silencing, since it presents unpaired bases that are in theory available for hybridization by complementary single-stranded nucleic acids (170). Through hybridization to complementary mRNA sequences, antisense oligonucleotides can inhibit mRNA expression and translation, thereby inducing a blockade in the transfer of genetic information from DNA to protein. Two major mechanisms are known to mediate the action of antisense oligonucleotides. The first involves heteroduplex-mediated induction of RNaseH, which digests the RNA component of the hybrid. The second mechanism is based on the steric blockade of mRNA, causing physical inhibition of RNA splicing and translation (170, 171). DNA antisense technology has been used for several years in various target validation projects (172, 173). Ribozymes are naturally occurring catalytically active nucleic acids, which have their own built-in RNA cleavage activity for site-specific cleavage of target mRNAs (170, 174, 175). Unfortunately, their application in target validation is limited because of their lack of substrate specificity (174, 176). Small RNAs mediating RNA interference (RNAi) appear to be better options than the antisense- and ribozyme-based techniques for gene silencing (168, 174, 177). Generally speaking, there are two basic methods for RNAi (177). The first involves cytoplasmic delivery of short double-stranded RNA oligonucleotides specific for a particular target gene (also called short interfering RNA or siRNA). This method mimics the active intermediate of the naturally occurring endogenous RNAi mechanism. The second method entails nuclear delivery of gene expression cassettes that express a short hairpin RNA (also called shRNA). This method is also similar to a naturally occurring RNAi mechanism mediated by microRNA (miRNA).

It should be recognized that significant challenges remain at this point for RNAi to become a routine and reliable tool for in vivo target validation. However, RNAi has been used successfully in animal experiments (177, 178). For instance, adenoviral vector-based shRNA technology was used to knockdown in the livers of *ob/ob* mice the expression of a putative target for metabolic syndrome, mitochondrial glycerol-3-phosphate acyltransferase (mtGPAT1) (179). This approach resulted in a significantly reduced expression of mtGPAT1 mRNA in the liver of *ob/ob* mice and a dramatic and dose-dependent reduction in mtGPAT1 activity. This decreased activity was associated with a reduction in hepatic diacylglycerol, hepatic free fatty acid, plasma cholesterol, and plasma glucose, confirming that inhibition of hepatic mtGPAT1 could correct obesity and related disorders. A similar approach was successfully used to validate stearoyl-CoA desaturase-1 (Scd1) as a target for metabolic syndrome (180).

We have been exploring the use of RNAi to predict potential target-related toxicity and investigate the on-target and off-target effects of some experimental compounds, but our experience is still too limited to draw clear conclusions regarding the value of this approach. Several key issues must be addressed for RNAi to become a useful tool in toxicological evaluation of drug targets. In particular, the potency and selectivity of siRNA must be better understood and controlled. To ensure that a phenotype is primarily caused by gene silencing, in vitro investigations often use as controls several independent RNAi reagents, including scrambled or nonsense shRNA sequences. This is often not feasible in animal experiments for cost and logistical reasons. RNAi reagents are indeed costly, which limits most in vivo studies to mice, a species that is not optimal for toxicogenomics evaluations. Furthermore, there are several important caveats to this approach (166). First, gene knockdown via RNAi is functionally distinct from pharmacological inhibition with a compound, and therefore may not completely mirror the action of a compound. For instance, a specific kinase inhibitor would target the kinase domain of the enzyme, while RNAi would result in lack of expression of the enzyme, thereby shutting down activities associated with domains other than the kinase domain (174). Second, the tissue distribution of RNAi reagents may be very different from that of a compound. Third, RNAi may induce classic antiviral responses, such as activation of interferon and specific kinase signaling pathways (174). Fourth, this technology is still relatively new, and other unrecognized effects of siRNA on cells may exist. For instance, a recent study investigated the long-term effects of sustained high-level shRNA expression in the livers of adult mice through intravenous infusion with an optimized shRNA delivery vector based on duplex-DNA-containing adeno-associated virus type 8 (AAV8). In this study, 49 distinct AAV-shRNA vectors, unique in length and sequence and directed against six targets, were evaluated (181). Thirty-six vectors resulted in dose-dependent liver injury, with 23 ultimately causing death. Pathological findings included severe ascites, widespread subcutaneous edema, elevated total and direct serum bilirubin levels, elevated serum ALT and AST levels, decreased serum total protein, albumin, and globulin, and chronic hepatic changes characterized by multifocal hepatocyte necrosis and regeneration with presence of large hepatocytes with open-faced nuclei. These unexpected side effects were associated with down-regulation of liver-derived miRNAs, indicating a possible competition of these miRNAs with shRNAs for cellular factors required for processing of various small RNAs. Further in vitro and in vivo studies demonstrated that one such factor was the nuclear karyopherin exportin-5. These findings indicate that controlling intracellular shRNA expression levels is imperative.

It should be mentioned that RNAi-based approaches have already been successfully applied in mechanistic investigations of toxicity. For instance, the toxicity of acetaminophen has been studied with a glutathione (GSH)-depletion experimental rat model constructed by knocking down hepatic γ-glutamylcysteine synthetase (γ-GCS) with a shRNA (182). This resulted in an 80% decrease in hepatic GSH levels for at least 2 weeks. In this GSH knockdown rat model, acetaminophen-induced hepatotoxicity was shown to be significantly potentiated compared with normal rats, confirming the central protective role of GSH in acetaminophen hepatotoxicity.

5.6. PREDICTING SPECIES-SPECIFIC TOXICITY

Generally speaking, preclinical species are very useful in predicting and characterizing toxic changes that may occur in humans. We have mentioned on several occasions that there is overall a good correlation between preclinical findings and adverse events observed in the clinic (12). However, not all toxicological changes occurring in preclinical species are relevant to humans because of species differences in cell biology, physiology, or responses to changes induced by compounds (86). Evaluating the relevance of toxic changes observed preclinically to the clinical situation is not a trivial exercise, and it typically requires a thorough understanding of the molecular mechanism whereby the toxicity occurs. In the majority of cases, establishing the relevance of preclinical toxic changes requires a significant investigational effort. Sometimes this is not even possible, and therefore preclinical data need to be considered by default relevant to humans. Since chemically induced rat-specific changes are commonly encountered in routine toxicology studies, toxicogenomics can be extremely useful to proactively address them.

Peroxisome proliferation represents a classic, and probably the best-studied, example of a rat-specific change with no relevance to humans. It is caused by exposure to a pleiotropic class of chemicals called peroxisome proliferators, which includes diverse groups of chemicals, for example, the fibrate class of lipid-lowering drugs, the phthalate ester plasticizers and herbicides, and various natural products (80). These compounds induce peroxisome proliferation in rodent liver, a change characterized by an increased volume and density of peroxisomes, organelles involved in fatty acid oxidation and lipid metabolism. Upon chronic administration, these compounds cause hepatomegaly, hepatocellular hypertrophy, and eventually hepatic neoplasms in rats (183). The fact that these chemicals are non-genotoxic animal carcinogens stimulated a large interest in understanding their mode of action to better assess the human relevance of the animal tumors associated with them.

There are marked species differences in the response to peroxisome proliferators, with mice and rats being highly responsive in contrast to humans (184). This differential species response correlates directly with the number of hepatic peroxisome proliferator-activated receptors α (PPAR-α), the nuclear hormone receptor that mediates most of the effects of peroxisome proliferators and that is an obligatory factor in peroxisome proliferation in rodent hepatocytes (80, 184). PPAR-α is expressed in human liver at only 5–10% of rodent liver levels, suggesting that humans are at minimal or no risk of developing hepatic tumors on chronic exposure to peroxisome proliferators. Mechanistic studies using rodent tissues have led to several postulated mechanisms of action and have identified important causal or associated events, such as activation of PPAR-α, perturbation of cell proliferation and apoptosis, and hepatocyte oxidative stress. By interrogating the entire transcriptome during the process of carcinogenesis, toxicogenomics has furthered the understanding of the molecular mechanisms associated with the various effects of several peroxisome proliferators (185–187). Thus it confirmed the mechanisms whereby certain compounds cause rodent hepatomegaly and hepatic carcinogenesis, improving the overall process of risk assessment. In addition, a better comprehension of the early transcriptomic responses and biological pathways associated with PPAR-α activation can lead to an early identification of compounds associated with peroxisome proliferation (Fig. 5.7).

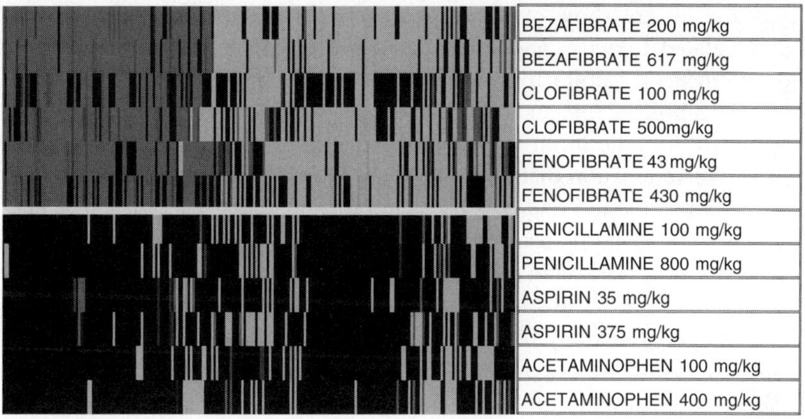

Figure 5.7 Heat map from an agglomerative hierarchical clustering analysis illustrating gene expression changes in the liver of rats following treatment with prototypical peroxisome proliferators. Male rats were treated orally and daily with various fibrates (bezafibrate, clofibrate, fenofibrate) at 2 doses for 5 days (the low dose representing a multiple of the therapeutic dose and the high dose representing a maximum tolerated dose). Fibrates are prototypical peroxisome proliferators. In addition, rats were treated with several non-peroxisome proliferators (penicillamine, aspirin, acetaminophen) using a similar dose selection. Shown here are the genes that were regulated at a P value <0.01 and with at least a two-fold change. Green indicates down-regulation, while red indicates up-regulation. Each treatment group corresponds to three animals that were pooled in silico. Note that the three peroxisome proliferators altered the expression levels of large numbers of genes. These differentially expressed genes can be used to easily identify test agents that act as peroxisome proliferators. See color insert.

While peroxisome proliferation is a prevalent change identified in preclinical rodent studies, other toxic changes have been detected whose relevance to humans needs to be assessed. Cyclosporine-induced nephrotoxicity is an illustration of an approach that can be used to explain findings of unclear toxicological relevance (188, 189). Rats treated with cyclosporine A develop renal tubular calcification due to accumulation of calcium in tubules associated with a marked down-regulation of calbindin-D28kDa, a calcium binding protein. In contrast, cyclosporine A does not regulate calbindin-D28kDa expression in the kidneys of dogs and monkeys, two species resistant to cyclosporine-mediated renal toxicity. Overall, these findings suggest that calbindin-D28kDa down-regulation represents a marker of species-related toxicity that can be exploited to better assess risks for humans.

5.7. EVALUATION OF IDIOSYNCRATIC TOXICITY WITH TOXICOGENOMICS

Idiosyncratic toxic responses present a major challenge to toxicologists and to the pharmaceutical industry in general. Drugs that cause these types of adverse events typically do not show evidence of toxicity in animals, generally do not exhibit dose-related toxicity, and cause low rates of injury in humans, with a typical occurrence of 1:5000 to 1:100,000 patients (190). These reactions are thought to reflect host factors and individual susceptibility, giving rise to the term **idiosyncratic** (or specific to an individual). Thus an idiosyncratic drug reaction (IDR) is an adverse reaction that does not occur in most people within the range of doses used clinically and does not involve the known pharmacological effects of a drug (191). Other terms used concurrently with IDR are **type B reactions**, **hypersensitivity reactions**, and **allergic reactions** (191).

There are many well-known examples of drugs that cause idiosyncratic toxicity, especially hepatotoxicity or drug-induced liver injury (DILI). For instance, troglitazone (Rezulin) was approved by the FDA in 1997 for the treatment of type 2 diabetes mellitus. Reviews of the clinical trials conducted before approval did not reveal any cases of liver failure, although a small proportion of troglitazone-treated subjects experienced slight elevations of serum ALT (192, 193). After its approval, several cases of acute liver failure associated with troglitazone were reported, ultimately leading to its market withdrawal in 2000 (194–196). Follow-up studies suggested a rate of severe liver injury around 1 in 3000–10,000 (197). Likewise, perhexiline, a drug widely used in Europe for the treatment of angina pectoris, produced an alcoholic-like cirrhosis after weeks to months of usage (198). Other examples of drugs causing IDRs include bromfenac (withdrawn in 1998 because of liver toxicity), fenfluramine (withdrawn in 1997 because of heart valve disease), fialuridine (acute liver failure and metabolic acidosis), and cerivastatin (withdrawn in 2001 because of rhabdomyolysis) (190, 192).

With such a low incidence, the chances of encountering IDRs in clinical trials are extremely low. Thus these events are mostly observed after approval of the drug, when epidemiological patterns of adverse reactions finally become evident. Such a late identification occurs after large investments in R&D and poses a significant risk to the public health, including fatalities. For instance, adverse drug reactions in the United Kingdom have been shown to be responsible for more than 6% of hospital admissions, with a mortality rate of approximately 2% (199). Although only approximately 5% of these adverse reactions were idiosyncratic, other studies suggest a much higher fraction of IDRs (200). Furthermore, IDRs represent a major cause of market withdrawal or black box warning (33). Clearly, an enhanced understanding of IDR would be beneficial to the pharmaceutical industry and to public health as a whole. The current testing paradigm does not predict IDRs, making this type of toxicity especially challenging to deal with. Efforts have been made to predict the risk of such reactions and to identify possible signals of IDRs during clinical trials and postmarketing surveillance (191). However, results have been less than satisfactory. Our inability to predict the occurrence of IDRs significantly increases the cost and uncertainty of drug development.

Several mechanisms have been hypothesized to explain the development of idiosyncratic drug toxicity, and multiple mechanisms of IDRs are likely to exist.

These include the formation of reactive metabolites in certain individuals because of the presence of genetic polymorphisms in drug-metabolizing enzymes, the development of immune-mediated responses to the drug or one of its metabolites, a synergistic effect of concurrent low-level inflammatory reactions, and changes in mitochondrial function and integrity (201–206). There is indeed substantial experimental evidence that these mechanisms mediate IDRs for certain drugs; however, these mechanisms also occur with drugs that do not appear to cause IDRs. Therefore, it is currently impossible to proactively identify compounds that may lead to idiosyncratic responses in a large patient population. This has prompted several investigators to take advantage of genomics tools to improve our understanding of IDRs. They have used animal models, as well as in vitro systems, the latter being covered in Chapter 6. In this section, we focus on idiosyncratic hepatic reactions as an illustration.

Given the limited opportunities to conduct investigations in humans with compounds inducing IDRs, various preclinical animal models have been proposed to study the mechanism of idiosyncratic hepatotoxicity (191, 207). These animal models were created under the assumption that IDRs are not human specific and can be re-engineered in animals. This is probably correct for some types of IDRs with defined, not species-specific mechanisms, but is completely unknown for others. Several preclinical models utilize specific strains or species that have been shown to reproduce drug-induced toxic changes with features similar to various IDRs. For instance, halothane hepatitis is considered to be a result of an idiosyncratic autoimmune reaction induced by the formation of neoantigens that have been generated by covalent binding of halothane biotransformation intermediates (208). While most animal species do not develop a hepatitis reaction after exposure to halothane, treatment of guinea pigs with halothane results in an immune response with evidence of Kupffer cell involvement. Similarly, the Brown Norway rat has been commonly used as an animal model of idiosyncratic toxicity, especially for the study of suspected immune-mediated idiosyncratic reactions. After exposure to nevirapine or D-penicillamine, Brown Norway rats develop toxic changes similar to those seen in humans (209, 210). Other models consist of standard laboratory species that are somehow modified to reflect situations or conditions that may lead to IDRs.

An example of these models is the lipopolysaccharide (LPS)-potentiated rat model developed by the laboratory of Dr. R. A. Roth at Michigan State University. This is an example of a two-hit model for idiosyncratic toxicity, in which compounds with potential idiosyncratic liabilities require another underlying factor (such as alcohol intake or concurrent infection) to cause toxicity. In this model, rats are coadministered a single dose of compound and LPS. Several compounds have been tested in this model. It is unfortunate that some of the compounds initially used were not ideal test compounds. For instance, ranitidine (a drug available over the counter), chloropromazine, and aflatoxin B1 have been studied extensively in this model, and most would agree that these drugs, albeit hepatotoxic, do not classify as DILI-causing agents (211–214). Therefore, we will not comment on the gene expression studies conducted with these compounds, since these data may not be relevant to idiosyncratic hepatotoxicity (215). Nevertheless, it is important to recognize that there are many possible mechanisms of DILI, and that the LPS concept may be relevant, because LPS can potentiate the hepatic toxicity of various compounds. Therefore, this model was further evaluated with novel compounds with proven idiosyncratic hepatotoxicity association, including trovafloxacin and diclofenac.

Our laboratory has used the quinolone antibiotic trovafloxacin as a tool compound for DILI. Quinolones are antibacterial agents that act by inhibiting bacterial DNA gyrase and DNA topoisomerase IV (216). Compounds of this class are generally well tolerated in humans, except for trovafloxacin (217). Before its regulatory approval in 1997, trovafloxacin had been tested in over 7000 patients with no evidence of hepatic failures or deaths. Once launched, over two million people were prescribed trovafloxacin and 150 cases of liver toxicity have been reported, including 14 cases of acute liver failure, four cases requiring liver transplants, and five fatalities (218). This has led to severe restrictions on its use, with prescriptions exclusively limited to life-threatening situations (217, 218). The mechanism underlying this adverse effect has not yet been determined, and no evidence of hepatic injury has been found in animals. To confirm that the normal laboratory rat does not recapitulate this hepatic toxicity, we treated male Sprague-Dawley rats orally with trovafloxacin (200 mg/kg/day) for 7 days. We used various parameters of toxicity, including serum chemistry, histopathology, and transcriptomic analysis of the liver, but did not detect any signal consistent with hepatic injury, thereby confirming that the normal laboratory rat is not susceptible to trovafloxacin-induced hepatic injury (219).

In contrast, in the LPS-potentiated rat model, coadministration of non-hepatotoxic doses of LPS and trovafloxacin resulted in a unique form of hepatotoxicity, characterized by elevations of serum ALT activity and histological evidence of multifocal, randomly distributed small foci of coagulative hepatocellular necrosis (220). None of these changes was observed with either agent alone. To ensure that this hepatic injury did not simply reflect an uncharacterized synergistic effect between LPS and quinolone compounds, we also demonstrated that hepatic injury does not occur after cotreatment with LPS and levofloxacin, a fluoroquinolone compound with no known human idiosyncratic hepatic liability. On the basis of these encouraging results suggesting that the LPS rat model may be relevant to human DILI, we evaluated the hepatic gene expression profiles in an attempt to understand the exact molecular cascade associated with this hepatic toxicity (220). Liver gene expression analysis identified unique changes induced by the cotreatment with LPS and trovafloxacin. These changes included an up-regulation of chemokines suggesting hepatic accumulation and activation of polymorphonuclear neutrophils (PMNs). Accumulation of PMNs and elevations of serum cytokine-induced neutrophil chemoattractant (CINC)-1 and macrophage inflammatory protein (MIP)-2 concentrations were demonstrated by immunohistochemical evaluation of the liver and by ELISA, respectively. These findings led us to hypothesize that PMNs may play a crucial role in the development of the toxicity. Indeed, this role was confirmed by demonstrating that prior depletion of PMNs with PMN antiserum considerably attenuates the liver injury (220). In summary, this study provided evidence that the LPS rat model could represent a useful animal model of DILI, and showed that gene expression analysis is a valuable method that can rapidly generate hypotheses on the mechanism of toxicity and can discover specific transcript changes that can be explored as biomarkers for idiosyncratic hepatic toxicity.

These positive results encouraged us to pursue our investigation of this rat LPS model. In a subsequent study, we determined whether diclofenac would trigger a similar type of toxicity in this model (221). Diclofenac is a non-steroidal anti-inflammatory drug that has caused rare, but sometimes serious hepatotoxicity in humans (222). The pathogenesis of diclofenac-induced hepatotoxicity is largely unknown, although several mechanisms have been postulated, including reactive drug metabolite, oxidative stress, mitochondrial injury, and immune-modulated hypersensitivity (222–225). Similar to trovafloxacin, a non-hepatotoxic dose of diclofenac (20 mg/kg), when coadministered with low doses of LPS, resulted in increased serum ALT activity and histopathological changes in the liver within 6 hours. Gene expression analysis demonstrated unique gene modulation in the cotreated group with evidence of up-regulation of multiple genes related to inflammation, cell death, and stress. As in the case of trovafloxacin, prior PMN depletion with PMN antiserum protected rats against liver injury, demonstrating the role of PMNs in the mechanism of the LPS/diclofenac toxicity. Diclofenac is known to induce hepatic toxicity in rats at high doses, and indeed with higher doses (100 mg/kg), we were able to demonstrate hepatotoxicity with diclofenac alone. Interestingly, these higher doses of diclofenac are also known to be associated with gastrointestinal injury and endotoxin translocation from the intestine to the liver, suggesting that diclofenac toxicity may be related to a partial loss of intestinal integrity leading to increased intestinal permeability to LPS or other products from the normal gut flora (226–228). Consistent with this hypothesis, sterilization of the gut with oral treatment with a mixture of polymyxin B (150 mg/kg/day) and neomycin (450 mg/kg/day) for 4 days alleviated the hepatotoxicity induced by high doses of diclofenac. In conclusion, this study further demonstrated that the LPS rat model can mimic, to some extent, the clinical situation and can be used as a basis to create preclinical models predictive of human IDRs. Most importantly, because of their global coverage of molecular changes, large-scale genomic profiles represent ideal tools to investigate these models to confirm similarity in responses between compounds and decipher their mechanisms, thus facilitating the assessment of risks to humans.

5.8. CONCLUSION

In this chapter, we have reviewed several applications of large-scale gene expression profiling during the toxicological evaluation of experimental pharmaceutical compounds in animal studies. In this in-depth review, we have tried to illustrate the potential of toxicogenomics, when used in a predictive approach, to significantly impact the productivity of discovery organizations by enabling earlier and more precise toxicological evaluations. In particular, adding multigene biomarkers to the current toolbox of the toxicologist should lead to smaller, more cost-effective, and robust rat studies requiring less labor and lower quantities of API. Today, it is too early to draw conclusions on the full impact of predictive toxicogenomics, but the evidence reviewed in this chapter suggests that this technology will improve the discipline of toxicology and risk assessment. With the expansion of public and private repositories of gene expression profiles and with the help of advanced computational tools, numerous multigene markers of various toxic states are also likely to become available in the near future, further promoting the application of toxicogenomics in various institutions. However, to be used in regulatory preclinical studies or in a clinical setting, these novel markers will need to be validated in prospective, well-controlled studies across multiple institutions with well-established standards for all steps in the process (229). The examples used in this chapter are a good illustration of the scale of the efforts required before enough confidence is generated for their widespread use. These validations will require access to large amounts of diverse reference data generated through consistent experimental procedures. Such large endeavors can only be successfully achieved by combining resources and expertise from various parts of the scientific community (230). Various efforts have already been initiated in the form of industry consortia. The Predictive Safety Testing Consortium between the C-Path Institute (www.c-path.org) and several large pharmaceutical companies is an example of a consortium formed to share proprietary markers or laboratory methods to improve the prediction and monitoring of toxicity.

The impact of toxicogenomics on mechanistic toxicology is easier to recognize, since mechanistic investigations applications may be more visible in organizations, especially when dealing with toxicity issues that hamper the advancement of highly visible projects. In our opinion, toxicogenomics is the most effective approach available today to the toxicologist to rapidly investigate the mechanism of most toxic changes. The objective of a typical mechanistic toxicogenomic study is not to gain an immediate understanding of the mechanism, but rather to rapidly generate relevant molecular data that can be used to formulate hypotheses. These hypotheses can then be quickly confirmed or refuted in orthogonal studies. Once mechanisms of toxicity are understood, risks can be better assessed and compounds can be either moved forward or terminated on the basis of more reliable information. Furthermore, this mechanistic information can be used to set up appropriate counterscreens to select backup compounds unlikely to have similar safety liabilities.

As indicated in Chapter 4, the reliability and precision of microarray platforms in the context of biomarkers has recently been confirmed by a consortium led by the U.S. Food and Drug Administration (231). However, it should be recognized that these analytical tools have still not reached a stage at which standardized, reproducible, and high-throughput applications of diagnostic assays would be feasible. However, given the pace of technological improvements, this should occur in the very near future for such species as humans, rats, and mice. Unfortunately, the progress is much slower in developing experimental platforms for other species commonly used in toxicological assessment (such as dogs). In our experience, toxicity detected in dogs represents a significant cause of preclinical attrition, and any improvement in our ability to predict and characterize toxicity in this species could have a very substantial impact on toxicology.

In this chapter, we have focused on specific applications for which we had enough published or internal data to illustrate the strengths and limitations of the approach. It is noteworthy, however, that gene expression profiling is also relevant to other areas of toxicology. For instance, toxicogenomics may be applied to gain mechanistic insights into immunotoxic processes or even predict immunotoxicity (232, 233). Similarly, transcriptomic studies may be useful in deciphering the mechanisms of teratogenicity of experimental compounds and detecting chemically induced birth defects (234). Others have also successfully utilized this technology to investigate drug-induced muscle toxicity (27). In fact, virtually any aspect of toxicology can theoretically be interrogated at the transcriptomic level. It remains to be seen where the use of gene expression profiling will provide enough value to justify its use. Most likely, this would be determined by the particular drug discovery setting. Therefore, one could argue that developing the necessary skills and tools to use toxicogenomics in various tissues represents a useful investment for the future.

REFERENCES

1. Service RF. Surviving the blockbuster syndrome. *Science* 2004;303:1796–1799.
2. Kola I, Landis J. Can the pharmaceutical industry reduce attrition rates? *Nat Rev Drug Discov* 2004;3:711–715.
3. Prentis RA, Lis Y, Walker SR. Pharmaceutical innovation by the seven UK-owned pharmaceutical companies (1964–1985). *Br J Clin Pharmacol* 1988;25:387–396.
4. Roberts SA. High-throughput screening approaches for investigating drug metabolism and pharmacokinetics. *Xenobiotica* 2001;31:557–589.
5. Kramer JA, Sagartz JE, Morris DL. The application of discovery toxicology and pathology towards the design of safer pharmaceutical lead candidates. *Nat Rev Drug Discov* 2007;6:636–649.
6. Ulrich RG, Friend SH. Toxicogenomics and drug discovery: will new technologies help us produce better drugs? *Nat Rev* 2002;1:84–88.
7. Yang Y, Blomme EA, Waring JF. Toxicogenomics in drug discovery: from preclinical studies to clinical trials. *Chem Biol Interact* 2004;150:71–85.
8. Lovett RA. Toxicogenomics. Toxicologists brace for genomics revolution. *Science* 2000;289:536–537.
9. Larson JL. Toxicity evaluations: ICH guidelines and current practice. In: Rogge MC, Taft DR, editors. *Preclinical Drug Development*. Boca Raton, FL: Taylor & Francis Group, 2005;349–414.
10. Knight A. Systematic reviews of animal experiments demonstrate poor human clinical and toxicological utility. *Altern Lab Anim* 2007;35:641–659.
11. Greaves P, Williams A, Eve M. First dose of potential new medicines to humans: how animals help. *Nat Rev Drug Discov* 2004;3:226–236.

12. Olson H, Betton G, Robinson D, Thomas K, Monro A, Kolaja G, Lilly P, Sanders J, Sipes G, Bracken W, Dorato M, Van Deun K, Smith P, Berger B, Heller A. Concordance of the toxicity of pharmaceuticals in humans and in animals. *Regul Toxicol Pharmacol* 2000;32:56–67.
13. Jacobson-Kram D, Contrera JF. Genetic toxicity assessment: employing the best science for human safety evaluation. Part I: Early screening for potential human mutagens. *Toxicol Sci* 2007;96:16–20.
14. Kirkland D, Aardema M, Henderson L, Muller L. Evaluation of the ability of a battery of three in vitro genotoxicity tests to discriminate rodent carcinogens and non-carcinogens. I. Sensitivity, specificity and relative predictivity. *Mutat Res* 2005;584:1–256.
15. Dorato MA, Engelhardt JA. The no-observed-adverse-effect-level in drug safety evaluations: use, issues, and definition(s). *Regul Toxicol Pharmacol* 2005;42:265–274.
16. Stevens JL. Future of toxicology—mechanisms of toxicity and drug safety: where do we go from here? *Chem Res Toxicol* 2006;19:1393–1401.
17. Spielmann H. Predicting the risk of developmental toxicity from in vitro assays. *Toxicol Appl Pharmacol* 2005;207:375–380.
18. Williams GM. Mechanisms of chemical carcinogenesis and application to human cancer risk assessment. *Toxicology* 2001;166:3–10.
19. Kramer JA, Curtiss SW, Kolaja KL, Alden CL, Blomme EA, Curtiss WC, Davila JC, Jackson CJ, Bunch RT. Acute molecular markers of rodent hepatic carcinogenesis identified by transcription profiling. *Chem Res Toxicol* 2004;17:463–470.
20. Monro A. How useful are chronic (life-span) toxicology studies in rodents in identifying pharmaceuticals that pose a carcinogenic risk to humans? *Adverse Drug React Toxicol Rev* 1993;12:5–34.
21. Robinson DE, MacDonald JS. Background and framework for ILSI's collaborative evaluation program on alternative models for carcinogenicity assessment. International Life Sciences Institute. *Toxicol Pathol* 2001;29 Suppl:13–19.
22. Gintant GA, Su Z, Martin RL, Cox BF. Utility of hERG assays as surrogate markers of delayed cardiac repolarization and QT safety. *Toxicol Pathol* 2006;34:81–90.
23. Searfoss GH, Ryan TP, Jolly RA. The role of transcriptome analysis in pre-clinical toxicology. *Curr Mol Med* 2005;5:53–64.
24. Suter L, Babiss LE, Wheeldon EB. Toxicogenomics in predictive toxicology in drug development. *Chem Biol* 2004;11:161–171.
25. Martin RL, Su Z, Limberis JT, Palmatier JD, Cowart MD, Cox BF, Gintant GA. In Vitro preclinical cardiac assessment of tolterodine and terodiline: multiple factors predict the clinical experience. *J Cardiovasc Pharmacol* 2006;48:199–206.
26. Blomme EA, Ciurlionis R, Marsh KC, Waring JF, Yang Y. Evaluation of the effects of serial phlebotomy on the transcriptome of major tissues and on the response to toxicants in rats. *Toxicol Lett* 2008;176:138–148.
27. Foster WR, Chen SJ, He A, Truong A, Bhaskaran V, Nelson DM, Dambach DM, Lehman-McKeeman LD, Car BD. A retrospective analysis of toxicogenomics in the safety assessment of drug candidates. *Toxicol Pathol* 2007;35:621–635.
28. Hardisty JF, Brix AE. Comparative hepatic toxicity: prechronic/chronic liver toxicity in rodents. *Toxicol Pathol* 2005;33:35–40.
29. Nunez M, Soriano V. Hepatotoxicity of antiretrovirals: incidence, mechanisms and management. *Drug Saf* 2005;28:53–66.
30. Lewis JH. Drug-induced liver disease. *Med Clin North Am* 2000;84:1275–1311.
31. Kaplowitz N. Drug-induced liver injury. *Clin Infect Dis* 2004;38 Suppl 2:S44–S48.
32. Schuster D, Laggner C, Langer T. Why drugs fail—a study on side effects in new chemical entities. *Curr Pharm Des* 2005;11:3545–3559.

33. Lasser KE, Allen PD, Woolhandler SJ, Himmelstein DU, Wolfe SM, Bor DH. Timing of new black box warnings and withdrawals for prescription medications. *JAMA* 2002;287: 2215–2220.
34. Boone L, Meyer D, Cusick P, Ennulat D, Bolliger AP, Everds N, Meador V, Elliott G, Honor D, Bounous D, Jordan H. Selection and interpretation of clinical pathology indicators of hepatic injury in preclinical studies. *Vet Clin Pathol* 2005;34:182–188.
35. O'Brien PJ, Slaughter MR, Polley SR, Kramer K. Advantages of glutamate dehydrogenase as a blood biomarker of acute hepatic injury in rats. *Lab Anim* 2002;36:313–321.
36. Malarkey DE, Johnson K, Ryan L, Boorman G, Maronpot RR. New insights into functional aspects of liver morphology. *Toxicol Pathol* 2005;33:27–34.
37. Dai X, He YD, Dai H, Lum PY, Roberts CJ, Waring JF, Ulrich RG. Development of an approach for ab initio estimation of compound-induced liver injury based on global gene transcriptional profiles. *Genome Inform* 2006;17:77–88.
38. Ruepp S, Boess F, Suter L, de Vera MC, Steiner G, Steele T, Weiser T, Albertini S. Assessment of hepatotoxic liabilities by transcript profiling. *Toxicol Appl Pharmacol* 2005;207:161–170.
39. Ganter B, Tugendreich S, Pearson CI, Ayanoglu E, Baumhueter S, Bostian KA, Brady L, Browne LJ, Calvin JT, Day GJ, Breckenridge N, Dunlea S, Eynon BP, Furness LM, Ferng J, Fielden MR, Fujimoto SY, Gong L, Hu C, Idury R et al. Development of a large-scale chemogenomics database to improve drug candidate selection and to understand mechanisms of chemical toxicity and action. *J Biotechnol* 2005;119:219–244.
40. Blomme EA, Yang Y, Brennan R, Jeffy B. Development and evaluation of gene expression-based signatures predictive and diagnostic of bile duct hyperplasia in rats. *Vet Pathol* 2007;44:781.
41. Fielden MR, Eynon BP, Natsoulis G, Jarnagin K, Banas D, Kolaja KL. A gene expression signature that predicts the future onset of drug-induced renal tubular toxicity. *Toxicol Pathol* 2005;33:675–683.
42. Waring JF, Jolly RA, Ciurlionis R, Lum PY, Praestgaard JT, Morfitt DC, Buratto B, Roberts C, Schadt E, Ulrich RG. Clustering of hepatotoxins based on mechanism of toxicity using gene expression profiles. *Toxicol Appl Pharmacol* 2001;175:28–42.
43. Werner M, Costa MJ, Mitchell LG, Nayar R. Nephrotoxicity of xenobiotics. *Clin Chim Acta* 1995;237:107–154.
44. Davis JW, Kramer JA. Genomic-based biomarkers of drug-induced nephrotoxicity. *Expert Opin Drug Metab Toxicol* 2006;2:95–101.
45. Loeb WF. The measurement of renal injury. *Toxicol Pathol* 1998;26:26–28.
46. Thukral SK, Nordone PJ, Hu R, Sullivan L, Galambos E, Fitzpatrick VD, Healy L, Bass MB, Cosenza ME, Afshari CA. Prediction of nephrotoxicant action and identification of candidate toxicity-related biomarkers. *Toxicol Pathol* 2005;33:343–355.
47. Omenn GS. Assessment of human cancer risk: challenges for alternative approaches. *Toxicol Pathol* 2001;29 Suppl:5–12.
48. Greim H, Gelbke HP, Reuter U, Thielmann HW, Edler L. Evaluation of historical control data in carcinogenicity studies. *Hum Exp Toxicol* 2003;22:541–549.
49. Guengerich FP. Roles of cytochrome P-450 enzymes in chemical carcinogenesis and cancer chemotherapy. *Cancer Res* 1988;48:2946–2954.
50. Michel C, Roberts RA, Desdouets C, Isaacs KR, Boitier E. Characterization of an acute molecular marker of nongenotoxic rodent hepatocarcinogenesis by gene expression profiling in a long term clofibric acid study. *Chem Res Toxicol* 2005;18:611–618.
51. Shibanuma M, Kuroki T, Nose K. Isolation of a gene encoding a putative leucine zipper structure that is induced by transforming growth factor beta 1 and other growth factors. *J Biol Chem* 1992;267:10219–10224.

52. Kester HA, Blanchetot C, den Hertog J, van der Saag PT, Van der BB. Transforming growth factor-beta-stimulated clone-22 is a member of a family of leucine zipper proteins that can homo- and heterodimerize and has transcriptional repressor activity. *J Biol Chem* 1999;274:27439–27447.
53. Iida M, Anna CH, Gaskin ND, Walker NJ, Devereux TR. The putative tumor suppressor Tsc-22 is downregulated early in chemically induced hepatocarcinogenesis and may be a suppressor of Gadd45b. *Toxicol Sci* 2007;99:43–50.
54. Kawamata H, Nakashiro K, Uchida D, Hino S, Omotehara F, Yoshida H, Sato M. Induction of TSC-22 by treatment with a new anti-cancer drug, vesnarinone, in a human salivary gland cancer cell. *Br J Cancer* 1998;77:71–78.
55. Nakashiro K, Kawamata H, Hino S, Uchida D, Miwa Y, Hamano H, Omotehara F, Yoshida H, Sato M. Down-regulation of TSC-22 (transforming growth factor beta-stimulated clone 22) markedly enhances the growth of a human salivary gland cancer cell line in vitro and in vivo. *Cancer Res* 1998;58:549–555.
56. Iida M, Anna CH, Hartis J, Bruno M, Wetmore B, Dubin JR, Sieber S, Bennett L, Cunningham ML, Paules RS, Tomer KB, Houle CD, Merrick AB, Sills RC, Devereux TR. Changes in global gene and protein expression during early mouse liver carcinogenesis induced by non-genotoxic model carcinogens oxazepam and Wyeth-14,643. *Carcinogenesis* 2003;24:757–770.
57. Ellinger-Ziegelbauer H, Stuart B, Wahle B, Bomann W, Ahr HJ. Characteristic expression profiles induced by genotoxic carcinogens in rat liver. *Toxicol Sci* 2004;77:19–34.
58. Ellinger-Ziegelbauer H, Stuart B, Wahle B, Bomann W, Ahr HJ. Comparison of the expression profiles induced by genotoxic and nongenotoxic carcinogens in rat liver. *Mutat Res* 2005;575:61–84.
59. Ellinger-Ziegelbauer H, Gmuender H, Bandenburg A, Ahr HJ. Prediction of a carcinogenic potential of rat hepatocarcinogens using toxicogenomics analysis of short-term in vivo studies. *Mutat Res* 2008;637:23–39.
60. Fielden MR, Brennan R, Gollub J. A gene expression biomarker provides early prediction and mechanistic assessment of hepatic tumor induction by nongenotoxic chemicals. *Toxicol Sci* 2007;99:90–100.
61. Nie AY, McMillian M, Parker JB, Leone A, Bryant S, Yieh L, Bittner A, Nelson J, Carmen A, Wan J, Lord PG. Predictive toxicogenomics approaches reveal underlying molecular mechanisms of nongenotoxic carcinogenicity. *Mol Carcinog* 2006;45:914–933.
62. Peters EH, Rojas-Caro S, Brigell MG, Zahorchak RJ, des Etages SA, Ruppel PL, Knight CR, Austermiller B, Graham MC, Wowk S, Banks S, Madabusi LV, Turk P, Wilder D, Kempfer C, Osborn TW, Willey JC. Quality-controlled measurement methods for quantification of variations in transcript abundance in whole blood samples from healthy volunteers. *Clin Chem* 2007;53:1030–1037.
63. Willey JC, Knight CR, Crawford EL, Olson DE, Hammersly J, Yoon Y, Sharief I. Use of standardized reverse transcription-polymerase chain reaction and the standardized expression measurement center in multi-institutional trials to develop meaningful lung cancer classification based on molecular genetic criteria. *Chest* 2004;125:155S–156S.
64. Willey JC, Crawford EL, Knight CR, Warner KA, Motten CA, Herness EA, Zahorchak RJ, Graves TG. Standardized RT-PCR and the standardized expression measurement center. *Methods Mol Biol* 2004;258:13–41.
65. Rockett JC, Burczynski ME, Fornace AJ, Herrmann PC, Krawetz SA, Dix DJ. Surrogate tissue analysis: monitoring toxicant exposure and health status of inaccessible tissues through the analysis of accessible tissues and cells. *Toxicol Appl Pharmacol* 2004;194:189–199.
66. Burczynski ME, Dorner AJ. Transcriptional profiling of peripheral blood cells in clinical pharmacogenomic studies. *Pharmacogenomics* 2006;7:187–202.

67. Rockett JC, Kavlock RJ, Lambright CR, Parks LG, Schmid JE, Wilson VS, Wood C, Dix DJ. DNA arrays to monitor gene expression in rat blood and uterus following 17beta-estradiol exposure: biomonitoring environmental effects using surrogate tissues. *Toxicol Sci* 2002;69: 49–59.
68. Baechler EC, Batliwalla FM, Karypis G, Gaffney PM, Ortmann WA, Espe KJ, Shark KB, Grande WJ, Hughes KM, Kapur V, Gregersen PK, Behrens TW. Interferon-inducible gene expression signature in peripheral blood cells of patients with severe lupus. *Proc Natl Acad Sci USA* 2003;100:2610–2615.
69. Aune TM, Maas K, Moore JH, Olsen NJ. Gene expression profiles in human autoimmune disease. *Curr Pharm Des* 2003;9:1905–1917.
70. Olsen NJ, Moore JH, Aune TM. Gene expression signatures for autoimmune disease in peripheral blood mononuclear cells. *Arthritis Res Ther* 2004;6:120–128.
71. Tang Y, Gilbert DL, Glauser TA, Hershey AD, Sharp FR. Blood gene expression profiling of neurologic diseases: a pilot microarray study. *Arch Neurol* 2005;62:210–215.
72. Burczynski ME, Peterson RL, Twine NC, Zuberek KA, Brodeur BJ, Casciotti L, Maganti V, Reddy PS, Strahs A, Immermann F, Spinelli W, Schwertschlag U, Slager AM, Cotreau MM, Dorner AJ. Molecular classification of Crohn's disease and ulcerative colitis patients using transcriptional profiles in peripheral blood mononuclear cells. *J Mol Diagn* 2006;8:51–61.
73. Scherzer CR, Eklund AC, Morse LJ, Liao Z, Locascio JJ, Fefer D, Schwarzschild MA, Schlossmacher MG, Hauser MA, Vance JM, Sudarsky LR, Standaert DG, Growdon JH, Jensen RV, Gullans Sr. Molecular markers of early Parkinson's disease based on gene expression in blood. *Proc Natl Acad Sci USA* 2007;104:955–960.
74. Borovecki F, Lovrecic L, Zhou J, Jeong H, Then F, Rosas HD, Hersch SM, Hogarth P, Bouzou B, Jensen RV, Krainc D. Genome-wide expression profiling of human blood reveals biomarkers for Huntington's disease. *Proc Natl Acad Sci USA* 2005;102:11023–11028.
75. Dressman HK, Muramoto GG, Chao NJ, Meadows S, Marshall D, Ginsburg GS, Nevins JR, Chute JP. Gene expression signatures that predict radiation exposure in mice and humans. *PLoS Med* 2007;4:e106.
76. Whitney AR, Diehn M, Popper SJ, Alizadeh AA, Boldrick JC, Relman DA, Brown PO. Individuality and variation in gene expression patterns in human blood. *Proc Natl Acad Sci USA* 2003;100:1896–1901.
77. Kim SJ, Dix DJ, Thompson KE, Murrell RN, Schmid JE, Gallagher JE, Rockett JC. Effects of storage, RNA extraction, genechip type, and donor sex on gene expression profiling of human whole blood. *Clin Chem* 2007;53:1038–1045.
78. Larrey D, Tinel M, Pessayre D. Formation of inactive cytochrome P-450 Fe(II)-metabolite complexes with several erythromycin derivatives but not with josamycin and midecamycin in rats. *Biochem Pharmacol* 1983;32:1487–1493.
79. Amacher DE, Schomaker SJ, Boldt SE, Mirsky M. The relationship among microsomal enzyme induction, liver weight, and histological change in cynomolgus monkey toxicology studies. *Food Chem Toxicol* 2006;44:528–537.
80. Klaunig JE, Babich MA, Baetcke KP, Cook JC, Corton JC, David RM, DeLuca JG, Lai DY, McKee RH, Peters JM, Roberts RA, Fenner-Crisp PA. PPARalpha agonist-induced rodent tumors: modes of action and human relevance. *Crit Rev Toxicol* 2003;33:655–780.
81. Waring JF, Gum R, Morfitt D, Jolly RA, Ciurlionis R, Heindel M, Gallenberg L, Buratto B, Ulrich RG. Identifying toxic mechanisms using DNA microarrays: evidence that an experimental inhibitor of cell adhesion molecule expression signals through the aryl hydrocarbon nuclear receptor. *Toxicology* 2002;181–182:537–550.
82. Wolfgang GH, Robertson DG, Welty DF, Metz AL. Hepatic and adrenal toxicity of a novel lipid regulator in beagle dogs. *Fundam Appl Toxicol* 1995;26:272–281.

83. Hamadeh HK, Bushel PR, Jayadev S, Martin K, DiSorbo O, Sieber S, Bennett L, Tennant R, Stoll R, Barrett JC, Blanchard K, Paules RS, Afshari CA. Gene expression analysis reveals chemical-specific profiles. *Toxicol Sci* 2002;67:219–231.
84. Kemper B. Regulation of cytochrome P450 gene transcription by phenobarbital. *Prog Nucleic Acid Res Mol Biol* 1998;61:23–64.
85. Denison MS, Nagy Sr. Activation of the aryl hydrocarbon receptor by structurally diverse exogenous and endogenous chemicals. *Annu Rev Pharmacol Toxicol* 2003;43:309–334.
86. Guerreiro N, Staedtler F, Grenet O, Kehren J, Chibout SD. Toxicogenomics in drug development. *Toxicol Pathol* 2003;31:471–479.
87. Mimura J, Fujii-Kuriyama Y. Functional role of AhR in the expression of toxic effects by TCDD. *Biochim Biophys Acta* 2003;1619:263–268.
88. Nebert DW, Dalton TP, Okey AB, Gonzalez FJ. Role of aryl hydrocarbon receptor-mediated induction of the CYP1 enzymes in environmental toxicity and cancer. *J Biol Chem* 2004;279:23847–23850.
89. Whitlock JP, Jr. Induction of cytochrome P4501A1. *Annu Rev Pharmacol Toxicol* 1999; 39:103–125.
90. Hu W, Sorrentino C, Denison MS, Kolaja K, Fielden MR. Induction of cyp1a1 is a nonspecific biomarker of aryl hydrocarbon receptor activation: results of large scale screening of pharmaceuticals and toxicants in vivo and in vitro. *Mol Pharmacol* 2007;71:1475–1486.
91. Ohno M. Genetic and pharmacological basis for therapeutic inhibition of beta- and gamma-secretases in mouse models of Alzheimer's memory deficits. *Rev Neurosci* 2006; 17:429–454.
92. Citron M. Strategies for disease modification in Alzheimer's disease. *Nat Rev Neurosci* 2004; 5:677–685.
93. Czirr E, Leuchtenberger S, Dorner-Ciossek C, Schneider A, Jucker M, Koo EH, Pietrzik CU, Baumann K, Weggen S. Insensitivity to Abeta42-lowering nonsteroidal anti-inflammatory drugs and gamma-secretase inhibitors is common among aggressive presenilin-1 mutations. *J Biol Chem* 2007;282:24504–24513.
94. Beher D, Graham SL. Protease inhibitors as potential disease-modifying therapeutics for Alzheimer's disease. *Expert Opin Investig Drugs* 2005;14:1385–1409.
95. Tomita T, Iwatsubo T. gamma-secretase as a therapeutic target for treatment of Alzheimer's disease. *Curr Pharm Des* 2006;12:661–670.
96. Evin G, Sernee MF, Masters CL. Inhibition of gamma-secretase as a therapeutic intervention for Alzheimer's disease: prospects, limitations and strategies. *CNS Drugs* 2006;20:351–372.
97. Milano J, McKay J, Dagenais C, Foster-Brown L, Pognan F, Gadient R, Jacobs RT, Zacco A, Greenberg B, Ciaccio PJ. Modulation of notch processing by gamma-secretase inhibitors causes intestinal goblet cell metaplasia and induction of genes known to specify gut secretory lineage differentiation. *Toxicol Sci* 2004;82:341–358.
98. Searfoss GH, Jordan WH, Calligaro DO, Galbreath EJ, Schirtzinger LM, Berridge BR, Gao H, Higgins MA, May PC, Ryan TP. Adipsin, a biomarker of gastrointestinal toxicity mediated by a functional gamma-secretase inhibitor. *J Biol Chem* 2003;278:46107–46116.
99. Nair S, Xu C, Shen G, Hebbar V, Gopalakrishnan A, Hu R, Jain MR, Liew C, Chan JY, Kong AN. Toxicogenomics of endoplasmic reticulum stress inducer tunicamycin in the small intestine and liver of Nrf2 knockout and C57BL/6J mice. *Toxicol Lett* 2007;168:21–39.
100. Shen G, Xu C, Hu R, Jain MR, Nair S, Lin W, Yang CS, Chan JY, Kong AN. Comparison of (−)-epigallocatechin-3-gallate elicited liver and small intestine gene expression profiles between C57BL/6J mice and C57BL/6J/Nrf2 (−/−) mice. *Pharm Res* 2005;22:1805–1820.
101. Shen G, Xu C, Hu R, Jain MR, Gopalkrishnan A, Nair S, Huang MT, Chan JY, Kong AN. Modulation of nuclear factor E2-related factor 2-mediated gene expression in mice liver and small intestine by cancer chemopreventive agent curcumin. *Mol Cancer Ther* 2006;5:39–51.

102. Zhang JN, Liu XG, Zhu M, Chiu FC, Li RC. Assessment of presystemic factors on the oral bioavailability of rifampicin following multiple dosing. *J Chemother* 1998;10: 354–359.
103. Graham RA, Tyler LO, Krol WL, Silver IS, Webster LO, Clark P, Chen L, Banks T, LeCluyse EL. Temporal kinetics and concentration-response relationships for induction of CYP1A, CYP2B, and CYP3A in primary cultures of beagle dog hepatocytes. *J Biochem Mol Toxicol* 2006;20:69–78.
104. Guth BD. Preclinical cardiovascular risk assessment in modern drug development. *Toxicol Sci* 2007;97:4–20.
105. Wallace KB, Hausner E, Herman E, Holt GD, MacGregor JT, Metz AL, Murphy E, Rosenblum IY, Sistare FD, York MJ. Serum troponins as biomarkers of drug-induced cardiac toxicity. *Toxicol Pathol* 2004;32:106–121.
106. York M, Scudamore C, Brady S, Chen C, Wilson S, Curtis M, Evans G, Griffiths W, Whayman M, Williams T, Turton J. Characterization of troponin responses in isoproterenol-induced cardiac injury in the Hanover Wistar rat. *Toxicol Pathol* 2007;35:606–617.
107. Anderson TD, Van Alstine WG, Ficken MD, Miskimins DW, Carson TL, Osweiler GD. Acute monensin toxicosis in sheep: light and electron microscopic changes. *Am J Vet Res* 1984;45:1142–1147.
108. Robert J. Preclinical assessment of anthracycline cardiotoxicity in laboratory animals: predictiveness and pitfalls. *Cell Biol Toxicol* 2007;23:27–37.
109. Singal PK, Iliskovic N. Doxorubicin-induced cardiomyopathy. *N Engl J Med* 1998;339: 900–905.
110. Minotti G, Recalcati S, Menna P, Salvatorelli E, Corna G, Cairo G. Doxorubicin cardiotoxicity and the control of iron metabolism: quinone-dependent and independent mechanisms. *Methods Enzymol* 2004;378:340–361.
111. Minotti G, Menna P, Salvatorelli E, Cairo G, Gianni L. Anthracyclines: molecular advances and pharmacologic developments in antitumor activity and cardiotoxicity. *Pharmacol Rev* 2004;56:185–229.
112. Kalivendi SV, Konorev EA, Cunningham S, Vanamala SK, Kaji EH, Joseph J, Kalyanaraman B. Doxorubicin activates nuclear factor of activated T-lymphocytes and Fas ligand transcription: role of mitochondrial reactive oxygen species and calcium. *Biochem J* 2005;389:527–539.
113. Robert J. Long-term and short-term models for studying anthracycline cardiotoxicity and protectors. *Cardiovasc Toxicol* 2007;7:135–139.
114. Chaiswing L, Cole MP, St Clair DK, Ittarat W, Szweda LI, Oberley TD. Oxidative damage precedes nitrative damage in adriamycin-induced cardiac mitochondrial injury. *Toxicol Pathol* 2004;32:536–547.
115. Jang YM, Kendaiah S, Drew B, Phillips T, Selman C, Julian D, Leeuwenburgh C. Doxorubicin treatment in vivo activates caspase-12 mediated cardiac apoptosis in both male and female rats. *FEBS Lett* 2004;577:483–490.
116. Gu YG, Weitzberg M, Clark RF, Xu X, Li Q, Lubbers NL, Yang Y, Beno DW, Widomski DL, Zhang T, Hansen TM, Keyes RF, Waring JF, Carroll SL, Wang X, Wang R, Healan-Greenberg CH, Blomme EA, Beutel BA, Sham HL et al. *N*-{macrosleftcurlybrace3-[2-(4-alkoxyphenoxy)thiazol-5-yl]-1-methylprop-2-ynyl}carboxy derivatives as acetyl-coA carboxylase inhibitors—improvement of cardiovascular and neurological liabilities via structural modifications. *J Med Chem* 2007;50:1078–1082.
117. Tong L, Harwood HJ, Jr. Acetyl-coenzyme A carboxylases: versatile targets for drug discovery. *J Cell Biochem* 2006;99:1476–1488.
118. Ruderman N, Prentki M. AMP kinase and malonyl-CoA: targets for therapy of the metabolic syndrome. *Nat Rev Drug Discov* 2004;3:340–351.

119. Abu-Elheiga L, Almarza-Ortega DB, Baldini A, Wakil SJ. Human acetyl-CoA carboxylase 2. Molecular cloning, characterization, chromosomal mapping, and evidence for two isoforms. *J Biol Chem* 1997;272:10669–10677.
120. Mao J, Chirala SS, Wakil SJ. Human acetyl-CoA carboxylase 1 gene: presence of three promoters and heterogeneity at the 5′-untranslated mRNA region. *Proc Natl Acad Sci USA* 2003;100:7515–7520.
121. Abu-Elheiga L, Matzuk MM, Kordari P, Oh W, Shaikenov T, Gu Z, Wakil SJ. Mutant mice lacking acetyl-CoA carboxylase 1 are embryonically lethal. *Proc Natl Acad Sci USA* 2005;102:12011–12016.
122. Oh W, Abu-Elheiga L, Kordari P, Gu Z, Shaikenov T, Chirala SS, Wakil SJ. Glucose and fat metabolism in adipose tissue of acetyl-CoA carboxylase 2 knockout mice. *Proc Natl Acad Sci USA* 2005;102:1384–1389.
123. Boekelheide K. Mechanisms of toxic damage to spermatogenesis. *J Natl Cancer Inst Monogr* 2005;6–8.
124. Creasy DM. Evaluation of testicular toxicity in safety evaluation studies: the appropriate use of spermatogenic staging. *Toxicol Pathol* 1997;25:119–131.
125. Comhaire FH, Mahmoud AM, Schoonjans F. Sperm quality, birth rates and the environment in Flanders (Belgium). *Reprod Toxicol* 2007;23:133–137.
126. Richburg JH, Johnson KJ, Schoenfeld HA, Meistrich ML, Dix DJ. Defining the cellular and molecular mechanisms of toxicant action in the testis. *Toxicol Lett* 2002;135:167–183.
127. Kartalou M, Essigmann JM. Recognition of cisplatin adducts by cellular proteins. *Mutat Res* 2001;478:1–21.
128. Petersen PM, Hansen SW, Giwercman A, Rorth M, Skakkebaek NE. Dose-dependent impairment of testicular function in patients treated with cisplatin-based chemotherapy for germ cell cancer. *Ann Oncol* 1994;5:355–358.
129. Prahalathan C, Selvakumar E, Varalakshmi P. Lipoic acid ameliorates adriamycin-induced testicular mitochondriopathy. *Reprod Toxicol* 2005;20:111–116.
130. El Demerdash FM, Yousef MI, Kedwany FS, Baghdadi HH. Cadmium-induced changes in lipid peroxidation, blood hematology, biochemical parameters and semen quality of male rats: protective role of vitamin E and beta-carotene. *Food Chem Toxicol* 2004;42:1563–1571.
131. Morris ID, Phillips DM, Bardin CW. Ethylene dimethanesulfonate destroys Leydig cells in the rat testis. *Endocrinology* 1986;118:709–719.
132. Delic JI, Hendry JH, Morris ID, Shalet SM. Serum androgen binding protein and follicle stimulating hormone as indices of Sertoli cell function in the irradiated testis. *Br J Cancer Suppl* 1986;7:105–107.
133. Stewart J, Turner KJ. Inhibin B as a potential biomarker of testicular toxicity. *Cancer Biomark* 2005;1:75–91.
134. Anawalt BD, Bebb RA, Matsumoto AM, Groome NP, Illingworth PJ, McNeilly AS, Bremner WJ. Serum inhibin B levels reflect Sertoli cell function in normal men and men with testicular dysfunction. *J Clin Endocrinol Metab* 1996;81:3341–3345.
135. Illingworth PJ, Groome NP, Byrd W, Rainey WE, McNeilly AS, Mather JP, Bremner WJ. Inhibin-B: a likely candidate for the physiologically important form of inhibin in men. *J Clin Endocrinol Metab* 1996;81:1321–1325.
136. Foresta C, Bettella A, Garolla A, Ambrosini G, Ferlin A. Treatment of male idiopathic infertility with recombinant human follicle-stimulating hormone: a prospective, controlled, randomized clinical study. *Fertil Steril* 2005;84:654–661.
137. Majumdar SS, Winters SJ, Plant TM. Procedures for the isolation and culture of Sertoli cells from the testes of infant, juvenile, and adult rhesus monkeys (*Macaca mulatta*). *Biol Reprod* 1998;58:633–640.

138. Kukucka MA, Misra HP. Isolation and culture of highly enriched populations of Leydig cells from guinea-pig (*Cavia porcellus*) testes. *Andrologia* 1994;26:217–224.
139. Murugesan P, Balaganesh M, Balasubramanian K, Arunakaran J. Effects of polychlorinated biphenyl (Aroclor 1254) on steroidogenesis and antioxidant system in cultured adult rat Leydig cells. *J Endocrinol* 2007;192:325–338.
140. Adachi T, Ono Y, Koh KB, Takashima K, Tainaka H, Matsuno Y, Nakagawa S, Todaka E, Sakurai K, Fukata H, Iguchi T, Komiyama M, Mori C. Long-term alteration of gene expression without morphological change in testis after neonatal exposure to genistein in mice: toxicogenomic analysis using cDNA microarray. *Food Chem Toxicol* 2004;42:445–452.
141. Adachi T, Koh KB, Tainaka H, Matsuno Y, Ono Y, Sakurai K, Fukata H, Iguchi T, Komiyama M, Mori C. Toxicogenomic difference between diethylstilbestrol and 17beta-estradiol in mouse testicular gene expression by neonatal exposure. *Mol Reprod Dev* 2004;67:19–25.
142. Moustafa GG, Ibrahim ZS, Hashimoto Y, Alkelch AM, Sakamoto KQ, Ishizuka M, Fujita S. Testicular toxicity of profenofos in matured male rats. *Arch Toxicol* 2007;81:875–81.
143. Tully DB, Bao W, Goetz AK, Blystone CR, Ren H, Schmid JE, Strader LF, Wood CR, Best DS, Narotsky MG, Wolf DC, Rockett JC, Dix DJ. Gene expression profiling in liver and testis of rats to characterize the toxicity of triazole fungicides. *Toxicol Appl Pharmacol* 2006;215:260–273.
144. Shultz VD, Phillips S, Sar M, Foster PM, Gaido KW. Altered gene profiles in fetal rat testes after in utero exposure to di(*n*-butyl) phthalate. *Toxicol Sci* 2001;64:233–242.
145. Rockett JC, Christopher LJ, Brian GJ, Krawetz SA, Hughes MR, Hee KK, Oudes AJ, Dix DJ. Development of a 950-gene DNA array for examining gene expression patterns in mouse testis. *Genome Biol* 2001;2:14.1–14.9.
146. Ryu JY, Lee BM, Kacew S, Kim HS. Identification of differentially expressed genes in the testis of Sprague-Dawley rats treated with di(*n*-butyl) phthalate. *Toxicology* 2007;234: 103–112.
147. Latini G, Del Vecchio A, Massaro M, Verrotti A, De Felice C. Phthalate exposure and male infertility. *Toxicology* 2006;226:90–98.
148. Mylchreest E, Cattley RC, Foster PM. Male reproductive tract malformations in rats following gestational and lactational exposure to di(*n*-butyl) phthalate: an antiandrogenic mechanism? *Toxicol Sci* 1998;43:47–60.
149. Saillenfait AM, Payan JP, Fabry JP, Beydon D, Langonne I, Gallissot F, Sabate JP. Assessment of the developmental toxicity, metabolism, and placental transfer of di-*n*-butyl phthalate administered to pregnant rats. *Toxicol Sci* 1998;45:212–224.
150. Lehmann KP, Phillips S, Sar M, Foster PM, Gaido KW. Dose-dependent alterations in gene expression and testosterone synthesis in the fetal testes of male rats exposed to di(*n*-butyl) phthalate. *Toxicol Sci* 2004;81:60–68.
151. Fukushima T, Yamamoto T, Kikkawa R, Hamada Y, Komiyama M, Mori C, Horii I. Effects of male reproductive toxicants on gene expression in rat testes. *J Toxicol Sci* 2005;30: 195–206.
152. Linder RE, Klinefelter GR, Strader LF, Suarez JD, Roberts NL, Dyer CJ. Spermatotoxicity of dibromoacetic acid in rats after 14 daily exposures. *Reprod Toxicol* 1994;8:251–259.
153. Linder RE, Klinefelter GR, Strader LF, Veeramachaneni DN, Roberts NL, Suarez JD. Histopathologic changes in the testes of rats exposed to dibromoacetic acid. *Reprod Toxicol* 1997;11:47–56.
154. Linder RE, Klinefelter GR, Strader LF, Suarez JD, Dyer CJ. Acute spermatogenic effects of bromoacetic acids. *Fundam Appl Toxicol* 1994;22:422–430.
155. Blomme EA, Ciurlionis R, Milicic I, Carr T, Whitney K, Miller J et al. Role of cytochrome P450C17α in dibromoacetic acid-induced testicular toxicity in rats (Abstract). *Proceedings of the 45th Annual Meeting of the Society of Toxicology*, 1546, 2006.

156. Dalla VL, Vianello S, Belvedere P, Colombo L. Rat cytochrome P450c17 gene transcription is initiated at different start sites in extraglandular and glandular tissues. *J Steroid Biochem Mol Biol* 2002;82:377–384.
157. Benson JD, Chen YN, Cornell-Kennon SA, Dorsch M, Kim S, Leszczyniecka M, Sellers WR, Lengauer C. Validating cancer drug targets. *Nature* 2006;441:451–456.
158. Su AI, Cooke MP, Ching KA, Hakak Y, Walker JR, Wiltshire T, Orth AP, Vega RG, Sapinoso LM, Moqrich A, Patapoutian A, Hampton GM, Schultz PG, Hogenesch JB. Large-scale analysis of the human and mouse transcriptomes. *Proc Natl Acad Sci USA* 2002;99:4465–4470.
159. Su AI, Wiltshire T, Batalov S, Lapp H, Ching KA, Block D, Zhang J, Soden R, Hayakawa M, Kreiman G, Cooke MP, Walker JR, Hogenesch JB. A gene atlas of the mouse and human protein-encoding transcriptomes. *Proc Natl Acad Sci USA* 2004;101:6062–6067.
160. Smitskamp-Wilms E, Giaccone G, Pinedo HM, van der Laan BF, Peters GJ. DT-diaphorase activity in normal and neoplastic human tissues; an indicator for sensitivity to bioreductive agents? *Br J Cancer* 1995;72:917–921.
161. Zappa F, Ward T, Pedrinis E, Butler J, McGown A. NAD(P)H:quinone oxidoreductase 1 expression in kidney podocytes. *J Histochem Cytochem* 2003;51:297–302.
162. Hanna WT, Krauss S, Regester RF, Murphy WM. Renal disease after mitomycin C therapy. *Cancer* 1981;48:2583–2588.
163. Zambrowicz BP, Turner CA, Sands AT. Predicting drug efficacy: knockouts model pipeline drugs of the pharmaceutical industry. *Curr Opin Pharmacol* 2003;3:563–570.
164. Anders HJ, Vielhauer V. Identifying and validating novel targets with in vivo disease models: guidelines for study design. *Drug Discov Today* 2007;12:446–451.
165. Lewandoski M. Conditional control of gene expression in the mouse. *Nat Rev Genet* 2001;2:743–755.
166. Singh M, Johnson L. Using genetically engineered mouse models of cancer to aid drug development: an industry perspective. *Clin Cancer Res* 2006;12:5312–5328.
167. Berg AL, Bohlooly Y. The program for phenotyping of genetically modified animals at AstraZeneca. *Exp Toxicol Pathol* 2006;57:383–384.
168. Semizarov D, Kroeger P, Fesik S. siRNA-mediated gene silencing: a global genome view. *Nucleic Acids Res* 2004;32:3836–45.
169. Semizarov D, Frost L, Sarthy A, Kroeger P, Halbert DN, Fesik SW. Specificity of short interfering RNA determined through gene expression signatures. *Proc Natl Acad Sci USA* 2003;100:6347–6352.
170. Cairns MJ, Saravolac EG, Sun LQ. Catalytic DNA: a novel tool for gene suppression. *Curr Drug Targets* 2002;3:269–279.
171. Dias N, Stein CA. Antisense oligonucleotides: basic concepts and mechanisms. *Mol Cancer Ther* 2002;1:347–355.
172. Honore P, Kage K, Mikusa J, Watt AT, Johnston JF, Wyatt JR, Faltynek CR, Jarvis MF, Lynch K. Analgesic profile of intrathecal P2X(3) antisense oligonucleotide treatment in chronic inflammatory and neuropathic pain states in rats. *Pain* 2002;99:11–19.
173. Dias N, Stein CA. Potential roles of antisense oligonucleotides in cancer therapy. The example of Bcl-2 antisense oligonucleotides. *Eur J Pharm Biopharm* 2002;54:263–269.
174. Chatterjee-Kishore M, Miller CP. Exploring the sounds of silence: RNAi-mediated gene silencing for target identification and validation. *Drug Discov Today* 2005;10:1559–1565.
175. Li QX, Tan P, Ke N, Wong-Staal F. Ribozyme technology for cancer gene target identification and validation. *Adv Cancer Res* 2007;96:103–143.
176. Rossi JJ. Ribozymes, genomics and therapeutics. *Chem Biol* 1999;6:R33–R37.
177. Lu PY, Xie F, Woodle MC. In Vivo application of RNA interference: from functional genomics to therapeutics. *Adv Genet* 2005;54:117–142.

178. Li L, Lin X, Staver M, Shoemaker A, Semizarov D, Fesik SW, Shen Y. Evaluating hypoxia-inducible factor-1alpha as a cancer therapeutic target via inducible RNA interference in vivo. *Cancer Res* 2005;65:7249–7258.
179. Xu H, Wilcox D, Nguyen P, Voorbach M, Suhar T, Morgan SJ, An WF, Ge L, Green J, Wu Z, Gimeno RE, Reilly R, Jacobson PB, Collins CA, Landschulz K, Surowy T. Hepatic knockdown of mitochondrial GPAT1 in ob/ob mice improves metabolic profile. *Biochem Biophys Res Commun* 2006;349:439–448.
180. Xu H, Wilcox D, Nguyen P, Voorbach M, Smith H, Brodjian S, Suhar T, Reilly RM, Jacobson PB, Collins CA, Landschulz K, Surowy TK. Hepatic knockdown of stearoyl-CoA desaturase 1 via RNA interference in obese mice decreases lipid content and changes fatty acid composition. *Front Biosci* 2007;12:3781–3794.
181. Grimm D, Streetz KL, Jopling CL, Storm TA, Pandey K, Davis CR, Marion P, Salazar F, Kay MA. Fatality in mice due to oversaturation of cellular microRNA/short hairpin RNA pathways. *Nature* 2006;441:537–541.
182. Akai S, Hosomi H, Minami K, Tsuneyama K, Katoh M, Nakajima M, Yokoi T. Knock down of gamma-glutamylcysteine synthetase in rat causes acetaminophen-induced hepatotoxicity. *J Biol Chem* 2007;282:23996–24003.
183. Cattley RC. Peroxisome proliferators and receptor-mediated hepatic carcinogenesis. *Toxicol Pathol* 2004;32 Suppl 2:6–11.
184. Holden PR, Tugwood JD. Peroxisome proliferator-activated receptor alpha: role in rodent liver cancer and species differences. *J Mol Endocrinol* 1999;22:1–8.
185. Kramer JA, Blomme EA, Bunch RT, Davila JC, Jackson CJ, Jones PF, Kolaja KL, Curtiss SW. Transcription profiling distinguishes dose-dependent effects in the livers of rats treated with clofibrate. *Toxicol Pathol* 2003;31:417–431.
186. Crunkhorn SE, Plant KE, Gibson GG, Kramer K, Lyon J, Lord PG, Plant NJ. Gene expression changes in rat liver following exposure to liver growth agents: role of Kupffer cells in xenobiotic-mediated liver growth. *Biochem Pharmacol* 2004;67:107–118.
187. Hamadeh HK, Bushel PR, Jayadev S, DiSorbo O, Bennett L, Li L, Tennant R, Stoll R, Barrett JC, Paules RS, Blanchard K, Afshari CA. Prediction of compound signature using high density gene expression profiling. *Toxicol Sci* 2002;67:232–240.
188. Grenet O, Varela MC, Staedtler F, Steiner S. The cyclosporine A-induced decrease in rat renal calbindin-D28kDa protein as a consequence of a decrease in its mRNA. *Biochem Pharmacol* 1998;55:1131–1133.
189. Varela MC, Arce A, Greiner B, Schwald M, Aicher L, Wahl D, Grenet O, Steiner S. Cyclosporine A-induced decrease in calbindin-D 28kDa in rat kidney but not in cerebral cortex and cerebellum. *Biochem Pharmacol* 1998;55:2043–2046.
190. Waring JF, Anderson MG. Idiosyncratic toxicity: mechanistic insights gained from analysis of prior compounds. *Curr Opin Drug Discov Devel* 2005;8:59–65.
191. Uetrecht J. Idiosyncratic drug reactions: current understanding. *Annu Rev Pharmacol Toxicol* 2007;47:513–539.
192. Watkins PB. Idiosyncratic liver injury: challenges and approaches. *Toxicol Pathol* 2005;33:1–5.
193. Watkins PB. Insight into hepatotoxicity: The troglitazone experience. *Hepatology* 2005;41: 229–230.
194. Gitlin N, Julie NL, Spurr CL, Lim KN, Juarbe HM. Two cases of severe clinical and histologic hepatotoxicity associated with troglitazone. *Ann Intern Med* 1998;129:36–38.
195. Vella A, de Groen PC, Dinneen SF. Fatal hepatotoxicity associated with troglitazone. *Ann Intern Med* 1998;129:1080.
196. Herrine SK, Choudhary C. Severe hepatotoxicity associated with troglitazone. *Ann Intern Med* 1999;130:163–164.

197. Knowler WC, Hamman RF, Edelstein SL, Barrett-Connor E, Ehrmann DA, Walker EA, Fowler SE, Nathan DM, Kahn SE. Prevention of type 2 diabetes with troglitazone in the Diabetes Prevention Program. *Diabetes* 2005;54:1150–1156.
198. Pessayre D, Bichara M, Degott C, Potet F, Benhamou JP, Feldmann G. Perhexiline maleate-induced cirrhosis. *Gastroenterology* 1979;76:170–177.
199. Pirmohamed M, James S, Meakin S, Green C, Scott AK, Walley TJ, Farrar K, Park BK, Breckenridge AM. Adverse drug reactions as cause of admission to hospital: prospective analysis of 18,820 patients. *BMJ* 2004;329:15–19.
200. Bousquet J, Demoly P, Hillaire-Buys D, Blayac JP, Godard P, Michel FB. [Drug allergies: a public health problem]. *Bull Acad Natl Med* 2000;184:737–750.
201. Williams DP, Park BK. Idiosyncratic toxicity: the role of toxicophores and bioactivation. *Drug Discov Today* 2003;8:1044–1050.
202. Uetrecht J. Prediction of a new drug's potential to cause idiosyncratic reactions. *Curr Opin Drug Discov Devel* 2001;4:55–59.
203. Knowles SR, Uetrecht J, Shear NH. Idiosyncratic drug reactions: the reactive metabolite syndrome. *Lancet* 2000;356:1587–1591.
204. Park BK, Kitteringham NR, Powell H, Pirmohamed M. Advances in molecular toxicology—towards understanding idiosyncratic drug toxicity. *Toxicology* 2000;153:39–60.
205. Roth RA, Luyendyk JP, Maddox JF, Ganey PE. Inflammation and drug idiosyncrasy–is there a connection? *J Pharmacol Exp Ther* 2003;307:1–8.
206. Liguori MJ, Waring JF. Investigations toward enhanced understanding of hepatic idiosyncratic drug reactions. *Expert Opin Drug Metab Toxicol* 2006;2:835–846.
207. Shenton JM, Chen J, Uetrecht JP. Animal models of idiosyncratic drug reactions. *Chem Biol Interact* 2004;150:53–70.
208. Furst SM, Luedke D, Gaw HH, Reich R, Gandolfi AJ. Demonstration of a cellular immune response in halothane-exposed guinea pigs. *Toxicol Appl Pharmacol* 1997;143:245–255.
209. Shenton JM, Teranishi M, Abu-Asab MS, Yager JA, Uetrecht JP. Characterization of a potential animal model of an idiosyncratic drug reaction: nevirapine-induced skin rash in the rat. *Chem Res Toxicol* 2003;16:1078–1089.
210. Masson MJ, Uetrecht JP. Tolerance induced by low dose D-penicillamine in the brown Norway rat model of drug-induced autoimmunity is immune-mediated. *Chem Res Toxicol* 2004;17:82–94.
211. Luyendyk JP, Shores KC, Ganey PE, Roth RA. Bacterial lipopolysaccharide exposure alters aflatoxin B(1) hepatotoxicity: benchmark dose analysis for markers of liver injury. *Toxicol Sci* 2002;68:220–225.
212. Luyendyk JP, Maddox JF, Cosma GN, Ganey PE, Cockerell GL, Roth RA. Ranitidine treatment during a modest inflammatory response precipitates idiosyncrasy-like liver injury in rats. *J Pharmacol Exp Ther* 2003;307:9–16.
213. Luyendyk JP, Maddox JF, Green CD, Ganey PE, Roth RA. Role of hepatic fibrin in idiosyncrasy-like liver injury from lipopolysaccharide-ranitidine coexposure in rats. *Hepatology* 2004;40:1342–1351.
214. Buchweitz JP, Ganey PE, Bursian SJ, Roth RA. Underlying endotoxemia augments toxic responses to chlorpromazine: is there a relationship to drug idiosyncrasy? *J Pharmacol Exp Ther* 2002;300:460–467.
215. Luyendyk JP, Mattes WB, Burgoon LD, Zacharewski TR, Maddox JF, Cosma GN, Ganey PE, Roth RA. Gene expression analysis points to hemostasis in livers of rats cotreated with lipopolysaccharide and ranitidine. *Toxicol Sci* 2004;80:203–213.
216. Drlica K, Zhao X. DNA gyrase, topoisomerase IV, and the 4-quinolones. *Microbiol Mol Biol Rev* 1997;61:377–392.

217. Ball P, Mandell L, Niki Y, Tillotson G. Comparative tolerability of the newer fluoroquinolone antibacterials. *Drug Safety* 1999;21:407–421.
218. Bertino J, Jr., Fish D. The safety profile of the fluoroquinolones. *Clin Ther* 2000;22:798–817.
219. Liguori MJ, Anderson LM, Bukofzer S, McKim J, Pregenzer JF, Retief J, Spear BB, Waring JF. Microarray analysis in human hepatocytes suggests a mechanism for hepatotoxicity induced by trovafloxacin. *Hepatology* 2005;41:177–186.
220. Waring JF, Liguori MJ, Luyendyk JP, Maddox JF, Ganey PE, Stachlewitz RF, North C, Blomme EA, Roth RA. Microarray analysis of lipopolysaccharide potentiation of trovafloxacin-induced liver injury in rats suggests a role for proinflammatory chemokines and neutrophils. *J Pharmacol Exp Ther* 2005;316:1080–7.
221. Deng X, Stachlewitz RF, Liguori MJ, Blomme EA, Waring JF, Luyendyk JP, Maddox JF, Ganey PE, Roth RA. Modest inflammation enhances diclofenac hepatotoxicity in rats: role of neutrophils and bacterial translocation. *J Pharmacol Exp Ther* 2006;319:1191–1199.
222. Boelsterli UA. Diclofenac-induced liver injury: a paradigm of idiosyncratic drug toxicity. *Toxicol Appl Pharmacol* 2003;192:307–322.
223. Masubuchi Y, Nakayama S, Horie T. Role of mitochondrial permeability transition in diclofenac-induced hepatocyte injury in rats. *Hepatology* 2002;35:544–551.
224. Cantoni L, Valaperta R, Ponsoda X, Castell JV, Barelli D, Rizzardini M, Mangolini A, Hauri L, Villa P. Induction of hepatic heme oxygenase-1 by diclofenac in rodents: role of oxidative stress and cytochrome P-450 activity. *J Hepatol* 2003;38:776–783.
225. Greaves RR, Agarwal A, Patch D, Davies SE, Sherman D, Reynolds N, Rolles K, Davidson BR, Burroughs AK. Inadvertent diclofenac rechallenge from generic and non-generic prescribing, leading to liver transplantation for fulminant liver failure. *Eur J Gastroenterol Hepatol* 2001;13:71–73.
226. Kim JW, Jeon WK, Yun JW, Park DI, Cho YK, Sung IK, Sohn CI, Kim BI, Yeom JS, Park HS, Kim EJ, Shin MS. Protective effects of bovine colostrum on non-steroidal anti-inflammatory drug induced intestinal damage in rats. *Asia Pac J Clin Nutr* 2005;14:103–107.
227. Banks AT, Zimmerman HJ, Ishak KG, Harter JG. Diclofenac-associated hepatotoxicity: analysis of 180 cases reported to the Food and Drug Administration as adverse reactions. *Hepatology* 1995;22:820–827.
228. Seitz S, Boelsterli UA. Diclofenac acyl glucuronide, a major biliary metabolite, is directly involved in small intestinal injury in rats. *Gastroenterology* 1998;115:1476–1482.
229. Ludwig JA, Weinstein JN. Biomarkers in cancer staging, prognosis and treatment selection. *Nat Rev Cancer* 2005;5:845–856.
230. Fostel JM. Future of toxicogenomics and safety signatures: balancing public access to data with proprietary drug discovery. *Pharmacogenomics* 2007;8:425–430.
231. Shi L, Reid LH, Jones WD, Shippy R, Warrington JA, Baker SC, Collins PJ, de Longueville F, Kawasaki ES, Lee KY, Luo Y, Sun YA, Willey JC, Setterquist RA, Fischer GM, Tong W, Dragan YP, Dix DJ, Frueh FW, Goodsaid FM, et al. The MicroArray Quality Control (MAQC) project shows inter- and intraplatform reproducibility of gene expression measurements. *Nat Biotechnol* 2006;24:1151–1161.
232. Baken KA, Pennings JL, Jonker MJ, Schaap MM, de Vries A, Van Steeg H, Breit TM, Van Loveren H. Overlapping gene expression profiles of model compounds provide opportunities for immunotoxicity screening. *Toxicol Appl Pharmacol* 2007;226:46–59.
233. Baken KA, Vandebriel RJ, Pennings JL, Kleinjans JC, Van Loveren H. Toxicogenomics in the assessment of immunotoxicity. *Methods* 2007;41:132–141.
234. Kultima K, Nystrom AM, Scholz B, Gustafson AL, Dencker L, Stigson M. Valproic acid teratogenicity: a toxicogenomics approach. *Environ Health Perspect* 2004;112:1225–1235.

Chapter 6

Toxicogenomics: Applications in In Vitro Systems

6.1. INTRODUCTORY REMARKS ON IN VITRO TOXICOLOGY

The toxicity of compounds is ultimately assessed in standardized animal toxicology studies, and as we have mentioned on several occasions in this book, these non-clinical toxicology studies have been relatively robust to predict potential adverse events in humans (1). However, animal studies, even innovative studies enabled by technologies such as toxicogenomics, require gram amounts of compounds and are resource-intensive in terms of both time and cost. In addition, animal welfare issues represent a strong incentive to reduce testing in animals as much as possible. Alternatives to animal tests in toxicology are highly desirable, and the concept of reducing animal testing is known in the scientific community as the 3 Rs principles, which stand for reduce, refine, and replace (2). Albeit sound in principle, these alternative methods are only useful if they generate reliable data that can be used to infer the effects of compounds in humans. In other words, they must maintain our ability to predict the presence or absence of specific toxicities. In the last three decades, a significant effort has been made toward the development of alternatives to animal testing in toxicology. These initiatives have led to the development of several in vitro methods that have even gained regulatory acceptance for the classification of toxic agents, such as severe eye irritants, or for the prediction of specific end points, such as skin corrosivity and phototoxicity (3). While these examples are mostly useful in the fields of worker safety, food, chemicals, or cosmetics, several tests (such as the in vitro genotoxicity assays) are also relevant to the development of therapeutic agents.

Genomics in Drug Discovery and Development, by Dimitri Semizarov and Eric Blomme

In vitro approaches have been very successful in predicting with good accuracy the pharmacokinetics and metabolism of compounds. Therefore, they are now widely used throughout the industry, predominantly at the lead optimization stage of drug discovery. While it is beyond the scope of this chapter to discuss in depth what is known as in vitro ADME assays (where ADME stands for absorption, distribution, metabolism, and excretion), it is worth discussing them briefly, as general principles relevant to in vitro toxicology can be learned from their success and implementation. Frontline, automated, and miniaturized ADME screening assays are now commonly utilized in all discovery organizations (4, 5). The main contributors of suboptimal ADME properties are metabolic stability (leading to rapid clearance from the body) and drug-drug interactions (DDI) due to inhibition or induction of cytochrome P450 monooxygenases (CYP enzymes) (6, 7). Therefore, in vitro ADME assays have been developed to optimize these parameters and rapidly assess other important physicochemical attributes, such as solubility and protein binding. Over the years, these assays have evolved considerably, with improvements in automation, miniaturization, compound requirements, and data quality, and their predictive value has contributed to minimizing the use of animal testing and to decreasing late-stage attrition.

In contrast to in vitro ADME screens and despite numerous initiatives and significant investments, robust in vitro toxicology approaches for use in lead optimization are still limited in number. In this chapter, we provide an overview of the current status of in vitro toxicology methodologies frequently used in testing of therapeutic agents. This overview provides a framework for a better understanding of the value that toxicogenomics may create to advance this subdiscipline of toxicology. Next, various applications of toxicogenomics to in vitro systems are discussed in detail. We conclude the chapter by providing an outlook on the use of in vitro toxicogenomics in drug discovery.

6.2. OVERVIEW OF CURRENT APPROACHES TO IN VITRO TOXICOLOGY

In contrast to animal testing mandated by regulations to guarantee minimum safety standards for the testing and licensing of drugs, few in vitro tests are required by regulations (8). For small molecules, these in vitro tests include the various genotoxicity assays and the hERG assay. The hERG assay is an electrophysiological test that measures the effects of compounds on the hERG (human-ether-à-go-go-related gene) current, a cardiac potassium channel that plays a critical role in defining ventricular repolarization (9). This assay is conducted on all experimental small molecules before clinical trials, since several drugs associated with a rare, but potentially lethal ventricular arrhythmia called torsades de pointes, have been linked to delayed cardiac repolarization due to blockade of hERG current (9).

The paucity of in vitro required regulatory tests illustrates the challenges associated with the development and validation of in vitro assays predictive of toxic outcomes in humans. However, while in vitro approaches may not be sufficient to evaluate with high confidence the potential hazards and risks to human beings, like the in vitro ADME screens discussed above, they may have useful applications during the prioritization of compounds at the lead optimization stage of drug discovery. Indeed, while most assays are not sufficient to fully comprehend all toxic properties of compounds, when applied judiciously they can reduce testing in animals. Moreover, because they have higher throughput than animal studies and are compound-sparing (with requirements in the milligram range in contrast to the gram range), they can be performed at much earlier stages and can represent a powerful tool during lead optimization. Finally, in vitro assays can represent useful methods for mechanistic investigations to elucidate mechanisms of toxicity and efficient approaches to conduct structure-toxicity relationship studies for specific toxic end points known to be associated with a particular chemical series (Fig. 6.1). For instance, it is well accepted that in vitro methods are very appropriate for conducting mechanistic studies and that their role can be quite valuable in the screening of a series of structurally related chemicals when at least one of the chemicals has a known toxic effect in vivo (10). In vitro assays have definite advantages, but also important limitations. Below, we discuss their characteristics in more details.

With the exception of the regulated in vitro tests for genotoxicity and hERG interactions that are used in screening and GLP formats by virtually all discovery organizations, there is no industry standard for the use of in vitro toxicology systems used for the profiling of compounds during lead optimization. Therefore, there exists a wide range of approaches and strategies, with some companies

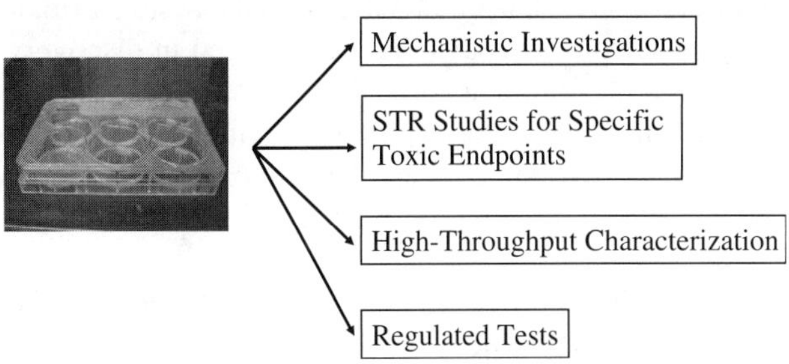

Figure 6.1 Applications of in vitro toxicology assays in drug discovery. In vitro assays represent useful methods for investigations designed to elucidate mechanisms of toxicity. If a relevant toxic end point is identified, in vitro assays can be used to conduct structure-toxicity relationship (STR) studies during lead optimization to develop alternative structures not associated with the toxicity. In general, in vitro toxicology assays are not sufficient to fully comprehend all toxic properties of compounds. However, when applied judiciously, they can reduce testing in animals, and because of their high throughput and low compound requirements, they are ideally suited for early characterization of compounds, especially during lead optimization. Finally, in vitro tests, such as the in vitro genotoxicity assays, may be required by regulations.

using a large battery of customized in vitro tests and others seldom interrogating toxicity with in vitro methodologies. Methods may or may not be published and are usually not standardized, and without personal experience in applying them, it is difficult to comment on them beyond the published data. Models may range from simple methods designed to evaluate non-specific end points, such as cytotoxicity, to more elaborate methods evaluating a series of mechanistically relevant end points. For instance, in vitro assays have recently been developed using high-content screening (HCS) to assess simultaneously a series of toxicologically relevant end points. HCS is a recent advance in the automation of quantitative fluorescence microscopy and image analysis, and in the application of fluorescent multiprobe technology (11). HCS is a multiplexed, functional cell-based screening approach that can be used in different drug discovery-related applications, such as the study of cell signaling and cellular morphology, as well as in toxicological assessment. It enables the monitoring of live cells for multiple cellular biomarkers indicative of toxicologically relevant processes (12). Validation results of a novel HCS-based assay have recently been reported (12). This assay uses HepG2 cells cultured for 3 days in 96-well plates and loaded with four fluorescent dyes to monitor calcium influx (fluo-4 AM), mitochondrial membrane potential (TMRM), DNA content (Hoechst 33342), and plasma membrane permeability (TOTO-3). Compared to conventional in vitro cytotoxicity assays, this assay showed better sensitivity and specificity. It would be unrealistic to expect such assays to be specific for hepatotoxicity because they are being conducted with a liver cell line. Thus, not surprisingly, several non-hepatotoxic compounds that tested positive in this assay were known to produce serious toxic changes in other organs. Because of their richer contents and mechanistically related end points, these novel assays may represent significant advances in predicting human specific toxicity.

Table 6.1 provides a non-exhaustive list of in vitro toxicity systems other than genotoxicity and safety pharmacology assays that can be used in discovery toxicology. For simplicity, these assays are categorized under broad categories: cytotoxicity, apoptosis, mitochondrial function/respiration, lipid accumulation, phospholipidosis, oxidative stress, embryo toxicity, and hematotoxicity. For more details, readers are referred to several reviews or investigations of the topic (13–20).

The major limitation of in vitro systems is their inability to recapitulate the overall complexity of the living organism, which limits their potential to detect toxic events requiring multicellular interactions or membrane polarity. There have been multiple attempts to address these limitations, such as the concurrent use of several cell types with differentiated functions (for instance, the co-culture of hepatocytes with Kupffer cells) or the use of more refined culture conditions and extracellular matric (21, 22). However, it is unreasonable to expect that even elaborate in vitro systems will recapitulate the complex endocrine and paracrine interactions that occur in tissues. Additionally, most in vitro systems are also typically short-lived and therefore may not be used to detect, identify, or interrogate chronic effects (23). This is compounded by the fact that long-term cultivation of

Table 6.1 In Vitro Toxicity Assays Used in Pharmaceutical Discovery

Assay Types	Examples
Cytotoxicity assays	MTT or MTS assays (standard colorimetric assays for measuring cell proliferation or cytotoxicity) Other ATP detection systems (e.g., CyQUANT) LDH release assay
Mitochondrial membrane potential assays	JC-1 or rhodamine 123-based mitochondrial membrane potential assay
Mitochondrial respiration assay (oxygraph assay)	Use of isolated mitochondria to evaluate the potential of compounds to interfere with mitochondrial function
Apoptosis assays	Detection of DNA fragmentation (TUNEL assay) Caspase activity assays Annexin 5 binding assays
Cytoplasmic lipid accumulation (lipidosis)	Niles Red or Oil Red O assays
Phospholipidosis assays	Fluorescent cell-based assays using dyes that bind specifically to phospholipids (e.g., NBD-PE, β-BODIPY C12-HPC, Lipitox®)
Oxidative stress assays	Assays based on fluorescent dyes (e.g., Amplex Red) GSH depletion ATP depletion
Embryo toxicity (development toxicity) assays	Whole embryo cultures Stem cell assays Micromass test
Hematotoxicity	Clonogenic assays

primary cell/tissue cultures or even cell lines is associated with significant loss of function (24).

Another major shortcoming of in vitro tests is their inability to predict safety margins. Virtually every molecule can cause deleterious effects on cells if dosed at sufficiently high concentrations. These high concentrations may be completely irrelevant to the in vivo situation based on the anticipated therapeutic doses and the resulting systemic exposures. Systemic blood exposures based on pharmacokinetics studies and quantified by the area under the curve (AUC, or the area under the curve in a plot of concentration of drug in plasma against time) are used to calculate the safety margins associated with a compound. By definition, for a compound the safety margin is the ratio between the blood exposure at the No Adverse Effect Level (NOAEL) and the blood exposure at the efficacious level. The use of blood exposures is necessary, since blood can easily be accessed and monitored in clinical trials, and preclinical safety data can consequently be easily bridged with clinical data. However, blood concentrations do not necessarily reflect compound concentrations in various tissues and organs. Compounds frequently accumulate in specific tissues at levels higher than those in blood for various reasons, and therefore predictions of safety margins based on toxic concentrations in vitro are typically very inaccurate. For instance, in an internal retrospective study, we have evaluated the correlation between blood AUC_{0-24h} and in vitro concentrations for a series of structurally diverse compounds shown to induce phospholipidosis in vitro and in vivo. These data indicated a total lack of correlation, supporting the fact that in vitro concentrations could not be used to predict the blood exposures at which this well-defined toxic end point would occur.

While this statement is appropriate when trying to differentiate between compounds from diverse chemical structures and with different physicochemical properties, it is important to stress that the correlation is much more robust when evaluating chemically related compounds. For these compounds, a more similar body distribution can be expected, and therefore, in such instances, it is more likely that the in vitro concentrations at which toxic end points are detected correlate with the blood concentrations at which the toxicity occurs in vivo. Therefore, while the absolute therapeutic margins may be unclear for a series of compounds, the relative risk for these compounds can be reliably estimated, making in vitro tests appropriate for rank-ordering and prioritization of compounds. Finally, it should also be mentioned that a reasonably fair correlation has been demonstrated between in vitro basal cytotoxicity and acute toxicity in humans and animals (25). For instance, studies have shown correlation of over 70% between in vitro basal cytotoxicity and rodent Lethal Dose, 50% values (LD_{50} values, or the median lethal doses of toxic agents required to kill 50% of the members of a tested population; in the past, LD_{50} has been used frequently as a quantitative descriptor of the acute toxicity of a test toxic agent) or human lethal blood concentrations (26). Therefore, while for the subacute and chronic toxic changes most commonly associated with therapeutic agents the complex relationship between tissue toxicokinetics and specific organ toxicity limits one's ability to develop a robust predictor of toxic levels in vivo, in acute toxicity testing in vitro methods may be reliable or even officially accepted (26). These data suggest that simple and robust in vitro testing strategies could be used for prediction of human acute systemic toxicity and could even eventually replace the current animal acute studies used for regulatory purposes.

Except for the regulated in vitro tests, we have indicated the lack of real standardization of in vitro methods in the industry. Pharmaceutical R&D organizations use a wide range of cell types and experimental methods to interrogate specific end points. The choice of a cell system depends on personal experience, the availability of historical data, or scientific reasons (proactive screens vs. mechanistic investigations). For instance, we were interested in gaining a better understanding of the cardiac toxicity observed in rats with a particular compound. As shown in Table 6.2, the compound induced cell death at very different concentrations in two cell systems, indicating that the selection of the cell system would have important bearing on prioritization of compounds emerging from the same chemical series. When prospectively evaluating compounds, it is not possible to predict the target organ of toxicity, and therefore testing must be done in a cell system that is not necessarily derived from the tissue ultimately

Table 6.2 In Vitro Cytotoxicity in Primary Rat Cardiomyocytes and Hepatocytes

Compound	Cardiomyocytes TC_{50} (μM)	Hepatocytes TC_{50} (μM)
Abbott 1	32	170
Amitriptyline	61	50
Clomipramine	21	30
Doxorubicin	34	20

TC_{50}: toxic concentration; 24-hour concentration resulting in 50% cell death.

affected in vivo. The assumption here is that the vast majority of toxic end points can occur in any cell type, and that differences observed in vivo may reflect the biodistribution of compounds, rather than tissue-specific reactions. Exceptions to this rule include toxicity related to cell functions or due to specific reactive metabolites; however, it is generally accepted that identifying general markers of toxicity (such as DNA damage, apoptosis, or oxidative stress) is feasible with cell types of various origins.

Since the liver represents a common, if not the most common, target organ of toxicity in both preclinical and clinical studies, cell cultures derived from the liver represent the model of choice (27, 28). In vitro systems derived from the liver encompass simple systems, such as liver cell lines or primary hepatocyte preparations, but also more complex systems, which may have more relevance but are also more resource-intensive and have lower throughput. Isolated perfused livers or liver slices represent two examples of these more elaborate systems. In contrast to monolayer cultures, isolated perfused livers and liver slices maintain intact cellular interactions and spatial arrangements, and allow for long-term studies (29, 30). Furthermore, these models are considered useful for studying toxic effects on the biliary system because they contain phenotypically and functionally intact biliary epithelial cells (31). Nevertheless, these models are more appropriate to evaluate compounds suspected to be associated with specific toxic end points or for mechanistic investigations, while monolayer cultures are more typically used in prospective evaluation of compounds.

Liver cell lines are particularly advantageous since they are readily available and cost-effective and generally yield reproducible results over time. However, it should be recognized that liver cell lines are quite different from liver or primary hepatocytes in terms of function and phenotype. Of particular importance is the fact that liver cell lines express very low or undetectable levels of phase I metabolizing enzymes, at both the RNA and protein levels (28, 32). For instance, toxicology studies often employ the HepG2 cell line derived in 1979 from a human hepatocarcinoma (33). These cells express lower levels of the major phase I drug metabolizing enzymes, especially the cytochrome P450 enzymes (CYPs), than normal liver (34, 35). These differences have important implications, since HepG2 cells are unlikely to generate the same quantity or types of metabolites as seen in the liver or with primary hepatocytes. For instance, recent studies have demonstrated differences in cytotoxicity between primary human hepatocytes and

HepG2 cells likely related to differences in metabolism (28, 36, 37). Therefore, the use of hepatocytes offers clear advantages, particularly the opportunity to assess the toxicity associated with certain metabolites without prior metabolic activation or transfection of CYP enzymes (28).

6.3. TOXICOGENOMICS IN IN VITRO SYSTEMS: TECHNICAL CONSIDERATIONS

As already indicated, the liver is a frequent target organ of toxicity in preclinical toxicology studies. Not surprisingly, the vast majority of toxicogenomics efforts have focused on liver-derived in vitro systems. However, there are also large numbers of published data in cell types derived from other organs. In other words, gene expression profiling in vitro should not be viewed as limited to the prediction of hepatotoxicity. In light of our personal experience and published reports, some general statements can be made regarding toxicogenomics studies in cell cultures.

6.3.1. Reproducibility

There is ample experimental evidence that genomics technologies used in the context of in vitro experiments are reproducible and robust enough to predict or characterize the toxic properties of compounds. This is best illustrated by the recently released findings of an interlaboratory collaboration involving four pharmaceutical companies (38). In this collaborative effort, gene expression profiles induced in rat primary hepatocytes by a 24-hour incubation with a prototypical hepatotoxic compound (methapyrilene) correctly classified the test article with an independently generated in vitro database. This indicates that the interlaboratory and interplatform reproducibility of in vitro toxicogenomics tests is sufficient to enable the use of these methodologies for toxicological profiling. As expected, when different microarray platforms are used, some degree of site-to-site variability can be expected; however, there was good concordance in the results when the data were analyzed in the context of affected biological processes.

6.3.2. Genomic Classifiers

Several studies have evaluated the possibility of generating predictive genomic classifiers for in vitro characterization of compounds. These studies have clearly demonstrated that indeed specific mechanisms of toxicity can be identified with gene expression profiles generated from in vitro systems (39–41). In other words, given a sufficiently robust repository of reference profiles, it is feasible to generate multigene classifiers that can be used for in vitro toxicological profiling of compounds. Several examples are used below in this chapter to illustrate how these classifiers have been generated and used during lead optimization. However, in contrast to the experience with profiles generated from in vivo toxicology studies (discussed in Chapter 5), it has been more challenging to interpret transcriptomic data through clustering of expression profiles with reference data or through extensive pathway analyses. These challenges may be due to several reasons. One potential explanation is that the high concentrations used in in vitro experiments may produce a level of noise that may obscure relevant mechanisms of action.

6.3.3. Testing Concentrations

The selection of appropriate testing concentrations is a recurring topic of intense debate for any in vitro approach, and, not surprisingly, it also represents a critical issue for in vitro toxicogenomics. We have already discussed the challenge of correlating concentrations used in vitro with exposures in vivo, especially when comparing compounds from different chemical structures and with differing physicochemical properties. Let us illustrate this point with a classic toxicant, carbon tetrachloride (CCl_4), which is commonly used in hepatotoxicity studies. When administered as a single dose to animals, carbon tetrachloride induces severe liver injury characterized by centrolobular necrosis within 24–48 hours (42). Yet the in vitro cytotoxicity of carbon tetracholoride is remarkably low (less than 20% cell death at concentrations over 700 μM based on internal experimental data), probably because of its extreme lipophilicity. Strikingly, exposure of primary hepatocytes to extremely high concentrations of carbon tetrachloride also results in a very small number of differentially expressed genes. There are multiple other examples, for which the lack of in vivo–in vitro correlation is more difficult to explain.

So what should the appropriate dosing paradigm be for in vitro toxicogenomics? Ideally, this paradigm should maximize the ability to generate consistent and robust signals for hazard identification and facilitate comparisons among compounds for rank-ordering purposes. It is clear that there is currently no consensus regarding the optimal testing concentrations for in vitro toxicogenomics assessment. However, in our experience robust and reproducible gene expression profiles can only be obtained at concentrations high enough to cause some detectable phenotypic changes in cells, such as cytotoxicity. This reproducibility is critical to developing robust predictive signatures and to characterizing the toxicological profiles of compounds. As discussed below, in our laboratory we routinely characterize compounds in a primary rat hepatocyte model at concentrations sufficient to cause death of 20% of cells after a 24-hour exposure. For many toxicologically relevant end points, we have shown that lower concentrations markedly decrease the sensitivity and the utility of the assay (Fig. 6.2). This observation has been confirmed by others (personal communication). For instance, we will briefly discuss data from a large comprehensive study, showing that inclusion of gene expression profiles induced at low concentrations decreased the accuracy of genomic classifiers (43).

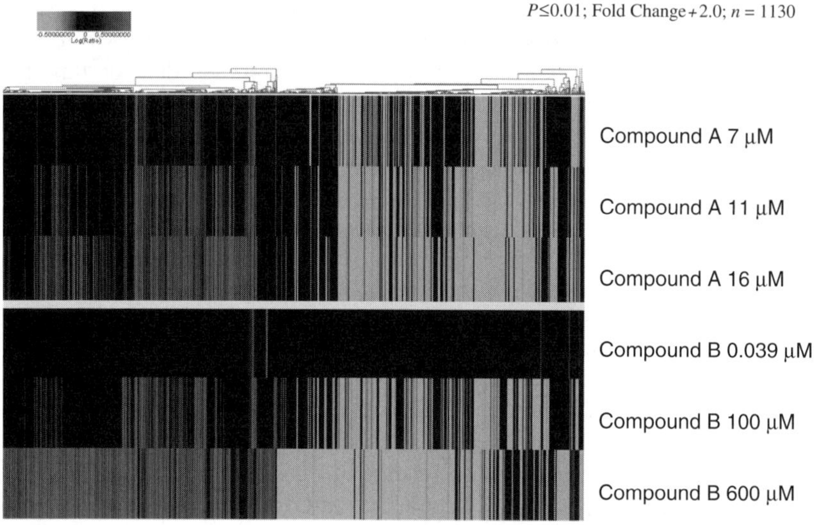

Figure 6.2 Heat map illustrating the transcriptomic effect on primary rat hepatocytes by treatment with two experimental compounds (compounds A and B). In this experiment, primary rat hepatocytes were treated with increasing doses of two experimental compounds. For compound A, 11 μM represents the concentration causing approximately 20% cell death after a 24-hour exposure, while the cytotoxic concentration for compound B was estimated to be around 200 μM. The lowest concentrations used for both compounds are 10-fold multiples of efficacious in vitro concentrations. Note that at concentrations lower than cytotoxic doses, limited gene expression changes are detected. In our experience, robust and reproducible gene expression profiles can only be obtained at concentrations high enough to cause some detectable cytotoxic changes in cells. In our laboratory, compounds are characterized in primary rat hepatocytes at concentrations sufficient to cause death of 20% of cells after a 24-hour exposure. Genes shown ($n = 1130$) are genes that were up- or down-regulated by at least twofold with a P value < 0.01. Green indicates down-regulation, while red indicates up-regulation. See color insert.

6.3.4. Throughput and Cost

While there is ample evidence that genomic classifiers can be developed in in vitro systems, the cost of the procedure may limit the practical use of such assays in a discovery setting. In addition to their costs, microarray platforms remain resource-intensive and have inadequate throughput despite rapid improvements in automation. In lead optimization, assays must be accurate and cost-effective and have appropriate throughput. Furthermore, a useful in vitro toxicogenomics assay must simultaneously interrogate not a single, but several, relevant toxicological end points. These requirements are currently being addressed. Several groups have shown that signatures derived from microarray data can reliably be transferred to more cost-effective platforms with appropriate throughput, such as RT-PCR-based assays (41, 44). Furthermore, the following sections demonstrate that, indeed, gene expression signatures can be developed for multiple, independent end points with a single or at most a few, concentrations.

6.4. PROOF-OF-CONCEPT STUDIES USING PRIMARY RAT HEPATOCYTES

As already alluded to, a significant number of published in vitro toxicogenomics studies have used rat liver cells. The rationale is that the liver is a common target organ of toxicity, that hepatocytes are commonly used for in vitro toxicology, and that these systems typically retain sufficient metabolic functions to generate relevant metabolites that may play central roles in toxicity. In addition, several groups were initially interested in developing in vivo–in vitro correlations for toxicogenomics, and the availability of public and private repositories of gene expression profiles generated in rat liver made a rat hepatic-based in vitro system quite attractive.

Early studies were mostly designed to characterize at the transcriptomic level several commonly used liver-based in vitro systems. For example, rat livers, rat liver slices, primary rat hepatocytes cultured on a collagen monolayer or a collagen sandwich, and two rat liver cell lines (BRL3A and NRL clone 9 cells) were compared at the transcriptomic level (32). These results confirmed that liver slices were more similar to intact rat livers at the transcriptomic level compared to primary hepatocytes and that the liver cell lines were quite different from intact livers, in particular in the transcript expression levels of phase I metabolizing enzymes. These results are consistent with our experience with other cell types, such as HepG2 cells.

A few years ago, we began using primary rat hepatocytes cultured on collagen in our laboratory as the primary system to evaluate the use and application of in vitro toxicogenomics (39, 45). Our initial efforts were focused on evaluating whether genomics studies in this in vitro system would be useful to identify mechanisms of toxicity. In a study in which 15 prototypical hepatotoxicants (carbon tetrachloride, allyl alcohol, aroclor 1254, methotrexate, diquat, carbamazepine, methapyrilene, arsenic, diethylnitrosamine, monocrotaline, dimethyl-formamide, amiodarone, indomethacin, etoposide, and 3-methylcholanthrene) were characterized in primary rat hepatocytes, compounds with similar mechanisms of toxicity were found to induce shared expression profiles (39, 45). Hence, an unsupervised hierarchical clustering was sufficient to distinguish the various classes of hepatotoxicants included in the study (39). In addition, there was a significant correlation between the genes regulated in vivo and in vitro for some of the prototypical toxic molecules used in this study.

These results have been confirmed, reproduced, and expanded in studies conducted by independent groups. For instance, despite some biological variability and technological problems, it was demonstrated that gene expression profiling can distinguish two mechanistically unrelated classes of toxicants (cytotoxic non-steroidal anti-inflammatory drugs or NSAIDs and DNA-damaging agents) based solely on clustering analysis of genes differentially induced or repressed in cultured cells during exposure to these compounds (40). Further refinement of the clustering strategy with a computer-based optimization algorithm yielded even better results and demonstrated that genes that ultimately best discriminated between DNA damage and NSAIDs were involved in such diverse processes as DNA repair, xenobiotic metabolism, transcriptional activation, structural maintenance, cell cycle control, signal transduction, and apoptosis (40).

Additional independent studies using different platforms and different experimental procedures have confirmed that gene expression profiling in in vitro systems can be used to classify compounds according to their mechanisms of toxicity and have demonstrated the value of this approach to further our understanding of toxic mechanisms (30, 46, 47). It is impossible to review all these studies in detail here, but generally speaking, they have demonstrated that classification can be achieved by using a relatively small number of genes and that the resulting multigene classifiers can be transferred to RT-PCR-based platforms without significant loss of performance.

With these early data showing evidence that one may be able to construct gene expression-based classifiers, it became clear that repositories of reference gene expression profiles would be necessary. Therefore, we generated a reference database in collaboration with Iconix Biosciences (Mountain View, CA) by profiling a large number of primary rat hepatocyte cultures exposed for 24 hours to compounds at concentrations resulting in 20% cell death. The compounds were selected based on a variety of criteria, such as mechanisms of toxicity or chemical and pharmacological diversity. The concentrations resulting in 20% cell death were determined beforehand in primary rat hepatocytes. This compendium of expression profiles was then used to develop predictive signatures for several toxicological end points. We used different algorithms to develop these signatures and ultimately selected the methods that resulted in signatures that would meet our preset expectations. In particular, two of these signatures were recently released (41). These two signatures were designed to classify two well-characterized classes of toxic agents, the aryl hydrocarbon receptor (AhR) agonists and the peroxisome proliferator-activated receptor-α (PPAR-α) agonists. The testing set consisted of gene expression profiles induced by prototypical AhR agonists, such as 3-methylcholanthrene, aroclor, and β-napthoflavone, as well as by various PPAR-α agonists, such as bezafibrate, clofibrate, and WY-14643. Gene expression profiles after exposure to a variety of negative controls (i.e., compounds known not to agonize the AhR and PPAR-α nuclear receptors) were included as well. With linear discriminant analysis (LDA) coupled with a permutation-based t-test, gene expression signatures were generated that were able to classify compounds according to a discriminant score. The final gene signatures consisted of eight genes for AhR agonism and 11 genes for PPAR-α agonism. The performance of the two signatures was then estimated in a limited forward validation step using profiles induced by additional reference compounds and demonstrated perfect accuracy. Of particular interest is the fact that we were able to successfully transfer these microarray-based signatures to a real-time RT-PCR platform without loss of performance (41). Using slightly different algorithms, we have since expanded this work to generate multigene classifiers for additional toxicological mechanistic classes, such as phospholipidosis, oxidative stress, mitochondrial toxicity, and DNA damage.

Other groups have reported their experience with in vitro toxicogenomics. For instance, a comprehensive rat-based toxicogenomics study was recently published in which the authors explored novel toxicogenomics approaches to predict the hepatotoxic potential of compounds. The ultimate objective of the study of Martin et al. was to establish a genomics-based screen that could be used in the early stages of drug discovery (43). Gene expression profiles were generated in vitro with primary rat hepatocytes and 86 compounds, as well as in rat livers obtained from single-dose, 24-hour studies evaluating over 100 compounds. For the in vitro approach, high concentrations known to cause approximately 10% cell death were used, an approach similar to the experimental design adopted by our laboratory. In addition, the authors used a lower concentration selected as the maximum dose at which no significant cytotoxicity was observed. The time course included 3-, 6-, and 24-hour time points. These expression profiles were obtained from the commercial ToxExpress database® (Gene Logic, Inc., Gaithersburg, MD, USA). To generate gene-based classification rules, the authors used a stochastic gradient boosting machine learner and estimated classification error with a modified bootstrap estimate of true error. With this approach, robust classification rules were constructed for both the in vitro and the in vivo systems. In contrast to our approach, which was designed to detect specific mechanism-based end points, the authors of this study were interested in distinguishing hepatotoxicants from nonhepatotoxicants with a single classifier. The in vitro and in vivo classifiers consisted of 173 and 168 genes, respectively. Their results were particularly interesting for several reasons that we carefully review here.

- **Comparisons between the in vitro and in vivo classifiers.** There was little overlap between the in vivo and in vitro classifiers, although both contained a fairly large number of genes. Indeed, only five genes were common to both classifiers, and none of them was ranked in the top 30 of either classifier, as measured by relative importance. These results are consistent with our experience and that of others and clearly demonstrate how in vitro and in vivo systems differ in terms of response and how difficult it will be to understand in vitro–in vivo correlations for toxicology assays (48).
- **Evaluations of accuracy associated with different time points.** When the 3- and 6-hour time points were considered instead of the 24-hour treatment, a decreased accuracy of the assay was observed. These results are also consistent with our experience. As we constructed our database, we also profiled hepatocyte cultures after a 16-hour exposure to various compounds. This shorter exposure period was associated with decreased reproducibility and repeatibilty of experiments and higher variability in response to similar treatments. Overall, these data suggest that longer exposure periods are likely to be more reliable than shorter periods.

- **Comparisons of testing concentrations.** In the study, the inclusion of the low-dose data also resulted in decreased accuracy of the gene expression-based classifier. These results confirm that interrogation of toxicity in vitro with gene expression profiling benefits from using concentrations sufficiently high to induce some low level of cytotoxicity.

Overall, our data and those of Martin et al. confirmed that gene expression profiling in the rat hepatocyte model represents a robust approach to predict hepatotoxicity when sufficient training data are available to build classifiers. Similar to the repositories of gene expression profiles used to construct in vivo multigene classifiers, these training data must cover a sufficiently wide range of mechanisms of toxicity and chemical structures, and the accuracy of these genomics classifiers ultimately must be estimated with a testing set composed of gene expression profiles induced in the same cell system by a range of compounds naive to the training set, but also reflecting the chemical space of the user.

6.5. USE OF GENE EXPRESSION PROFILING TO ASSESS GENOTOXICITY

We briefly review the required genetic toxicological assessments for pharmaceuticals in Chapter 5. As a brief reminder, two in vitro tests (an Ames test for mutagenicity and a mammalian cell-based assay for chromosomal damage) are usually required. Of particular interest in the industry is the realization that based on data accumulated over the past 15 years, the current in vitro genotoxicity assays using mammalian cells (mammalian mutation and/or chromosomal damage assays) provide a limited insight into genotoxic mechanisms and result in a high rate of positive results (49, 50). Many of these positive results obtained in vitro have not been confirmed in in vivo genotoxicity tests and/or in carcinogenicity studies. Hence, the majority of positive genotoxicity findings for marketed drugs with negative carcinogenicity data have been observed in these in vitro mammalian cell assays (51). This retrospective evaluation of the in vitro mammalian cell-based genotoxicity tests demonstrates their low specificity and the need to develop approaches enabling mechanism based risk assessment of genotoxicity results.

Recent data have shown that compounds can induce chromosomal damage in mammalian cells through indirect mechanisms, for example, by causing nucleotide pool imbalance, inhibiting protein or DNA synthesis, suppressing DNA repair, overloading oxidative defense mechanisms, or disrupting chromosome segregation (52). In addition, experimental conditions such as extreme changes in pH and osmolarity, excessive cytotoxicity, or endolytic DNA degradation can contribute to these false positive results (52). For all these mechanisms, the dose response is typically non-linear with a threshold effect (52). Consequently, the understanding of the genotoxic mechanism of action is considered essential in the improvement of the risk assessment for humans, and gene expression analysis may represent a useful tool to gain a better understanding of these genotoxic mechanisms.

The objective of initial toxicogenomics studies was not to find a replacement for the current battery of genetic toxicology tests but to determine whether gene expression profiling could differentiate between DNA-reactive and non-DNA-reactive compounds that both result in positive signals in in vitro mammalian cell-based assays. This realistic objective was initially addressed by collaborative studies conducted under the auspices of the Health and Environmental Sciences Institute (HESI) Committee on the Application of Genomics in Mechanism-Based Risk Assessment from the International Life Science Institute (ILSI) (49, 50). This consortium was composed of representatives from the pharmaceutical industry, academic institutions, and regulatory authorities. Several studies have demonstrated that gene expression profiling can indeed provide some useful mechanistic insight regarding mechanisms of genetic toxicity that would allow a better risk assessment of genotoxicants (50, 53). These results were recently reviewed by Thybaud et al. (52). Rather than going into the specifics of all published studies, we concentrate here on the major findings and conclusions from these studies.

6.5.1. Toxicogenomics Can Differentiate Genotoxic Carcinogens from Nongenotoxic Carcinogens

We covered this topic in Chapter 5 when discussing the potential of toxicogenomics to predict the outcome of rodent carcinogenicity studies. As a brief reminder, using rat studies of short duration (up to 14 days) to evaluate several rodent genotoxic and non-genototoxic hepatic carcinogens, gene expression profiling of livers demonstrated that the two classes of agents modulate different pathways. This difference can be used to differentiate between these two mechanisms of hepatic carcinogenesis (54, 55).

6.5.2. Toxicogenomics Can Differentiate DNA-Reactive from Non-DNA-Reactive Compounds Positive in In Vitro Mammalian Cell-Based Genotoxicity Assays

Data from several studies have demonstrated differences in gene expression profiles between DNA-reactive and non-DNA-reactive compounds. Hence, in various cell types (p53-deficient mouse lymphoma cells L5178Y/TK+/−, TK6 cells), DNA-reactive compounds regulated genes involved in cell cycle regulation, DNA repair, apoptosis, and cellular signaling that were distinct from those affected by non-DNA-reactive agents (49). For instance, a panel of DNA-reactive genotoxic agents (methyl methanesulfonate, mitomycin C, cisplatin) and compounds known to act through non-DNA-reactive mechanisms (etoposide, hydroxyurea, paclitaxel) were evaluated in L5178Y cells. Clear differences in the mRNA expression levels of 43 marker genes were found between the two classes of molecules after a 4-hour exposure (56). Likewise, by using a different cell type (human lymphoblastoid TK6 cells) and 14 anticancer agents, genotoxic compounds could be classified according to their direct or indirect interaction with DNA, but also according to their mechanism of action by using a set of 28 genes (57). In other words, irrespective of the experimental conditions used, there is strong evidence that gene expression signatures or pathway analysis can be used to differentiate DNA-reactive from non-DNA-reactive compounds. These data support toxicogenomics as a viable follow-up experimental approach to evaluate compounds with positive findings in standard in vitro mammalian cell-based genetic toxicity assays.

6.5.3. Toxicogenomics Assays May Be Less Sensitive than the Standard Battery of In Vitro Genetic Toxicity Tests

We have provided several examples demonstrating that gene expression profiling is typically more sensitive in detecting pharmacologically or toxicologically significant responses. Therefore, it was initially anticipated that toxicogenomics would also prove more sensitive than the current genetic toxicity tests. In fact, existing evidence points toward the opposite. Several studies have shown that the gene expression changes induced by genotoxicants were not as sensitive as the end points provided by the conventional genetic toxicity assays (52, 58). To some extent, for in vitro tests, this should not come as a surprise. We have already mentioned that high concentrations of test compounds are required to generate robust and reproducible gene expression profiles. Furthermore, gene expression profiles reflect transcriptomic changes occurring in entire cell populations, while the current in vitro genetic toxicity tests are designed to detect rare events occurring in a small minority of cells (52). These rare events occurring in a small fraction of cells are likely to trigger specific transcriptomic responses, but these responses may be difficult, if not impossible, to detect because they are diluted by the normal transcriptomic status of the rest of the cell population. The existing evidence positions toxicogenomics as a useful adjunct to the standard battery of genetic toxicology assays.

However, it is noteworthy that we are far from seeing this technology as a commonly accepted approach to validating positive findings from in vitro mammalian cell-based tests. Like any novel technology, substantial experimental data are needed to fully comprehend the predictive value and limitations of toxicogenomics. These data are especially critical to convince the conservative membership of the toxicology community. Moreover, while a significant number of studies have been published, it is difficult to interpret these data given the wide variety of experimental procedures followed by investigators. Protocols have ranged from short-term to long-term exposures with or without a recovery period and have utilized different cell types and various custom-made or commercially available platforms (52). Furthermore, data have been analyzed with a diversity of approaches and software packages (52). Comparing results from these studies represents a daunting challenge. Standardization of experimental protocols will be required to comprehend the performance of toxicogenomics in the assessment of genotoxicants. Such an endeavor will likely be successful only in the form of a consortium formed by major stakeholders from industry, academia, and regulatory authorities.

6.6. APPLICATION OF GENE EXPRESSION PROFILING FOR IN VITRO DETECTION OF PHOSPHOLIPIDOSIS

Phospholipidosis refers to the excessive cytoplasmic accumulation of phospholipids, a normal cellular component. Morphologically by light microscopy, phospholipidosis is characterized by various levels of cytoplasmic vacuolation in a wide range of tissues, and this microscopic vacuolation is caused by the presence of membranous lamellar inclusions called lamellar bodies, that can be detected at the ultrastructural level (Fig. 6.3) (59). Drug-induced phospholipidosis is typically caused by cationic amphiphilic drugs (59, 60). These compounds

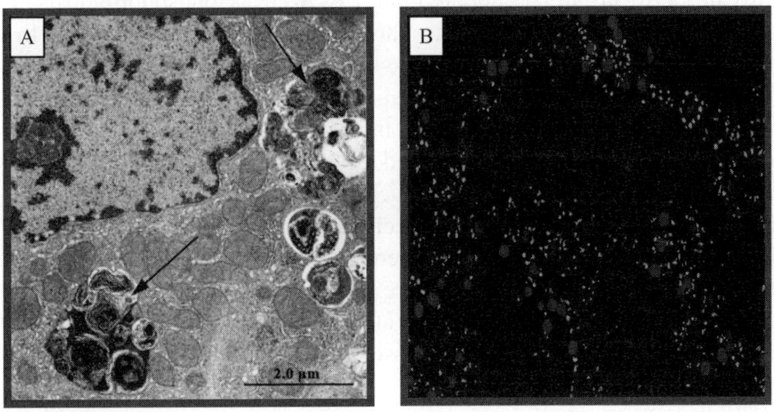

Figure 6.3 Drug-induced phospholipidosis. A. Electron microscopic evaluation of the liver of a rat treated for 2 weeks with an experimental compound inducing hepatic phospholipidosis. Phospholipidosis is the excessive cytoplasmic accumulation of phospholipids, a normal cellular component. Histologically, it is characterized by various levels of cytoplasmic vacuolation in a wide range of tissues. This vacuolation is due to membranous lamellar inclusions called lamellar bodies, detectable by electron microscopy evaluation (arrow). B. Detection of phospholipidosis with fluorescent microscopy. Screening in vitro approaches for phospholipidosis use fluorescent dyes or fluorescence-labeled phospholipids in cultures of hepatocytes or HepG2 cells. Illustrated here are primary rat hepatocytes exposed for 24 hours to amiodarone, a drug known to induce phospholipidosis in rats. A fluorescent probe (BODOPY-C_{12}-HPC) was used to detect the cytoplasmic lamellar bodies (green granules). (Courtesy of Abbott Department of Exploratory and Investigative Technologies.) See color insert.

contain a hydrophobic ring structure and a hydrophilic side chain with a primary or substituted nitrogen group that is charged at physiological pH (59, 60). These types of compounds typically penetrate the brain very well, such that phospholipidosis is not an infrequent finding when developing compounds to treat neurological diseases. The most prevailing hypothesis regarding the mechanism of phospholipidosis is that compounds bind endogenous phospholipids, thereby creating complexes that cannot be degraded by cellular phospholipases (59, 61). There is also evidence that some compounds may directly inhibit endogenous phospholipase activity, resulting in a decreased turnover of phospholipids, or impact the synthesis and metabolism of phospholipids (61, 62). Phospholipidosis can theoretically be found in any tissue, its tissue distribution can be quite different across animal species, and the tissue affected probably reflects the tissue distribution of the compounds (63). Finally, this change is known to be reversible, although this process may span several weeks in preclinical species (61, 63).

It is unclear whether phospholipidosis can lead to tissue injury and whether it represents a significant hazard for humans. Hence, there are over 50 marketed drugs known to cause phospholipidosis at least preclinically (61). The problem resides in the fact that it has proven extremely difficult to show a clear link between the presence of phospholipidosis and tissue injury. For instance, gentamycin can cause proximal tubular injury, as well as phospholipidosis in the kidney, but it is unclear whether phospholipidosis is the main cause of the tubular injury or just a coincidental change (64). Nevertheless, in some instances, there seems to be a link. Hence, phospholipidosis has been associated with amiodarone-induced pulmonary toxicity or myopathy and chloroquine-induced myopathy and retinopathy (65–67). In addition, human genetic conditions resulting in lysosomal storage diseases, such as Niemann–Pick disease, can prove fatal, suggesting that excessive intracellular accumulation of endogenous phospholipids may represent an adverse event in humans (68). Because of the uncertainty associated with this change, several companies have established in silico or in vitro screening assays in an attempt to eliminate at early stages compounds likely to induce phospholipidosis in vivo, at least at low safety margins (69, 70). Most of these in vitro approaches use fluorescent dyes or fluorescently labeled phospholipids in cultures of hepatocytes or HepG2 cells (Fig. 6.3) (62, 71–72). Improvements of automated fluorescent microscopy have contributed to a significant improvement in throughput, such that these approaches can easily be used during lead optimization (71).

As you may recall, phospholipidosis is an end point for which our laboratory has generated a multigene classifier using primary rat hepatocytes. In general, this classifier performs as well as our in vitro fluorescent dye system, but with the advantage of using only one concentration of test compounds. Others have also successfully used toxicogenomics to detect phospholipidosis in vitro. For instance, one report used gene expression profiling of HepG2 cells exposed to 12 phospholipidosis-inducing compounds to identify 17 candidate marker genes that could be used in an in vitro screen for drug-induced phospholipidosis (73). Using real-time PCR, these investigators confirmed that 12 of those 17 gene markers showed significant concordance with the phenotype of phospholipidosis and that the magnitude of the mRNA expression changes for these markers correlated well with the phenotypic change. Of particular interest from a logistical view is the fact that these investigators also reported the successful transfer of this multigene classifier to a 96-well plate genomics-based platform to increase the throughput (74).

This set of genes was independently validated by two other groups using the same cell line. In the first independent validation study, using this multigene marker, 11 of 12 compounds known to induce phospholipidosis were identified as positive and all the negative compounds (5 of 5) were confirmed as negative (75). The same investigators conducted a second experiment where they tested 26 novel compounds; 14 of 15 compounds known to induce phospholipidosis and all the negative controls (6 compounds) were correctly classified (75). In the second published validation study, the investigators also evaluated the utility of this gene panel to predict the phospholipidosis-inducing potential of compounds (76). They evaluated 10 known phospholipidosis inducers and four negative compounds and quantified gene expression changes by quantitative RT-PCR. Generally, the negative control compounds had little or no effect on the expression of any of the genes in the panel. By contrast, the known inducers of phospholipidosis caused substantial changes in the expression of these genes, and seven of 10 were correctly classified in two separate experiments. Interestingly, the three false-negative compounds were classified as positive when testing concentrations were increased, while higher concentrations did not result in the misclassification of a negative control compound. This finding is consistent with our experience and with what has just been discussed with toxicogenomics assays to detect genotoxic end points: Genomic classifiers for in vitro systems may not prove more sensitive than biochemical assays, and therefore high concentrations are required. As stated before, in our experience one needs to treat cells at concentrations resulting in low levels of cytotoxicity. This high testing concentration apparently does not result in lower specificity of well-constructed gene expression-based signatures.

Altogether, these three independent evaluations represent an important step for the field of in vitro toxicogenomics for two major reasons. First, they demonstrated the reproducibility and good accuracy of genomics-based in vitro assays developed with appropriate methods. Second, by using standardized protocols (similar multigene classifier, same cell line, similar treatment protocols), they generated meaningful results that can be used to properly evaluate a proposed assay in terms of utility and performance at different sites and with novel compounds. Hopefully, others will use similar approaches to evaluate additional genomics-based in vitro assays, such that performance can be properly understood and the technology can be transferred to other laboratories. This can only be achieved if the appropriate scientific journals do not negate publication of these studies because of their lack of mechanistic insight and if methodologies are sufficiently described.

6.7. TOXICOGENOMICS IN ASSESSMENT OF IDIOSYNCRATIC HEPATOTOXICITY

In Chapter 5, we discuss idiosyncratic toxicity and explain how gene expression profiling can be applied to further our understanding of the mechanisms behind these toxic reactions. As a brief reminder, idiosyncratic toxicity refers to the toxic events occurring in humans that are not considered to be related to the pharmacology of the drug and that occur unexpectedly in a small proportion of treated patients, often in a non-dose-dependent manner (77–79). These toxic reactions are typically not predicted in animal studies during preclinical safety evaluations. In addition, since these toxic events occur in a small proportion of patients, they are typically not observed during the early phases of clinical trials.

There are likely several mechanisms of idiosyncratic toxicity (77, 80). Therefore, these toxic responses are better addressed on a compound-by-compound basis. In addition, studying these toxic events is an extremely challenging task because it is unethical to test potentially harmful compounds in humans in a targeted manner and because there is a paucity of exposed human tissues given the rarity of these events. If these toxic reactions are indeed host-specific, animal models may prove of little or no value, and investigations in human cells may well represent the only approach to understand the molecular basis of idiosyncratic drug reactions (IDRs). This rationale has motivated several groups to initiate investigations of IDRs in human cells with gene expression technologies. As expected, since the liver is a common target organ of idiosyncratic toxicity, most of these investigations have selected hepatocytes or liver-derived cell lines.

Our laboratory has been particularly interested in developing a better understanding of the molecular mechanisms of hepatoxicity caused by trovafloxacin (Trovan). As you may recall, trovafloxacin is a quinolone antibacterial agent that inhibits bacterial DNA gyrase and DNA topoisomerase IV (81). Since its regulatory approval in 1997, treatment with trovafloxacin has been associated with 150 cases of liver toxicity, including 14 cases of acute liver failure and five deaths (82). Because of this hepatotoxicity, the use of trovafloxacin is now restricted to life-threatening situations. In Chapter 5, we summarize our studies in the rat-lipopolysaccharide (LPS) model. However, our earlier efforts were focused on experiments using primary human hepatocytes. Our initial objective was to determine whether we could use gene expression profiles to differentiate trovafloxacin from other quinolone agents not associated with IDR. In this study, human hepatocytes from four different donors were treated with six quinolone agents (trovafloxacin, levofloxacin, grepafloxacin, gatifloxacin, ciprofloxacin, and clinafloxacin) to generate gene expression profiles using microarrays (83). Treatment with trovafloxacin resulted in far more gene expression changes than treatment with the other compounds (83). When these gene expression changes were analyzed in the context of biological pathways, it was apparent that trovafloxacin affected several critical pathways, suggesting an effect of the compound on mitochondrial function, RNA processing, transcription, and inflammation. In particular, trovafloxacin regulated the expression of a number of mitochondrial genes that were not affected by the other quinolones (83).

We have since then reported the results of a comparative toxicogenomics study that evaluated the differences and similarities in mRNA expression between HepG2 cells and human hepatocytes before and after exposure to a panel of fluoroquinolone compounds (84). Interestingly, while trovafloxacin treatment resulted in many gene expression differences between HepG2 cells and human hepatocytes, there were also a number of trovafloxacin-induced commonalities, including regulation of genes involved in RNA processing and mitochondrial function. These findings suggest that trovafloxacin-associated hepatotoxicity may not be the result of a human-specific metabolite given the low metabolizing capabilities of HepG2 cells. While it is difficult at this point to determine whether these transcriptomic changes are related to the hepatotoxicity associated with trovafloxacin, they nevertheless provide relevant data to formulate novel hypotheses regarding the mechanism of trovafloxacin-induced hepatotoxicity. This initial investigation was also interesting, since in parallel we conducted a short-term (7-day) repeat-dose toxicology study with male Sprague-Dawley rats and levofloxacin (600 mg/kg/day) and trovafloxacin (200 mg/kg/day). In this study, we did not see any evidence of hepatotoxicity as assessed by conventional end points (serum chemistry and histopathology). Furthermore, a microarray analysis of the rat livers did not identify unique gene expression changes induced by trovafloxacin compared to levofloxacin, and the biological pathways regulated in human hepatocytes did not appear to be affected in the rat. These results strongly suggest that, indeed, the toxicity associated with trovafloxacin in humans may be related to a host-specific reaction that may be impossible to detect in a preclinical species.

Troglitazone (Rezulin) is another tool compound frequently used to investigate the mechanisms of idiosyncratic hepatotoxicity. Troglitazone is a thiazolidinedione PPAR-γ agonist that is used for the treatment of type II diabetes because of its ability to reduce glucose and lipid levels in diabetic patients (85). Treatment with troglitazone resulted in hepatotoxicity in a small percentage of patients, leading to its removal from the market (79, 86). Several groups have evaluated the gene expression profiles associated with troglitazone in primary human hepatocytes or other culture systems, such as rat hepatocytes or HepG2 cells (87–89). In these studies, troglitazone could be distinguished at the transcriptomic level from the other control approved thiazolidinedione compounds, rosiglitazone (Avandia), and pioglitazone (Actos). For instance, with an approach similar to the one we used for trovafloxacin, human hepatocytes were treated with troglitazone, rosiglitazone, and pioglitazone. Troglitazone resulted in a large number of gene expression changes that were not observed with the other two thiazolidinedione compounds (87). Furthermore, the unique gene expression changes suggested potential mechanisms associated with the IDR. Another study evaluated the effects of these three compounds in rats, rat hepatocytes, and the clone 9 rat liver cell line by monitoring changes in expression levels of multiple preselected genes related to xenobiotic metabolism, cell proliferation, DNA damage, oxidative stress, apoptosis, and inflammation (89). Compared to the other two compounds, troglitazone had a significant impact on many of the pathways monitored in vitro, although no major perturbation was detected in vivo. Finally, in a separate study, the hepatotoxic effects of five PPAR-γ agonists (ciglitazone, pioglitazone, rosiglitazone, troglitazone, and JTT-501) were also investigated at the transcriptomic level with rat primary hepatocytes and HepG2 cells (88). As reported by others, the gene expression profiles induced by troglitazone and ciglitazone could be clearly distinguished from those induced by pioglitazone and rosiglitazone. Pathways impacted were related to cell death, apoptosis, and proliferation. This study also demonstrated a good agreement between chemical structural information and the clustering of compounds based on cytotoxicity or gene expression data, suggesting a strong relationship between chemical structure and biological end points. In many aspects, these results are in line with previous observations of the effects of troglitazone in cell culture. An elegant in vitro study investigating the effects of troglitazone on mitochondrial function and viability in HepG2 cells showed that troglitazone treatment decreases cellular ATP levels and mitochondrial membrane potential in HepG2 cells and that, consequently, mitochondrial dysfunction may be, at least in part, the cause of the idiosyncratic toxicity induced by troglitazone (90).

Obviously, it would be premature to draw any definitive conclusions from these preliminary results. In fact, even if these results represent a useful advance in generating novel or more specific hypotheses regarding the mechanisms of toxicity of specific IDRs, the reality is that it may be extremely difficult to design appropriate experiments to refute or confirm such hypotheses. Without confirmatory data, it may be hard to foresee the combined use of human cell cultures and gene expression profiling as a robust system to identify early IDR signals, especially considering the multiplicity of mechanisms involved in these toxic events. Nevertheless, if viewed as proof-of-concept, these results indicate that the application of gene expression profiling to human hepatocytes can distinguish some compounds associated with idiosyncratic hepatotoxicity (trovafloxacin, troglitazone) from structurally related compounds not associated with hepatotoxicity. More compounds now need to be profiled in the human hepatocyte system to evaluate whether similar differences can be found with compounds from different pharmacological and chemical classes. Once these data become available, it will be easier to judge the potential use of this approach in drug discovery and development.

6.8. DO PERIPHERAL BLOOD MONONUCLEAR CELLS REPRESENT A USEFUL ALTERNATIVE IN VITRO MODEL?

For any in vitro system, understanding the in vitro–in vivo correlation is quite challenging. This is especially true for in vitro systems using human cells, since the rare availability of tissues collected from patients exposed to experimental compounds limits one's ability to put results obtained in vitro into perspective.

Given this limitation, one must admit that toxicogenomics studies in human cell culture are difficult to validate. For this reason, we have in the last few years started to evaluate the suitability of cultured human peripheral blood mononuclear cells (PBMCs) as an alternative model to interrogate efficacy and toxicity end points with gene expression profiling. PBMCs represent an attractive system, since they are easily accessible in the clinic and can be cultured ex vivo, at least for short periods. As such, they would be suitable to generate sufficient data that could help understand in vitro–in vivo correlations for various toxicologic end points. Furthermore, PBMCs can be viewed as a circulating organ that is typically exposed to high and monitorable levels of drugs. This is in sharp contrast to what is seen with other organs, where local exposure to drugs may be challenging to estimate. The exposure of PBMCs to compounds is usually well understood because pharmacokinetics studies are conducted during clinical trials. Therefore, one could envision that the observed effects of compounds on these cells may be used to better predict therapeutic indexs for experimental compounds. In addition, as discussed in Chapter 5, blood cells could potentially be utilized as a surrogate target cell type that would serve as a sentinel by monitoring perturbations throughout the organism (91, 92). Hence, comparing the transcriptomic responses of PBMCs in vivo and ex vivo may be useful to distinguish direct compound-related changes from changes associated to other perturbations.

While we have not generated enough data to fully comprehend the utility of this approach, our current data indicate that human PBMCs represent a viable in vitro system for toxicogenomics. We have shown that PBMCs collected from different donors respond similarly to the same compound at the transcriptomic level, implying that donor-to-donor variability does not affect the reproducibility of the assay. In addition, pilot experiments showed that gene expression responses from multiple donors correlate well with the compound structure-activity classes evaluated, suggesting that gene expression signatures could be derived for various end points. Finally, we have shown that in this system transcriptomic responses induced by prototypical compounds, such as glucocorticoid receptor agonists, are very consistent with their known in vivo pharmacological effects. Our results are in agreement with previously published reports. For instance, PBMCs have been used to interrogate the pharmacological properties of an engineered form of interleukin-2 (IL-2) and compare them with the properties of a recombinant IL-2 (93). In another study, PBMCs were exposed to cigarette smoke condensate to identify novel biomarkers that could be used in population studies designed to monitor early biological effects caused by cigarette smoke (94).

Overall, these limited, yet promising data indicate that human PBMCs cultured ex vivo can be utilized in a toxicity assay in combination with gene expression profiling. This system has the advantage of being quite relevant to humans, and could be used to bridge the gap between in vitro and in vivo data. For instance, it may represent the source of new, compound-specific biomarkers in the form of gene expression-based signatures that could be applied in the clinic to monitor toxic events with improved sensitivity. The validity of this strategy can be confirmed in preclinical species and ultimately significantly expand our current battery of toxicity biomarkers. It is too early to comment on the likelihood of success of this approach; however, in the very near future, sufficient data should become available to better understand whether the ex vivo PBMC system has any useful applications in toxicology.

6.9. CURRENT AND FUTURE USE OF IN VITRO TOXICOGENOMICS

While more work is required for in vitro toxicogenomics to become a routine tool in discovery organizations, this area is developing at a rapid pace, and the aforementioned examples have generated substantial evidence that this tool could be used to improve the toxicological characterization and prioritization of compounds in discovery, especially at the lead optimization stage. Moreover, in vitro toxicogenomics is a useful alternative to investigate mechanisms of toxicity or generate compound-specific signatures that could be used as biomarkers of toxicity at both the preclinical and clinical stages. Since it is difficult to generalize on mechanistic investigations and compound-specific effects, we concentrate here on the required attributes and necessary improvements that must occur for toxicogenomics to have practical applications in compound optimization in discovery.

6.9.1. Improved Gene Expression Platforms

The majority of approaches have used gene expression microarrays. These analytical tools are still costly, have a limited throughput, and are too noisy to be efficiently used in lead optimization. Microarray-based tools are needed to construct optimal multigene classifiers; however, once these classifiers are developed, they must be transferred to platforms better suited for lead optimization. In this chapter, we have mentioned several examples where investigators were able to successfully transfer their gene panels to different platforms platforms, such as the Taqman Low Density Array (Applied Biosystems, Foster City, CA) or ArrayPlate (High Throughput Genomics, Inc., Tucson, AZ). In addition, independent investigators have successfully used these multigene panels at different sites. Altogether, this suggests that if classifiers are composed of genes with sufficient amplitude of regulation, the transferability to more cost-effective platforms should not represent a limiting factor. Several reports suggest that customized focused platforms may represent an attractive alternative (95). With the rapid technological improvements observed in the last few years, appropriate analytical methods suitable for targeted approaches should soon become available.

6.9.2. Standardization of Protocols and Experimental Approaches

Since published studies with similar objectives or addressing similar issues have used a wide variety of experimental approaches, analytical systems, and analysis tools, their results have only limited generalizability. In this context, generalizability refers to the ability of the investigator to generalize the findings beyond the training set used in the study (43). This is particularly evident for studies evaluating the use of gene expression profiling to differentiate DNA-reactive from non-DNA-reactive compounds (52). These early studies were nonetheless critical proof-of-concept experiments, and their encouraging results should lead to confirmatory studies performed by independent groups. As indicated earlier, this has already been done for the initial gene panel to detect phospholipidosis in vitro (74–76). In that particular instance, the multigene marker was relatively small and the experimental protocol was not too resource-intensive, so that independent investigators were able to provide the forward validation evidence of this predictive genomics-based assay. However, other assays may require far more resources and be technically more challenging, such that consortia of organizations would be better suited to conduct these forward validation efforts. Recent trends suggest that this is likely to happen, at least for some end points of interest.

6.9.3. Performance Accuracy

Useful assays in lead optimization should have well-understood performance characteristics, such that decisions can be made with sufficient confidence. In particular, most discovery scientists agree that early toxicology assays should have a high specificity (i.e., a low rate of false-positive signals). The assumption is that the opportunity cost of erroneously eliminating an otherwise good compound far exceeds the expense incurred in subsequent preclinical tests (96). In fact, one may argue that the challenge in early discovery may not only be to terminate compounds early, but to pick the best compounds. Obviously, no assay can have a perfect specificity; however, machine learning algorithms can be adjusted such that specificity is improved at the expense of sensitivity. This balance between specificity and sensitivity needs to be established based on the stage at which the test is likely to be used and the proposed use of the assay. If used in a prioritization exercise for a specific chemical series, confirmatory biochemical assays may be warranted to confirm a signal, such that perfect specificity may not be required. For instance, if during the screening of a chemical series, several members show positive signals with our phospholipidosis signature, we would attempt to confirm this finding with an independent fluorescent dye assay for selected representative molecules of the series.

6.9.4. Battery of Gene Expression Signatures

In their comparison between a gene expression-based and a fluorescent dye-based (LipidTox™, Invitrogen, Carlsbad, CA) assay to predict phospholipidosis in vitro, Nioi et al. showed some evidence that the fluorescent dye-based assay was less time consuming, more sensitive, and higher in throughput than the gene expression-based assay. They concluded that LipidTox™ should be the method of choice for screening compounds for phospholipidosis (76). While this conclusion is correct when specific gene expression signatures are used in isolation, it is debatable when several specific end points can be interrogated simultaneously with a battery of multigene classifiers. In this latter approach, one could classify compounds for several end points by using a single cell culture and at most a few concentrations of the compound. This methodology, especially when an appropriate analytical platform is utilized, may prove to be more economical, higher in throughput, and easier to interpret than an approach based on conducting multiple alternative biochemical assays concurrently at multiple concentrations. In addition, there are toxicological mechanisms that cannot be easily detected with conventional methods, and gene expression-based signatures may fill this gap. Therefore, we believe that for the in vitro toxicogenomics approach to be viable, a sufficient number of multigene classifiers must be constructed for useful toxicological end points.

6.9.5. Clear, Actionable Data Points

During lead optimization perhaps more than during any other stages of drug discovery and development, clear data points need to be generated for efficient communication, since these data points are not always transferred face to face and require rapid actions. As stated by Martin et al., most discovery scientists, especially medicinal chemists, who represent the main customer of toxicity data generated during lead optimization, do not tend to think in terms of gene expression, gene lists, or other genomics concepts (43). Communicating highly complex data with no clear interpretation is likely to generate confusion and misleading views of the results. The scientists involved in these genomics assessments need to communicate clear, actionable data to their customers. Genomics-based signatures or expression levels of specific genes represent clear data points that can easily be understood by most project stakeholders, and are our preferred way to communicate results from in vitro toxicogenomics screens. Likewise, correlation results from comparisons to gene expression profiles induced by reference toxicants are easily understood by most scientists; however, as mentioned earlier, these correlation analyses have been somewhat disappointing in in vitro systems in comparison to what has been experienced in vivo. Therefore, we do not recommend including them in communications.

With our current battery of gene expression signatures generated in a rat hepatocyte model, we have started to use toxicogenomics to evaluate compounds in vitro during lead optimization. Our current assays are not applied to all internal programs, nor are they utilized for all compounds generated in a project. In fact, one may argue that toxicity screens are best used at a stage when viable compounds optimized for potency and selectivity have been identified. Moreover, ADME screens are rather robust and, at this point, probably represent a more efficient approach to prioritizing compounds early. Finally, our current experimental protocol does not have the cost-effectiveness or throughput necessary to justify

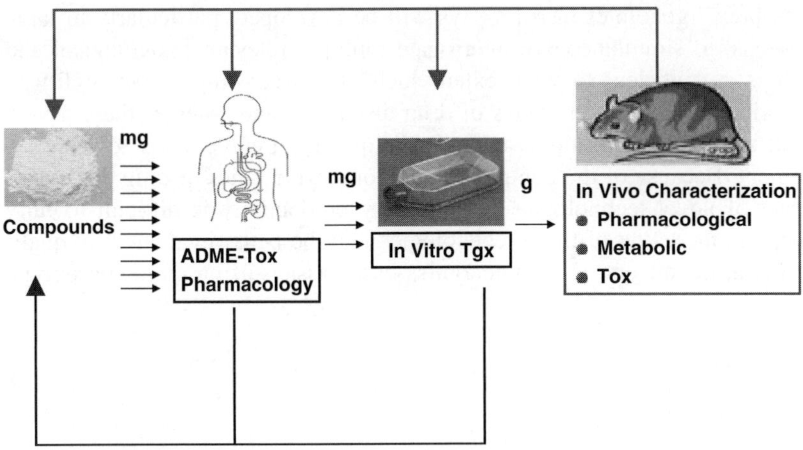

Figure 6.4 Proposed use of in vitro toxicogenomics (Tgx) at the lead optimization stage of drug discovery. In vitro Tgx can be used as a proactive screen during lead optimization in the form of refined and specific genomics signatures interrogated with economical and high-throughput platforms. Toxicity screens are best used at a stage when viable compounds optimized for potency and selectivity have been identified. Therefore, they should preferentially be used after the more robust and high-throughput ADME-Tox and pharmacology screens, which are more efficient to prioritize compounds early. Ideally, these toxicity screens can be used as an additional testing layer before in vivo characterization of compounds. Since lead optimization is an iterative process, where rapid feedback is critical, these in vitro toxicogenomics assays also offer the potential to rapidly screen backup compounds for specific toxicological liabilities identified in vivo.

large-scale applications. Nevertheless, with future technical improvements, we believe that solutions in terms of customized targeted platforms and automation will become available to start using in vitro toxicogenomics at a larger scale. Figure 6.4 represents an attempt to summarize our current vision of how in vitro toxicogenomics could be used effectively at the lead optimization stage.

6.10. CONCLUSIONS

Data generated so far in our laboratory or published elsewhere suggest that toxicogenomics represent a feasible approach to screening in vitro compounds early in the discovery process, likely at the lead optimization stage. However, more work and technical improvements are required for in vitro toxicogenomics to become an integral component of the battery of routine assays used in drug discovery. More cost-effective and flexible analytical gene expression platforms with appropriate throughput will be needed for toxicogenomics to have practical applications for the toxicological screening of compounds. Current data suggest that microarray-generated signatures can easily be transferred to other platforms, and therefore it is reasonable to envision that reliable and economical in-house or "home brew" genomics-based assays will be developed, particularly in large organizations, to simultaneously interrogate multiple relevant toxicological end points. It is still unclear to what extent such assays can impact our ability to reduce attrition rates at later stages of drug discovery, and whether these assays will result in a significant improvement in the quality of compounds advanced to development. Because of the complexity and long cycle times of drug discovery, the impact of novel technologies on discovery productivity is difficult to fully ascertain, and the value of these technologies may be better evaluated by qualitative criteria. In concept, refined genomics-based assays using more appropriate

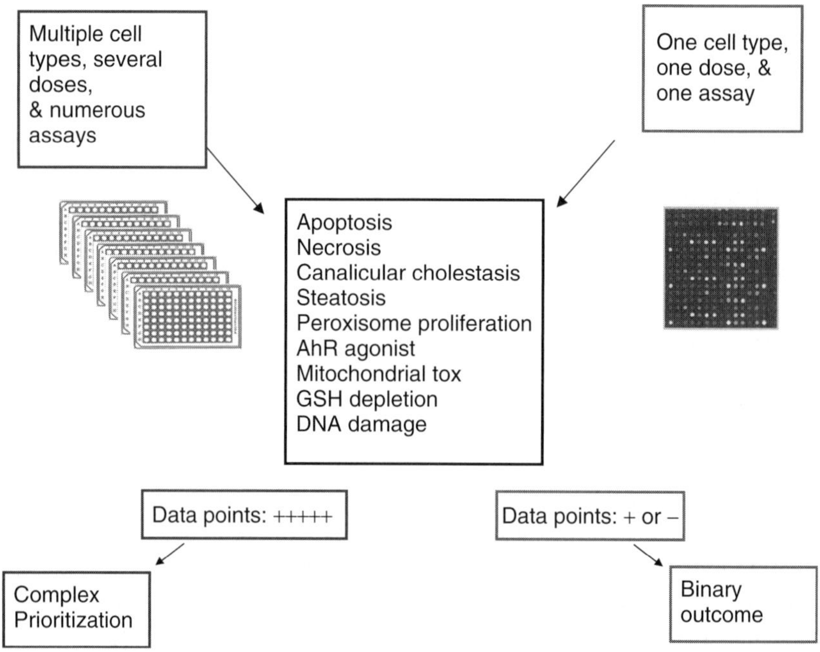

Figure 6.5 Advantages of in vitro toxicogenomics. In contrast to traditional in vitro toxicology screens (left side), in vitro toxicogenomics (right side) offers clear potential economical and throughput advantages. First, it enables the simultaneous evaluation of multiple toxicological end points in one cell type at a single (or at most few) concentration of test articles. This is in contrast to traditional in vitro tests, where several cell types may be required for various end points and which typically evaluate a dose response. Second, robust gene expression classifiers may be developed for end points that have not been regularly screened previously or can only be evaluated with labor-intensive assay formats (such as mitochondrial toxicity, for instance). Third, focused in vitro toxicogenomics assays generate binary assignments (positive or negative) for various end points, in contrast to a complex data set more challenging to interpret with traditional biochemical tests. This binary outcome has a clear advantage in lead optimization, since at this stage, data need to be generated rapidly and communicated clearly.

gene expression platforms have potential economical and throughput advantages, since they would enable the simultaneous evaluation of multiple relevant end points in one cell type and at a single, or at most, a few concentrations of test articles (Fig. 6.5). In addition, it is likely that robust gene expression classifiers may be developed for end points that have not been regularly screened previously or must be evaluated with labor-intensive assay formats. However, the use of more focused platforms will limit the data generated to simple binary assignments (positive vs. negative) for various end points. This binary outcome has a clear advantage in lead optimization, since at this stage data need to be rapidly generated and communicated, often through inclusion in complex databases integrating a multitude of parameters evaluated for novel compounds. In addition, these end points must be mined effectively to enable the development of in silico models for prediction of toxicity end points. However, gene expression signatures may not capture mechanisms beyond the end points generated, and may not provide much mechanistic clarity.

In contrast to the proactive screens used during lead optimization, different approaches will be required for mechanistic applications, where high-content microarray platforms may still represent the analytical methodology of choice. Indeed, in these mechanistic investigations, speed and cost are not as important as data content, since they would be used in a retroactive mode with limited sets of compounds. Genetic toxicology issues may fall into this category, although more work is needed to fully understand the value of toxicogenomics to improve the risk assessment of compounds positive in in vitro genetic toxicity assays.

REFERENCES

1. Olson H, Betton G, Robinson D, Thomas K, Monro A, Kolaja G, Lilly P, Sanders J, Sipes G, Bracken W, Dorato M, Van Deun K, Smith P, Berger B, Heller A. Concordance of the toxicity of pharmaceuticals in humans and in animals. *Regul Toxicol Pharmacol* 2000;32:56–67.
2. Flecknell P. Replacement, reduction and refinement. *ALTEX* 2002;19:73–78.
3. Liebsch M, Spielmann H. Currently available in vitro methods used in the regulatory toxicology. *Toxicol Lett* 2002;127:127–134.
4. Riley RJ, Kenna JG. Cellular models for ADMET predictions and evaluation of drug-drug interactions. *Curr Opin Drug Discov Devel* 2004;7:86–99.
5. Kassel DB. Applications of high-throughput ADME in drug discovery. *Curr Opin Chem Biol* 2004;8:339–345.
6. Riley RJ, Kenna JG. Cellular models for ADMET predictions and evaluation of drug-drug interactions. *Curr Opin Drug Discov Devel* 2004;7:86–99.
7. Jenkins KM, Angeles R, Quintos MT, Xu R, Kassel DB, Rourick RA. Automated high throughput ADME assays for metabolic stability and cytochrome P450 inhibition profiling of combinatorial libraries. *J Pharm Biomed Anal* 2004;34:989–1004.
8. Gruber FP, Hartung T. Alternatives to animal experimentation in basic research. *ALTEX* 2004;21 Suppl 1:3–31.

9. Gintant GA, Su Z, Martin RL, Cox BF. Utility of hERG assays as surrogate markers of delayed cardiac repolarization and QT safety. *Toxicol Pathol* 2006;34:81–90.
10. Spielmann H. Predicting the risk of developmental toxicity from in vitro assays. *Toxicol Appl Pharmacol* 2005;207:375–380.
11. Rausch O. High content cellular screening. *Curr Opin Chem Biol* 2006;10:316–320.
12. O'Brien PJ, Irwin W, Diaz D, Howard-Cofield E, Krejsa CM, Slaughter MR, Gao B, Kaludercic N, Angeline A, Bernardi P, Brain P, Hougham C. High concordance of drug-induced human hepatotoxicity with in vitro cytotoxicity measured in a novel cell-based model using high content screening. *Arch Toxicol* 2006;80:580–604.
13. Xu JJ, Diaz D, O'Brien PJ. Applications of cytotoxicity assays and pre-lethal mechanistic assays for assessment of human hepatotoxicity potential. *Chem Biol Interact* 2004;150:115–128.
14. Suter W. Predictive value of in vitro safety studies. *Curr Opin Chem Biol* 2006;10:362–366.
15. Liebsch M, Spielmann H. Currently available in vitro methods used in the regulatory toxicology. *Toxicol Lett* 2002;127:127–134.
16. Guengerich FP, MacDonald JS. Applying mechanisms of chemical toxicity to predict drug safety. *Chem Res Toxicol* 2007;20:344–369.
17. Flick B, Klug S. Whole embryo culture: an important tool in developmental toxicology today. *Curr Pharm Des* 2006;12:1467–1488.
18. Genschow E, Spielmann H, Scholz G, Seiler A, Brown N, Piersma A, Brady M, Clemann N, Huuskonen H, Paillard F, Bremer S, Becker K. The ECVAM international validation study on in vitro embryotoxicity tests: results of the definitive phase and evaluation of prediction models. European Centre for the Validation of Alternative Methods. *Altern Lab Anim* 2002;30:151–176.
19. Pessina A, Malerba I, Gribaldo L. Hematotoxicity testing by cell clonogenic assay in drug development and preclinical trials. *Curr Pharm Des* 2005;11:1055–1065.
20. Negro GD, Bonato M, Gribaldo L. In Vitro bone marrow granulocyte-macrophage progenitor cultures in the assessment of hematotoxic potential of the new drugs. *Cell Biol Toxicol* 2001;17:95–105.
21. LeCluyse E, Madan A, Hamilton G, Carroll K, DeHaan R, Parkinson A. Expression and regulation of cytochrome P450 enzymes in primary cultures of human hepatocytes. *J Biochem Mol Toxicol* 2000;14:177–188.
22. LeCluyse EL, Fix JA, Audus KL, Hochman JH. Regeneration and maintenance of bile canalicular networks in collagen-sandwiched hepatocytes. *Toxicol in vitro* 2000;14:117–132.
23. Amin K, Ip C, Jimenez L, Tyson C, Behrsing H. In Vitro detection of differential and cell-specific hepatobiliary toxicity induced by geldanamycin and 17-allylaminogeldanamycin using dog liver slices. *Toxicol Sci* 2005;87:442–450.
24. Pfaller W, Balls M, Clothier R, Coecke S, Dierickx P, Ekwall B, Hanley BA, Hartung T, Prieto P, Ryan MP, Schmuck G, Sladowski D, Vericat JA, Wendel A, Wolf A, Zimmer J. Novel advanced in vitro methods for long-term toxicity testing: the report and recommendations of ECVAM workshop 45. European Centre for the Validation of Alternative Methods. *Altern Lab Anim* 2001;29:393–426.
25. Bernauer U, Oberemm A, Madle S, Gundert-Remy U. The use of in vitro data in risk assessment. *Basic Clin Pharmacol Toxicol* 2005;96:176–181.

26. Clemedson C, Blaauboer B, Castell J, Prieto P, Risteli L, Vericat JA, Wendel A. ACuteTox—optimation and pre-validation of an in vitro test strategy for predicting human acute toxicity. *ALTEX* 2006;23 Suppl:254–258.
27. Ballet F. Hepatotoxicity in drug development: detection, significance and solutions. *J Hepatol* 1997;26 Suppl 2:26–36.
28. Dambach DM, Andrews BA, Moulin F. New technologies and screening strategies for hepatotoxicity: use of in vitro models. *Toxicol Pathol* 2005;33:17–26.
29. Gomez-Lechon MJ, Ponsoda X, Bort R, Castell JV. The use of cultured hepatocytes to investigate the metabolism of drugs and mechanisms of drug hepatotoxicity. *Altern Lab Anim* 2001;29:225–231.
30. Harries HM, Fletcher ST, Duggan CM, Baker VA. The use of genomics technology to investigate gene expression changes in cultured human cells. *Toxicol in vitro* 2001;15:399–405.
31. Vickers AE, Fisher RL. Organ slices for the evaluation of human drug toxicity. *Chem Biol Interact* 2004;150:87–96.
32. Boess F, Kamber M, Romer S, Gasser R, Muller D, Albertini S, Suter L. Gene expression in two hepatic cell lines, cultured primary hepatocytes, and liver slices compared to the in vivo liver gene expression in rats: possible implications for toxicogenomics use of in vitro systems. *Toxicol Sci* 2003;73:386–402.
33. Aden DP, Fogel A, Plotkin S, Damjanov I, Knowles BB. Controlled synthesis of HBsAg in a differentiated human liver carcinoma-derived cell line. *Nature* 1979;282:615–616.
34. Guillouzo A. Liver cell models in in vitro toxicology. *Environ Health Perspect* 1998;106 Suppl 2:511–532.
35. Rodriguez-Antona C, Donato MT, Boobis A, Edwards RJ, Watts PS, Castell JV, Gomez-Lechon MJ. Cytochrome P450 expression in human hepatocytes and hepatoma cell lines: molecular mechanisms that determine lower expression in cultured cells. *Xenobiotica* 2002;32:505–520.
36. Westerink WM, Schoonen WG. Cytochrome P450 enzyme levels in HepG2 cells and cryopreserved primary human hepatocytes and their induction in HepG2 cells. *Toxicol in vitro* 2007;21:1581–91.
37. Westerink WM, Schoonen WG. Phase II enzyme levels in HepG2 cells and cryopreserved primary human hepatocytes and their induction in HepG2 cells. *Toxicol in vitro* 2007;21:1592–602.
38. Beekman JM, Boess F, Hildebrand H, Kalkuhl A, Suter L. Gene expression analysis of the hepatotoxicant methapyrilene in primary rat hepatocytes: an interlaboratory study. *Environ Health Perspect* 2006;114:92–99.
39. Waring JF, Ciurlionis R, Jolly RA, Heindel M, Ulrich RG. Microarray analysis of hepatotoxins in vitro reveals a correlation between gene expression profiles and mechanisms of toxicity. *Toxicol Lett* 2001;120:359–68.
40. Burczynski ME, McMillian M, Ciervo J, Li L, Parker JB, Dunn RT2, Hicken S, Farr S, Johnson MD. Toxicogenomics-based discrimination of toxic mechanism in HepG2 human hepatoma cells. *Toxicol Sci* 2000;58:399–415.
41. Yang Y, Abel SJ, Ciurlionis R, Waring JF. Development of a toxicogenomics in vitro assay for the efficient characterization of compounds. *Pharmacogenomics* 2006;7:177–186.
42. Rao KS, Recknagel RO. Early onset of lipoperoxidation in rat liver after carbon tetrachloride administration. *Exp Mol Pathol* 1968;9:271–278.

43. Martin R, Rose D, Yu K, Barros S. Toxicogenomics strategies for predicting drug toxicity. *Pharmacogenomics* 2006;7:1003–1016.
44. Sawada H, Taniguchi K, Takami K. Improved toxicogenomic screening for drug-induced phospholipidosis using a multiplexed quantitative gene expression ArrayPlate assay. *Toxicol in vitro* 2006;20:1506–1513.
45. Waring JF, Ciurlionis R, Jolly RA, Heindel M, Gagne G, Fagerland JA, Ulrich RG. Isolated human hepatocytes in culture display markedly different gene expression patterns depending on attachment status. *Toxicol in vitro* 2003;17:693–701.
46. de Longueville F, Surry D, Meneses-Lorente G, Bertholet V, Talbot V, Evrard S, Chandelier N, Pike A, Worboys P, Rasson JP, Le Bourdelles B, Remacle J. Gene expression profiling of drug metabolism and toxicology markers using a low-density DNA microarray. *Biochem Pharmacol* 2002;64:137–49.
47. Morgan KT, Ni H, Brown HR, Yoon L, Qualls CW, Jr., Crosby LM, Reynolds R, Gaskill B, Anderson SP, Kepler TB, Brainard T, Liv N, Easton M, Merrill C, Creech D, Sprenger D, Conner G, Johnson PR, Fox T, Sartor M et al. Application of cDNA microarray technology to in vitro toxicology and the selection of genes for a real-time RT-PCR-based screen for oxidative stress in Hep-G2 cells. *Toxicol Pathol* 2002;30:435–51.
48. Jessen BA, Mullins JS, De Peyster A, Stevens GJ. Assessment of hepatocytes and liver slices as in vitro test systems to predict in vivo gene expression. *Toxicol Sci* 2003;75:208–222.
49. Newton RK, Aardema M, Aubrecht J. The utility of DNA microarrays for characterizing genotoxicity. *Environ Health Perspect* 2004;112:420–422.
50. Aubrecht J, Caba E. Gene expression profile analysis: An emerging approach to investigate mechanisms of genotoxicity. *Pharmacogenomics* 2005;6:419–428.
51. Snyder RD, Green JW. A review of the genotoxicity of marketed pharmaceuticals. *Mutat Res* 2001;488:151–169.
52. Thybaud V, Le Fevre AC, Boitier E. Application of toxicogenomics to genetic toxicology risk assessment. *Environ Mol Mutagen* 2007;48:369–379.
53. Dickinson DA, Warnes GR, Quievryn G, Messer J, Zhitkovich A, Rubitski E, Aubrecht J. Differentiation of DNA reactive and non-reactive genotoxic mechanisms using gene expression profile analysis. *Mutat Res* 2004;549:29–41.
54. Ellinger-Ziegelbauer H, Stuart B, Wahle B, Bomann W, Ahr HJ. Characteristic expression profiles induced by genotoxic carcinogens in rat liver. *Toxicol Sci* 2004; 77: 19–34.
55. Ellinger-Ziegelbauer H, Stuart B, Wahle B, Bomann W, Ahr HJ. Comparison of the expression profiles induced by genotoxic and nongenotoxic carcinogens in rat liver. *Mutat Res* 2005;575:61–84.
56. Hu T, Gibson DP, Carr GJ, Torontali SM, Tiesman JP, Chaney JG, Aardema MJ. Identification of a gene expression profile that discriminates indirect-acting genotoxins from direct-acting genotoxins. *Mutat Res* 2004;549:5–27.
57. Le Fevre AC, Boitier E, Marchandeau JP, Sarasin A, Thybaud V. Characterization of DNA reactive and non-DNA reactive anticancer drugs by gene expression profiling. *Mutat Res* 2007;619:16–29.
58. Akerman GS, Rosenzweig BA, Domon OE, McGarrity LJ, Blankenship LR, Tsai CA, Culp SJ, MacGregor JT, Sistare FD, Chen JJ, Morris SM. Gene expression profiles and genetic damage in benzo(*a*)pyrene diol epoxide-exposed TK6 cells. *Mutat Res* 2004;549:43–64.

59. Halliwell WH. Cationic amphiphilic drug-induced phospholipidosis. *Toxicol Pathol* 1997; 25:53–60.
60. Reasor MJ, Kacew S. Drug-induced phospholipidosis: are there functional consequences? *Exp Biol Med* 2001;226:825–830.
61. Reasor MJ, Hastings KL, Ulrich RG. Drug-induced phospholipidosis: issues and future directions. *Expert Opin Drug Saf* 2006;5:567–583.
62. Casartelli A, Bonato M, Cristofori P, Crivellente F, Dal Negro G, Masotto I, Mutinelli C, Valko K, Bonfante V. A cell-based approach for the early assessment of the phospholipidogenic potential in pharmaceutical research and drug development. *Cell Biol Toxicol* 2003;19:161–176.
63. Rudmann DG, McNerney ME, VanderEide SL, Schemmer JK, Eversole RR, Vonderfecht SL. Epididymal and systemic phospholipidosis in rats and dogs treated with the dopamine D3 selective antagonist PNU-177864. *Toxicol Pathol* 2004;32:326–332.
64. Laurent G, Kishore BK, Tulkens PM. Aminoglycoside-induced renal phospholipidosis and nephrotoxicity. *Biochem Pharmacol* 1990;40:2383–2392.
65. Carella F, Riva E, Morandi L, Cappiello E, Mangoni A. Myopathy during amiodarone treatment: a case report. *Ital J Neurol Sci* 1987;8:605–608.
66. Riva E, Marchi S, Pesenti A, Bizzi A, Cini M, Veneroni E, Tavbani E, Boeri R, Bertani T, Latini R. Amiodarone induced phospholipidosis. Biochemical, morphological and functional changes in the lungs of rats chronically treated with amiodarone. *Biochem Pharmacol* 1987;36:3209–3214.
67. Finbloom DS, Silver K, Newsome DA, Gunkel R. Comparison of hydroxychloroquine and chloroquine use and the development of retinal toxicity. *J Rheumatol* 1985;12:692–694.
68. McGovern MM, Aron A, Brodie SE, Desnick RJ, Wasserstein MP. Natural history of Type A Niemann-Pick disease: possible endpoints for therapeutic trials. *Neurology* 2006;66:228–232.
69. Pelletier DJ, Gehlhaar D, Tilloy-Ellul A, Johnson TO, Greene N. Evaluation of a published in silico model and construction of a novel Bayesian model for predicting phospholipidosis inducing potential. *J Chem Inf Model* 2007;47:1196–1205.
70. Monteith DK, Morgan RE, Halstead B. In Vitro assays and biomarkers for drug-induced phospholipidosis. *Expert Opin Drug Metab Toxicol* 2006;2:687–696.
71. Schurdak ME, Vernetti LA, Abel SJ, Thiffault C. Adaptation of an in vitro phospholipidosis assay to an automated image analysis system. *Toxicol Mech Method* 2007;17:77–86.
72. Kasahara T, Tomita K, Murano H, Harada T, Tsubakimoto K, Ogihara T, Ohnishi S, Kakinuma C. Establishment of an in vitro high-throughput screening assay for detecting phospholipidosis-inducing potential. *Toxicol Sci* 2006;90:133–141.
73. Sawada H, Takami K, Asahi S. A toxicogenomic approach to drug-induced phospholipidosis: analysis of its induction mechanism and establishment of a novel in vitro screening system. *Toxicol Sci* 2005;83:282–292.
74. Sawada H, Taniguchi K, Takami K. Improved toxicogenomic screening for drug-induced phospholipidosis using a multiplexed quantitative gene expression ArrayPlate assay. *Toxicol in vitro* 2006;20:1506–1513.
75. Atienzar F, Gerets H, Dufrane S, Tilmant K, Cornet M, Dhalluin S, Ruty B, Rose G, Canning M. Determination of phospholipidosis potential based on gene expression analysis in HepG2 cells. *Toxicol Sci* 2007;96:101–114.

76. Nioi P, Perry BK, Wang EJ, Gu YZ, Snyder RD. In Vitro detection of drug-induced phospholipidosis using gene expression and fluorescent phospholipid based methodologies. *Toxicol Sci* 2007;99:162–173.
77. Uetrecht J. Idiosyncratic drug reactions: current understanding. *Annu Rev Pharmacol Toxicol* 2007;47:513–539.
78. Liguori MJ, Waring JF. Investigations toward enhanced understanding of hepatic idiosyncratic drug reactions. *Expert Opin Drug Metab Toxicol* 2006;2:835–846.
79. Waring JF, Anderson MG. Idiosyncratic toxicity: mechanistic insights gained from analysis of prior compounds. *Curr Opin Drug Discov Devel* 2005;8:59–65.
80. Knowles SR, Uetrecht J, Shear NH. Idiosyncratic drug reactions: the reactive metabolite syndrome. *Lancet* 2000;356:1587–1591.
81. Drlica K, Zhao X. DNA gyrase, topoisomerase IV, and the 4-quinolones. *Microbiol Mol Biol Rev* 1997;61:377–392.
82. Bertino J, Jr., Fish D. The safety profile of the fluoroquinolones. *Clin Ther* 2000;22:798–817.
83. Liguori MJ, Anderson LM, Bukofzer S, McKim J, Pregenzer JF, Retief J, Spear BB, Waring JF. Microarray analysis in human hepatocytes suggests a mechanism for hepatotoxicity induced by trovafloxacin. *Hepatology* 2005;41:177–186.
84. Liguori MJ, Blomme EA, Waring JF. Trovafloxacin-induced gene expression changes in liver-derived in vitro systems: comparison of primary human hepatocytes to HepG2 cells. *Drug Metab Dispos* 2008;36:223–233.
85. Gorson DM. Effect of troglitazone in type 2 diabetes mellitus. *N Engl J Med* 1998;339:406.
86. Watkins PB. Insight into hepatotoxicity: the troglitazone experience. *Hepatology* 2005; 41:229–230.
87. Kier LD, Neft R, Tang L, Suizu R, Cook T, Onsurez K, Tiegler K, Sakai Y, Ortiz M, Nolan T, Sankar U, Li AP. Applications of microarrays with toxicologically relevant genes (tox genes) for the evaluation of chemical toxicants in Sprague Dawley rats in vivo and human hepatocytes in vitro. *Mutat Res* 2004;549:101–113.
88. Guo L, Zhang L, Sun Y, Muskhelishvili L, Blann E, Dial S, Shi L, Schroth G, Dragan YP. Differences in hepatotoxicity and gene expression profiles by anti-diabetic PPAR gamma agonists on rat primary hepatocytes and human HepG2 cells. *Mol Divers* 2006;10:349–360.
89. Vansant G, Pezzoli P, Saiz R, Birch A, Duffy C, Ferre F, Monforte J. Gene expression analysis of troglitazone reveals its impact on multiple pathways in cell culture: a case for in vitro platforms combined with gene expression analysis for early (idiosyncratic) toxicity screening. *Int J Toxicol* 2006;25:85–94.
90. Tirmenstein MA, Hu CX, Gales TL, Maleeff BE, Narayanan PK, Kurali E, Hart TK, Thomas HC, Schwartz LW. Effects of troglitazone on HepG2 viability and mitochondrial function. *Toxicol Sci* 2002;69:131–138.
91. Rockett JC, Burczynski ME, Fornace AJ, Herrmann PC, Krawetz SA, Dix DJ. Surrogate tissue analysis: monitoring toxicant exposure and health status of inaccessible tissues through the analysis of accessible tissues and cells. *Toxicol Appl Pharmacol* 2004;194:189–199.
92. Burczynski ME, Dorner AJ. Transcriptional profiling of peripheral blood cells in clinical pharmacogenomic studies. *Pharmacogenomics* 2006;7:187–202.
93. Steppan S, Kupfer K, Mayer A, Evans M, Yamasaki G, Greve JM, Eckart MR, Cassell DJ. Genome wide expression profiling of human peripheral blood mononuclear cells stimulated with BAY 50–4798, a novel T cell selective interleukin-2 analog. *J Immunother* 2007;30:150–168.

94. Van Leeuwen DM, Gottschalk RW, Van Herwijnen MH, Moonen EJ, Kleinjans JC, Van Delft JH. Differential gene expression in human peripheral blood mononuclear cells induced by cigarette smoke and its constituents. *Toxicol Sci* 2005;86:200–210.
95. Zidek N, Hellmann J, Kramer PJ, Hewitt PG. Acute hepatotoxicity: a predictive model based on focused illumina microarrays. *Toxicol Sci* 2007;99:289–302.
96. Barros SA. The importance of applying toxicogenomics to increase the efficiency of drug discovery. *Pharmacogenomics* 2005;6:547–550.

Chapter 7

Germ Line Polymorphisms and Drug Response

7.1. INTRODUCTION TO GERM LINE POLYMORPHISMS

Germ line polymorphisms are genetic variants present in an appreciable fraction of individuals in a population. Unlike somatic mutations in tumors considered in Chapter 3, germ line polymorphisms are present in all cells of an organism. They may represent various sequence variations or deletions, but the most frequent form of genetic polymorphism is a single nucleotide variant. A **single nucleotide polymorphism** (SNP; Fig. 7.1A) is defined as a genomic locus in which two or more alternative bases occur at a frequency of more than 1% of the population. SNPs represent the most common form of variation in the human genome. Other common types of human polymorphisms include microsatellites (or tandem repeats), which represent multiple copies of repeated DNA sequences in a region 0.1–10 kb in length. Microsatellites consist of repeated short sequences up to 4 nucleotides. Table 7.1 presents a glossary of key terms related to germ line polymorphisms.

The functional role of many known human genetic polymorphisms is not known. Depending on whether they are located in the coding or noncoding sequence, polymorphisms may affect protein composition and activity, gene transcription or translation, or mRNA stability (1). If a polymorphism is found in a gene that plays a role in a drug's mechanism or its metabolism, it may affect the efficacy or toxicity of the drug. In this case, the presence of the polymorphism would correlate with the response to the drug and thus, if measured before administration of therapy, might serve as a predictor of efficacy or toxicity

Genomics in Drug Discovery and Development, by Dimitri Semizarov and Eric Blomme

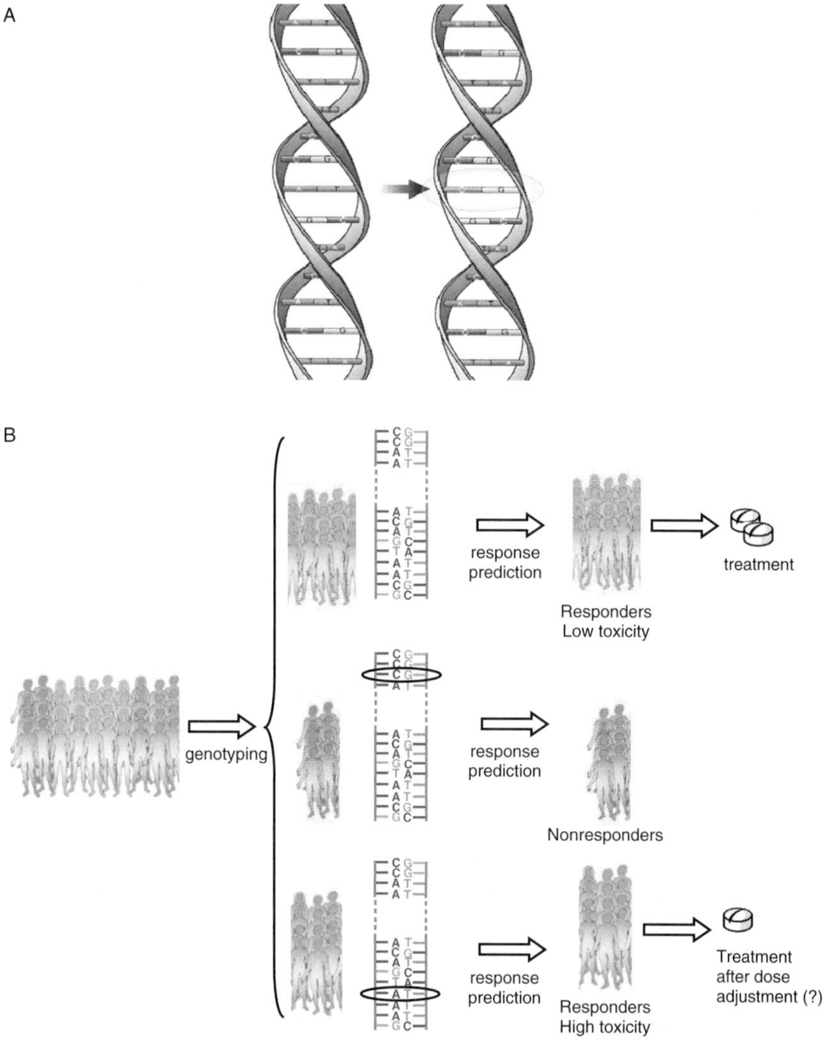

Figure 7.1 A) A single-base substitution in the DNA sequence (e.g. A:T → G:C) is the most common form of polymorphism in the human genome. B) If associated with drug efficacy or toxicity, SNPs can be used to predict drug response and select patients for therapy. See color insert.

(Fig. 7.1B). Strictly speaking, there is no reason to separate polymorphic variants from mutations while considering their role as genomic biomarkers: In oncology, for example, many mutations are as frequent as polymorphisms when a diseased population is considered. However, traditionally polymorphisms have been considered separately, and pharmacogenetic studies use different approaches and rely on separately established databases. Therefore, we chose to devote a separate chapter to polymorphic variants as predictive markers of drug response.

Table 7.1 Glossary of Pharmacogenetic Terms

Term	Definition
Single nucleotide polymorphism (SNP)	A genomic locus in which 2 or more alternative bases occur at a frequency of more than 1% of the population
Microsatellites (or tandem repeats)	Multiple copies of repeated DNA sequences in a region 0.1–10 kb in length. Microsatellites consist of repeated short sequences up to 4 nucleotides in length.
Allele	An alternate form of a gene
Haplotype	A tightly linked group of genetic markers, which are often inherited as a unit because of their close proximity
Linkage disequilibrium	The property of 2 polymorphic loci in a population such that the polymorphic states at the 2 loci are not independent of one another, and as a result the state of the polymorphism at 1 locus has a higher probability of being associated with a particular state at the second locus
Pharmacogenetics	Study of the relationships between the genetics of an individual and drug response
Pharmacogenomics	Study of the relationship between the genomic profile of an individual and response to drugs. The term "genomic profile" includes the SNP genotype, gene copy number pattern, and the gene expression pattern.
Toxicogenomics	Study of the genomics correlates of adverse effects to drugs
Association study	A study looking for increased occurrence of a polymorphism in the group with a phenotype vs. the group that does not display this phenotype
Aneuploidy	The presence of an abnormal number of chromosomes in the cell
Double minute	Acentric, extrachromosomally amplified chromatin, which usually contains a particular chromosomal segment or gene; common in cancer cells
Segmental duplications	Segments of chromosomal DNA that are >1 kb in size and have >90% intercopy sequence identity (also called low-copy repeats or duplicons)
Copy number variant (CNV)	A segment of DNA that is 1 kb or larger and is present at a variable copy number in comparison with a reference genome. Classes of CNVs include insertions, deletions and duplications.
Copy number polymorphism	A copy number variant that occurs in >1% of the population.
Inversion	A segment of DNA that is reversed in orientation with respect to the rest of the chromosome. Pericentric inversions include the centromere, whereas paracentric inversions do not.

Studies have shown that that the contribution of genetics to the variability in drug disposition and effects may range from 20% to 95% (2). Although the response of an individual to a therapeutic agent depends on numerous nongenetic factors such as age, lifestyle, disease subtype, coexisting diseases, and comedication, it has been clearly demonstrated that interindividual variability in drug response is determined by polymorphisms in genes coding for drug targets, drug metabolism enzymes, or drug transporters. Traditionally, genetic polymorphisms in drug-metabolizing enzymes have received the most study. They may affect both the pharmacological benefit of the drug and its toxicity by altering the available drug concentrations. If the polymorphism results in a decrease in function of the drug-metabolizing enzyme, the patient will be exposed to higher drug concentrations, resulting in increased toxicity. Indeed, it was recently reported that approximately 100,000 deaths and more than 2 million hospitalizations annually in the United States occur because of unforeseen adverse reactions to properly prescribed medications (3). On the contrary, if the genetic variant leads to increased enzyme activity, the patient will be undertreated and will not receive the desired therapeutic benefit. Another obvious source of potential genetic predictors of drug response is polymorphisms in drug targets. In this case, the more relevant correlate is drug efficacy, although the toxicological response to the drug may also be affected if on-target toxicity is substantial. As the drug discovery paradigm shifts toward targeted therapeutics, prospective identification of potential polymorphisms in drug targets becomes a critical activity that may affect the clinical development process. Finally, the search for predictive polymorphisms can be expanded to genes whose protein products are not directly affected by the drug, but which are involved in the disease mechanism. In this case, polymorphisms may affect drug response either by indirectly affecting the expression or function of the drug target or by altering the course of the disease.

The majority of this chapter is a collection of case studies in several therapeutic areas, namely oncology, inflammation, virology, and neuroscience. Most of the published studies on polymorphisms and drug response deal with established marketed drugs. However, since the main focus of this book is on application of genomics in drug discovery, we will attempt to emphasize the key learnings from the drug discovery perspective and analyze potential strategies for early identification of polymorphisms as predictors of drug response.

7.2. POLYMORPHISMS AND DRUG RESPONSE IN ONCOLOGY

Genetic polymorphisms have been extensively studied in oncology because of the well-established observation that cancer risk strongly depends on the genetic background of the individual. Additionally, in cancer drug discovery it is particularly important to understand the pharmacogenetic associations between the genotype and the drug response, because many chemotherapeutic agents used in the treatment of cancer are characterized by a narrow therapeutic index.

7.2.1. UGT1A1 Polymorphism and Response to Irinotecan

Perhaps the best-studied and most advanced example of a polymorphism as a predictor of drug effect involves irinotecan, an approved therapeutic agent used in the treatment of various solid tumors such as colon and lung cancers. To exert its therapeutic effect, irinotecan must be converted to its active form, SN-38, which suppresses tumorigenesis by inhibiting topoisomerase I (4). The metabolism of SN-38 is controlled in part by a hepatic enzyme, UDP glucuronosyltransferase 1A1 (UGT1A1). The enzyme inactivates SN-38 by converting it to a glucuronide, which is then eliminated in urine and bile (5). The response to irinotecan has been associated with significant interpatient variability. It has also been determined that the rate of SN-36 inactivation varies up to 50-fold among patients, mostly because of interindividual variation in the UGT1A1 expression (6, 7). In light of these observations, it was suggested that the variation in response to irinotecan is related to the UGT1A1 status of the patient. Specifically, the dose-limiting toxicities of irinotecan, namely, diarrhea and leukopenia, were associated with UGT1A1 expression. In turn, it was established that the expression of the gene is determined by genetic variation in its promoter sequence. The UGT1A1 enzyme is encoded by the *UGT1* locus on chromosome 2, which may contain at least 12 alternative versions of exon 1, each controlled by its own promoter (8). The variability in the levels of UGT1A1 is determined by differences in the number of TA repeats in the UGT1A1 promoter (the transcription factor IID binds to these TA repeats). The majority of people have six TA repeats, but some have seven, and this increase results in lower expression levels of UGT1A1 (9, 10). This condition is called Gilbert syndrome and is characterized by mild chronic hyperbilirubinemia. One consequence of the reduced UGT1A1 expression in patients with Gilbert syndrome is a relatively lower level of SN-38 inactivation and hence increased accumulation of this active form of irinotecan (6, 7). Thus when patients with a polymorphism in the UGT1A1 gene promoter are treated with the standard, nonoptimized dose of irinotecan, they are exposed to much higher levels of the active metabolite and therefore more frequently develop diarrhea and leukopenia relative to patients with the wild-type UGT1A1 gene (11).

Thus it has been established that the UGT1A1 genotype may serve as a predictor of irinotecan toxicity. This is one of the best-studied examples of human genetic polymorphisms as predictive markers for therapeutic agents. The high degree of validation of the marker and the clear mechanistic link between the UGT1A1 genotype and metabolism of irinotecan has prompted the inclusion of a special warning in the prescription label for irinotecan regarding a lower initial dose of the drug for patients with the variant UGT1A1 allele.

7.2.2. FGFR4 Polymorphism and Response to Chemotherapy

Identification of a polymorphism that affects the efficacy of anticancer drugs is equally important as discovery of variants predictive of increased toxicity. There are multiple striking examples of SNPs as biomarkers predictive of efficacy. Notably, some of these represent SNPs located in genes with known association with cancer. For instance, fibroblast growth factor receptor 4 (FGFR4) belongs to a group of four closely related receptors, which, together with their more than 20 known ligands, are involved in the control of cell growth, differentiation, and migration (12). It is known that FGFs and FGFRs mediate tumorigenesis by affecting angiogenesis or directly stimulating cell growth (12). Several years ago, a variant allele of the FGFR4 gene was discovered that contains a G to A conversion (13). This nucleotide change leads to a substitution of arginine for glycine at position 388 in the transmembrane domain of the receptor. The Arg388 allele was detected in several tumor types and cancer cell lines, as well as in the germ line of cancer patients and healthy individuals. Sequencing analysis of the FGFR4 gene in three geographically separated populations revealed an SNP frequency of approximately 50%. The putative involvement of FGFR4 in carcinogenesis prompted sequence analysis of the FGFR4 allele in two patient populations with different types of cancer (13). In a group of 84 breast cancer patients, homo- or heterozygous carriers of the Arg388 allele had a significantly reduced disease-free survival time (P value $= 0.01$) within a median follow-up of 62 months. In a population of 82 colorectal cancer patients, the presence of the Arg388 allele of FGFR4 correlated with early manifestation of lymph node metastases and advanced tumor node metastasis (TNP).

Consistent with the notion that mechanistic explanation is an important part of biomarker discovery and validation, the clinical observations described above prompted an in vitro study to elucidate the effects of the Arg388 allele of FGFR4 on breast cancer cells. It was found that MDA-MB-231 mammary tumor cells expressing FGFR4 Arg388 have increased motility relative to cells expressing the wild-type FGFR4. It was concluded that the FGFR4 Arg388 allele is innocuous in normal individuals but represents a marker of significantly accelerated disease progression in breast cancer patients (13).

Later it was demonstrated that high expression of the variant Arg388 allele of FGFR4 is associated with poor clinical outcome in head and neck squamous cell carcinoma (14). Additionally, a correlation between the presence of the Arg388 polymorphic variant and disease outcome has been detected in adenocarcinoma of the lung (15). These observations further confirm the clinical relevance of the Arg388 FGFR4 variant to cancer and further validate it as a genomic stratification biomarker predictive of poor disease outcome. Obviously, the value of the marker would be much greater if it could be used to predict patient response to therapy. Fortunately, breast cancer represents a disease well suited for testing hypotheses on predictive value of biomarkers, as therapeutic benefits from adjuvant therapy have been clearly demonstrated (16).

The question of predictive power of the Arg388 allele of FGFR4 in breast cancer chemotherapy was promptly addressed with a large patient cohort (n = 372) subjected to adjuvant therapy and a long follow-up (94.5 months median) (17). In parallel, HER2 status was analyzed in the same patients by using the immunohistochemistry-based test. In 51% of patients, homo- or heterozygous Arg388 allele of FGFR4 was detected. No correlation was found between the presence of the FGFR4 Arg388 marker and the HER2 status. In node-negative patients, the FGFR4 genotype did not correlate with disease outcome. However, in node-positive patients the FGFR4 Arg388 marker was significantly associated with poor disease-free survival (P value = 0.02) and overall survival (P value = 0.04). In the adjuvant therapy setting, this association is likely to be due to poor response to therapy. An additional statistical procedure, a multivariate proportional hazards model, indicated that Arg388 FGFR4 carriers receive approximately half as much benefit from adjuvant systemic therapy as carriers of the wild-type FGFR4 allele. In summary, the Arg388 FGFR4 marker was initially discovered as a variant allele in some breast cancer cell lines and tumors, then found to be a frequent variant in a larger population of breast cancer patients, associated with poor outcome in breast and other cancers, and finally established as a predictive marker of efficacy for adjuvant breast cancer therapy. It thus represents a remarkable example of biomarker development starting from the discovery of an SNP in a gene with mechanistic connection to cancer and ending with a predictive marker validated in a large cohort of relevant patients.

7.2.3. Mdr-1 Polymorphism and Response to Paclitaxel

In the study described above, a polymorphism was selected for clinical investigation based on its localization in a cancer-related gene and the known overexpression of this gene in breast cancer cells. A different approach to identification of polymorphisms predictive of drug response involves study of variants in the genes known to be associated with drug resistance. For example, a transport protein called P-glycoprotein, encoded by the mdr-1 gene, is known to cause multidrug resistance. It is a 170-kDa plasma membrane protein that functions as an ATP-driven drug export pump. Recently, different polymorphisms in the mdr-1 gene have been discovered, and their correlation with the response to drugs that are P-glycoprotein substrates is being investigated. In particular, the allelic frequencies at polymorphic sites G2677T/A and C3435T in ovarian cancer patients were studied in patients with good or poor response to paclitaxel in combination with carboplatin in order to evaluate their predictive power (18). Paclitaxel (Taxol) is a broad-spectrum cytotoxic drug commonly used to treat breast, ovarian, and lung cancer. Resistance is a major obstacle to the successful treatment of patients, and therefore mechanisms of paclitaxel resistance are subjects of intense scrutiny.

In the aforementioned study, 53 patients were included in the study, of which 28 had been relapse-free for at least 1 year (good prognosis group) and 25 had progressive disease or relapsed within 12 months (poor prognosis group). The genotypes of each of the known SNPs in the mdr-1 gene were determined with pyrosequencing. A missense SNP (G2677T/A) was found to correlate with the response to paclitaxel. A statistically significant correlation was found between the homozygous variant and response to paclitaxel (Fisher's Exact test P value < 0.05). Nine of the 28 cases with a good response were homozygous variants, compared with 2 of the 25 cases with a poor response. The frequency of the T and A alleles in the good-response group of patients was also significantly higher than in the poor-response group (32 of 56 compared with 18 of 50; Fisher's Exact test P value < 0.05). The effect of increasing number of mutated alleles (G/G < G/T < T/T or T/A) on the drug response was also significant (χ^2-test for linear-by-linear association; P value $= 0.03$). These data indicate that ovarian cancer patients with homozygous variants at the G2677T/A SNP in the mdr-1 gene are more likely to respond to paclitaxel treatment. The presence of two variant alleles is a predictive factor for paclitaxel response.

A mechanistic explanation would strongly support further evaluation of the G2677T/A SNP as a stratification marker for paclitaxel treatment. One possible mechanistic connection between the polymorphism and the effect of the drug could be a reduced efflux of paclitaxel from the tumor cells or a decreased elimination from the body, resulting in higher plasma concentrations. In this case, however, one would expect a higher incidence of adverse reactions to paclitaxel, which was not the case in this study. The functional role of the G2677T/A SNP and its mechanistic connection with the response to paclitaxel thus remain to be elucidated.

7.2.4. DPD Polymorphisms and Response to 5-Fluorouracil

A common cancer therapeutic, 5-fluorouracil is widely used in the treatment of gastrointestinal malignancies as well as breast and head and neck cancers. Both the efficacy and toxicity of 5-fluorouracil are determined by the rate of its anabolism into cytotoxic nucleotides. The drug is converted to an inactive compound, 5,6-dihydro-5-fluorouracil, by an enzyme called dihydropyrimidine dehydrogenase (DPD) (19). DPD is the first and rate-limiting enzyme in the catabolism of uracil and thymine (20). Consistent with the mechanism of 5-fluorouracil inactivation, it has been demonstrated that reduced DPD activity increases the half-life and toxicity of 5-fluorouracil (21). A severe toxic reaction to 5-fluorouracil has been described in cancer patients, which is manifested by neutropenia, neurological symptoms, and mucositis and can sometimes lead to death. It was shown that this syndrome is caused by mutations in the DPD gene resulting in a significant decrease in DPD activity (22–24). According to population studies, approximately 3% of the population is estimated to have reduced catalytic activity of DPD (25). Among patients with severe toxicity caused by 5-fluorouracil, the fraction of individuals with decreased (< 70%) DPD activity in peripheral blood mononuclear cells ranges from 36% to 59% (26, 27). The most frequent variant found in patients with partial or complete DPD deficiency is a G → A substitution within the 50-splicing site of intron 14 (often referred to as the exon 14–skipping mutation, *DPYD*2A*) (28). The substitution leads to loss of exon 14, resulting in a truncation of the protein product by 55 amino acids and consequent loss of enzyme activity. The frequency of this variant allele in Caucasian populations is 0.9% (29, 30), bordering the threshold set by the definition of SNPs (frequency $\geq$ 1%). The frequency of *DPYD*2A* in other ethnic groups remains to be determined. The data accumulated so far on the relationship between the *DPYD*2A* genotype and the 5-fluorouracil-related toxic reactions are controversial. A number of studies reveled a strong concordance, with approximately 50% of patients with toxic reactions carrying the heterozygous *DPYD*2A* allele (27, 30). However, others found no genotypic correlates for reduced DPD activity and susceptibility to toxic reaction to 5-fluorouracil (31). Currently, genotyping tests for *DPYD* variants have low sensitivity in identifying patients potentially susceptible to the toxic reaction; therefore, no test has yet been validated for use in the clinic (28).

7.2.5. TPMT Variants and Response to Thiopurines

Another noteworthy example of polymorphic markers in cancer therapy is related to a group of thiopurine agents frequently used in the treatment of acute lymphoblastic leukemia, which includes azathioprine, 6-mercaptopurine, and 6-thioguanine. These compounds represent prodrugs that are metabolized into therapeutically active forms, 6-thioguanine nucleoside 5′-triphosphates. They can also be inactivated through *S*-methylation, a reaction catalyzed by thiopurine *S*-methyltransferase (TPMT), an enzyme critical for the metabolism of many heterocyclic sulfhydryl compounds. Thus both the efficacy and toxicity of thiopurine drugs depend on the enzymatic activity of TPMT. Patients with genetically predetermined low or undetectable TPMT activity develop severe myelosuppression when treated with commonly used doses of these agents, whereas individuals with high levels of enzyme activity are potentially undertreated (32). Approximately 89% of Caucasian subjects are homozygous for the inherited trait of high TPMT activity, approximately 11% are heterozygous and have intermediate activity, and 0.3% are homozygous for the trait of very low or absent activity (33). The low enzyme activity results from accelerated degradation of the protein produced by the polymorphic alleles, containing nonsynonymous SNPs in the open reading frame (34). Polymorphism of the TPMT locus has been detected in a number of different populations, including Caucasians, Asians, Africans, and African Americans.

Nine polymorphic TPMT alleles have been detected so far, including three variant alleles [*TPMT*2* (G238C), *TPMT*3A* (G460A and A719G), and *TPMT*3C* (A719G)], which account for 80–95% of the cases of low and intermediate enzyme activity (28). It has been found that *TPMT*3A* is the most frequent variant allele in Caucasians, while *TPMT*3C* is the most frequent polymorphic allele in Asians and populations of African descent. Another type of polymorphism, namely, a polymorphic tandem repeat, was recently identified within a GC-rich area in the 5′-flanking region of the human TPMT gene (35). The repeat elements were either 17 or 18 base pairs in length. Allele lengths varied from three to nine repeats. In Caucasians, the most frequent alleles contain four or five repeat elements. In a large study that associated TPMT activity with genotype, the TPMT activity in erythrocytes was inversely related to the sum of the number of repeat elements on the two alleles in each DNA sample (35). This was an extremely interesting observation with potentially important consequences for thiopurine drug therapy, but as we have already emphasized in this chapter, association studies of this type require in vitro functional validation to provide the mechanistic connection between the polymorphism and the drug action. Toward this goal, transient expression was used to demonstrate that the increased number of repeats reduces the reporter gene expression (36), suggesting that the variable number of tandem repeats controls the level of TPMT activity. However, the effect of the tandem repeats on the TMPT enzyme activity was significantly smaller than that of the SNPs discussed above. Additionally, it was shown that *VNTR**5 and *TPMT***3A* are in linkage disequilibrium (37).

Thus the link between the polymorphisms in the TMPT gene and enzyme activity has been established. The remaining challenge was to correlate TMPT enzyme activity with the toxicity of thiopurine drugs. A number of clinical observations have been published that support the correlation between the genetically predetermined TPMT deficiency and hematologic toxicity of thiopurines, including severe myelosuppression (38–40). This prompted a comprehensive assessment of the TPMT phenotype and thiopurine metabolism in patients displaying excessive toxicity while receiving mercaptopurine or azathioprine (41). Deficiency in TMPT activity or heterozygosity for the variant allele was sixfold more frequent among patients developing dose-limiting hematopoietic toxicity from therapy containing thiopurines. However, it was possible to adjust the dose so that TPMT-deficient and heterozygous patients could be treated with thiopurines, without acute dose-limiting toxicity. The pharmacology of 6-mercaptopurine was studied in 180 patients who achieved remission on a protocol composed of weekly methotrexate and daily oral 6-mercaptopurine given for 2.5 years. It was shown that, at conventional doses of 6-mercaptopurine, patients heterozygous for the TPMT variant accumulate approximately twofold more thioguanine nucleotides in their erythrocytes relative to homozygous wild-type individuals. This difference in thioguanine nucleotide accumulation resulted in a fivefold higher frequency of dose-limiting toxicity in TPMT heterozygotes compared with the wild-type patients. Patients with the wild-type TPMT alleles tolerated the conventional dose of 6-mercaptopurine during 84% of scheduled therapy compared with 65% for heterozygous patients and only 7% in completely TPMT-null patients (42).

The TPMT genetic marker has been associated not only with the toxicity of thiopurine drugs but also with their therapeutic efficacy. Multivariate analysis has revealed that in children with acute lymphoblastic leukemia decreased TPMT activity correlated with better outcome, implying that children with high TPMT enzymatic activity derive less therapeutic benefit from the drug (43).

Thus the TMPT polymorphism represents an interesting type of genomic marker that predicts both efficacy and toxicity for an entire class of anticancer drugs, and may be used to determine the optimal dose. An accurate diagnostic test would enable patient stratification for therapy and dose selection. Two techniques may be utilized to prospectively identify TPMT-deficient and heterozygous patients: measurement of TPMT enzymatic activity in erythrocytes or TPMT genotyping. The utility of the enzymatic activity test is limited by the fact that leukemia patients often receive allogeneic erythrocyte transfusions. In contrast, genotyping-based diagnostics can be extremely accurate and useful in identifying TPMT-deficient individuals (42). The existing tests employ polymerase chain reaction-based methods, but as more TPMT polymorphisms are validated as predictors of thiopurine efficacy and toxicity, high-throughput array-based genomic technologies may become useful in patient screening.

7.2.6. MTHFR Polymorphisms and Response to Chemotherapy

In some cases, polymorphisms in one gene may be associated with the response to a broad range of anticancer drugs. One of the best studied examples of this phenomenon is the gene encoding 5,10-methylenetetrahydrofolate reductase (MTHFR), an enzyme that catalyzes the conversion of 5,10-methylenetetrahydrofolate to 5-methyltetrahydrofolate, a compound mediating the remethylation of homocysteine to methionine. A change in MTHFR function would alter the intracellular concentration of folates and consequently affect the efficacy and toxicity of antifolate therapeutics, such as methotrexate. More research is needed to establish a firm correlation between polymorphic variants of the MTHFR gene and the enzyme activity of the protein product. Currently, one polymorphism, a $C \rightarrow T$ substitution at position 677, has been associated with the enzymatic activity (homozygotes show activity reduced by 30% relative to heterozygotes) (44). The importance of correlating this polymorphism with drug response is underscored by its high frequency in various populations: It has been detected in 24–40% of Caucasians, 26–37% of Japanese, 11% of African Americans, and 6.6% of Africans. In the Caucasian population, 10–12% of individuals are homozygous for the T allele (45).

It has been demonstrated that in bone marrow transplant patients treated with methotrexate, the TT genotype at position 677 is associated with poor hematologic recovery and increased risk of oral mucositis relative to the CC genotype (46). Since MTHFR activity may affect the toxicity of chemotherapy and because the SNP at position 677 of the MTHFR gene may alter the enzyme activity, genotyping the MTHFR gene in chemotherapy patients represents a potential strategy for identifying biomarkers for a wide spectrum of drugs. Indeed, it has been shown that five of six breast cancer patients who developed severe toxicity in the first cycle of adjuvant chemotherapy with a combination of cyclophosphamide, methotrexate, and 5-fluorouracil had the variant TT genotype (47). To the contrary, the TT genotype was associated with lower toxicity relative to the CT or CC genotypes in a phase I clinical trial of a thymidylate synthase inhibitor, raltitrexed, in combination with irinotecan (48). A potential explanation for this phenomenon is the potential competition between the intracellular folates and raltitrexed for binding to thymidylate synthase (48).

Obviously, more genotyping data are needed to establish firm associations between polymorphisms in the MTHFR gene, but this is clearly an area of great interest, given the high potential utility of a biomarker predictive of toxicity for a wide spectrum of antifolate, fluoropyrimidine, and related therapies. A clear mechanistic connection to drug activity is a factor supporting the use of MTHFR polymorphisms as predictive markers of drug toxicity and efficacy. Additionally, it has been demonstrated that the $C \rightarrow T$ polymorphism at position 677 of MTHFR reduces the risk of cancers, in which folate metabolism plays an important role, namely colon cancer (49) and childhood acute lymphocytic leukemia (ALL) (50). Interestingly, the protective effect of the polymorphism was specific to ALL, and not observed in acute myeloid leukemia (AML), consistent with the fact that methotrexate is effective in ALL, but not AML. This differential effect of the polymorphism in ALL may also justify routine genotyping of the patients in clinical trials for ALL drugs to identify possible associations with the drug response.

7.2.7. Tandem Repeat Polymorphisms in the TS Gene and Response to Drugs Targeting Thymidylate Synthase

Polymorphisms in drug targets are obvious biomarker candidates that have a high probability of affecting the drug mechanism and therefore need to be explored as predictors of drug response. In oncology, several polymorphic drug targets have been explored as genomic markers, of which thymidylate synthase has received the most study. Thymidylate synthase provides thymidylate for DNA biosynthesis by catalyzing the conversion of deoxyuridine monophosphate (dUMP) to deoxythymidine monophosphate (dTMP) (51). Because thymidylate synthase plays a central rate-limiting role in DNA synthesis, it has been actively explored and utilized in cancer therapy. Its inhibition leads to suppression of deoxythymidine triphosphate synthesis, followed by chromosome degradation and cell death. The thymidylate synthase inhibitor raltitrexed is commonly used in the treatment of metastatic colorectal cancer. Inhibition of thymidylate synthase is part of the cytotoxic mechanism of 5-fluorouracil, a drug already discussed in the context of dihydropyrimidine dehydrogenase polymorphisms.

When a polymorphism in the drug target is explored as a possible genetic marker of drug response, two possible questions may be asked: Does the polymorphism affect the expression of the gene? Does the polymorphism result in an amino acid change that may affect the binding of the drug? In the case of thymidylate synthase the first question is particularly relevant because the levels of this enzyme vary significantly in different tumors (52–55). In this case, to firmly establish the mechanistic relationship between a polymorphism and the drug response, two connections need to be confirmed: the association between the polymorphic variant and the expression of the gene and the correlation between protein level in the tumor and the patient response to the drug. The association between the polymorphism and gene expression can be established in model systems such as cell lines or in patient tumors. The thymidylate synthase gene contains a variable number of tandem repeat sequences (28 base pairs) in the 5′-untranslated region (56). Two, three, four, five, and nine repeats have been found in different ethnic groups (56–58). For instance, double (2R) and triple (3R) repeats are mainly found in individuals of Japanese origin (56). For the 3R allele, significant ethnic differences in frequency have been reported (58).

The correlation between the polymorphism and gene expression was established by studying the thymidylate synthase protein abundance in 92 colorectal cancer tissues by the fluoro-dUMP binding assay (57). These values were correlated with the genotypes of the samples determined by a PCR-based assay. It was found that cancer tissues with the 3R/3R genotype have a significantly higher thymidylate synthase protein expression than those with the 2R/3R genotype. As the thymydylate synthase mRNA was measured in the same samples by real-time QPCR, no relation was found between the mRNA expression and the genotype. These data suggest that the genotype-specific variation in thymidylate synthase expression in colon cancer samples is caused by differences at the translational level. These observations from clinical samples were confirmed in an in vitro system, in which thymidylate synthase 5′-untranslated region-luciferase reporter constructs were transfected into HeLa cells. The RNA with the three-repeat sequence was translated three to four times more efficiently than that with two-repeat sequence, confirming the mechanism elucidated in tumor samples (57). The relationship between the thymidylate synthase genotype and the expression of the gene has also been demonstrated in other studies (59).

The association between thymidylate synthase gene expression and disease prognosis and drug response has also been established for several tumor types. Initial studies have demonstrated that high levels of thymidylate synthase expression correlate with poor prognosis in breast cancer (60), gastric cancer (61), and colorectal cancer (62, 63). This has set the stage for investigations of the efficacy/expression and efficacy/genotype relationships for several drugs targeting thymidylate synthase. Overexpression of thymidylate synthase has been associated with poor response to 5-fluorouracil and related folate-based compounds in vitro and in vivo (64–67).

Direct correlation of the thymidylate synthase genotype with drug response has also supported the tandem repeat polymorphism as a predictor of drug efficacy. In a study of 24 patients with metastatic colorectal cancer treated with 5-fluorouracil, the frequencies of the low-expressing 2R/2R genotype were 40% among responders and only 20% among nonresponders (68). A clear correlation between the thymidylate synthase genotype and drug response was observed in another set of 50 colorectal cancer patients: The response rate to 5-fluorouracil was 9%, 15%, and 50% in patients with the 3R/3R, 2R/3R, and 2R/2R genotypes, respectively (68). The survival benefit from 5-fluorouracil adjuvant therapy was lower for high-expressing 3R/3R patients than for 2R/2R or 2R/3R individuals, when compared with surgery alone (69). Finally, intriguing correlations were also observed in other cancers. As metatrexate was administered to children with acute lymphoblastic leukemia, patients with the three-repeat thymidylate synthase variant demonstrated shorter event-free survival than those with the 2*R*/2*R* or 2*R*/3R genotypes (*P* value = 0.005). Thus the data accumulated to date indicate that tandem repeat polymorphisms in the thymidylate synthase gene affect the expression of the gene and thus determine the response to an entire class of chemotherapeutic agents in a wide range of cancers. This undoubtedly justifies large-scale prospective genotyping of patients who would receive drugs targeting thymidylate synthase. Moreover, in drug discovery determination of the tandem repeat polymorphisms would likely produce novel important correlations if genotyping were performed in preclinical and clinical studies for novel agents with the relevant mechanism of action. This and other case studies reviewed in this chapter strongly support the idea of **routine genotyping of the drug targets for all drug discovery programs**. Potential approaches to early discovery of marker polymorphisms in drug targets and other mechanism-related proteins are discussed at the end of this chapter.

7.2.8. Use of Cancer Cell Lines to Identify Predictive SNPs

Analysis of the case studies reviewed above reveals one of the greatest challenges in pharmacogenetic studies: the requirement of a large number of patients for whom the response data need to be collected. Indeed, to achieve statistical significance, associations between the desired response and the phenotype must be studied in a large population of patients. Because of the difficulties in recruiting sufficient numbers of patients or other logistical issues, many pharmacogenetic studies provide intriguing hints at genotype/response associations but do not yield statistically significant correlations. One possible solution to this problem would be to establish reliable correlations in preclinical models and then design clinical trials based on the pharmacogenetic data obtained. Most preclinical systems in oncology are reproducible with respect to drug sensitivity, and for many cancer types multiple cell lines are available, so that a sufficient number of responses can be obtained.

A recent study (70) utilized the NCI-60 panel, the collection of 60 human tumor cell lines assembled by the NCI in the early 1990s to serve as a primary screen for the antiproliferative activity of thousands of compounds (71). This panel is ideally suited for a proof-of-concept in vitro pharmacogenetic study, because cytotoxicity data have been already accumulated for several hundred compounds, including the core anticancer drugs with a known mechanism of action. To establish the feasibility of using the NCI-60 panel for pharmacogenetic studies, three genes were selected whose polymorphisms have already been associated with the tumor response to alkylating agents: two DNA repair enzymes belonging to the nucleotide excision repair (NER) pathway, *ERCC2* (XPD) and *ERCC5* (XPG), and the drug detoxification enzyme *GSTP1*. These polymorphic variants were chosen because they are frequent and are known to correlate with cancer risk and drug efficacy. Additionally, the expression of *ERCC2* had been shown to be associated with response to therapeutics in the NCI-60 panel (72); *ERCC5* deficiency had been found to occur in cell lines selected for resistance to a novel alkylating agent, ecteinascidin-743 (73). Finally, GSTP1, like other GSTs, is an established factor in cytotoxicity of cancer drugs (74). As various polymorphisms in these three genes were determined in the NCI-60 panel, it was demonstrated that a Lys751 → Gln polymorphism in the *ERCC2* gene is strongly associated with the cytotoxicity of many anticancer drugs, especially the spindle poisons, whereas an Asp1104 → *His* polymorphism in ERCC5 and a Ile105 → Val substitution in the *GSTP1* gene correlate with the in vitro cytotoxicity of several other classes of drugs (70). These results have established the possibility of correlating polymorphisms with drug sensitivity in a characterized panel of model cell lines. It remains to be seen whether the associations observed in the panel will reproduce in patients, but these data are encouraging also because they open the way to more extensive genotyping of the NCI-60 cell lines. One could envision that a genome-wide analysis of SNPs as predictors of drug response in the NCI-60 panel would generate important hypotheses on genotype/response associations.

7.3. POLYMORPHISMS AND RESPONSE TO ANTICOAGULANTS

Discovery of polymorphisms as stratification biomarkers has also been productive in therapeutic areas other than oncology. One of the most common therapies in clinical practice is oral anticoagulation, which is used to prevent and treat patients with arterial and venous thrombosis. However, this therapy is often complicated by bleeding, emphasizing the importance of anticoagulant dose optimization. In anticoagulation therapy with warfarin, two particular challenges, namely, the narrow therapeutic range of the drug and large interpatient variability in the dose required, have stimulated interest in possible genetic determinants of warfarin efficacy. Warfarin works by inhibiting vitamin K epoxide reductase (VKOR), preventing vitamin K recycling and the subsequent gamma-carboxylation of clotting factors. Recently a number of common polymorphisms have been identified in the gene coding for VKOR (*VKORC1*) and associated with variable warfarin dose requirements.

Initially an attempt was made to correlate the *VKORC1* haplotypes, VKORC1 mRNA concentrations in liver, and the required warfarin dose in African American, European American, and Asian American populations (75).

Ten common noncoding SNPs were identified in the VKORC1 gene. Low-dose and high-dose haplotype groups were identified; *VKORC1* haplotype groups A and B explained approximately 25% of the variance in dose. Asian Americans had a higher proportion of group A haplotypes, and African Americans a higher proportion of group B haplotypes. Thus it was demonstrated that *VKORC1* haplotypes can be used to stratify patients into low-, intermediate-, and high-dose warfarin groups and may explain differences in dose requirements among patients of different ancestries. A relationship was established between the gene variants and the VKORC1 mRNA expression, implying that the molecular mechanism of the warfarin dose response is regulated at the transcriptional level.

In a separate study, 147 patients were followed from the start of warfarin therapy to investigate whether *VKORC1* gene mutations affected doses of drug prescribed to acquire the target anticoagulation intensity. Two common polymorphisms were identified, an 1173C → T transition in intron 1 and a 3730G → A transition in the 3′-UTR. The 1173TT genotype was associated with a dose of warfarin significantly lower than that for the CC or CT genotypes. The 3730G → A polymorphism also correlated with the average dose of warfarin prescribed: patients carrying the GG genotype received a significantly lower average daily dose. No mechanistic explanation for the effects of the polymorphism on warfarin sensitivity has been found. The 1173C → T polymorphism did not affect the splicing of VKORC1 mRNA (76).

The 3730G → A polymorphism was also detected in an independent study of patients receiving warfarin (77), supporting its association with the sensitivity to the drug. Additionally, a warfarin-relevant SNP was detected in the promoter region of the VKORC1 gene (1639G → A). Its homozygous form (genotype AA) was detected in 11 of 11 warfarin-sensitive patients. In the resistant patients ($n = 5$), the 1639 genotype was either AG or GG. This association was tested in a separate validation set of randomly selected patients receiving warfarin. In this population, the AA genotype at position 1639 also had a lower dose than the AG/GG genotype (P value < 0.0001). As interethnic variability in dose requirements is an important factor in warfarin treatment, the frequencies of the AA, AG, and GG genotypes at the 1639 polymorphic site were compared in Chinese and Caucasian patients receiving warfarin (77). These frequencies differed significantly (79.7%, 17.6%, and 2.7% in Chinese patients vs. 14%, 47%, and 39%, respectively, in Caucasians; P value < 0.0001), implying that the differences in warfarin sensitivity between the ethnicities can be explained at least partly by the VKORC1 1639 genotype. To determine the mechanism whereby the 1639 polymorphism in VKORC1 affects sensitivity to warfarin, the relationship between the 1639 genotype and the VKORC1 promoter activity was explored. The polymorphic site represents the second nucleotide of an E-box (CANNTG), and therefore the 1639 polymorphism could change the promoter activity by altering the E-box consensus sequence. To explore this possibility, the VKORC1 promoter encompassing the 1639 A or G was PCR-amplified from patients with the AA or GG genotype at 1639 and cloned. The promoter activity was then tested by luciferase assays in a human hepatoma cell line (the highest level of VKORC1 gene expression is found in the liver). The 1639 G VKORC1 promoter displayed an ~44% higher luciferase activity compared with the 21639 A promoter, thus confirming the hypothesis that the 1639 polymorphism affects warfarin sensitivity by altering VKORC1 gene promoter activity.

The G/A SNP at the position 1639 was also assessed in a large study aimed at assessing the combined contribution of polymorphisms in genes whose products are involved in the pharmacodynamics and pharmacokinetics of warfarin, including VKORC1, the coagulation factors, and *CYP2C9* (78). Because of the differences in nucleotide numbering (i.e., numbering based on translation start site rather than based on the *VKORC1* reference sequence), this SNP was referred to as 3673. Among various genetic and nongenetic factors, the *VKORC1* 3673 G/A polymorphism was the strongest predictor of warfarin dose in a population of therapeutically stable patients receiving warfarin. Subjects with the A/A and G/A genotypes required 50.4% and 29.4% lower weekly doses of warfarin, respectively, compared with those with the wild-type G/G genotype. Linear regression analysis revealed that the *VKORC1* genotype is responsible for the largest portion of dose variability.

The example of warfarin response biomarkers convincingly demonstrates the utility of polymorphisms in drug targets in patient stratification. In this case, researchers specifically searched for polymorphisms in the target of the drug, while known SNPs in the drug metabolism genes (such as *CYP2C9*) had already been associated with the warfarin dose requirements. As discovery of targeted agents becomes the prevailing paradigm in drug discovery, one can argue that all discovery programs should be accompanied by complete resequencing of the gene coding for the drug target in relevant populations, followed by genotype/response association studies in clinical trials. This approach would enable early identification of polymorphisms correlating with response to the clinical candidate, facilitating early patient stratification and guided enrollment in late-phase trials.

7.4. POLYMORPHISMS IN NEUROSCIENCE

In many disorders, however, the current understanding of molecular mechanisms is insufficient to focus SNP identification efforts on defined molecular targets. Nevertheless, progress in identification of SNP biomarkers has been achieved even in such complex disorders as depression and schizophrenia (79). Several hypotheses exist on the mechanism of depression. One of these, the monoamine theory, suggests that depressive symptoms result from disruptions in the serotonergic and noradrenergic systems. Therefore, the serotonin transporter (SERT) has become a pharmacological target for treating depression. This has prompted significant interest in genetic polymorphisms in the *SERT* gene as possible correlates of efficacy for antidepressants. Initially, a polymorphism was discovered in the promoter region of the human *SERT* gene that resulted in a deletion of a 44-base pair repeat (80). Follow-up studies using luciferase-promoter fusion constructs in JAR cells demonstrated that the promoter with the deletion is three times less potent than the intact promoter.

Moreover, levels of PKC- and cAMP-activated transcription were significantly lower for the shorter variant promoter (80). As the frequencies of the variants were determined in a population of 505 subjects, it was found that 19% of the individuals were homozygous for the Deleted/Deleted (D/D) genotype, whereas 49% were Deleted/Intact (D/I) and the remaining subjects were Intact/Intact (I/I) (81). Lymphocytes from individuals with different variants of the SERT promoter were used to determine the promoter transcriptional activity. The SERT protein expression was increasing in the following sequence: I/I → I/D → D/D, implying higher activity of the promoter with the deletion (81).

Because the mechanistic studies described above supported the hypothesis that the SERT promoter polymorphisms are important factors in depression, their association with the efficacy of antidepressants became a subject of active research. Initially, the genotype of the SERT promoter was determined in 51 depression patients and correlated with the efficacy of fluoxetine on various symptoms of depression (82). It was shown that patients with the D/D genotype of the SERT promoter have a higher Hamilton depression score after treatment with either fluoxetine or placebo, indicating poor response to the drug. An independent study of Chinese patients with depression demonstrated that individuals with the I/I genotype of the SERT promoter have a greater decrease in the Hamilton score after fluoxetine therapy, compared with patients with I/D and D/D genotypes (83). Additionally, the I/I patients responded to the drug more quickly than their counterparts with the variant genotypes.

In biomarker studies, validation of a marker in an independent cohort of patients is an important step. In the case of the SERT promoter polymorphism, the association of the variant promoter with drug response was confirmed in a separate study using a different antidepressant (paroxetine), adding validity to the correlation (84). In a population of elderly patients with depression, individuals with the I/I genotype of the SERT promoter showed lower Hamilton depression scores compared with their counterparts with the variant genotypes, implying a faster onset of the therapeutic effect. Importantly, a negative control arm was included in this study, in which patients were treated with nortriptyline, an antidepressant not selective to SERT. In this group, there was no difference in the timing of drug response between the patients with normal and variant SERT promoter genotypes, suggesting that the correlation is specific to SERT-selective antidepressants and thus supporting the mechanism-based relationship. A faster rate of onset for response in patients with the I/I SERT promoter was also observed in a study of 176 elderly patients treated with another relevant antidepressant, sertraline, whereas no differences in the response were seen in the placebo group (85). Although the data from the aforementioned studies are convincing, more research is necessary to correlate the SERT promoter genotype, SERT mRNA and protein expression, protein function, and the mechanism of the antidepressants. A strong mechanistic link between these and other polymorphisms and the response to antidepressants would facilitate the assessment of predictive value of the polymorphism-based biomarkers.

Identification of polymorphisms affecting therapy with antischizophrenia therapeutics is an equally important task, because schizophrenia patients exhibit a high degree of variability in drug response. Pharmacological treatment of this disease typically includes typical and atypical antipsychotic drugs targeting dopamine (DA) and serotonin (5-hydroxytryptamine, or 5-HT) G protein-coupled receptors (GPCRs). Unfortunately, 30–60% of schizophrenia patients do not respond to atypical antipsychotic therapy (86), while those who respond frequently experience side effects that affect compliance and/or adverse events that may be life threatening (87). This variability has stimulated interest in polymorphisms in the relevant drug target genes that may be associated with the response to antipsychotics. Several preliminary studies have been run to investigate the relationship between known SNPs in the 5-HT2a receptor, but their clinical relevance remains to be established (87, 88). A recent in vitro pharmacology study examined the affinity to antipsychotics and functional activity for the known variant of 5-HT2a (88). It was shown that the effect of individual SNPs were drug-specific and that none of the known SNPs was associated with the effects of all the drugs. This finding underscores the complexity of biomarker identification for psychotropic drugs and implies the necessity to design separate clinical studies to address particular drug/polymorphism combinations.

7.5. POLYMORPHISMS AND DRUG RESPONSE IN IMMUNOLOGY

Rheumatoid arthritis (RA) is a chronic inflammatory disease characterized by progressive destruction of synovium and cartilage in the joints. A critical factor in the pathogenesis of this disease is a cytokine called tumor necrosis factor (TNF)-α. Several therapeutics have recently been developed and approved that target TNF-α, including etanercept, infliximab, and adalimumab. The introduction of these biologicals represents a breakthrough in the treatment of RA, as these agents effectively slow the progression of the disease. However, only approximately 60% of patients respond to etanercept, infliximab, and adalimumab. Additionally, these biologics are very expensive, and the treatment is typically lifelong. Therefore, there is a clear need for reliable stratification markers for selecting patients most likely to benefit from therapy with TNF-α antagonists.

This has prompted a search for polymorphisms within the TNF-α gene and other genes related to the mechanism of TNF-α antagonists (89). Several polymorphisms have been identified within the coding sequence and the promoter of the TNF gene. The best-studied TNF polymorphic sites are at positions −308 and −238 in the promoter of the gene and at position +489 in an intron of the gene (89). A number of studies attempted to correlate the presence of these polymorphisms with TNF-α gene expression and the clinical course and severity of RA, but the results were somewhat contradictory (89). Since promoter sequence variation may affect binding of transcription factors and consequently the expression levels of the gene, particularly strong focus has been on the SNPs at positions −308 and −238. In particular, the −308 SNP was correlated to the TNF-α production in 57 healthy individuals (90). The production of TNF-α after stimulation with lipopolysaccharides was evaluated with a whole blood cell culture model. In the studied population, there were 41 TNF1 homozygotes and 16 TNF1/TNF2 heterozygotes. The production of TNF-α *ex vivo* after lipopolysaccharide stimulation of whole blood cell culture was significantly higher among TNF2 carriers than among TNF1 homozygotes [929 pg/ml (480–1473 pg/ml) vs. 521 pg/ml (178–1307 pg/ml); P value < 0.05], implying that the polymorphism at position −308 is associated with TNF-α production in healthy individuals. The intronic SNP at position +489 has also been tested for association with the clinical course of RA (91). Genotyping of the TNF-α gene by polymerase chain reaction-restriction fragment length polymorphism was performed for a total of 163 patients, of which 66 had severe disease, and 67 healthy individuals. The severe disease was defined as active disease nonresponsive to disease-modifying antirheumatic drug combination therapy. The SNP at position +489 was found to correlate with the severity of the disease. The AA genotype at this position was very rare in patients with moderate to severe disease (91). In addition to the aforementioned SNPs, the TNF locus also contains other types of polymorphisms. In particular, several so-called DNA microsatellites have been mapped to the TNF gene. These are repeat A and T sequences that reside in noncoding sequences and have no assigned function. However, DNA microsatellites are highly polymorphic, and the number of microsatellites may affect DNA folding and conformation and thus alter the expression level of the gene. In vitro studies have demonstrated that microsatellites in the TNF locus may influence TNF-α production by peripheral blood monocytes. Specifically, two microsatellites (TNFd and TNFa2) have been shown to correlate with increased production of TNF-α in vitro, while microsatellite TNFa6 has been associated with low concentrations of TNF-α (92).

As polymorphisms in the TNF receptor genes may also affect the severity of RA, a number of studies have addressed genetic variation in TNFRSF1A and TNFRSF1B. A nonsynonymous SNP has been identified in exon 6 of the TNFRSF1B gene (T → G at codon 196) that results in the substitution of arginine for methionine in an extracellular domain of the receptor (93). The polymorphism is associated with increased levels of interleukin-6 and altered membrane receptor shedding and ligand binding.

Building on the reported associations of polymorphisms with the clinical course of RA, a number of studies have been run to evaluate known SNPs as predictive biomarkers of response to TNF-α antagonists. The aforementioned SNP at position −308 was tested for predictive associations in a population of 59 patients with RA (94). The patients were classified into two categories by the genotype: A-allele carriers (including A/A and A/G genotypes) and non-A-allele carriers (G/G genotype). After 22 weeks of treatment with infliximab, the clinical response to the drug was evaluated by measuring the standard RA disease activity score (DAS, composite score that includes clinical and laboratory parameters). In the A-allele carrier group, 42% of individuals showed an improvement in DAS of 1.2 (P value = 0.009), while in the non-A-allele carriers, 81% of patients showed an equivalent improvement. Average improvement in DAS in 28 joints (DAS28) was 1.24 in the A/A and A/G patients and 2.29 in the G/G patients (P value = 0.029), a clinically significant difference. Thus the results obtained suggest that the G/G allele at position −308 of the TNF-α gene is a predictive marker of response to infliximab.

To elucidate the mechanistic basis for this correlation between the −308 TNF-α genotype and the response to infliximab, the relationship between the polymorphism and TNF-α levels was examined in patients with RA. Groups of 10 individuals with the G/A genotype at position −308 of TNF-α and 10 individuals with the G/G genotype were treated with infliximab (95). There was no difference in drug response between the two genotypically different groups when standard disease activity parameters such as the ACR20 and 50 response rates were measured. In view of previous findings on the increase of TNF-α levels after administration of an anti-TNF-α monoclonal antibody (95), the TNF-α levels were measured in the groups under study. It was found that treatment with infliximab increased the TNF-α levels in both groups, but this increase correlated with the ACR50 response only in individuals with the G/A genotype (P value < 0.03). This led to the hypothesis that that the −308 polymorphism could affect the response to infliximab by altering the levels of circulating TNF-α (95).

Polymorphisms in the TNF-α receptor genes can also affect the response to TNF-α antagonists. No conclusive results have been reported on the predictive power of the known TNF-α receptor polymorphisms, but preliminary studies indicate that a correlation may exist for the 196 T/G polymorphism of TNFRSF1B (96). A cohort of 175 Caucasian patients with RA was genotyped, which included 97 individuals with mild to moderate disease and 78 individuals with severe disease, who did not respond to methotrexate. In the latter group, 66 patients were treated with TNF-α antagonists (infliximab or etanercept) and monitored by measuring the DAS. The genotypic distribution in this group was as follows: TT (38 patients), TG (22 patients), and GG (6 patients). The frequency of the GG genotype at position 196 was higher in individuals with severe disease (6.4%) than in patients with mild to moderate disease (3.1%). The difference was not statistically significant, implying that the genotype may not correlate with the clinical course of the disease. However, when associations with TNF-α antagonist therapy were explored, it was found that patients with the TT variant responded better to the drugs than patients with the TG or GG genotypes. The difference in response, as measured by the change in the DAS, was most pronounced after 12 weeks of therapy (odds ratio 5.1; confidence interval 1.3–19.96; P value = 0.03), suggesting that the presence of a G allele is associated with poor response to TNF-α antagonists. The results reported suggest that the TT polymorphism may serve as a stratification marker for response to TNF-α antagonists, but the association with drug response needs to be tested and validated in larger populations of patients.

If therapy with TNF-α antagonists is placed in a broader context of treating an inflammatory disease, it seems logical that polymorphic variants of genes relevant to the inflammation processes may also affect the response to anti-TNF-α drugs. In particular, genes encoding cytokine and cytokine receptors appear to be good candidates for predicting disease progression and therapy response. In a recent study (97), 123 patients with RA were genotyped to determine the presence of the following polymorphisms:

- TNF-α gene: G/G, A/G, and A/A at position −308
- IL10 gene: G/G, A/G, and A/A at position −1087
- TGF-β1 gene: G/G and G/C (codon 25)
- IL1 receptor antagonist intron 2: A1/A1, A1/A2, A2/A2, A1/A4, and A2/A4 (variable nucleotide tandem repeats)

The patients were treated with etanercept, and the response to the drug was evaluated by measuring the ACR parameters of disease activity and the DAS28. In this population, the evaluation identified 99 patients as responders and 24 as nonresponders. Assessment of the frequencies for each of the individual polymorphisms has revealed no significant differences between the responder and nonresponder groups. However, following the logic of evaluating a composite panel of polymorphisms in relevant genes involved in inflammatory processes, the aforementioned variants can be considered in combinations and the resulting composite markers can be assessed for associations with the drug response.

When the individual polymorphisms were assessed in combinations, one particular genotype (TNF −308 G/G–IL10 −1087 G/G) was associated with response to the drug (P value < 0.05). There was only one patient with this genotype that did not respond to etanercept, but continued treatment for 3 more months induced response. This genotype can be characterized as a "weak inflammatory response" marker because patients with this combination of variant alleles had lower predisposition to inflammatory response (89, 97). In contrast, patients with the "strong inflammatory response" genotype (C allele in codon 25 of TGFB1 and A2 allele in intron 2 of IL1RN) showed poor response to the therapy (P value < 0.05). Thus analysis of composite polymorphic markers associated with inflammation demonstrated that patients genetically predisposed to weak immune response are more likely to benefit from etanercept than those predisposed to strong immune response. This approach represents a potentially very powerful strategy in the discovery of polymorphisms as patient stratification biomarkers. If one can identify combinations of polymorphisms associated with a particular phenotype, a relationship with the drug's mechanism can be explored, with the eventual goal of identifying a composite biomarker predictive of response to the therapeutic. Alternatively, a more empirical approach would involve screening patients for all known polymorphisms in the genes relevant to the disease mechanism, followed by statistical analysis to assess all possible combinations of individual polymorphisms for associations with sensitivity or resistance to the drug. With the recent improvement in the throughput of the genotyping technologies and the easy access to the information on polymorphisms in the human genome, this biomarker identification strategy can significantly advance the drug development process. Currently, hundreds of thousands of SNPs can be profiled on one microarray in dozens of samples in less than a week, enabling generation of extremely complex patterns of potential markers. The key step in the subsequent analysis would be to correctly identify SNPs relevant to the disease process and filter out less relevant genes. Multiple statistical approaches common for gene expression analysis, copy number profiling, and genotyping can be then used to identify patterns of individual SNPs that may predict response to the drug. If such composite biomarkers are identified early in the clinical trials, they can be validated and used to guide selection of patients for late-stage trials who are more likely to respond to the drug, thus improving the response rates and reducing the duration and cost of clinical trials. Additionally, retrospective analysis of samples from failed clinical trials may resurrect failed drugs by clearly defining populations in which the drug has produced sufficiently high response rates.

Because of the complexity of the inflammation mechanisms, a comprehensive approach to targeted marker discovery should involve searching for polymorphisms in multiple inflammation-related genes, including the genes coding for major histocompatibility complex (MHC) components. An important argument for possible associations between MHC alleles and susceptibility to TNF-α antagonists is the close proximity of the TNF locus to the HLA B and HLA DR genes on chromosome 6. Additionally, a strong correlation has been established between specific HLA DRB1 alleles (referred to as shared epitope alleles) and susceptibility to RA and its severity (98). Therefore, polymorphisms in the MHC genes have been actively explored as potential markers of sensitivity to TNF-α antagonists. Initially, PCR- and hybridization-based methods were applied to genotype 78 patients with RA undergoing treatment with infliximab (99). The list of determined polymorphisms included HLA-DRB1, HLA-DQA1, and HLA-DQB1 alleles, a trinucleotide repeat polymorphism within the MHC class I chain-related gene A (*MICA*), TNF microsatellites a through e, D6S273, HLA-B-associated transcript 2 (BAT2), and D6S2223. The TNFd microsatellite has been included because of its association with high concentrations of TNF-α in vitro. Additionally, the microsatellites profiled had either been associated with susceptibility to RA or correlated with SNPs in the promoter region of the TNF-α gene. For instance, the TNFa6;b5;c1;d3;e3 haplotype confers the risk of an increased susceptibility to RA (100), and the TNFa2;b3;c1;d1;e3 haplotype is linked to the −308 TNF polymorphism (101). The BAT2 and D6S273 are HLA class III microsatellites, whereas the D6S2223 is a microsatellite marker located telomeric to the HLA class I genes. χ^2-Tests were performed to compare allele proportions between responder and nonresponder patients.

Patients were categorized into a responder group based on either a 25% improvement in the DAS or a 50% improvement in at least two of the following criteria: number of swollen or painful joints affected, analytic data, erythrocyte sedimentation rate and/or C-reactive protein level, and scores on the HAQ and VAS. To detect linkage disequilibrium between pairs of markers, 342 healthy control subjects were also profiled. When single alleles were tested for association with response to the drug, no correlations were seen. However, assessment of microsatellite haplotypes, particularly in combinations, revealed possible predictive markers. Specifically, the combination of D6S273_4 and BAT2_2 was detected in 46% of responders and only 11% of nonresponders (*P* value = 0.001), implying that it may serve as a predictive marker of infliximab sensitivity. Two potential explanations were offered for this association. First, the two polymorphic markers may be on the same haplotype, and this associated haplotype may carry an unknown gene associated with the response to the drug. Alternatively, each of the two microsatellite alleles may represent a marker of a distinct neighboring gene. Indeed, if this were the case, a combination of these two genes (either in *cis* or in *trans* orientation) would provide a new phenotype, potentially with a different drug sensitivity. The TNFa11;b4 minihaplotype was more frequent in drug responders (41% vs. 16% in nonresponders; *P* value = 0.01), while the D6S273_3 allele was less frequent (32% in responders vs. 56% in nonresponders; *P* value = 0.04). The frequencies of the MICA and D6S2223 polymorphisms and the HLA-DRB1, HLA-DQA1, and HLA-DQB1 alleles did not reveal any statistically significant differences between the responder and nonresponder groups (99).

Recently, a number of putative genetic predictors of response were tested in a large prospective trial of TNF-α antagonists (102). A total of 457 patients with early RA (disease duration of <3 years) were enrolled in a randomized controlled trial comparing weekly methotrexate and two dosages of etanercept (10 mg twice weekly and 25 mg twice weekly). The response was defined as a 50% decrease in ACR50 after 12 months of treatment. The ACR50 parameter is a categorical measure standardized by the American College of Rheumatology that indicates whether or not a subject has achieved a 50% improvement in a set of standardized outcomes. The patients were genotyped for HLA-DRB1 alleles as well as for polymorphisms in the following genes: *TNF*, *LTA*, *TNFRSF1A*, *TNFRSF1B*, *FCGR2A*, *FCGR3A*, and *FCGR3B*. To identify potential correlations between the polymorphisms and the response to treatment, univariate and multivariate statistic analyses were performed. In the multivariate analyses the following covariates were used: sex, ethnicity, age, disease duration, and baseline levels of the rheumatoid factor and the number of tender and swollen joints. Patients were genotyped for specific HLA-DRB1 alleles, the shared epitope (SE) alleles, and categorized as having 0, 1, or 2 copies of the SE. The SNPs at positions −308, −238, and +488 of the TNF-α gene and positions +249, +365, and +720 of the LTA gene were determined. These six SNPs were selected for analysis because they are known to define haplotypes spanning the region. Additionally, the proximity of TNF-α and *LTA* genes to the HLA-DRB1 locus potentially associates polymorphic variants in these genes with response to etanercept, which binds to both TNF and LTA. The *LTA–TNF* region (TNFa, b, c, d, and

e) contains five microsatellites, which were also genotyped. Since variation in the TNF-α receptors may affect the activity of TNF-α antagonists, SNPs in the genes encoding TNF receptors TNFRI (*TNFRSF1A*; positions −609, −580, −383) and TNFRII (*TNFRSF1B*; 1 SNP at amino acid 196) were also determined. Finally, etanercept–TNF complexes are degraded through the Fc receptor (FcR) pathway, and FcR may be associated with the activity of immunoglobulin-based biologics (89). Therefore, several polymorphisms in the FcR gene were genotyped, namely, *FCGR2A* H131/R131, *FCGR3A* 176F/176V (nucleotide 449 G/T), and *FCGR3B* NA1/NA2. It was shown that the HLA-DRB1 alleles, but not SNPs in TNF, LTA, TNFRSF1A, TNFRSF1B, and FcR genes, correlate with response to etanercept treatment. Response to the drug was better in patients with two copies of the SE alleles, compared with individuals without or with one a copy of the allele (OR = 4.3; 95% *CI* = 1.8–10.3).

The statistical analysis was restricted to a subset of 224 Caucasian patients. Haplotypes defined by the six LTA–TNF SNPs and DRB1 alleles were deduced for the 16 most common DRB1 alleles. Two particular haplotypes, each representing approximately 4% of all haplotypes, were associated with a high percentage of reponders to treatment (61% and 76% of patients showed an ACR50 response at 12 months) in all treatment groups combined. These two haplotypes were very similar with respect to the six *LTA–TNF* alleles. One of these haplotypes included HLA–DRB1*0404, and the other included DRB1*0101, both of which encode the shared epitope.

It was thus demonstrated that polymorphisms in the MHC region, such as the number of copies of HLA-DRB1 SE alleles inherited, and specific haplotypes spanning the HLA-DRB1 region and SNPs in the LTA–TNF region, can be used to predict response to the TNF-α antagonist etanercept (102). The association is, however, limited to the Caucasian population. Nevertheless, the results of this study demonstrate that genetic predictors of response may be found among polymorphic variants of genes not directly implicated in the mechanism of the drug but related to the disease mechanism through the function of their protein products or through genetic links.

7.6. POLYMORPHISMS AND RESPONSE TO ANTIVIRAL AGENTS

7.6.1. Anti-HIV Drugs

Pharmacological treatment of the HIV infection consists of long-term administration of various combinations of antiretroviral medications. The currently used drugs have reduced HIV mortality and improved the quality of life for AIDS patients, but their use has been associated with significant toxicity. Because toxic responses are often variable and correlate with with the patient's genotype, multiple studies have been conducted to identify predictors of toxicity for antiretroviral agents used in HIV management. Additionally, a number of polymorphisms have been correlated with the risk of HIV infection and the rate of disease progression, stimulating interest in genetic correlates of drug efficacy.

Variants of the major histocompatibility complex (MHC) genes associated with hypersensitivity to abacavir are good examples of polymorphisms predictive of toxicity of antiretroviral therapy. Abacavir is a nucleoside-based inhibitor of HIV reverse transcriptase that displays potent activity against HIV type 1. It is usually well tolerated, but up to 9% of Caucasian patients develop acute idiosyncratic reactions to the drug, potentially with a fatal outcome (103). If drug administration is discontinued the symptoms typically improve within 24 hours, but rechallenge with the drug usually leads to recurrence of the hypersensitivity symptoms within hours, and the risk of life-threatening hypotension increases significantly. Since predisposition to this hypersensitivity is heritable, a search was initiated for genetic factors that may predispose an individual to this type of drug hypersensitivity. Because evidence existed in favor of the involvement of MHC-restricted presentation of drugs in similar drug sensitivity reactions, initial studies interrogated polymorphic variants of MHC for possible associations with abacavir hypersensitivity. The MHC region was genotyped in a cohort of 200 Australian patients with HIV exposed to abacavir (104). Hypersensitivity to the drug was detected in 18 patients, while 15 other individuals exhibited some symptoms but did not meet the criteria for hypersensitivity. The remaining 167 patients were considered abacavir-tolerant because they did not develop the hypersensitivity reaction after more than 6 weeks of exposure to abacavir. Several polymorphic MHC variants were identified in this patient cohort and associated with hypersensitivity to abacavir. Specifically, HLA-B*5701 was present in 14 (78%) of the 18 hypersensitive patients, but only in four (2%) of the 167 tolerant patients [odds ratio 117 (95% confidence interval 29–481), P value < 0.0001]. A combination of HLA-DR7 and HLA-DQ3 was found in 13 (72%) hypersensitive and five (3%) tolerant patients [73 (20–268), P value < 0.0001]. A combination of HLAB*5701, HLA-DR7, and HLA-DQ3 was detected in 13 (72%) hypersensitive patients and none of the tolerant patients [822 (43,675), P value < 0.0001]. Within the entire abacavir-exposed cohort ($n = 200$), the presence of HLA-B*5701, HLA-DR7, and HLA-DQ3 had a positive predictive value for hypersensitivity of 100% and a negative predictive value of 97%.

In an independent study, 85 clinical trial subjects with abacavir hypersensitivity and 115 control patients who had tolerated abacavir for at least 6 weeks were assessed for multiple markers in the vicinity of HLA-B that had previously been associated with abacavir hypersensitivity (105). The patients were matched whenever possible for age, sex, race, CD4 count, and treatment status (naive or experienced). The HLA-B57 variant was detected in 39 (46%) of 84 patients, but only in four (4%) of 113 controls (P value < 0.0001). Recursive partitioning and conditional logistic regression have revealed that the presence of the HLA-B57 genotype is the most significant predictor of hypersensitivity to abacavir (P value < 0.00001).

Finally, fine recombinant genetic mapping was used in an expanded study of a cohort of 248 Western Australian HIV patients exposed to abacavir, of which 18 (7.3%) were hypersensitive to the drug and 230 were fully tolerant. It was shown that a combination of the HLA-B*5701 genotype and a haplotypic polymorphism of Hsp70-Hom is an excellent predictor of abacavir hypersensitivity. The statistical data obtained indicate that the prospective use of a genetic test to interrogate HLA-B*5701 alone or in combination with the Hsp70-Hom M493T variant in the cohort under study would decrease the occurrence of abacavir hypersensitivity from 8% to 0.4%. The use of a test for HLA-B*5701 alone would result in inappropriately denied access to the drug for 1.6% of the tested population. Assessment of HLA-B*5701 and the Hsp70-Hom M493T variants would reduce this percentage to 0.4%. These data indicate that prospective genetic testing for the relevant polymorphic variants would be highly predictive of abacavir hypersensitivity and thus effectively stratify patients for therapy with the drug, in particular in populations of European descent, in which the HLA-B*5701 allele is relatively common.

As the study of the MHC genotypes as predictive markers for abacavir hypersensitivity was pioneered in Australia, genetic testing before prescribing abacavir has quickly become common in that country. After the discovery of the predictive power of the HLA-B*5701 marker in a study population in 2001, a prospective testing strategy was initiated to stratify patients into low-risk and high-risk groups on the basis of predicted probability of developing hypersensitivity to abacavir (<1% and >70%, respectively) (104). Abacavir prescription was avoided in HLAB*5701–positive patients in Western Australia from 2002 onward (106). The cost-effectiveness of prospective HLAB*5701 genotyping was determined with a probabilistic decision analytic model that incorporated the cost of managing abacavir hypersensitivity as well as the cost of alternative antiretroviral medications for patients with high probability of abacavir hypersensitivity (107). By utilizing Monte Carlo simulations, the model clearly demonstrated that prospective genotype testing followed by stratification for abacavir therapy is a cost-effective use of health care resources.

In summary, the use of the newly discovered genomic markers has substantially reduced the incidence of hypersensitivity reactions to abacavir (106), thus establishing a successful precedent for introduction of a polymorphism into clinical use as a genomic stratification biomarker. While in this particular case the discovery was made when the drug was already in widespread clinical use, one of the key biomarker discovery studies was conducted with samples from the GlaxoSmithKline database of clinical trials for abacavir, suggesting that the drug development process can be optimized by stratifying patients based on potential toxicity.

Another advanced example of polymorphisms as genetic markers in HIV therapy is the association between a well-known UDP glucuronosyltransferase 1A1 (UGT1A1) polymorphism and hyperbilirubinemia induced by protease inhibitors. The defect in bilirubin conjugating activity known as Gilbert syndrome was introduced earlier in this chapter, when polymorphic markers for anticancer drug efficacy were discussed. It is caused by a TA insertion into the sequence of the UGT1A1 promoter (9). The polymorphic genotype associated with Gilbert syndrome causes a 50% reduction in the bilirubin conjugating activity (10). As HIV protease inhibitors atazanavir and indinavir compete with bilirubin for binding to UGT1A1, they frequently cause unconjugated hyperbilirubinemia, which occasionally leads to discontinuation of treatment because of jaundice. It was hypothesized that this side effect is caused by indinavir-mediated suppression of UGT and hence would be most pronounced in individuals with reduced hepatic enzyme levels (Gilbert syndrome patients) (108). This hypothesis was tested in a rat model of UGT deficiency and in HIV-infected patients with and without Gilbert syndrome. It was shown that indinavir competitively inhibits UGT enzymatic activity and induces hepatic bilirubin UGT mRNA and protein expression (108). In heterozygous rats, oral indinavir increased plasma bilirubin levels more significantly than in wild-type rats. The findings from animal studies were extended to humans by analyzing a series of HIV-infected patients treated with indinavir. A total of 15 male patients receiving the drug were screened for the TATA box polymorphism in the UGT1A1 gene, and the genotype was correlated with the serum bilirubin concentration before and after the treatment with indinavir. In patients with Gilbert syndrome, treatment with indinavir resulted in an approximately fivefold higher increase in serum bilirubin than in patients with the wild-type UGT1A1 genotype. The data revealed a strong association between the presence of the polymorphic allele and the risk of developing hyperbilirubinemia from indinavir treatment, suggesting that a genetic test for the UGT1A polymorphism could be used to stratify patients for treatment with indinavir.

Thus predictive markers have been identified for several anti-HIV therapeutics of different classes. Notably, some of these are already used in clinical practice to guide treatment decisions. While many of the associations discovered for currently marketed drugs remain to be validated in larger patient cohorts, the reviewed studies have important implications for drug discovery. The examples described above underscore the importance of prospective genotyping of drug targets and all disease-related genes at early stages of drug discovery and development, as an early identification of polymorphic variants affecting drug efficacy or toxicity could substantially improve the drug development process.

7.6.2. Interferon Therapy in Hepatitis B Treatment

Interferon, together with nucleotide analogs, is currently the mainstay of hepatitis B treatment. It is estimated that interferon therapy controls infection in only about one-third of chronically infected individuals (109). This variability has prompted researchers to investigate the genetic basis for response to interferon. A novel approach to identification of polymorphic markers predictive of drug response has been used recently to identify markers of interferon response (110). Multiple differences between the genotypes of interferon responders and nonresponders were identified on a global scale by utilizing advanced statistical procedures, such as hierarchical clustering and principal component analysis. These methods are reviewed in Chapter 2 devoted to genomics technologies. A total of 221 polymorphic STR markers on 12 different chromosomes were genotyped, and 13 of them were found to have different frequencies in the responder and nonresponder populations. Hierarchical clustering and principal component analysis were performed to identify populations of patients with different STR marker profiles. While this is an early work and the identified markers need to be explored in different patient cohorts, it offers a new approach to identification of polymorphisms as genomic biomarkers. As the development of genotyping technologies, in particular microarray-based genotyping, facilitates large-scale identification of polymorphisms on a genome-wide scale, the search for associations between polymorphic variants and drug response will increasingly rely on advanced statistical methods that correlate composite patterns of polymorphisms with a predefined class label, such as sensitivity to a drug or the presence of a toxic reaction.

7.7. GENE COPY NUMBER POLYMORPHISMS

The overwhelming majority of the case studies reviewed in this chapter deal with SNPs as predictors of drug response. Until recently, they were considered the dominant form of genetic variation in humans. However, the rapid development of high-throughput genomic technologies in the first years of the twenty-first century has enabled discovery of a new form of variation in the human genome, namely, copy number polymorphisms. This term refers to microscopic and submicroscopic variants, which include deletions, duplications, insertions, and inversions. In the past 2 years, a significant amount of evidence has been accumulated that indicates that structural variants may cover millions of nucleotides within the human genome (111–120).

The scale and importance of copy number variation became apparent several years ago, when two initial breakthrough studies utilized representational oligonucleotide microarray analysis (ROMA) (117) and CGH (118) to identify multiple regions of genomic gains and losses in normal individuals. The ROMA study involved germ line DNA from 20 individuals and detected 221 copy number variants (CNVs) covering approximately 44 Mb of the genome. The individuals under study differed on average by 11 copy number polymorphisms (CNPs), while the average length of a CNV region was 465 kb. The variants were widely distributed throughout the genome. Regions of copy number variation covered 70 different genes, many of which are involved in regulation of cell growth or regulation of metabolism. Since it has been known for a long time that alterations in the dosage of certain genes cause profound phenotypic changes, this study suggested an intriguing possibility that a significant part of the diversity in the human population can be explained by copy number variation (117). In the context of drug discovery, the most important aspect of this finding is the variation in the copy number of genes involved in the disease mechanisms. For example, the screen detected triplication of the neuropeptide-Y4 receptor (PPYR1), a gene that is implicated in the regulation of food intake and body weight, suggesting a possible relationship between the presence of a CNP and the development of obesity.

The second of the two initial studies used CGH to analyze genomes of 55 unrelated inviduals, including 39 karyotypically normal persons and 16 patients with previously characterized chromosomal aberrations (118). Despite the differences in the methods used, this screen revealed a similar statistics of copy number variation in the genome: 255 widely distributed genomic clones showed variation, with an average of 12.4 CNVs per individual. Because of the limited resolution of the CGH method applied, the precise length of the CNVs could not be measured, but it was estimated that the individual regions of variation covered up to 2 Mb of genomic sequence. The most common CNV was detected in nearly half of the individuals studied and included the amylase alpha 1a and alpha 1b loci on 1p13.3. Both losses and gains of this region were identified, in 24% and 26% of cases, respectively. The alterations in the gene dosage were confirmed by using FISH and Q-PCR; the length of the region of variation was found to vary among individuals in the range of 150–425 kb. It is noteworthy that because of the limited content of the CGH array, only 12% of the genome was interrogated, suggesting that application of more advanced CGH tools would enable discovery of more recurrent CNVs.

Indeed, considering the known relationship between gene copy number and expression, one can envision that discovery of CNPs in genes related to disease or controlling drug metabolism in germ line DNA would explain a significant portion of the observed variation in drug response in the human population. When responses to therapeutic agents are reliably associated with CNPs, simple FISH- or QPCR-based tests would be used to select patients for therapy. This makes the task of complete mapping of the copy number variation in the human genome an extremely important task in pharmacogenetic research. Toward this goal, a first-generation CNV map of the human genome was constructed in 2006 (120) by profiling 270 individuals from four populations of European, African, and Asian origin comprising the HapMap collection (discussed in more detail in Section 7.8.4). Germ line DNA from lymphoblastoid cell lines was profiled for copy number changes with two independent methodologies, high-resolution CGH on SNP genotyping arrays and traditional BAC array CGH (both are reviewed comprehensively in Chapter 1). Analysis of data obtained from the two platforms yielded a total of 1447 CNVs covering 360 Mb, or 12% of the genome. These polymorphisms contained hundreds of genes, functional elements, and segmental duplications. It was demonstrated that CNVs encompass more sequence content that SNPs (120), implying the crucial role of copy number variation in genetic diversity and evolution. Additionally, one can speculate that copy number changes may potentially make a greater contribution to genetic diversity than SNPs because changes in gene expression that they may cause have more potential to affect the phenotype than substitution of a single nucleotide.

The discovery of this fundamentally new mechanism that represents a key driver of interindividual variation in humans will undoubtedly transform the field of pharmacogenomics. In the context of drug discovery, the most important next step is to determine possible associations between CNPs and mechanisms of disease. This would enable identification of CNPs potentially affecting response to therapeutics and would facilitate their testing as potential predictors of response in the clinic. So far, preliminary observations have been made on copy number variation for several genes of potential disease relevance. A CNV was found that encompassed the LPA apolipoprotein A gene, suggesting a possible association with susceptibility to atherosclerosis (120). Another CNV was suggested as a possible factor in susceptibility to psoriasis because it covers the PSORS1 locus on chromosome 6p21.3. Because this first genome-wide CNV mapping study was conducted on HapMap donors, for whom no phenotypic information is available, it was not possible to derive any correlations between CNVs and disease susceptibility and outcome. However, the first map of copy number variation may now be used to investigate associations between specific CNVs and disease outcome and response to therapeutic agents. The CNVs identified in the HapMap samples can now be mapped to intracellular pathways, and those relevant to the target disease can be explored for associations with the drug response by using QPCR or other techniques for copy number measurement. However, given the limited set of reference samples profiled, the 1500 CNPs detected (120) are likely to be just the tip of the iceberg.

Germ line variation in copy number of genes encoding drug-metabolizing enzymes is an area of special interest for drug discovery research, as alterations in the copy number of these genes may lead to changes in their expression and consequently to increases or decreases in enzyme activity, altering the metabolism of the drugs processed by the enzyme. For example, a recent study demonstrated germ line copy number variation for the SULT1A1 gene, which encodes an enzyme catalyzing sulfate conjugation of a wide spectrum of natural and synthetic compounds (121). Analysis of germ line DNA from 362 Caucasian and 99 African American individuals by quantitative PCR revealed the presence of one to five copies of the gene. In the Caucasian group, 5% of subjects carried a single copy of SULT1A1 and 26% had three or more copies, while in the African American cohort, 63% of individuals carried three or more copies. As the SULT1A1 activity was analyzed in human platelet and liver samples, the variability in the enzyme activity was best explained by differences in the gene copy number when all sources of genetic variability were considered (P value < 0.0001), implying that any variation in drug metabolism by SULT1A1 would most likely be due to the presence of a CNP.

As sufficient knowledge on human CNPs is accumulated, we believe that the most informative methodology for finding correlates of drug response would involve a genome-wide high-resolution scan of germ line DNA for CNPs. Application of advanced statistical methods would then permit identification of CNPs or composite CNP patterns associated with response to the therapeutic. A significant advantage of genome-wide profiling is that it would enable discovery of new polymorphisms not represented in the original HapMap data set. Additionally, it would eliminate the subjectivity associated with the selection of genes and pathways relevant to the disease. In Chapter 3 devoted to the genomic biomarkers, we discuss the potential value of copy number profiling of tumors from clinical trial subjects as a strategy in biomarker identification in oncology. While this is clearly the most direct and informative approach for identification of markers predictive of drug response, frozen tumor tissue from clinical trials is rarely available for CGH analysis. In contrast, blood samples are almost always available in amounts sufficient for genome-wide copy number analysis. Furthermore, analysis of blood DNA by high-density SNP genotyping arrays will produce information on both the genotype and the copy number profile of the patients. Thus, as we expand our knowledge of copy number variation in germ line DNA, it is likely that comprehensive copy number profiling will become standard in all clinical trials and the newly discovered CNPs in the human genome will be used as predictors of drug response.

7.8. CONCLUSION: APPROACHES TO IDENTIFICATION OF POLYMORPHISMS AS PREDICTORS OF DRUG RESPONSE

It is obvious from the numerous case studies reviewed in this chapter that genetic polymorphisms are a major determinant of the interindividual variability in response to drugs. Given the current advanced state of genomic technologies and the knowledge of the sequence of the human genome, the rate of discovery of pharmacogenetic biomarkers of drug response is primarily determined by the degree of understanding of the molecular mechanisms of pathogenic processes and the pharmacological effects of drugs. The polymorphisms reviewed in this chapter will likely be further pursued both as stratification markers for the existing therapies described here and as predictors of response for novel therapies directed at the same drug targets or targeting the same disease mechanisms. However, they undoubtedly represent only a tip of the iceberg, and it is likely that numerous other polymorphisms in a large spectrum of disease-related genes will be identified as predictive markers of drug response. Their number will grow exponentially as both additional disease-causing genes and additional polymorphisms in known genes are identified. Several methodologies exist for the discovery of polymorphic variants predictive of drug response.

7.8.1. Candidate Gene Approach

The first and the most straightforward methodology, referred to as the candidate gene approach, involves genotyping of candidate genes selected a priori based on their relevance to the drug's mechanism of action or its toxicological profile.

This approach may entail comprehensive genotype analysis of known disease targets, other disease-causing genes, as well as genes involved in drug transport, metabolism, or toxicity. The initial studies can be done in a small group of patients, and the mechanisms can be elucidated in established disease models such as cultured cells and animals or in toxicological models such as cultured hepatocytes or laboratory animals.

In the context of drug discovery, this work can be initiated at the lead discovery and lead optimization stages by prospective identification of the genes related to the mechanism of the drug and, if possible, other genes driving the disease (Fig. 7.2). The genes identified would be genotyped in an ethnically diverse population to determine all existing polymorphic variants. Alternatively, one of the existing databases of human polymorphisms can be used. Once sufficient knowledge is accumulated on the sequence diversity of the relevant genes, mechanistic studies can be initiated. Previously established disease models can be used at this stage, such as cell lines. The main goal at this stage is to determine whether any possible connection may exist between the polymorphic variants and the sensitivity of the model systems to the lead compound. An important aspect is the mechanism whereby the polymorphism affects the drug response. As demonstrated by the case studies reviewed in this chapter, most commonly this occurs through the effects of the polymorphism on the expression of the gene. Alternatively, polymorphic variants of the target gene may produce proteins that have different affinity for the drug candidate, thus causing the variability in the drug sensitivity. In either case, the mechanistic studies should yield "validated" polymorphisms that are likely to affect drug response in preclinical and clinical studies through a known mechanism and filter out genotypes that do not correlate with drug sensitivity in model systems (Fig. 7.2).

Depending on the relative timing of the compound optimization and biomarker discovery programs, the lead compound may be ready for first-in-human clinical trials by the time a list of candidate polymorphisms is assembled. In this case, patients enrolled into phase I and phase II trials should be genotyped for the variants selected. As efficacy is determined in phase II (and possibly even in phase I), associations between the polymorphisms and drug response can be determined in humans, leading to the next stage of biomarker discovery. The correlation between the presence of specific polymorphisms and the drug efficacy and toxicity can then be validated in phase III trials. Additional support can be obtained from independent studies correlating the same polymorphism with disease development in a general patient population with the target disease not receiving the drug. In many cases, such studies can be easier to set up, and they may be very useful if the statistical power of phase II studies (with possible addition of data from phase I) is not sufficient to identify associations between the genotype and the response significant enough to justify selecting patients for phase III. If the clinical trials are successful, regulatory approval will be sought for the use of the drug in a preselected population. In this case, a validated and approved genotyping test must be available for the drug to be approved. The example of irinotecan and the UGT1A1 polymorphism comprehensively

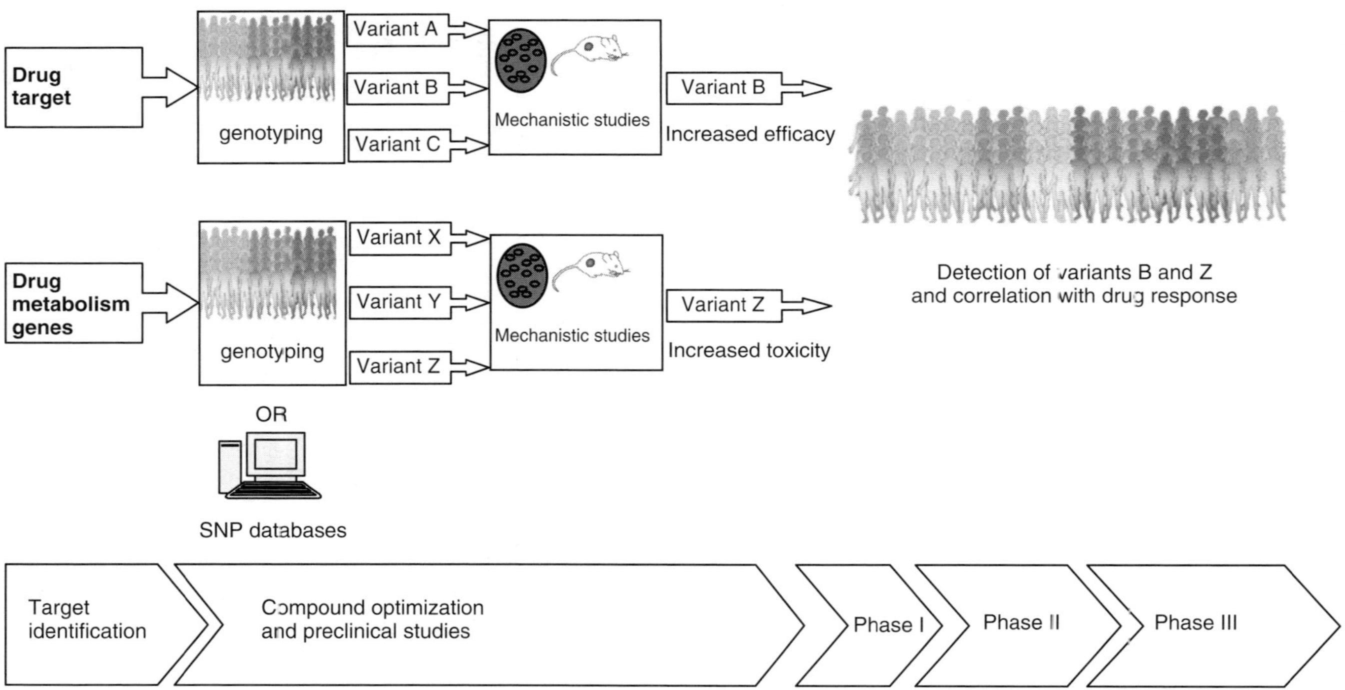

Figure 7.2 Candidate gene approach to pharmacogenetics studies in drug discovery. Drug target and known genes involved in drug metabolism are genotyped in an ethnically diverse population to identify existing variants. Alternatively, human SNP databases can be used. Once the spectrum of variation is defined for the candidate genes, possible associations with drug activity are explored in model systems, such as cell lines or animal models. The polymorphisms that affect the drug's mechanism in model systems are then studied in humans as clinical trials are initiated. See color insert.

reviewed in this chapter illustrates the potential impact of pharmacogenetic stratification of patients for therapy. In the case of irinotecan, the UGT1A1 diagnostic test was designed and validated when the drug was already on the market, prompting the change of the label for the drug. However, we expect that as pharmacogenetic profiling and patient stratification gains acceptance in the pharmaceutical industry, more drugs will be developed for specific preselected populations, necessitating parallel development and validation of diagnostic genotyping tests. The specifics of the drug/diagnostics codevelopment process are discussed elsewhere in this book.

7.8.2. Genome-wide Approach

The second methodology involves comprehensive high-throughput profiling of thousands of SNPs without a preconceived hypothesis followed by statistical analysis to correlate either single SNPs or composite patterns of polymorphisms with the predefined phenotype, primarily with response to the therapeutic candidate. In a drug development setting, this process may be initiated as soon as the lead compound enters clinical trials (Fig. 7.3A). High-throughput microarray-based genotyping of patients receiving the drug in phase II would yield genome-wide SNP profiles, which can be analyzed for associations with the response. Statistical procedures used for this analysis are similar to those applied for analysis of other composite biomarkers, such as gene expression profiles and patterns of gene copy number abnormalities (reviewed in Chapter 1). If associations are found that lead to hypotheses on possible response predictors, the genome-wide profiles may be reduced to a more focused set of SNPs that will be interrogated in phase III. Alternatively, the routine genome-wide genotyping can be continued in phase III to generate larger data sets with greater numbers of patients that will provide higher statistical power. One shortcoming of this approach is that polymorphic biomarkers are identified relatively late in drug development, and even if the data from the clinical trials provide sufficient statistical power to associate the presence of the marker with response to the drug, the drug will likely enter the approval process without a companion diagnostic test. More likely, additional studies will be required after preliminary associations are identified in phase III, and if the drug is approved genotyping of patients receiving the drug will continue to validate polymorphic biomarkers identified in clinical trials. Can biomarker programs targeted at polymorphisms be initiated earlier to enable co-development of the drug and a companion diagnostic test? Some of the case studies reviewed in this chapter indicate that polymorphic variants may produce in vitro results consistent with their in vivo effects, implying that if search for polymorphic variants is initiated at the preclinical stage in model systems the drug would enter the clinic with a specific biomarker to be validated. For example, the polymorphic FGFR4 allele that was associated with poor outcome in breast cancer and poor response to adjuvant therapy conferred a more oncogenic phenotype to breast cancer cells,

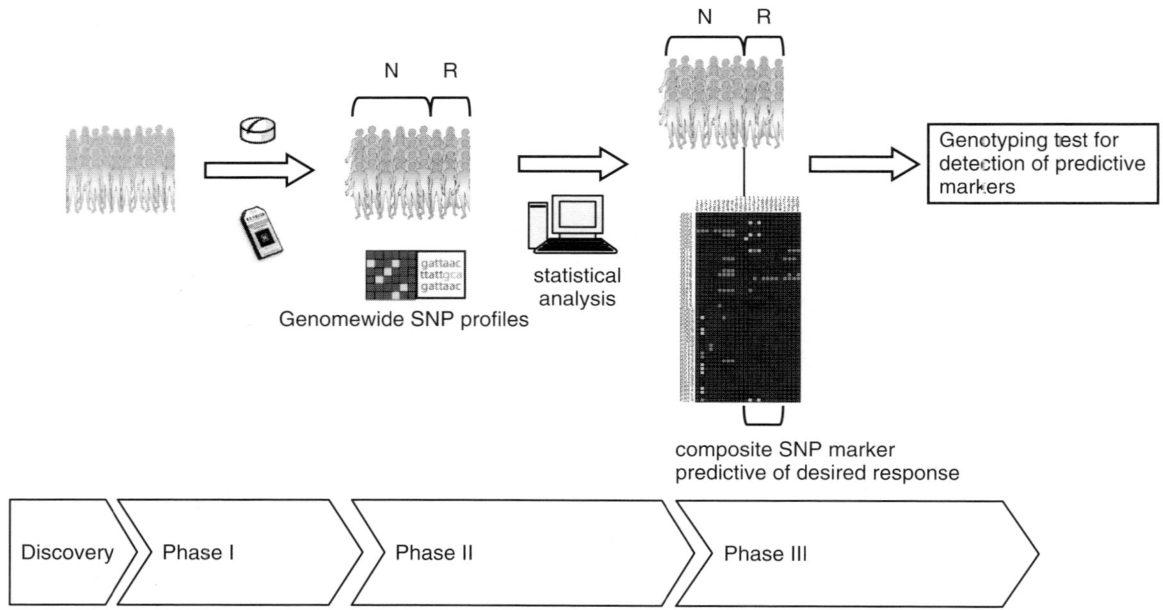

Figure 7.3 A) Genome-wide apprcach to identification of polymorphic markers of drug response. Patients enrolled in the clinical trial are genotyped with a high-throughput method, such as microarray-based genotyping. After the response data have been collected, the genome-wide SNP profiles are correlated with the drug response to identify SNP profiles associated with the desired outcome (good efficacy or low toxicity). Once composite markers predictive of response are identified, a genotyping test may be designed to interrogate these markers in an independent cohort of patients. B) Early application of the genome-wide approach in drug discovery. Preclinical drug screening is coupled with high-throughput genotyping of the preclinical models to identify polymorphisms associated with sensitivity to the drug. Once composite SNP markers are identified in preclinical models, they can be explored in clinical trials, and if correlation with response is confirmed, a diagnostic genotyping test can be developed. See color insert.

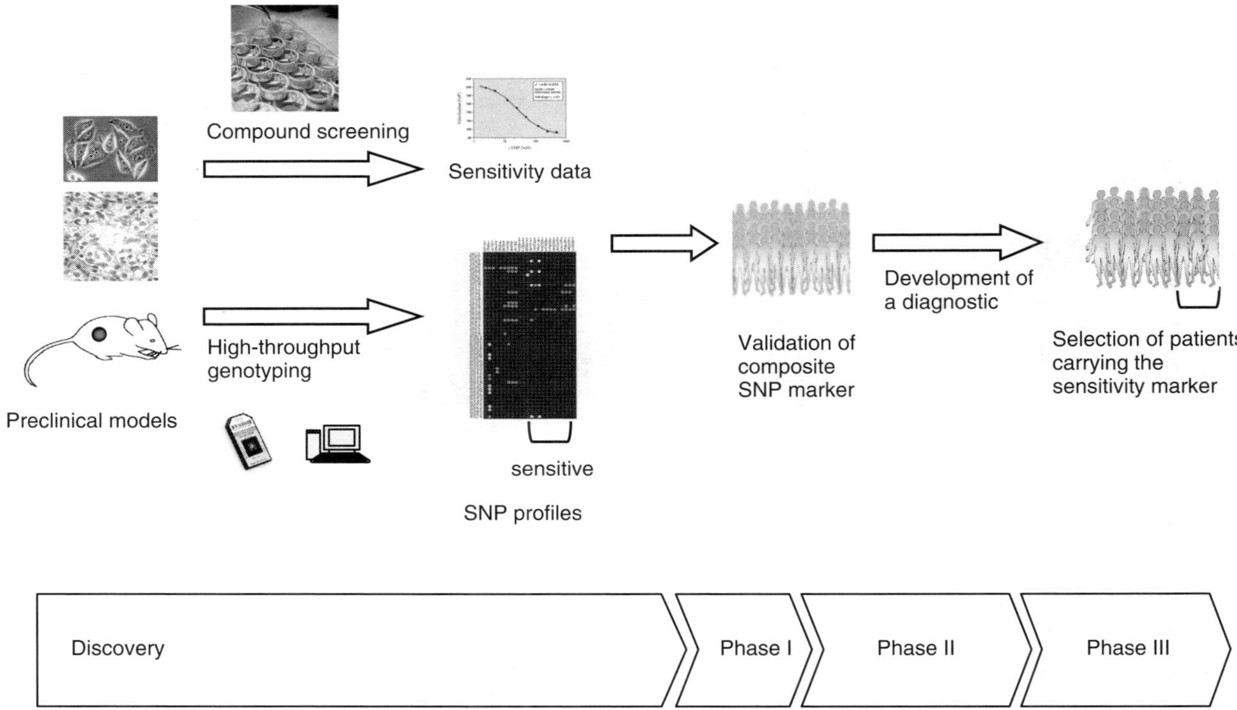

Figure 7.3 (Continued)

enabling mechanistic confirmation in vitro. Transient expression of the TPMT variant was used to demonstrate that the variable number tandem repeat controls the level of enzyme activity. In vitro studies with the TNF-α gene demonstrated that the presence of certain microsatellites influences TNF-α production.

Thus the search for polymorphisms predictive of drug response may be initiated in discovery by coupling the compound testing with comprehensive SNP profiling of model systems, as illustrated in Figure 7.3B. Today, this is often done for other types of biomarkers, such as RNA markers; for example, model systems used for drug screening are routinely profiled with gene expression microarrays. The concept of early identification of polymorphic variants predictive of response by high-throughput profiling of disease models still remains to be validated, but the potential value of early stratification decisions undoubtedly justifies continued efforts in this area.

7.8.3. Pathway Approach

The third methodology, referred to as the pathway approach (122), is based on the notion that most disease-modifying and drug-processing enzymes function in complex networks sharing common regulatory mechanisms. Consequently, it is unlikely that an isolated polymorphism with modest effects on enzyme function will individually determine the drug response, while a combination of variants affecting the same disease-relevant pathway may significantly perturb the molecular mechanism of the therapeutic agent and thus alter the drug response (Fig. 7.4). For example, the patient response to antifolates and thymidylate synthase inhibitors is affected by a number of polymorphisms in the genes that constitute the folate metabolism pathway (122). While the studies of associations between individual polymorphisms and the response to these drugs have provided the foundation for pharmacogenetic exploration, it is clear that drug response is controlled by a composite genotype comprised of polymorphic variants in several genes that determine the enzyme levels and the effective drug concentrations. To be able to examine composite genetic markers as predictors of response, one would need to build pathway maps relevant to the metabolism and on-target mechanism of the drug and systematically explore existing combinations of variants in the context of drug sensitivity in vitro or in vivo. For the antifolate drugs cited above and, more generally, for drugs that mimic natural enzyme substrates, the relevant pathways have been extensively studied, so the remaining challenge is to compile the information on individual polymorphisms and rationally examine their combinations as predictors of drug response. However, for many other drugs the precise intracellular pathways are not known and hence the individual variants contributing to drug sensitivity remain to be identified. One possible approach to elucidation of the relevant biological pathways is through gene expression studies. Microarray analysis has enabled high-throughput analysis of gene expression patterns associated with drug treatments, while advanced bioinformatics approaches facilitate mapping of expression signatures to intracellular pathways, and thus permit analysis of pathways affected by a drug. The

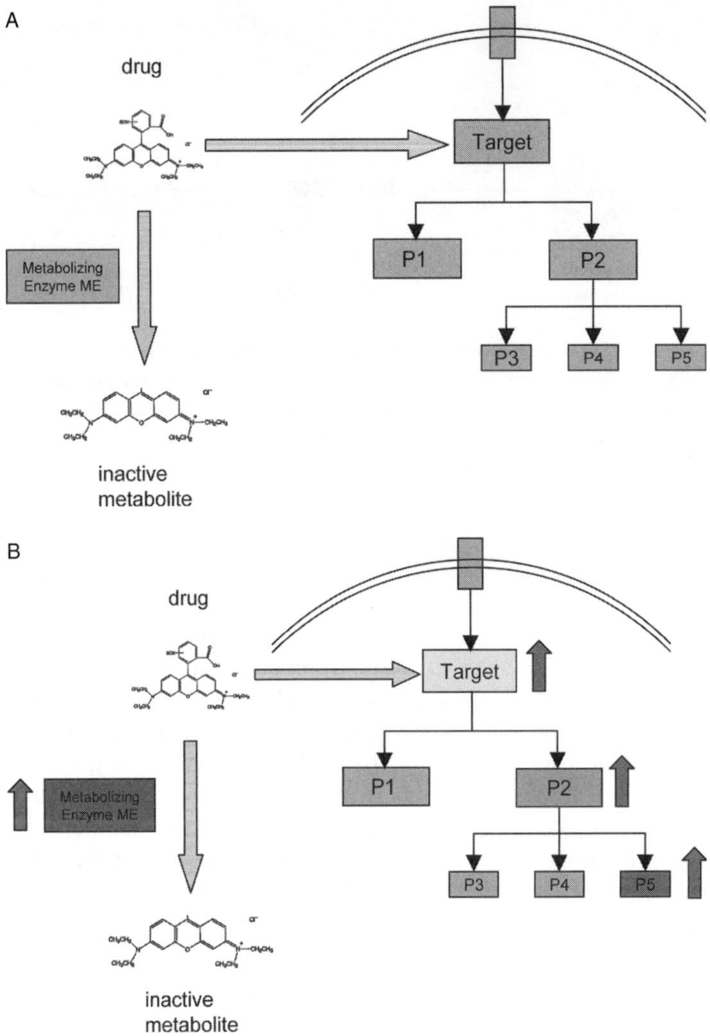

Figure 7.4 The pathway controlling a drug's mechanism may be affected by several polymorphisms. A) The target signals through proteins P1 and P2. Protein P2 induces P3, P4, and P5, which are implicated in the disease phenotype. The drug is converted into an inactive metabolite by a metabolizing enzyme ME. The dose of the drug is optimized for the most frequent genotype or the "average" individual. B) Polymorphic variants exist for the target, as well as SNPs in proteins P2, P5, and ME. Each of the polymorphisms causes an increase in expression or activity of the respective protein. When present individually in a patient, none of these SNPs significantly affects the efficacy of the drug, because the corresponding changes in expression are small. However, when all these SNPs are present in the same individual, their combined effect on the drug efficacy is significant. The higher concentrations of the target and its downstream modulators as well as the higher activity of the drug metabolizing enzyme contribute to the lower efficacy of the drug. Therefore, patients with this combination of variants may not respond to the drug. See color insert.

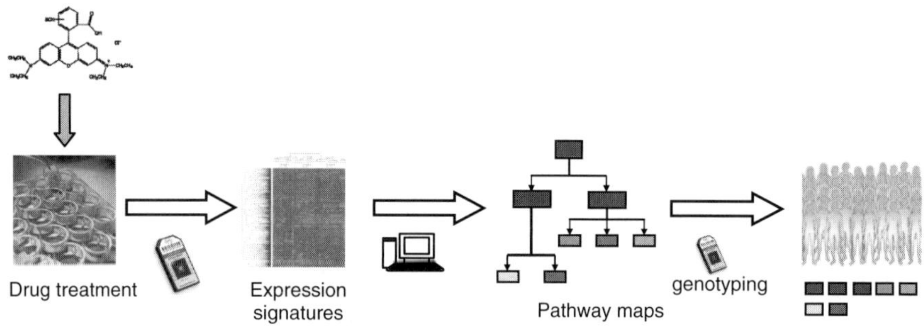

Figure 7.5 Use of gene expression microarrays to facilitate the pathway approach in pharmacogenetic studies. To elucidate the pathway involved in the drug's mechanism, cells are treated with the drug at different time points and profiled with expression microarrays. The resulting gene expression signatures are mapped to intracellular pathways. Genes involved in the activated pathways are then scanned for polymorphisms. The polymorphisms identified are considered candidate markers for drug efficacy because they may affect the drug's mechanism. See color insert.

applications of microarrays to study effects of drugs in model systems and the bioinformatics approaches to pathway analysis are covered in Chapter 2. Thus one can envisage identification of composite polymorphic markers by successive application of gene expression microarrays and SNP genotyping technologies (Fig. 7.5). First, gene expression analysis can be applied to identify the networks of genes activated by the drug and map them to intracellular pathways. Genotyping microarrays or alternative genotyping technologies can then be utilized to detect all existing SNPs in the genes that compose the pathways relevant to the drug mechanism.

The resulting composite SNP markers may have greater predictive power because they account for associations between the relevant proteins in the drug mechanism. If drug metabolism pathways are considered in addition to the mechanism-related pathways, then the composite biomarkers will also account for the available concentrations of the drug and its pharmacodynamics. Thus this hypothesis-driven approach may provide greater insights into the interindividual variability in drug response and yield biomarkers with a higher predictive power. Importantly, this analysis can be initiated early in drug discovery with disease models and thus may yield biomarker candidates before the initiation of clinical trials for the drug. However, the approach requires advance knowledge of the drug mechanism and involves significant amount of data analysis, thus necessitating substantial bioinformatics expertise. Nevertheless, the early applications of the approach in marker discovery for antifolate drugs and thymidylate synthase inhibitors (122) have yielded promising results, and one can certainly envisage that it will be increasingly adopted by the drug discovery community.

In summary, there are three possible approaches to identification of polymorphisms predictive of drug response. The so-called candidate gene approach implies a priori knowledge of the drug's pharmacology and toxicology and involves targeted analysis of individual polymorphisms in the genes previously implicated in the mechanism of action or the toxicological mechanism of the drug. In contrast, the genome-wide strategy does not rely on any previous knowledge of the drug mechanism or disease pathogenesis. Instead, it entails a whole-genome SNP screen to determine individual SNPs or composite SNP patterns that are associated with the desired response to the drug (high efficacy and low toxicity). Finally, the third, pathway-based approach relies on knowledge of the intracellular pathways responsible for metabolism, on-target mechanism of action, and toxicological properties of the therapeutic candidate. It recognizes the fact that the pharmacological effects of drugs are determined by a complex network of genes whose products interact with each other. Therefore, its distinguishing feature is the focus on composite patterns of polymorphisms in related genes, rather than individual genetic variants.

7.8.4. Use of Model Systems in Identification of Predictive Pharmacogenetic Markers

In the section devoted to genome-wide methodologies (Section 7.8.2) we initiated a discussion of possible in vitro pharmacogenetics methodologies aimed at **early identification of polymorphic variants predictive of drug efficacy**. Discovery of such markers before the initiation of clinical trials, or even generation of specific hypotheses for testing in the clinic, is likely to deliver substantial value in the drug discovery process. Therefore, we briefly cover additional tools available for **in vitro pharmacogenetics methodologies** and review selected published work in this area.

Cell-based assays also have other important advantages as tools for pharmacogenetic marker discovery. They enable an unlimited number of drug treatment experiments in a controlled environment, where the influence of environmental factors is eliminated. Even more importantly, they permit testing of drug effects on cells without the variability associated with drug delivery and metabolism, a factor that frequently complicates pharmacogenetic association studies. Additionally, cell lines derived from family members may facilitate traditional pharmacogenetics linkage studies, which are otherwise unfeasible in human clinical trials. This is an advantage particularly important in oncology, as tumors rarely arise at the same time in members of the same family and chemotherapy cannot be administered to healthy subjects, rendering traditional familial genetics methods inapplicable in oncology. The need for characterized and annotated tools for in vitro pharmacogenetics studies prompted the creation of a panel of pedigree cell lines of the Centre d'Etude du Polymorphisme Humain (CEPH) (123). They represent lymphoblastoid lines derived from approximately 50 three-generation pedigrees and transformed with the Epstein-Barr virus to enable continuous growth in culture. The CEPH panel of cell lines is available to the scientific community through the Coriell Institute of Medical Research (http://www.coriell.org).

The genotypic information on all the lines is freely accessible through the Internet (http://www.cephb.fr/cephdb/ and www.hapmap.org) and contains both SNP and microsatellite data.

The task of creating an **annotated and standardized tool for pharmacogenomic marker discovery** has been facilitated by the **HapMap project** (www.hapmap.org), an international effort aimed at establishing a publicly available resource that could be used to define the haplotypes and SNPs across the genome associated with disease mechanisms and drug response. Thirty CEPH trios have already been profiled that include two parents and a child (124), and further additions to the database are planned (125).

Several studies have recently been published that demonstrate the potential value of phenotype/genotype correlations in CEPH cell lines in the discovery of genes associated with drug response (124, 126). In particular, an *ex vivo* familial genetics methodology was applied to CEPH cell lines to identify genetic determinants of cytotoxicity of chemotherapeutic agents (126). Cytotoxic effects of two mechanistically distinct drugs (5-fluorouracil and docetaxel) were shown to be heritable traits. The heritability values ranged from 0.26 to 0.65 for 5-fluorouracil and from 0.21 to 0.70 for docetaxel. The heritability of drug response was lowest at low drug concentrations, when these agents have a relatively small effect on cell viability. At higher doses, when these agents significantly affect cell survival, the heritability was much higher, implying that significant interindividual differences in drug response are more apparent at higher drug doses. These results suggest that inherited genetic polymorphisms are an important determinant of cytotoxicity for a broad range of chemotherapeutic agents (126). Genome-Wide linkage analysis was also used to map a quantitative trait locus influencing the cytotoxicity of 5-fluorouracil to chromosome 9q13–q22, and two quantitative trait loci influencing the cellular effects of docetaxel to chromosomes 5q11–21 and 9q13–q22. By narrowing the region of interest, a list of candidate genes affecting the response to chemotherapy agents can be assembled. Further studies are needed to identify the specific genes within the mapped quantitative trait loci regions. Possible approaches may include RNA interference (serial knockdown of genes in the region followed by measurement of phenotypic changes) and microarray-based expression analysis (comparison of baseline expression profiles of the sensitive and resistant cell lines). Finally, 5-fluorouracil and docetaxel were found to cause apoptotic cell death through caspase-3 cleavage in CEPH lines, indicating that cytotoxicity in these cells occurs through a mechanism similar to that in tumor cells. This study thus provides a widely applicable strategy for pharmacogenomic discovery using commonly available tools.

In an independent study, the degree of cisplatin sensitivity was estimated for lymphoblastoid cell lines derived from 10 CEPH pedigrees (124). The heritability for cisplatin sensitivity was estimated to be 0.47, implying a substantial genetic component of the response to chemotherapy. Linkage analysis detected the strongest signal on chromosome 1 at 44 cM. Susceptibility to cisplatin-induced cytotoxicity is likely due to multiple loci, with low locus-specific heritability contributing to the trait. Thus the studies mentioned above have demonstrated the feasibility of identifying markers predictive of drug response with the use of characterized human cell lines with known genotypes. These data are extremely encouraging as they open the way to high-throughput genome-wide studies of the pharmacogenetics of drug response in vitro (graphically outlined in Fig. 7.6).

One promising of area of research is the application of mouse models to determine pharmacogenetic associations in humans (122). Since drug response can be relatively quickly determined in mice at early discovery stages, the use of mouse models may accelerate the process of pharmacogenetic biomarker development and generate candidate markers to be tested and validated in clinical trials. The laboratory mouse has a high degree of similarity to humans in genome sequence and physiology. With well-defined inbred mouse strains, it is possible to discover genomic loci associated with a phenotype of interest (such as drug response), determine the gene location, and explore the polymorphisms that may affect the response to the therapeutic. In the oncology field, mouse genetics has been used to identify genes responsible for susceptibility to pulmonary fibrosis resulting from therapy with bleomycin (127). Genome-wide profiling of the offspring produced by breeding a sensitive and a resistant mouse strain revealed two loci that were associated with pulmonary fibrosis. The first region (located on chromosome 17 and containing the gene for TNF-α) had already been associated with pulmonary fibrosis unrelated to bleomycin. However, the second locus was unique to bleomycin-related fibrosis; it contained bleomycin hydrolase, an enzyme that detoxifies bleomycin. Functional studies of bleomycin hydrolase activity indicated that this enzyme modulates bleomycin-induced pulmonary fibrosis, implicating its gene as a candidate gene for the bleomycin-induced toxic effect. It should be mentioned that applications of mouse genetics have certain limitations, such as interspecies differences in drug metabolism. Multiple genes related to drug response are known to be polymorphic in both mice and humans, but differences exist between the two species in drug-metabolizing enzymes and drug receptors. To obviate this problem, "humanized" mice have been created that carry genes encoding human drug receptors and drug metabolism enzymes (128). Given the large amount of genomic information accumulated for the mouse and the relative availability of mouse-based disease models, application of mouse genetics to human pharmacogenetic marker discovery is likely to yield important results in the nearest future. Coordinated phenotyping efforts such as the Mouse Phenome Project (http://www.jax.org/phenome) have been initiated to collect diverse mouse phenotype data, including drug response. Thus, despite potential interspecies differences in drug response, discoveries made in mouse models may lead to important insights into the mechanism of drug response in humans and reveal novel polymorphisms predictive of drug response.

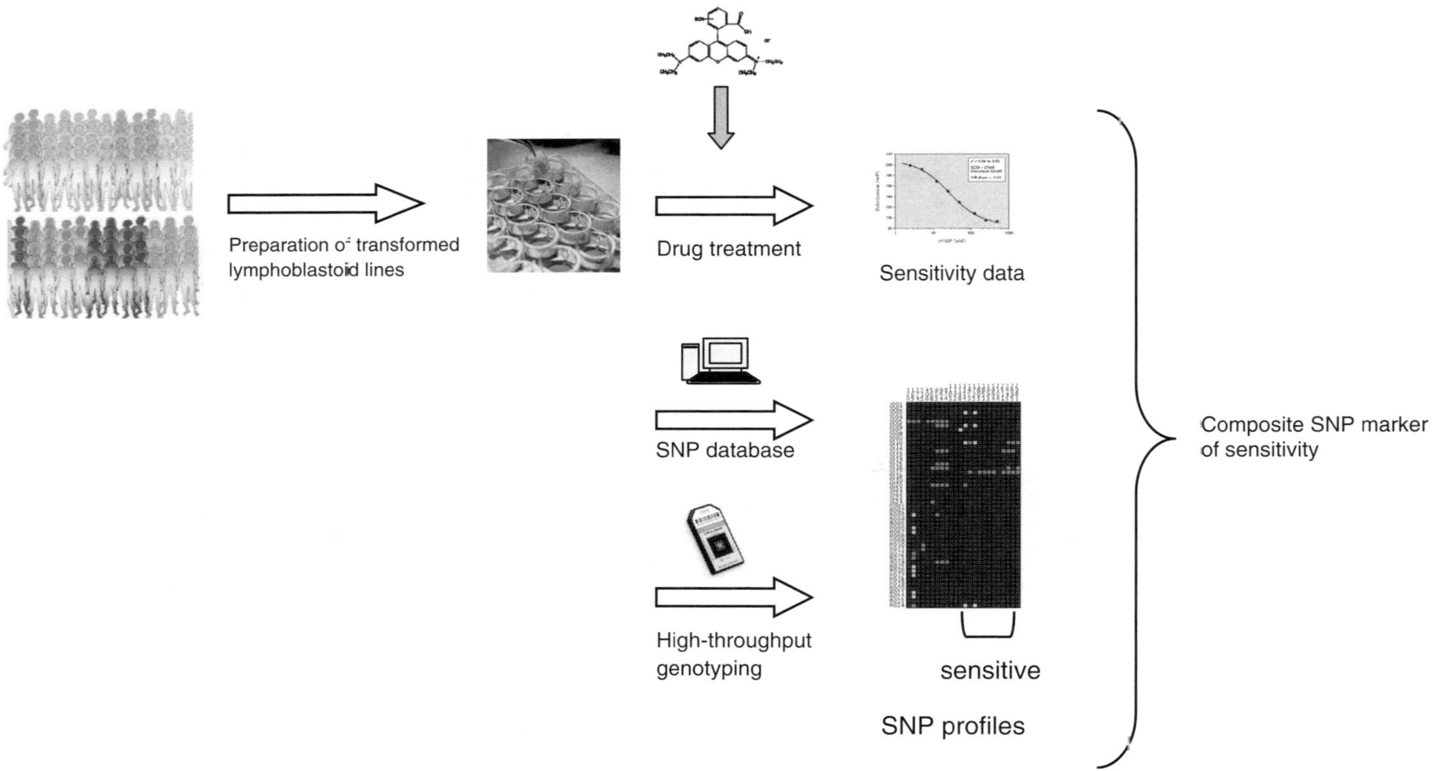

Figure 7.6 Use of cultured cell lines in pharmacogenetic studies in oncology to identify markers of sensitivity to anticancer drugs. Transformed lymphoblastoid cell lines are prepared from an ethnically diverse panel of individuals and cultured in vitro. The sensitivity of cell lines to drugs is determined with a cytotoxicity assay, and the genotypes of cells are obtained by high-throughput genotyping. See color insert.

7.8.5. Comparison of Methodologies in the Context of Drug Discovery

Naturally, there are advantages and disadvantages to each of the three reviewed approaches. The candidate gene approach has limited applicability because it requires substantial prior knowledge, but it is less resource-intensive because the number of target genes is limited. Typically, the list of candidate genes includes five to 100 genes most relevant to the drug's mechanism of action and metabolism. In many cases, this information is available from the literature. The list can be further prioritized, thus providing flexibility to the management, depending on the commitment to the pharmacogenetic program and the resources available. If the initial set of high-priority genes provides sufficient information on the variability in response and enables the identification of a predictive biomarker, the program can be refocused on validation of the biomarker discovered rather than further exploration of the candidate gene list. Of course, the most significant disadvantage of the candidate approach is that it may miss most predictive biomarkers if the initial knowledge of the disease mechanism is incomplete.

In contrast, the whole-genome approach is not based on any prior knowledge of the drug mechanism and therefore has the potential to identify all polymorphisms associated with the desired drug response. However, a critical prerequisite for a successful genome-wide analysis is the availability of the complete list of human polymorphisms. The current SNP databases continue to grow, and it is likely that in the near future most or all common SNPs will be identified. The next challenge will be to create a tool that would be able to interrogate the complete set of human polymorphisms. Currently, the most comprehensive tools interrogate over one million human SNPs. It is noteworthy, however, that the technology is developing so rapidly that the next generations of microarrays with higher SNP content will be available by the time this book is published. A significant challenge in implementing a genome-wide screen for predictive polymorphisms is the substantial sample size required to achieve statistical significance for composite patterns of polymorphisms. Finally, the genome-wide approach requires very significant resources, both intellectual and financial. Data analysis for millions of SNPs in hundreds of patients will necessitate significant investments in bioinformatics resources. However, as the genotyping technologies and data analysis software develop, the genotyping costs and resource requirements are likely to decrease, similar to what happened with high-throughput genome-wide analysis of gene expression in the first years of the twenty-first century.

Finally, the pathway-based approach may combine the advantages of the candidate gene and genome-wide methods: It is focused on a predefined set of relevant genes and therefore is less expensive and requires fewer resources, but at the same time it may yield composite polymorphic markers that account for the contribution of all relevant genes to the response to the drug. Thus, given the current state of the genotyping technologies, it may represent the optimal approach to discovery of SNPs predictive of drug response. However, it is very likely that the rapid development of genomic technologies and the bioinformatics tools will make genome-wide SNP profiling the method of choice for drug discovery organizations involved in genetic marker discovery.

BOX 7-1 *Approaches to Identification of Polymorphic Markers of Drug Response*

Candidate Gene Approach

A methodology involving genotyping of candidate genes selected a priori based on their relevance to the drug's mechanism of action or its toxicological profile. This approach may entail comprehensive genotype analysis of known disease targets, other disease-causing genes, as well as genes involved in drug transport, metabolism, or toxicity. In drug discovery, studies of this type may be initiated by prospective identification of the genes related to the mechanism of the drug with preclinical model systems.

Genome-wide Approach

A methodology that entails comprehensive high-throughput profiling of thousands of SNPs without a preconceived hypothesis, followed by statistical analysis to correlate composite patterns of polymorphisms with drug response. In a drug development setting, high-throughput microarray-based genotyping of patients receiving the drug in phases II and III would yield genome-wide SNP profiles, which can be analyzed for associations with the response. Unlike the candidate gene approach, the genome-wide methodology does not require prior knowledge of the efficacy or toxicity mechanisms for the drug.

Pathway Approach

A method that involves building pathway maps relevant to the metabolism and on-target mechanism of the drug, followed by a systematic exploration of the existing combinations of polymorphic variants in the context of drug sensitivity in vitro or in vivo. This approach is based on the notion that most disease-related and drug-metabolizing enzymes function in complex networks, which share common regulatory mechanisms. Genotyping of the chosen genes may be preceded by gene expression analysis aimed at the identification of relevant pathways. This approach requires advance knowledge of the drug mechanism and involves significant amount of data analysis, thus necessitating substantial bioinformatics expertise.

In conclusion, advances in target discovery and validation have enabled a substantial increase in the number of targeted drug candidates. This makes the early incorporation of pharmacogenomic studies into discovery research and clinical trials a more important and urgent task. The main goal of these studies would be to enable early identification of polymorphisms that are associated with high efficacy and low toxicity of the therapeutic candidate. In this chapter we have reviewed a substantial body of data on germ line polymorphisms as predictive biomarkers generated in diverse therapeutic areas, such as oncology, immunology, virology, and neuroscience. These data point to associations between the polymorphisms and the efficacy and toxicity of therapeutic agents as well as the outcome of the disease. However, despite the significant amounts of information generated, very few diagnostic tests exist that can predict the clinical efficacy and toxicity of drugs for the individual patient. The critical next step in the development of pharmacogenetics as a discipline will be to rapidly transfer the data accumulated into the clinic to enable individualized therapy. In drug discovery, the crucial next step would be to incorporate SNP profiling of relevant genes early into discovery and development programs, in order to generate biomarker candidates before the large phase III trials begin and thus enable rational patient selection. As genotyping technologies are developed that permit high-throughput genome-wide SNP profiling, pharmacogenetic programs in the pharmaceutical industry will likely expand to the whole-genome format. Undoubtedly, both the quantity and the quality of the pharmacogenetic data generated to date warrant an optimistic view of the future of personalized drug discovery.

REFERENCES

1. Lenz H-J. The use and development of germline polymorphisms in clinical oncology. *J Clin Oncol* 2004;22:2519–21.
2. Kalow W, Tang BK, and Endrenyi L. Hypothesis: comparisons of inter- and intra-individual variations can substitute for twin studies in drug research. *Pharmacogenetics* 1998;8:283–9.
3. Gandhi TK, Weingart SN, Borus J, Seger AC, Peterson J, Burdick E, Seger DL, Shu K, Federico F, Leape LL, and Bates DW. Adverse drug events in ambulatory care. *N Engl J Med* 2003;348:1556–64.
4. Kawato Y, Aonuma M, Hirota Y, Kuga H, and Sato K. Intracellular roles of SN-38, a metabolite of the camptothecin derivative CPT-11, in the antitumor effect of CPT-11. *Cancer Res* 1991;51:4187–91.
5. Gupta E, Lestingi TM, Mick R, Ramirez J, Vokes EE, and Ratain MJ. Metabolic fate of irinotecan in humans: correlation of glucuronidation with diarrhea. *Cancer Res* 1994;54:3723–5.
6. Iyer L, King CD, Whitington PF, Green MD, Roy SK, Tephly TR, Coffman BL, and Ratain MJ. Genetic predisposition to the metabolism of irinotecan (CPT-11). Role of uridine diphosphate glucuronosyltransferase isoform 1A1 in the glucuronidation of its active metabolite (SN-38) in human liver microsomes. *J Clin Invest* 1998;101:847–54.
7. Fisher MB, Vandenbranden M, Findlay K, Burchell B, Thummel KE, Hall SD, and Wrighton SA. Tissue distribution and interindividual variation in human UDP-glucuronosyltransferase activity: relationship between UGT1A1 promoter genotype and variability in a liver bank. *Pharmacogenetics* 2000;10:727–39.
8. Ritter JK, Chen F, Sheen YY, Tran HM, Kimura S, Yeatman MT, and Owens IS. A novel complex locus UGT1 encodes human bilirubin, phenol, and other UDP-glucuronosyltransferase isozymes with identical carboxyl termini. *J Biol Chem* 1992;267:3257–61.

9. Bosma PJ, Chowdhury JR, Bakker C, Gantla S, de Boer A, Oostra BA, Lindhout D, Tytgat GN, Jansen PL, Oude Elferink RP, and et al. The genetic basis of the reduced expression of bilirubin UDP-glucuronosyltransferase 1 in Gilbert's syndrome. *N Engl J Med* 1995;333:1171–5.
10. Raijmakers MT, Jansen PL, Steegers EA, and Peters WH. Association of human liver bilirubin UDP-glucuronyltransferase activity with a polymorphism in the promoter region of the UGT1A1 gene. *J Hepatol* 2000;33:348–51.
11. Ando Y, Saka H, Ando M, Sawa T, Muro K, Ueoka H, Yokoyama A, Saitoh S, Shimokata K, and Hasegawa Y. Polymorphisms of UDP-glucuronosyltransferase gene and irinotecan toxicity: a pharmacogenetic analysis. *Cancer Res* 2000;60:6921–6.
12. Burke D, Wilkes D, Blundell TL, and Malcolm S. Fibroblast growth factor receptors: lessons from the genes. *Trends Biochem Sci* 1998;23:59–62.
13. Bange J, Prechtl D, Cheburkin Y, Specht K, Harbeck N, Schmitt M, Knyazeva T, Muller S, Gartner S, Sures I, Wang H, Imyanitov E, Haring HU, Knayzev P, Iacobelli S, Hofler H, and Ullrich A. Cancer progression and tumor cell motility are associated with the FGFR4 Arg(388) allele. *Cancer Res* 2002;62:840–7.
14. Streit S, Bange J, Fichtner A, Ihrler S, Issing W, and Ullrich A. Involvement of the FGFR4 Arg388 allele in head and neck squamous cell carcinoma. *Int J Cancer* 2004;111:213–7.
15. Spinola M, Leoni V, Pignatiello C, Conti B, Ravagnani F, Pastorino U, and Dragani TA. Functional FGFR4 Gly388Arg polymorphism predicts prognosis in lung adenocarcinoma patients. *J Clin Oncol* 2005;23:7307–11.
16. Effects of chemotherapy and hormonal therapy for early breast cancer on recurrence and 15-year survival: an overview of the randomised trials. *Lancet* 2005;365:1687–717.
17. Thussbas C, Nahrig J, Streit S, Bange J, Kriner M, Kates R, Ulm K, Kiechle M, Hoefler H, Ullrich A, and Harbeck N. FGFR4 Arg388 allele is associated with resistance to adjuvant therapy in primary breast cancer. *J Clin Oncol* 2006;24:3747–55.
18. Green H, Soderkvist P, Rosenberg P, Horvath G, and Peterson C. mdr-1 single nucleotide polymorphisms in ovarian cancer tissue: G2677T/A correlates with response to paclitaxel chemotherapy. *Clin Cancer Res* 2006;12:854–9.
19. Heggie GD, Sommadossi JP, Cross DS, Huster WJ, and Diasio RB. Clinical pharmacokinetics of 5-fluorouracil and its metabolites in plasma, urine, and bile. *Cancer Res* 1987;47:2203–6.
20. Wasternack C. Degradation of pyrimidines and pyrimidine analogs—pathways and mutual influences. *Pharmacol Ther* 1980;8:629–51.
21. Baker SD, Diasio RB, O'Reilly S, Lucas VS, Khor SP, Sartorius SE, Donehower RC, Grochow LB, Spector T, Hohneker JA, and Rowinsky EK. Phase I and pharmacologic study of oral fluorouracil on a chronic daily schedule in combination with the dihydropyrimidine dehydrogenase inactivator eniluracil. *J Clin Oncol* 2000;18:915–26.
22. Tuchman M, Stoeckeler JS, Kiang DT, O'Dea RF, Ramnaraine ML, and Mirkin BL. Familial pyrimidinemia and pyrimidinuria associated with severe fluorouracil toxicity. *N Engl J Med* 1985;313:245–9.
23. Harris BE, Carpenter JT, and Diasio RB. Severe 5-fluorouracil toxicity secondary to dihydropyrimidine dehydrogenase deficiency. A potentially more common pharmacogenetic syndrome. *Cancer* 1991;68:499–501.
24. Wei X, McLeod HL, McMurrough J, Gonzalez FJ, and Fernandez-Salguero P. Molecular basis of the human dihydropyrimidine dehydrogenase deficiency and 5-fluorouracil toxicity. *J Clin Invest* 1996;98:610–5.
25. Milano G and Etienne MC. Dihydropyrimidine dehydrogenase (DPD) and clinical pharmacology of 5-fluorouracil (review). *Anticancer Res* 1994;14:2295–7.
26. Milano G, Etienne MC, Pierrefite V, Barberi-Heyob M, Deporte-Fety R, and Renee N. Dihydropyrimidine dehydrogenase deficiency and fluorouracil-related toxicity. *Br J Cancer* 1999;79:627–30.

27. Van Kuilenburg AB, Haasjes J, Richel DJ, Zoetekouw L, Van Lenthe H, De Abreu RA, Maring JG, Vreken P, and Van Gennip AH. Clinical implications of dihydropyrimidine dehydrogenase (DPD) deficiency in patients with severe 5-fluorouracil-associated toxicity: identification of new mutations in the DPD gene. *Clin Cancer Res* 2000;6:4705–12.
28. Nagasubramanian R, Innocenti F, and Ratain MJ. Pharmacogenetics in cancer treatment. *Annu Rev Med* 2003;54:437–52.
29. Van Kuilenburg AB, Muller EW, Haasjes J, Meinsma R, Zoetekouw L, Waterham HR, Baas F, Richel DJ, and Van Gennip AH. Lethal outcome of a patient with a complete dihydropyrimidine dehydrogenase (DPD) deficiency after administration of 5-fluorouracil: frequency of the common IVS14+1G>A mutation causing DPD deficiency. *Clin Cancer Res* 2001;7: 1149–53.
30. Raida M, Schwabe W, Hausler P, Van Kuilenburg AB, Van Gennip AH, Behnke D, and Hoffken K. Prevalence of a common point mutation in the dihydropyrimidine dehydrogenase (DPD) gene within the 5′-splice donor site of intron 14 in patients with severe 5-fluorouracil (5-FU)-related toxicity compared with controls. *Clin Cancer Res* 2001;7:2832–9.
31. Collie-Duguid ES, Etienne MC, Milano G, and McLeod HL. Known variant DPYD alleles do not explain DPD deficiency in cancer patients. *Pharmacogenetics* 2000;10:217–23.
32. Lennard L, Lilleyman JS, Van Loon J, and Weinshilboum RM. Genetic variation in response to 6-mercaptopurine for childhood acute lymphoblastic leukaemia. *Lancet* 1990;336:225–9.
33. Vuchetich JP, Weinshilboum RM, and Price RA. Segregation analysis of human red blood cell thiopurine methyltransferase activity. *Genet Epidemiol* 1995;12:1–11.
34. Tai HL, Krynetski EY, Schuetz EG, Yanishevski Y, and Evans WE. Enhanced proteolysis of thiopurine *S*-methyltransferase (TPMT) encoded by mutant alleles in humans (TPMT*3A, TPMT*2): mechanisms for the genetic polymorphism of TPMT activity. *Proc Natl Acad Sci USA* 1997;94:6444–9.
35. Spire-Vayron de la Moureyre C, Debuysere H, Mastain B, Vinner E, Marez D, Lo Guidice JM, Chevalier D, Brique S, Motte K, Colombel JF, Turck D, Noel C, Flipo RM, Pol A, Lhermitte M, Lafitte JJ, Libersa C, and Broly F. Genotypic and phenotypic analysis of the polymorphic thiopurine *S*-methyltransferase gene (TPMT) in a European population. *Br J Pharmacol* 1998;125:879–87.
36. Spire-Vayron de la Moureyre C, Debuysere H, Fazio F, Sergent E, Bernard C, Sabbagh N, Marez D, Lo Guidice JM, D'Halluin J C, and Broly F. Characterization of a variable number tandem repeat region in the thiopurine *S*-methyltransferase gene promoter. *Pharmacogenetics* 1999;9:189–98.
37. Yan L, Zhang S, Eiff B, Szumlanski CL, Powers M, O'Brien JF, and Weinshilboum RM. Thiopurine methyltransferase polymorphic tandem repeat: genotype-phenotype correlation analysis. *Clin Pharmacol Ther* 2000;68:210–9.
38. Evans WE, Horner M, Chu YQ, Kalwinsky D, and Roberts WM. Altered mercaptopurine metabolism, toxic effects, and dosage requirement in a thiopurine methyltransferase-deficient child with acute lymphocytic leukemia. *J Pediatr* 1991;119:985–9.
39. Schutz E, Gummert J, Mohr F, and Oellerich M. Azathioprine-induced myelosuppression in thiopurine methyltransferase deficient heart transplant recipient. *Lancet* 1993;341: 436.
40. Lennard L, Welch JC, and Lilleyman JS. Thiopurine drugs in the treatment of childhood leukaemia: the influence of inherited thiopurine methyltransferase activity on drug metabolism and cytotoxicity. *Br J Clin Pharmacol* 1997;44:455–61.
41. Evans WE, Hon YY, Bomgaars L, Coutre S, Holdsworth M, Janco R, Kalwinsky D, Keller F, Khatib Z, Margolin J, Murray J, Quinn J, Ravindranath Y, Ritchey K, Roberts W, Rogers ZR, Schiff D, Steuber C, Tucci F, Kornegay N, Krynetski EY, and Relling MV. Preponderance of thiopurine *S*-methyltransferase deficiency and heterozygosity among patients intolerant to mercaptopurine or azathioprine. *J Clin Oncol* 2001;19:2293–301.

42. Relling MV, Hancock ML, Rivera GK, Sandlund JT, Ribeiro RC, Krynetski EY, Pui CH, and Evans WE. Mercaptopurine therapy intolerance and heterozygosity at the thiopurine *S*-methyltransferase gene locus. *J Natl Cancer Inst* 1999;91:2001–8.
43. Relling MV, Hancock ML, Boyett JM, Pui CH, and Evans WE. Prognostic importance of 6-mercaptopurine dose intensity in acute lymphoblastic leukemia. *Blood* 1999;93:2817–23.
44. Frosst P, Blom HJ, Milos R, Goyette P, Sheppard CA, Matthews RG, Boers GJ, den Heijer M, Kluijtmans LA, van den Heuvel LP, et al. A candidate genetic risk factor for vascular disease: a common mutation in methylenetetrahydrofolate reductase. *Nat Genet* 1995;10:111–3.
45. Schneider JA, Rees DC, Liu YT, and Clegg JB. Worldwide distribution of a common methylenetetrahydrofolate reductase mutation. *Am J Hum Genet* 1998;62:1258–60.
46. Ulrich CM, Yasui Y, Storb R, Schubert MM, Wagner JL, Bigler J, Ariail KS, Keener CL, Li S, Liu H, Farin FM, and Potter JD. Pharmacogenetics of methotrexate: toxicity among marrow transplantation patients varies with the methylenetetrahydrofolate reductase C677T polymorphism. *Blood* 2001;98:231–4.
47. Toffoli G, Veronesi A, Boiocchi M, and Crivellari D. MTHFR gene polymorphism and severe toxicity during adjuvant treatment of early breast cancer with cyclophosphamide, methotrexate, and fluorouracil (CMF). *Ann Oncol* 2000;11:373–4.
48. Stevenson JP, Redlinger M, Kluijtmans LA, Sun W, Algazy K, Giantonio B, Haller DG, Hardy C, Whitehead AS, and O'Dwyer PJ. Phase I clinical and pharmacogenetic trial of irinotecan and raltitrexed administered every 21 days to patients with cancer. *J Clin Oncol* 2001;19:4081–7.
49. Ma J, Stampfer MJ, Giovannucci E, Artigas C, Hunter DJ, Fuchs C, Willett WC, Selhub J, Hennekens CH, and Rozen R. Methylenetetrahydrofolate reductase polymorphism, dietary interactions, and risk of colorectal cancer. *Cancer Res* 1997;57:1098–102.
50. Skibola CF, Smith MT, Kane E, Roman E, Rollinson S, Cartwright RA, and Morgan G. Polymorphisms in the methylenetetrahydrofolate reductase gene are associated with susceptibility to acute leukemia in adults. *Proc Natl Acad Sci USA* 1999;96:12810–5.
51. Rustum YM, Harstrick A, Cao S, Vanhoefer U, Yin MB, Wilke H, and Seeber S. Thymidylate synthase inhibitors in cancer therapy: direct and indirect inhibitors. *J Clin Oncol* 1997;15:389–400.
52. Nishimura R, Nagao K, Miyayama H, Matsuda M, Baba K, Matsuoka Y, Yamashita H, Fukuda M, Higuchi A, Satoh A, Mizumoto T, and Hamamoto R. Thymidylate synthase levels as a therapeutic and prognostic predictor in breast cancer. *Anticancer Res* 1999;19:5621–6.
53. Huang CL, Yokomise H, Kobayashi S, Fukushima M, Hitomi S, and Wada H. Intratumoral expression of thymidylate synthase and dihydropyrimidine dehydrogenase in non-small cell lung cancer patients treated with 5-FU-based chemotherapy. *Int J Oncol* 2000;17:47–54.
54. Salonga D, Danenberg KD, Johnson M, Metzger R, Groshen S, Tsao-Wei DD, Lenz HJ, Leichman CG, Leichman L, Diasio RB, and Danenberg PV. Colorectal tumors responding to 5-fluorouracil have low gene expression levels of dihydropyrimidine dehydrogenase, thymidylate synthase, and thymidine phosphorylase. *Clin Cancer Res* 2000;6:1322–7.
55. Yeh KH, Shun CT, Chen CL, Lin JT, Lee WJ, Lee PH, Chen YC, and Cheng AL. High expression of thymidylate synthase is associated with the drug resistance of gastric carcinoma to high dose 5-fluorouracil-based systemic chemotherapy. *Cancer* 1998;82:1626–31.
56. Kaneda S, Nalbantoglu J, Takeishi K, Shimizu K, Gotoh O, Seno T, and Ayusawa D. Structural and functional analysis of the human thymidylate synthase gene. *J Biol Chem* 1990;265:20277–84.
57. Kawakami K, Salonga D, Park JM, Danenberg KD, Uetake H, Brabender J, Omura K, Watanabe G, and Danenberg PV. Different lengths of a polymorphic repeat sequence in the thymidylate synthase gene affect translational efficiency but not its gene expression. *Clin Cancer Res* 2001;7:4096–101.

58. Marsh S, and McLeod HL. Thymidylate synthase pharmacogenetics in colorectal cancer. *Clin Colorectal Cancer* 2001;1:175–8; discussion 9–81.
59. Marsh S. Thymidylate synthase pharmacogenetics. *Invest New Drugs* 2005;23:533–7.
60. Pestalozzi BC, Peterson HF, Gelber RD, Goldhirsch A, Gusterson BA, Trihia H, Lindtner J, Cortes-Funes H, Simmoncini E, Byrne MJ, Golouh R, Rudenstam CM, Castiglione-Gertsch M, Allegra CJ, and Johnston PG. Prognostic importance of thymidylate synthase expression in early breast cancer. *J Clin Oncol* 1997;15:1923–31.
61. Lenz HJ, Leichman CG, Danenberg KD, Danenberg PV, Groshen S, Cohen H, Laine L, Crookes P, Silberman H, Baranda J, Garcia Y, Li J, and Leichman L. Thymidylate synthase mRNA level in adenocarcinoma of the stomach: a predictor for primary tumor response and overall survival. *J Clin Oncol* 1996;14:176–82.
62. Aschele C, Debernardis D, Casazza S, Antonelli G, Tunesi G, Baldo C, Lionetto R, Maley F, and Sobrero A. Immunohistochemical quantitation of thymidylate synthase expression in colorectal cancer metastases predicts for clinical outcome to fluorouracil-based chemotherapy. *J Clin Oncol* 1999;17:1760–70.
63. Kornmann M, Schwabe W, Sander S, Kron M, Strater J, Polat S, Kettner E, Weiser HF, Baumann W, Schramm H, Hausler P, Ott K, Behnke D, Staib L, Beger HG, and Link KH. Thymidylate synthase and dihydropyrimidine dehydrogenase mRNA expression levels: predictors for survival in colorectal cancer patients receiving adjuvant 5-fluorouracil. *Clin Cancer Res* 2003;9:4116–24.
64. Van Triest B, Pinedo HM, Van Hensbergen Y, Smid K, Telleman F, Schoenmakers PS, van der Wilt CL, Van Laar JAM, Noordhuis P, Jansen G, and Peters GJ. Thymidylate synthase level as the main predictive parameter for sensitivity to 5-fluorouracil, but not for folate-based thymidylate synthase inhibitors, in 13 nonselected colon cancer cell lines. *Clin Cancer Res* 1999;5:643–54.
65. Leichman C, Lenz H, Leichman L, Danenberg K, Baranda J, Groshen S, Boswell W, Metzger R, Tan M, and Danenberg P. Quantitation of intratumoral thymidylate synthase expression predicts for disseminated colorectal cancer response and resistance to protracted-infusion fluorouracil and weekly leucovorin. *J Clin Oncol* 1997;15:3223–9.
66. Johnston PG, Drake JC, Trepel J, and Allegra CJ. Immunological quantitation of thymidylate synthase using the monoclonal antibody TS 106 in 5-fluorouracil-sensitive and -resistant human cancer cell lines. *Cancer Res* 1992;52:4306–12.
67. Johnston PG, Lenz HJ, Leichman CG, Danenberg KD, Allegra CJ, Danenberg PV, and Leichman L. Thymidylate synthase gene and protein expression correlate and are associated with response to 5-fluorouracil in human colorectal and gastric tumors. *Cancer Res* 1995;55:1407–12.
68. Marsh S, McKay JA, Cassidy J, and McLeod HL. Polymorphism in the thymidylate synthase promoter enhancer region in colorectal cancer. *Int J Oncol* 2001;19:383–6.
69. Iacopetta B, Grieu F, Joseph D, and Elsaleh H. A polymorphism in the enhancer region of the thymidylate synthase promoter influences the survival of colorectal cancer patients treated with 5-fluorouracil. *Br J Cancer* 2001;85:827–30.
70. Le Morvan V, Bellott R, Moisan F, Mathoulin-Pelissier S, Bonnet J, and Robert J. Relationships between genetic polymorphisms and anticancer drug cytotoxicity vis-à-vis the NCI-60 panel. *Pharmacogenomics* 2006;7:843–52.
71. Monks A, Scudiero D, Skehan P, Shoemaker R, Paull K, Vistica D, Hose C, Langley J, Cronise P, Vaigro-Wolff A, et al. Feasibility of a high-flux anticancer drug screen using a diverse panel of cultured human tumor cell lines. *J Natl Cancer Inst* 1991;83:757–66.
72. Xu Z, Chen ZP, Malapetsa A, Alaoui-Jamali M, Bergeron J, Monks A, Myers TG, Mohr G, Sausville EA, Scudiero DA, Aloyz R, and Panasci LC. DNA repair protein levels vis-a-vis

anticancer drug resistance in the human tumor cell lines of the National Cancer Institute drug screening program. *Anticancer Drugs* 2002;13:511–9.

73. Takebayashi Y, Pourquier P, Zimonjic DB, Nakayama K, Emmert S, Ueda T, Urasaki Y, Kanzaki A, Akiyama SI, Popescu N, Kraemer KH, and Pommier Y. Antiproliferative activity of ecteinascidin 743 is dependent upon transcription-coupled nucleotide-excision repair. *Nat Med* 2001;7:961–6.
74. Ban N, Takahashi Y, Takayama T, Kura T, Katahira T, Sakamaki S, and Niitsu Y. Transfection of glutathione *S*-transferase (GST)-pi antisense complementary DNA increases the sensitivity of a colon cancer cell line to adriamycin, cisplatin, melphalan, and etoposide. *Cancer Res* 1996;56:3577–82.
75. Rieder MJ, Reiner AP, Gage BF, Nickerson DA, Eby CS, McLeod HL, Blough DK, Thummel KE, Veenstra DL, and Rettie AE. Effect of VKORC1 haplotypes on transcriptional regulation and warfarin dose. *N Engl J Med* 2005;352:2285–93.
76. D'Andrea G, D'Ambrosio RL, Di Perna P, Chetta M, Santacroce R, Brancaccio V, Grandone E, and Margaglione M. A polymorphism in the VKORC1 gene is associated with an interindividual variability in the dose-anticoagulant effect of warfarin. *Blood* 2005;105:645–9.
77. Yuan HY, Chen JJ, Lee MT, Wung JC, Chen YF, Charng MJ, Lu MJ, Hung CR, Wei CY, Chen CH, Wu JY, and Chen YT. A novel functional VKORC1 promoter polymorphism is associated with inter-individual and inter-ethnic differences in warfarin sensitivity. *Hum Mol Genet* 2005;14:1745–51.
78. Aquilante CL, Langaee TY, Lopez LM, Yarandi HN, Tromberg JS, Mohuczy D, Gaston KL, Waddell CD, Chirico MJ, and Johnson JA. Influence of coagulation factor, vitamin K epoxide reductase complex subunit 1, and cytochrome P450 2C9 gene polymorphisms on warfarin dose requirements. *Clin Pharmacol Ther* 2006;79:291–302.
79. White KJ, Walline CC, and Barker EL. Serotonin transporters: implications for antidepressant drug development. *AAPS J* 2005;7:E421–33.
80. Heils A, Teufel A, Petri S, Stober G, Riederer P, Bengel D, and Lesch KP. Allelic variation of human serotonin transporter gene expression. *J Neurochem* 1996;66:2621–4.
81. Lesch KP, Bengel D, Heils A, Sabol SZ, Greenberg BD, Petri S, Benjamin J, Muller CR, Hamer DH, and Murphy DL. Association of anxiety-related traits with a polymorphism in the serotonin transporter gene regulatory region. *Science* 1996;274:1527–31.
82. Rausch JL, Johnson ME, Fei YJ, Li JQ, Shendarkar N, Hobby HM, Ganapathy V, and Leibach FH. Initial conditions of serotonin transporter kinetics and genotype: influence on SSRI treatment trial outcome. *Biol Psychiatry* 2002;51:723–32.
83. Yu YW, Tsai SJ, Chen TJ, Lin CH, and Hong CJ. Association study of the serotonin transporter promoter polymorphism and symptomatology and antidepressant response in major depressive disorders. *Mol Psychiatry* 2002;7:1115–9.
84. Pollock BG, Ferrell RE, Mulsant BH, Mazumdar S, Miller M, Sweet RA, Davis S, Kirshner MA, Houck PR, Stack JA, Reynolds CF, and Kupfer DJ. Allelic variation in the serotonin transporter promoter affects onset of paroxetine treatment response in late-life depression. *Neuropsychopharmacology* 2000;23:587–90.
85. Durham LK, Webb SM, Milos PM, Clary CM, and Seymour AB. The serotonin transporter polymorphism, 5HTTLPR, is associated with a faster response time to sertraline in an elderly population with major depressive disorder. *Psychopharmacology (Berl)* 2004;174:525–9.
86. Arranz MJ and Kerwin RW. Neurotransmitter-related genes and antipsychotic response: pharmacogenetics meets psychiatric treatment. *Ann Med* 2000;32:128–33.
87. Malhotra AK, Murphy GM, Jr., and Kennedy JL. Pharmacogenetics of psychotropic drug response. *Am J Psychiatry* 2004;161:780–96.

88. Davies MA, Setola V, Strachan RT, Sheffler DJ, Salay E, Hufeisen SJ, and Roth BL. Pharmacologic analysis of non-synonymous coding h5-HT2A SNPs reveals alterations in atypical antipsychotic and agonist efficacies. *Pharmacogenomics J* 2006;6:42–51.
89. Ranganathan P. Pharmacogenomics of tumor necrosis factor antagonists in rheumatoid arthritis. *Pharmacogenomics* 2005;6:481–90.
90. Louis, Franchimont, Piron, Gevaert, Schaaf-Lafontaine, Roland, Mahieu, Malaise, Groote D, and Belaiche. Tumour necrosis factor (TNF) gene polymorphism influences TNF-α production in lipopolysaccharide (LPS)-stimulated whole blood cell culture in healthy humans. *Clin Exp Immunol* 1998;113:401–6.
91. Fabris M, Di PE, D'Elia A, Damante G, Sinigaglia L, and Ferraccioli G. Tumor necrosis factor-alpha gene polymorphism in severe and mild-moderate rheumatoid arthritis. *J Rheumatol* 2002;29:29–33.
92. Udalova IA, Nedospasov SA, Webb GC, Chaplin DD, and Turetskaya RL. Highly informative typing of the human TNF locus using six adjacent polymorphic markers. *Genomics* 1993;16:180–6.
93. Morita C, Horiuchi T, Tsukamoto H, Hatta N, Kikuchi Y, Arinobu Y, Otsuka T, Sawabe T, Harashima S, Nagasawa K, and Niho Y. Association of tumor necrosis factor receptor type II polymorphism 196R with systemic lupus erythematosus in the Japanese: molecular and functional analysis. *Arthritis Rheum* 2001;44:2819–27.
94. Mugnier B, Balandraud N, Darque A, Roudier C, Roudier J, and Reviron D. Polymorphism at position −308 of the tumor necrosis factor alpha gene influences outcome of infliximab therapy in rheumatoid arthritis. *Arthritis Rheum* 2003;48:1849–52.
95. Cuchacovich M, Ferreira L, Aliste M, Soto L, Cuenca J, Cruzat A, Gatica H, Schiattino I, Perez C, Aguirre A, Salazar-Onfray F, and Aguillon JC. Tumour necrosis factor-alpha (TNF-alpha) levels and influence of −308 TNF-alpha promoter polymorphism on the responsiveness to infliximab in patients with rheumatoid arthritis. *Scand J Rheumatol* 2004;33: 228–32.
96. Fabris M, Tolusso B, Di Poi E, Assaloni R, Sinigaglia L, and Ferraccioli G. Tumor necrosis factor-alpha receptor II polymorphism in patients from southern Europe with mild-moderate and severe rheumatoid arthritis. *J Rheumatol* 2002;29:1847–50.
97. Padyukov L, Lampa J, Heimburger M, Ernestam S, Cederholm T, Lundkvist I, Andersson P, Hermansson Y, Harju A, Klareskog L, and Bratt J. Genetic markers for the efficacy of tumour necrosis factor blocking therapy in rheumatoid arthritis. *Ann Rheum Dis* 2003;62:526–9.
98. Gregersen PK, Silver J, and Winchester RJ. The shared epitope hypothesis. An approach to understanding the molecular genetics of susceptibility to rheumatoid arthritis. *Arthritis Rheum* 1987;30:1205–13.
99. Martinez A, Salido M, Bonilla G, Pascual-Salcedo D, Fernandez-Arquero M, de Miguel S, Balsa A, de la Concha EG, and Fernandez-Gutierrez B. Association of the major histocompatibility complex with response to infliximab therapy in rheumatoid arthritis patients. *Arthritis Rheum* 2004;50:1077–82.
100. Mulcahy B, Waldron-Lynch F, McDermott MF, Adams C, Amos CI, Zhu DK, Ward RH, Clegg DO, Shanahan F, Molloy MG, and O'Gara F. Genetic variability in the tumor necrosis factor-lymphotoxin region influences susceptibility to rheumatoid arthritis. *Am J Hum Genet* 1996;59:676–83.
101. Waldron-Lynch F, Adams C, Amos C, Zhu DK, McDermott MF, Shanahan F, Molloy MG, and O'Gara F. Tumour necrosis factor 5′ promoter single nucleotide polymorphisms influence susceptibility to rheumatoid arthritis (RA) in immunogenetically defined multiplex RA families. *Genes Immun* 2001;2:82–7.
102. Criswell LA, Lum RF, Turner KN, Woehl B, Zhu Y, Wang J, Tiwari HK, Edberg JC, Kimberly RP, Moreland LW, Seldin MF, and Bridges SL, Jr. The influence of genetic variation in the

HLA-DRB1 and LTA-TNF regions on the response to treatment of early rheumatoid arthritis with methotrexate or etanercept. *Arthritis Rheum* 2004;50:2750–6.

103. Haas DW. Pharmacogenomics and HIV therapeutics. *J Infect Dis* 2005;191:1397–400.
104. Mallal S, Nolan D, Witt C, Masel G, Martin AM, Moore C, Sayer D, Castley A, Mamotte C, Maxwell D, James I, and Christiansen FT. Association between presence of HLA-B*5701, HLA-DR7, and HLA-DQ3 and hypersensitivity to HIV-1 reverse-transcriptase inhibitor abacavir. *Lancet* 2002;359:727–32.
105. Hetherington S, Hughes AR, Mosteller M, Shortino D, Baker KL, Spreen W, Lai E, Davies K, Handley A, Dow DJ, Fling ME, Stocum M, Bowman C, Thurmond LM, and Roses AD. Genetic variations in HLA-B region and hypersensitivity reactions to abacavir. *Lancet* 2002;359:1121–2.
106. Rauch A, Nolan D, Martin A, McKinnon E, Almeida C, and Mallal S. Prospective genetic screening decreases the incidence of abacavir hypersensitivity reactions in the Western Australian HIV cohort study. *Clin Infect Dis* 2006;43:99–102.
107. Hughes DA, Vilar FJ, Ward CC, Alfirevic A, Park BK, and Pirmohamed M. Cost-effectiveness analysis of HLA B*5701 genotyping in preventing abacavir hypersensitivity. *Pharmacogenetics* 2004;14:335–42.
108. Zucker SD, Qin X, Rouster SD, Yu F, Green RM, Keshavan P, Feinberg J, and Sherman KE. Mechanism of indinavir-induced hyperbilirubinemia. *Proc Natl Acad Sci USA* 2001;98:12671–6.
109. Delaney WEt, Locarnini S, and Shaw T. Resistance of hepatitis B virus to antiviral drugs: current aspects and directions for future investigation. *Antivir Chem Chemother* 2001;12:1–35.
110. Chen PJ, Lin CG, Lin FY, Chen E, and Wu LS. Genetic structural differences between responders and non-responders to interferon therapy for chronic hepatitis-B patients. *J Hum Genet* 2006;51:984–91.
111. Freeman JL, Perry GH, Feuk L, Redon R, McCarroll SA, Altshuler DM, Aburatani H, Jones KW, Tyler-Smith C, Hurles ME, Carter NP, Scherer SW, and Lee C. Copy number variation: new insights in genome diversity. *Genome Res* 2006;16:949–61.
112. Feuk L, Carson AR, and Scherer SW. Structural variation in the human genome. *Nat Rev Genet* 2006;7:85–97.
113. McCarroll SA, Hadnott TN, Perry GH, Sabeti PC, Zody MC, Barrett JC, Dallaire S, Gabriel SB, Lee C, Daly MJ, and Altshuler DM. Common deletion polymorphisms in the human genome. *Nat Genet* 2006;38:86–92.
114. Hinds DA, Kloek AP, Jen M, Chen X, and Frazer KA. Common deletions and SNPs are in linkage disequilibrium in the human genome. *Nat Genet* 2006;38:82–5.
115. Conrad DF, Andrews TD, Carter NP, Hurles ME, and Pritchard JK. A high-resolution survey of deletion polymorphism in the human genome. *Nat Genet* 2006;38:75–81.
116. Tuzun E, Sharp AJ, Bailey JA, Kaul R, Morrison VA, Pertz LM, Haugen E, Hayden H, Albertson D, Pinkel D, Olson MV, and Eichler EE. Fine-scale structural variation of the human genome. *Nat Genet* 2005;37:727–32.
117. Sebat J, Lakshmi B, Troge J, Alexander J, Young J, Lundin P, Maner S, Massa H, Walker M, Chi M, Navin N, Lucito R, Healy J, Hicks J, Ye K, Reiner A, Gilliam TC, Trask B, Patterson N, Zetterberg A, and Wigler M. Large-scale copy number polymorphism in the human genome. *Science* 2004;305:525–8.
118. Iafrate AJ, Feuk L, Rivera MN, Listewnik ML, Donahoe PK, Qi Y, Scherer SW, and Lee C. Detection of large-scale variation in the human genome. *Nat Genet* 2004;36:949–51.
119. Komura D, Shen F, Ishikawa S, Fitch KR, Chen W, Zhang J, Liu G, Ihara S, Nakamura H, Hurles ME, Lee C, Scherer SW, Jones KW, Shapero MH, Huang J, and Aburatani H. Genome-wide detection of human copy number variations using high-density DNA oligonucleotide arrays. *Genome Res* 2006;16:1575–84.

120. Redon R, Ishikawa S, Fitch KR, Feuk L, Perry GH, Andrews TD, Fiegler H, Shapero MH, Carson AR, Chen W, et al. Global variation in copy number in the human genome. *Nature* 2006;444:444–54.

121. Hebbring SJ, Adjei AA, Baer JL, Jenkins GD, Zhang J, Cunningham JM, Schaid DJ, Weinshilboum RM, and Thibodeau SN. Human SULT1A1 gene: copy number differences and functional implications. *Hum Mol Genet* 2007;16:463–70.

122. Ulrich CM, Robien K, and McLeod HL. Cancer pharmacogenetics: polymorphisms, pathways and beyond. *Nat Rev Cancer* 2003;3:912–20.

123. Dausset J, Cann H, Cohen D, Lathrop M, Lalouel JM, and White R. Centre d'etude du polymorphisme humain (CEPH): collaborative genetic mapping of the human genome. *Genomics* 1990;6:575–7.

124. Dolan ME, Newbold KG, Nagasubramanian R, Wu X, Ratain MJ, Cook EH, Jr., and Badner JA. Heritability and linkage analysis of sensitivity to cisplatin-induced cytotoxicity. *Cancer Res* 2004;64:4353–6.

125. The International HapMap Project. *Nature* 2003;426:789–96.

126. Watters JW, Kraja A, Meucci MA, Province MA, and McLeod HL. Genome-wide discovery of loci influencing chemotherapy cytotoxicity. *Proc Natl Acad Sci USA* 2004;101:11809–14.

127. Haston CK, Wang M, Dejournett RE, Zhou X, Ni D, Gu X, King TM, Weil MM, Newman RA, Amos CI, and Travis EL. Bleomycin hydrolase and a genetic locus within the MHC affect risk for pulmonary fibrosis in mice. *Hum Mol Genet* 2002;11:1855–63.

128. Kimura S and Gonzalez FJ. Applications of genetically manipulated mice in pharmacogenetics and pharmacogenomics. *Pharmacology* 2000;61:147–53.

Chapter 8

Pharmacogenetics of Drug Disposition

8.1. INTRODUCTION

Pharmacogenomics is the study of the human genome and its products (including RNA and protein) and their effect on drug response phenotypes; within a narrower scope, pharmacogenetics is the study of variations in DNA sequence and their effect on drug response phenotypes, including pharmacokinetics (PK) and pharmacodynamics. The evidence for variability in drug response, and an inheritable component to this variability, preceded the 1959 coining of the term "pharmacogenetics" by Friedrich Vogel (1). Early examples of pivotal discoveries in pharmacogenetics include the identification of polymorphic pathways that affect the disposition of drugs such as succinylcholine, isoniazid, sparteine, and debrisoquine (2–6).

Technological and informational advances in the past few decades, including increased knowledge of the human genome and critical advances in molecular genetic tools have provided the *how to* for further advances in pharmacogenetics. The promise of personalized medicine and encouragement in this direction from regulatory agencies such as the FDA have provided the *why* for advancing pharmacogenetics.

As mentioned elsewhere in this book, a concerning recent metric is the decrease in the success of pharmaceutical research and development. While productivity has increased in terms of the numbers of new compounds and drug targets identified (as measured by Investigational New Drug Applications to the FDA) and spending on biomedical research has increased, there has been a decrease in the number of New Drug Applications made to the FDA (7). So while the costs of drug development have steadily increased, the likelihood of

Genomics in Drug Discovery and Development, by Dimitri Semizarov and Eric Blomme

launching a new drug has decreased. The decreased success, despite increased spending, in developing new drugs may be partly due to a lack of understanding of the underlying variability in drug response or safety, and early incorporation of pharmacogenetic and pharmacogenomic technologies could aid successful clinical trials (7).

The fact that individuals respond differently to drug therapy has long been recognized—for a given dose of a given drug, a certain proportion of the population responds adequately, while others will manifest therapeutic failure (lack of efficacy) or adverse effects (Fig. 8.1). The goal of personalized medicine is to shift the health care and pharmaceutical industries from the "one size fits all" paradigm toward delivering the right dose of the right drug to the right patient every time. Typically, in the practice of medicine, the right drug and the right dose are reached by a trial-and-error approach or not at all. While the trial-and-error approach for dosing may be acceptable for certain types of drugs (e.g., those with wide therapeutic indices, prescribed for non-life-threatening illnesses), it could be inherently dangerous for others. The trial-and-error method of dosing for warfarin, for example, may result in increased risk of hemorrhage, and associated hospitalizations, especially during the early weeks of therapy (8). Identifying individuals who would respond to a particular drug or would be at risk for adverse events also promises to lower health care costs.

Drugs are developed and receive regulatory approval based on mean efficacy responses and mean tolerability. Drugs may fail in the development process when this *mean* response is not adequate. Identifying individual drug responses and

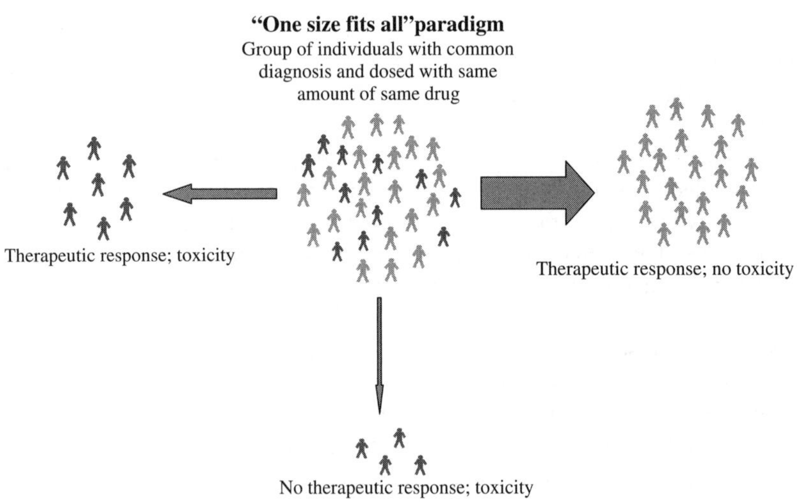

Figure 8.1 The goal of pharmacogenetics is to predict which individuals will respond well, will not respond, or will be at risk for toxicity, based on genetic variability; a goal of personalized medicine is to then use pharmacogenetic information to tailor drug therapy to each individual. See color insert.

subgroups of responders may "rescue" certain drugs that would otherwise fail. Predicting individual responses to individual drugs is the promise of personalized medicine and a *raison d'être* for pharmacogenetics. From an industry perspective, identifying responders and nonresponders and the right dose of a drug for a specific population can potentially shorten clinical trials or increase the likelihood of approval.

Genetic factors affecting drug efficacy are discussed in chapter 7; the focus of this chapter is on genetic factors that influence the PK of drugs, with particular emphasis on the value of pharmacogenetic studies in drug development.

8.2. GENES AND POLYMORPHISMS AFFECTING DRUG DISPOSITION

Identifying the sources of variability in drug disposition early in the development of a new chemical entity (NCE) enables improved strategies and decision-making for the development of the NCE. Pharmacogenetic-pharmacokinetic (PG-PK) analyses (Box 8-1) performed in the early phases of drug development can determine whether variability in the PK of the NCE can be attributed to underlying genetic differences between individuals. Examples of how PG-PK information can specifically influence drug development are provided in Section 8.4.

BOX 8-1 *Conducting a PG-PK Study*

Pharmacogenetic studies to investigate the genetic basis for variability in drug disposition are typically done in early clinical trials that are collecting PK data for a new chemical entity (NCE). A general overview of the process is described here.

- Appropriate language describing the pharmacogenetic substudy must be incorporated into the clinical study protocol and informed consent form.
  - The study protocol should include the logistics of sample collection (e.g., whether sampling is mandatory in the trial or optional, time of collection, type and amount of sample to be collected, and storage and shipping conditions) as well as a general description of the pharmacogenetic plan (e.g., which genes and what phenotypes). A whole blood sample is usually the tissue of choice for a pharmacogenetic study since it can easily be collected and yields quality DNA.
  - Each subject is required to sign an informed consent form, which conveys the pharmacogenetic study procedures and risks. The consent form should also inform subjects about how their samples will be utilized, and how they may withdraw from the study. A written consent is acquired prior to any study-related procedure.

- Genotyping can be performed on genomic DNA isolated from whole blood by a variety of methods.
- The pharmacogenetic data are related to PK phenotype (for e.g., **AUC**, $\mathbf{C_{max}}$, or **clearance**) with appropriate statistical methods, such as analysis of variation (ANOVA) or regression models.
- Any observed association should be replicated in a separate study to decrease the probability of false positive results.

Advances in sequencing and genotyping technologies have improved the ability to study genotype-phenotype correlations. These improvements in technology, combined with well-curated databases (Table 8.1), and in many instances (e.g., CYPs, UGTs, SULTs, NATs, SLC transporters, and more) the adoption of common, controlled nomenclature systems for genes and alleles, have facilitated the growth of pharmacogenetics (Box 8-2) (9–18)

Over 170 gene products including drug-metabolizing enzymes, drug transport proteins, serum binding proteins and regulatory factors, are estimated to affect drug disposition, and over half of the genes encoding these factors are polymorphic (19). The influence of polymorphisms in drug-metabolizing enzymes on the disposition of an NCE depends on several factors, including the functionality of the polymorphism, the importance of that enzyme in the NCE's metabolism, and the availability of alternative metabolic pathways for the NCE.

Various types of genetic sequence polymorphisms affect the expression or activity of proteins (Box 8-3). In general, where genotype-phenotype relationships are well established, three metabolizer phenotypes are identified and attributed to genetic polymorphisms: poor metabolizers (PM), who lack the functional enzyme (homozygous for nonfunctional or decreased function alleles), intermediate metabolizers (IM), who are heterozygous for a nonfunctional allele or homozygous for decreased function alleles, and extensive metabolizers (EM), who have two normal (usually designated *1) alleles; a fourth phenotype, ultrarapid metabolizers (UM), is often observed in individuals who have multiple copies of the functional gene. The EM phenotype is usually the most common phenotype observed in the population, and activity in these individuals is considered to be "normal." In certain cases (e.g., CYP2C19, CYP2D6), the IM phenotype may have a wide range of values and may not always be useful as an independent category.

Different metabolizer phenotypes result in interindividual variability in drug PK that may be reflected in drug effect. Figure 8.2 presents a theoretical strategy for use of PG-PK data for rational dose adjustments to limit genotype-based interindividual variability; such dosing regimens may be

Table 8.1 Useful Websites for PG-PK Research

Database	URL
Pharmacogenetic Research	
Genomics at FDA	http://www.fda.gov/cder/genomics/default.htm
HUGO Gene Nomenclature Committee database	http://www.genenames.org/cgi-bin/hgnc_search.pl
dbSNP	http://www.ncbi.nlm.nih.gov/projects/SNP/
PharmGKB	http://www.pharmgkb.org/
International Hapmap Project	http://www.hapmap.org
GeneCards	http://www.genecards.org
JSNP	http://snp.ims.u-tokyo.ac.jp/
Drug-Metabolizing Enzymes and Drug Transport Proteins	
Cytochrome P450 database	http://cpd.ibmh.msk.su
Home page of the Human Cytochrome P450 (CYP) Allele Nomenclature Committee	http://www.fda.gov/cder/genomics/default.htm
Cytochrome P450 Home Page	http://drnelson.utmem.edu/CytochromeP450.html
FMO3—Allellic Variant Database, University College London, UK	http://human-fmo3.biochem.ucl.ac.uk/Human_FMO3/
Consensus Arylamine *N*-Acetyltransferase (NAT) Gene Nomenclature	http://louisville.edu/medschool/pharmacology/NAT.html
UDP Glucuronosyltransferase Homepage	http://som.flinders.edu.au/FUSA/ClinPharm/UGT/
Cytosolic Sulfotransferase (SULT) Homepage	http://www.fccc.edu/research/labs/blanchard/sult/
UGT allele nomenclature home page	http://galien.pha.ulaval.ca/labocg/alleles/alleles.html
Organic Anion Transporting Polypeptides	http://www.kpt.unizh.ch/oatp/
Human ABC Transporters	http://nutrigene.4t.com/humanabc.htm
SLC Tables	http://www.bioparadigms.org/slc/intro.htm

especially valuable for drugs with narrow therapeutic windows (e.g., warfarin). The frequencies of variant alleles, and, consequently, frequencies of metabolizer phenotype, differ considerably among ethnic populations, making allele selection a critical consideration for any genotyping strategy (20).

Genetic variability in drug disposition is characterized best for drug-metabolizing enzymes, but in the past decade there has been increased focus on drug transport proteins as well. An overview of the pharmacogenetics of drug-metabolizing enzymes and drug transport proteins follows.

BOX 8-2 *Star Nomenclature*

Improvements in genetic technologies and a greater understanding of the molecular basis for variability in drug metabolism have resulted in a surge in the discovery of variant alleles that affect phenotype. The star nomenclature was developed as a system to organize, annotate, and standardize the genetic information being discovered in laboratories all over the world (169). Nomenclature committees for the CYPs and other drug-metabolizing enzymes were established and tasked with curating existing and newly discovered variants. This nomenclature system has been widely adopted by scientists and allows for a common, standardized basis for the exchange of information on genetic variants.

In the star-allele nomenclature, *1 is designated as the reference sequence against which all others are compared. Usually the *1 designation is given to the first allele sequenced, which is typically the most common allele in multiple ethnic populations, and usually also encodes for the protein with highest activity. However, this is not always the case, as illustrated by *CYP3A5*. In Caucasians, the allele frequency of *CYP3A5*1* is only approximately 4–7%. *CYP3A5*3*, a major contributor to the polymorphic expression of CYP3A5, is the most common allele, with an allele frequency of approximately 90%, (78, 170–173).

Subsequent to the designation of *1, a unique number is assigned when a novel polymorphism that affects expression, splicing, or results in an amino acid change is identified (e.g., *CYP2C9*2, CYP2C9*3*). If a nonfunctional variant is identified on an already existing allele then it is assigned a letter (e.g., *CYP2C9*2B*).

The star nomenclature is well established for several drug-metabolizing enzymes and drug transport proteins.

The star-allele nomenclature has served pharmacogenetics well over the past decade, but the large numbers of genetic variants being identified by high-throughput sequencing platforms may be straining the system. Limitations of the star-allele nomenclature tables include the lack of links to genome assemblies, limited phenotypic information for alleles, and the lack of information on frequencies and ethnic distributions (174).

BOX 8-3 *Types of Genetic Variations*

Various types of genetic sequence variations exist that can affect the expression or activity of proteins. Polymorphisms are DNA variants that occur with a frequency >1%. There are many different types of polymorphisms including nucleotide substitutions, insertions, deletions, and copy number variations. These variations are defined and discussed below.

- **Single nucleotide polymorphisms (SNP)**, which consist of a nucleotide substitution, are the most common type of genetic variation in the human genome. A SNP within the coding region of a gene can be synonymous (not causing an amino acid change) or nonsynonymous (causing an amino acid change). Nonsynonymous polymorphisms often alter the activity (usually a decrease) of the protein. For example, *CYP2C9*2*, which has an arginine to cysteine change at amino acid position 144, (commonly written as R144C) has decreased activity compared to the wild-type allele, *CYP2C9*1*. Since synonymous polymorphisms do not alter amino acid sequence they are generally not expected to affect protein activity. However, it is possible for synonymous polymorphisms to alter protein conformation and affect substrate binding (175–177). SNPs within promoter regions of genes can alter the transcription of the gene by interfering with transcription factor binding

- **Indels** are insertions or deletions of up to a few nucleotides. Depending on where they occur, they can have different effects on the expression or function of the protein. For example, the *UGT1A1*28* allele consists of a 2-base pair insertion within the TATA box of the gene (TA_7 instead of TA_6) and results in decreased expression of UGT1A1. If the insertion or deletion is within the coding region of the gene it can cause a shift in the reading frame, resulting in a truncated protein that may be nonfunctional (e.g., *CYP2A6*17*)
- **Copy number variations** result from deletion or amplification of genes. Copy number variations are detected in several genes encoding drug-metabolizing enzymes, including *CYP2A6, CYP2D6*, and *SULT1A1*. Gene deletions result in poor metabolizer (PM) phenotypes, while amplifications often result in ultrarapid metabolizer (UM) penotypes.

8.2.1. Drug-Metabolizing Enzymes

8.2.1.1. Cytochrome P450s

The cytochrome P450s (CYPs) are the most widely studied family of enzymes that bioactivate or metabolize most commonly used drugs. While there are currently 57 CYP genes identified in humans, only a small portion of these have a role in the metabolism of drugs. The remaining CYPs have key roles in cholesterol and steroid pathways (e.g., CYP7a, CYP17). The drug-metabolizing CYPs are primarily within the CYP1, CYP2, and CYP3 families. These CYPs exhibit broad and overlapping substrate specificities. Considerable interindividual variability is observed in the expression and activity of the CYPs, much of which is attributed to genetic sequence polymorphisms within the genes. The variability in the expression and activity of these CYPs influences the metabolism of several commonly prescribed drugs.

A complete listing of polymorphisms within the CYPs is available on the website "Home Page of the Human Cytochrome P450 (CYP) Allele Nomenclature Committee" (http://www.cypalleles.ki.se/). Key CYP subfamilies involved in drug metabolism are briefly discussed below.

CYP1A. The human CYP1A family comprises CYP1A1 and CYP1A2. These enzymes have a major role in the bioactivation of procarcinogens, including polycyclic aromatic hydrocarbons, heterocyclic aromatic amines,

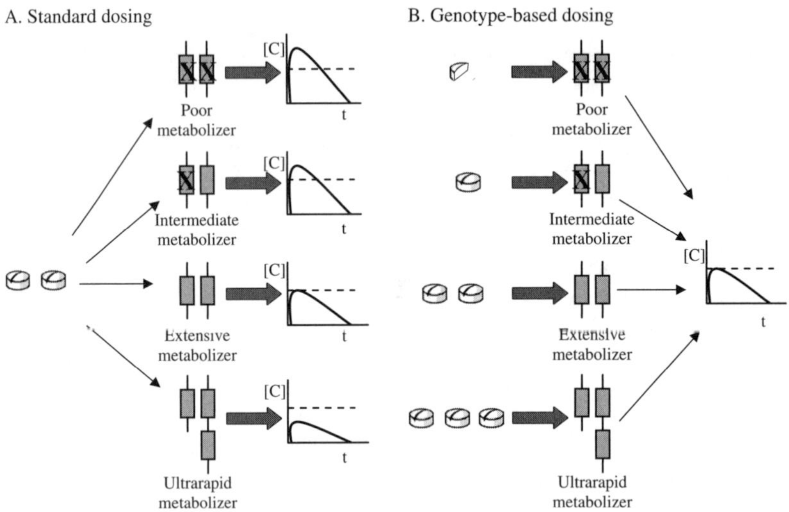

Figure 8.2 Strategy for genotype-based dosing. A) Standard, "one-size-fits-all" dosing results in different plasma levels in individuals with different metabolizer phenotypes. B) In genotype-based dosing, genotype/metabolizer phenotype is used to adjust the dose to achieve the desired plasma level. Adapted and reprinted from Biochim Biophys Acta, 1770, Kirchheiner and Seeringer, Clinical implications of pharmacogenetics of cytochrome P450 drug metabolizing enzymes, 489–494, 2007, with permission From Elsevier (59).

and aminoazo dyes, and are also involved in the metabolism of phenacetin and caffeine (21). While polymorphisms within CYP1A1 and CYP1A2 are not expected to affect drug metabolism in an appreciable way, inducibility of these enzymes by a variety of drugs and environmental pollutants does affect their activity, potentially leading to toxicity (discussed in Chapter 5). The CYP1A enzymes are highly inducible by environmental contaminants such as benzo[*a*]pyrene and 2,3,7,8-tetrachlorodibenzo-*p*-dioxin via an aromatic hydrocarbon receptor-mediated pathway. Additionally, although omeprazole is not a CYP1A substrate, it is known to induce CYP1A (22). Induction of CYP1A can potentially increase the risk of lung cancer in smokers because of increased bioactivation of polycyclic aromatic hydrocarbons. As such, CYP1A inducibility is an unwanted drug effect, and pharmaceutical companies routinely test compounds for CYP1A induction. A positive result may seriously influence whether a drug candidate will be pursued for further development.

CYP2A. The human CYP2A family comprises CYP2A6, CYP2A7, CYP2A13, and CYP2A18P. The two functional members of this family, CYP2A6 and CYP2A13, are expressed predominantly in the liver and respiratory tract, respectively. Of these, CYP2A6 is the primary contributor to drug metabolism.

Compared with other CYPs, CYP2A6 has relatively narrow substrate specificity. It is the principal CYP that metabolizes nicotine and is also involved in the metabolism of drugs such as coumarin, halothane, tegafur, and losigamone (23–26). *CYP2A6* is highly polymorphic, with over 30 alleles, including a gene deletion and duplication, currently identified and named (http://www.cypalleles.ki.se/cyp2a6.htm). Inactive alleles of *CYP2A6* include *CYP2A6*2* (does not incorporate haem), *CYP2A6*4* (gene deletion), *CYP2A6*5* (unstable protein), and *CYP2A6*20* (truncated protein) (27–33). Alleles with decreased activity include *CYP2A6*6, *7, *9, *10, *11, *12, *17, and *19* (34–39). Considerable interethnic variability is observed in the frequency of *CYP2A6* alleles. The combined frequencies of alleles with impaired or absent activity (*CYP2A6*2, *4, *5, *7, *9, *10, *11, *17, *19*, and **20*) are 9.1%, 21.9%, 42.9%, and 50.5% in white, black, Korean, and Japanese subjects, respectively (40). Reflecting the presence of functional polymorphisms and the interethnic differences in their frequencies, considerable interindividual and interethnic variability is observed in the metabolism of nicotine and coumarin (31, 41–43).

CYP2B. The human CYP2B subfamily consists of one functional gene, *CYP2B6*, and one pseudogene, *CYP2B7*. CYP2B6 is involved in the metabolism of several drugs including buproprion, efavirenz, nevirapine, propofol, and others (44). *CYP2B6* is highly polymorphic, with over 28 alleles currently defined (http://www.cypalleles.ki.se/cyp2b6.htm). While the molecular characterization of *CYP2B6* alleles lags behind other CYPs, thus far consistently replicated associations are reported between efavirenz exposure and *CYP2B6* genotype (45–48). Specifically, the $516G \rightarrow T$ polymorphism (which results in a change from glutamine to histidine at amino acid position 172), which is part of multiple alleles (including *CYP2B6*6, CYP2B6*7, CYP2B6*9*, and others), is associated with increased exposure to efavirenz. However, the metabolic phenotypes assigned to CYP2B6 (e.g., PM or EM) are not as clear as they are for other CYPs, and the effects of polymorphisms may be substrate specific (44). In addition to the variability conferred by polymorphisms, CYP2B6 is also inducible by various drugs including phenobarbital, cyclophosphamide, rifampin, efavirenz, carbamazepine, and others (44). The induction of CYP2B6, which is mediated by the nuclear receptors constitutive androstane receptor (CAR) and pregnane X receptor (PXR), also affects the PK of CYP2B6 substrates and, as such, confounds genotype-phenotype studies.

CYP2C. The CYP2C subfamily of CYPs includes CYP2C8, CYP2C9, CYP2C18, and CYP2C19. CYP2C9 and CYP2C19 are considered the clinically important members of the subfamily, though CYP2C8 also has a role in drug metabolism. CYP2C9 is one of the principal CYPs contributing to the metabolism of commonly used drugs (e.g., warfarin, phenytoin, tolbutamide, nonsteroidal anti-inflammatory drugs such as diclofenac and celecoxib, and others). While many SNPs have been detected in the *CYP2C9* gene (http://www.cypalleles.ki.se/cyp2c9.htm), two alleles, *CYP2C9*2* and *CYP2C9*3*, are typically responsible for the poor metabolism phenotype in individuals. These alleles each have a SNP that results in an amino acid change and decreased enzyme activity. There are at least 17 drugs for which an association exists between *CYP2C9*2* or *CYP2C9*3* genotype and pharmacokinetic parameters (49). Other impaired *CYP2C9* alleles such as *CYP2C9*6* and *CYP2C9*25* consist of a frame shift that results in absence of enzyme activity, but these alleles are rare. The importance of *CYP2C9* variability is reflected in the manufacturer's label for Celebrex®, which states, "Patients who are known or suspected to be P450 2C9 poor metabolizers based on a previous history should be administered celecoxib with caution as they may have abnormally high plasma levels due to reduced metabolic clearance" (http://www.fda.gov/cder/foi/label/2005/020998s018,019lbl.pdf). CYP2C19 has a major role in the metabolism of several commonly used drugs including proton pump inhibitors (e.g., omeprazole, lansoprazole), fluoxetine, sertraline, and nelfinavir). Several genetic variants result in decreased function, but most cases of PM phenotype can be attributed to *CYP2C19*2* and *CYP2C19*3*. The *CYP2C19*2* allele results in enzyme inactivation, while the *CYP2C19*3* allele results in a truncated protein. *CYP2C19* genotype is clinically relevant for the pharmacokinetics of several drugs including the proton pump inhibitors (omeprazole, lansoprazole, and pantoprazole), mephenytoin, and diazepam. For example, the AUCs of omeprazole, lansoprazole, and pantoprazole are approximately five-fold higher in CYP2C19 PMs than in CYP2C19 EMs (50). Furthermore, there is also an association between efficacy of omeprazole and *CYP2C19* genotype, with CYP2C19 PMs having higher cure rates [(51) and references therein].

CYP2D. The human CYP2D family comprises three highly homologous genes, *CYP2D6, CYP2D7*, and *CYP2D8*. While *CYP2D7* and *CYP2D8* are considered pseudogenes, CYP2D6 metabolizes a wide range of commonly prescribed drugs including codeine, dextromethorphan, tamoxifen, most tricyclic antidepressants, and many others. Genetic polymorphisms within the *CYP2D6* gene contribute to interindividual variability in the PK of drugs metabolized by this enzyme. Relationships between genotype and metabolizer phenotype for *CYP2D6* are well characterized, with individuals classified as PM, IM, EM, or UM based on genotype. Approximately 78 genetic variants of *CYP2D6* have been detected (http://www.cypalleles.ki.se/cyp2d6.htm). While some of these variants result in absent or inactive CYP2D6 protein (PM alleles), others demonstrate reduced catalytic activity (IM alleles).

The *CYP2D6*5* allele is the result of a large chromosomal deletion that results in the absence of the entire *CYP2D6* gene. Other PM alleles (e.g., *CYP2D6*3, CYP2D6*4, CYP2D6*6, CYP2D6*8*, and others) have either SNPs or single base insertions/deletions, which interfere with the reading frame or affect splicing, and result in CYP2D6 protein with no activity (52). Alleles such as *CYP2D6*10, *17, *36*, and **41* have decreased activity (IM alleles). In addition, alleles carrying multiple copies (ranging from 2 to 13) of the *CYP2D6* gene have been identified (53, 54). UM phenotype is often caused by amplification of functional *CYP2D6* alleles such as *CYP2D6*1, *2*, and others. Of note, the mere observation of *CYP2D6* amplification is not sufficient to classify an individual as a UM since amplification of a nonfunctional allele does not increase metabolic capacity (55).

CYP2D6 PMs have low CYP2D6 metabolic capacity and exhibit higher AUCs for CYP2D6 substrates, increased potential for drug-drug interactions and adverse reactions, and lower efficacy for drugs requiring activation by CYP2D6 (e.g., codeine, tramadol, tamoxifen) [(20, 56, 57) and references therein]. The UMs exhibit high metabolic activity, and these individuals may not achieve therapeutic efficacy at normal doses of drugs metabolized by CYP2D6. For example, the UM phenotype is 10-fold more common in nonresponders than in responders to antidepressant therapy (58). While individuals are classified as CYP2D6 IMs, this phenotype exhibits a wide range of activity that overlaps with EMs, and as such, this phenotype is not always useful. The clinically relevant differences are observed for CYP2D6 PMs and UMs versus EMs.

The influence of *CYP2D6* polymorphisms on the metabolism of antidepressant drugs has been extensively studied, and dose adjustments for CYP2D6 PMs and UMs are suggested, but prospective studies are required before such genotype-based dosing is utilized in antidepressant therapy (59, 60). Information on CYP2D6 metabolizer phenotype, as it pertains to drug PK, is found in the manufacturer's labels for several drugs including aripiprazole, thioridazine, atomoxetine, protryptyline, and others (http://www.fda.gov/cder/genomics/genomic-biomarkers-table.htm). While it is known that CYP2D6 metabolizer status affects the pharmacokinetics of several drugs, genotyping before therapy is not required. As with most of the CYPs, considerable interethnic differences in allele frequencies are observed for *CYP2D6*, and genotype selection is critical to pharmacogenetic studies (20, 61).

CYP3A. The human CYP3A subfamily comprises CYP3A4, CYP3A5, CYP3A7, and CYP3A43. CYP3A4 and CYP3A5 are the main members of this family. CYP3A7 is present in the fetus but not in most adults, and CYP3A43 is expressed at very low levels. CYP3A4 is the most important of all drug-metabolizing enzymes since it is involved in the metabolism of approximately 50% of currently used therapeutic agents (56, 62, 63). Interindividual variability in CYP3A activity is high. In vivo studies have shown 5- to 10-fold variability in the metabolism of CYP3A substrates, and in vitro studies show even greater (>30 fold) variability (64–67). Drug interactions with CYP3A are well characterized and can increase the range of variability to approximately 400-fold [(68) and references therein]. It is estimated that 70–90% of the variability in CYP3A activity may be attributable to genetic factors (69), but these factors remain largely unknown. While SNPs have been identified within the *CYP3A4* gene, they are generally of low frequency and cannot account for the observed variability. The continuous and unimodal distribution of CYP3A activity suggests that multiple genes are involved in its regulation, and that individual genetic factors play a small role (68). The *CYP3A4*1B* allele is the most extensively characterized of the *CYP3A4* alleles and has a SNP within the 5′-regulatory region. *CYP3A4*1B* has been associated with multiple disease states, but researchers have failed to show an effect on the metabolism of endogenous compounds or drugs [(70) and references therein]. In general, genotyping of known *CYP3A4* alleles is not expected to explain variability in drug metabolism, and the more clinically relevant issue is the regulation of *CYP3A4* expression. *CYP3A4* activity is highly inducible by a variety of ligands including rifampicin, phenobarbital, dexamethasone, and hyperforin. Induction of *CYP3A4*, which is mediated via the PXR pathway, can increase the metabolism of other drugs that are CYP3A substrates, making them ineffective. For example, St. John's wort, which is a potent inducer of CYP3A, can cause therapeutic failure of cyclosporine or HIV-protease inhibitors (also CYP3A substrates), if taken concurrently (71–73). Conversely, CYP3A inhibition, caused by drugs such as nitroimidazole antifungals, dilitiazem, and verapamil, as well as dietary components such as grapefruit juice, can increase the exposure to CYP3A substrates taken concurrently, leading to adverse events (65, 74, 75).

CYP3A5 is a highly polymorphic gene that is expressed in the livers of only 10–30% of Caucasians and Asians, but in more than 50% of African Americans (76–78). The most common allele in a number of ethnic populations is *CYP3A5*3*, in which a polymorphism creates a cryptic splice site resulting in an improperly spliced mRNA and a truncated protein (77, 78). The original CYP3A5 cDNA sequence came from a liver sample expressing CYP3A5, thus making it one of the few CYPs in which the *1 nomenclature is not designated to the most common allele (see also Box 8-2). The clinical relevance of polymorphic expression of *CYP3A5* is not readily apparent since most drugs metabolized by CYP3A5 are also metabolized by CYP3A4. However, *CYP3A5* status is associated with the steady-state PK of tacrolimus and has a dose-related effect on a ketolide antibiotic, suggesting that CYP3A5 status may only become an important factor under certain dosing conditions (79–82).

8.2.1.2. Flavin-Containing Monooxygenases

The flavin-containing monooxygenases (FMO) are a family of flavoproteins that catalyze the oxidation of a wide range of nucleophilic heteroatom-containing compounds. In humans there are six FMO isoforms, of which five (FMOs 1,2,3,4,6) are functional. Of these, FMO1 is the major fetal isoform, and FMO3 is the major adult isoform (83). FMO3 is the most abundantly expressed FMO in adult human liver, and as such is the most widely studied. FMO3 has a broad substrate specificity and is involved in the metabolism of commonly used drugs such as cimetidine, ranitidine, clozapine, methimazole, itopride, ketoconazole, tamoxifen, and sulindac sulfide (84–88). Numerous polymorphisms have been reported in the *FMO3* gene (http://human-fmo3.biochem.ucl.ac.uk/Human_−FM03/Tables/tableall.html), a few of which are associated with altered metabolism of various substrates, including sulindac, ranitidine, and benzydamine (89–92). The best-characterized phenotype of *FMO3* sequence variations is trimethylaminuria or "fish odor syndrome." Several genetic variants of the *FMO3* gene are associated with the inability to metabolize the tertiary amine trimethylamine (TMA), derived In vivo from the bacterial degradation of dietary choline, carnitine, and lecithin, which are found in various dietary sources including marine fish, beetroot, legumes, and others. Individuals with mutations in the *FMO3* gene have a diminished capacity to metabolize TMA. Accumulation of TMA in the body results in excretion of malodorous free amines in body secretions and the characteristic "fish odor."

8.2.1.3. Arylamine N-Acetyltransferases

The arylamine *N*-acetyltransferases (NATs) catalyze the transfer of an acetyl group from acetyl-CoA to the terminal nitrogen of substrates. Identification of the acetylation polymorphism is one of the early success stories of pharmacogenetics. In the early 1950s isoniazid was introduced as a treatment for tuberculosis. Although extremely effective, isoniazid use was also associated with peripheral nerve damage in a subset of individuals. Larger studies conducted to determine the cause of this toxicity led to the observation of a bimodal population distribution of isoniazid plasma levels following a standard dose of the drug. Biochemical studies identified acetyl-isoniazid as the main urinary metabolite in humans, and showed that individuals who excreted the largest amount of unchanged isoniazid and the least amount of acetyl-isoniazid were more likely to suffer from peripheral neuropathy. Family studies revealed that the half-life of isoniazid was an inherited trait. Further biochemical studies attributed the differences in isoniazid elimination to variability in NAT activity, and ultimately led to the identification of *NAT* polymorphisms affecting acetylation phenotype (93–97). The molecular genetic bases of the acetylation polymorphism are now well characterized, and individuals are classified as "slow" or "rapid" acetylators. Slow acetylators often experience adverse reactions in response to drugs such as isoniazid, sulfonamides, procainamide, and hydralazine, whereas rapid acetylators may not respond to isoniazid and hydralazine (93, 94, 98–100).

There are two functional NATs in humans—NAT1 and NAT2. Both *NAT1* and *NAT2* contain sequence polymorphisms affecting activity, and as an exception to standard nomenclature, the wild-type alleles associated with rapid acetylator phenotype are designated as NAT1*4 and NAT2*4 [see Box 8-2 and (101)]. While both human *NATs* contain sequence variations, the acetylator polymorphism observed for drugs (including isoniazid) is attributed to *NAT2*.

The *NAT2*5A, *5B, *6A, *7A*, and **13* alleles confer slow acetylator status, with *NAT2*5B* and *NAT2*6A* being the most common slow acetylator alleles in Caucasians and Asians, respectively.

8.2.1.4. UDP-Glucuronosyltransferases

The UDP-glucuronosyltransferases (UGTs) are a superfamily of enzymes that catalyze the transfer of the glucuronic group from uridine diphosphoglucuronic acid to a substrate, thereby increasing polarity and facilitating excretion of the substrate in bile or urine. UGTs have major roles in the conjugation of endogenous substances (including bilirubin, steroid and thyroid hormones, and biliary acids) as well as exogenous compounds (including drugs, dietary components, environmental contaminants, and chemical carcinogens) (102, 103).

The human *UGT* gene family is composed of *UGT1* and *UGT2*, with three subfamilies, *UGT1A, UGT2A*, and *UGT2B*. The major drug-metabolizing UGTs are described here.

UGT1A The entire UGT1A subfamily is derived from a single gene locus that consists of 17 exons encoding nine functional proteins (UGT1A1, and UGT1A3–1A10) and four pseudogenes (Fig. 8.3). Each UGT1A isoform consists of a unique exon 1, and common exons 2, 3, 4, and 5 downstream, such that all the UGT1A proteins have a common C-terminal cosubstrate binding domain and a unique N-terminal substrate binding region. An additional, recently identified exon (referred to as exon 5b) increases the genetic diversity at this locus by allowing for alternative splicing (104). Specifically, alternative splicing of exons 5a and 5b results in two isoforms of each UGT1A (the classical isoform 1 and the newly identified isoform 2), potentially increasing the number of proteins derived from this locus to 18. The newly identified isoforms 2 may act as negative modulators of their respective isoforms 1, but further investigations of their in vivo function are required (105).

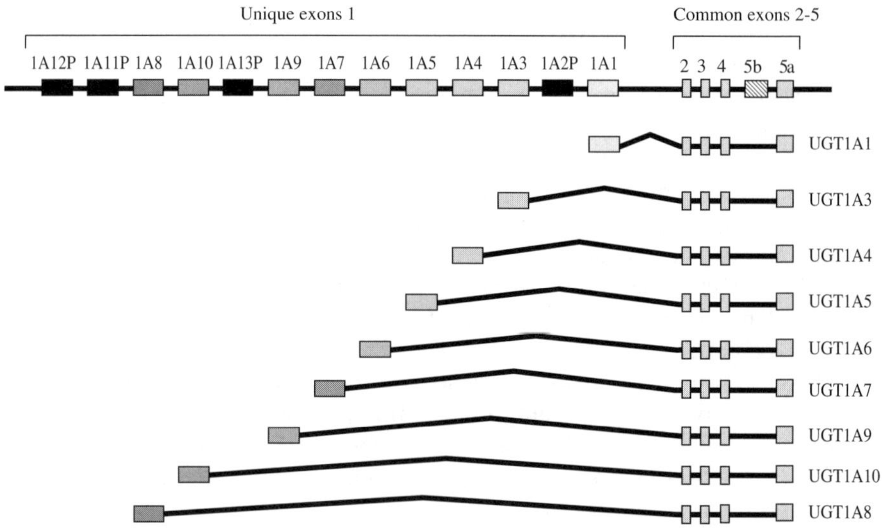

Figure 8.3 Schematic representation of the *UGT1A* locus and transcripts. The genomic structure includes 13 first exons, and four common exons (2–5). The black boxes represent pseudogenes; the hatched box represents the newly identified exon 5b. UGT1A transcripts generated from the 13 first exons and common exons 2, 3, 4, and 5a are shown. See color insert.

Within the *UGT1A* family, polymorphisms have been identified in *UGT1A1, UGT1A6, UGT1A7, UGT1A8*, and *UGT1A10* (106, 107). Of these, UGT1A1 and UGT1A6 are expressed in the liver and are the major contributors toward xenobiotic metabolism. The other UGT1A members may play roles in extrahepatic xenobiotic metabolism. UGT1A1 and UGT1A6 are discussed below.

UGT1A1 is the most extensively studied member of the UGT superfamily. It is the primary enzyme responsible for the conjugation of bilirubin and also has a role in conjugation of xenobiotics. Genetic alterations affecting *UGT1A1* expression or function are associated with severe familial forms, or milder, usually asymptomatic forms of unconjugated hyperbilirubinemia (Crigler-Najjar type I and II disorders and Gilbert syndrome, respectively) (108–113). A common polymorphism within the *UGT1A1* promoter is associated with *UGT1A1* expression level and glucuronidation capacity. This polymorphism consists of a variation in the number of "TA" dinucleotide repeats [A(TA)$_n$TAA...] in the TATA-box region of *UGT1A1* and results in four different alleles—*UGT1A1*1* ($n = 6$), *UGT1A1*33* ($n = 5$), *UGT1A1*28* ($n = 7$), and *UGT1A1*34* ($n = 8$). There is an inverse correlation between the number of "TA" dinucleotide repeats and level of UGT1A1 expression (111, 114, 115). The wild-type allele has six repeats, while the most common variant allele (*UGT1A1*28*) has seven repeats and is associated with decreased glucuronidation capacity for *UGT1A1* substrates. This polymorphism is discussed below (Section 8.3) in the context of irinotecan metabolism. The *UGT1A1*6* allele, comprising an amino acid change (Gly71Arg), is predominant in individuals of Japanese descent and is also associated with hyperbilirubinemia (116, 117).

UGT1A6 is involved in the metabolism of acetaminophen, aspirin, and other non-steroidal anti-inflammatory agents. The *UGT1A6*2* allele comprises two missense mutations (T181A and R184S) and has reduced catalytic activity toward some, but not all, UGT1A6 substrates (118). The observation of high linkage disequilibrium between the *UGT1A6*2* and *UGT1A1*28* alleles suggests an importance of haplotype structure and the need for caution when interpreting association results.

UGT2B7 is an important hepatic UGT that metabolizes several drugs, such as morphine, epirubicin, zidovudine, and gemfibrozil. A common polymorphism identified within the coding region of the *UGT2B7*2* allele results in an amino acid change (H268Y) that may affect glucuronidation activity in a substrate-specific manner (102). The clinical relevance of *UGT2B7*2* and other identified *UGT2B7* polymorphisms to drug metabolism remains unclear (102, 106, 107).

UGT2B15 is involved in the metabolism of various endogenous substances (e.g., steroid hormones) as well as clinical compounds (e.g., oxazepam). A common polymorphism that alters amino acid sequence (D85Y) has been detected in the *UGT2B15* gene, but its clinical relevance to drug metabolism is unclear (102).

8.2.1.5. Sulfotransferases

Two classes of sulfotransferases have been identified in vertebrates—the membrane-bound sulfotransferases found in the Golgi apparatus and the soluble, cytosolic forms (119). The membrane-bound sulfotransferases act on endogenous macromolecules and are not discussed further since they are not known to metabolize xenobiotics. The cytosolic sulfotransferases metabolize hormones as well as xenobiotics. The sulfotransferases catalyze the transfer of the sulfonyl group from 3′-phosphoadenosine 5′-phosphosulfate to the hydroxyl, sulfhydryl, amino, or N-oxide group of a variety of substrates.

The human cytosolic sulfotransferase (SULT) enzymes are part of an enzyme superfamily comprising 12 isoforms. Similar to the CYPs, the SULTs were originally named for the substrates they metabolized, but now follow a systematic nomenclature based on sequence homology (16). The human SULT1A family, which contributes significantly to drug metabolism, is discussed below.

SULT1A. The SULT1A family comprises SULT1A1, SULT1A2, and SULT1A3. SULT1A1 is the most abundant of the SULT1 enzymes in the liver and has broad substrate specificity, thus making it one of the primary SULTs for drug metabolism. SULT1A1 is involved in the metabolism of various hormones and xenobiotics (e.g., acetaminophen, minoxidil). Considerable interindividual variability is observed in SULT1A1 activity, and activity levels show familial segregation (120). The most common variant allele, *SULT1A1*2* (containing a nonsynonymous SNP, Arg213His), is associated with decreased SULT1A1 activity (121, 122). Two polymorphisms within the 5′-flanking region of the *SULT1A1* gene are also associated with variability in SULT1A1 activity. Since these polymorphisms are in linkage disequilibrium with the *SULT1A1*2* polymorphism, the effect of the *SULT1A1*2* polymorphism on activity may be attributed to the haplotype (123). The identified SNPs only account for a small portion of the SULT1A1 variability, and much of the variability in SULT1A1 activity may be attributable to a recently identified gene copy number variation. In a manner similar to *CYP2D6*, a *SULT1A1* gene deletion/duplication event results in individuals being "slow sulfators" or "rapid sulfators" (124). Thus genotyping for *SULT1A1* polymorphisms may generate spurious results if gene copy number is not taken into consideration. The clinical relevance of *SULT1A1* copy number polymorphisms remains to be determined.

8.2.2. Drug Transport Proteins

Drug transport proteins are expressed in many different tissues including liver, intestine, kidney, and brain, and are important modulators of drug absorption and disposition. Variability in the expression or function of drug transport proteins can affect the pharmacokinetic profile of a drug and hence its efficacy or adverse event profile. Drug transport proteins are generally classified into two major categories—the uptake transporters and the efflux transporters (Fig. 8.4). The uptake transporters belong to the solute carrier (SLC) transporter family, and allow the passage of small molecules across biological membranes along concentration gradients in a process known as "facilitated diffusion." Most efflux transporters belong to the ATP-binding cassette (ABC) transporter family and

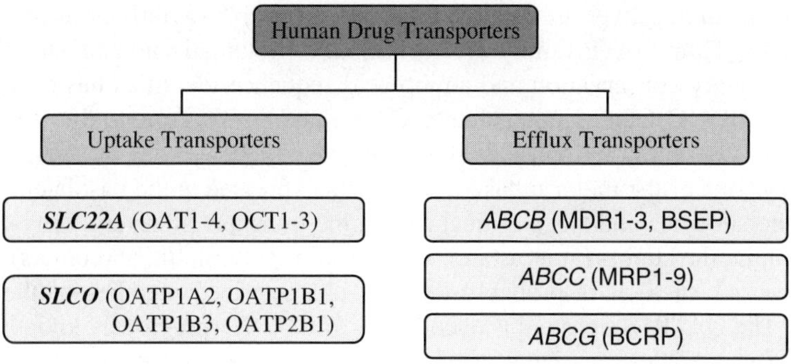

Figure 8.4 Human drug transporter families. Pharmacologically important members of the SLC and ABC-transporter families are listed [*gene* (proteins)]. OAT, organic anion transporter; OATP, organic anion transporting peptide; OCT, organic cation transporter; BSEP, bile salt export pump; MDR, multidrug resistance protein; MRP, multidrug resistance-associated protein; BCRP, breast cancer resistance protein.

utilize energy from ATP hydrolysis to transport substrates across biological membranes, either against or along concentration gradients, in a process known as "active transport." The balance between uptake and efflux transporters in different organs plays a crucial role in the drug disposition process. For example, the uptake transporters expressed on the sinusoidal membrane of the liver facilitate the extraction of drugs from the portal blood circulation into the hepatocyte, where the drug may be metabolized, and efflux transporters expressed on the apical membrane of the hepatocyte act to transport the drug out of the hepatocyte and into the bile (125).

While most of the drug transport proteins are known to contain polymorphisms, the genotype-phenotype relationships are not as well characterized as those for drug-metabolizing enzymes. However, as more information becomes available it is clear that polymorphisms affecting expression or activity of drug transport proteins contribute to the interindividual variability observed in drug disposition. The SLC and ABC transporter families are discussed briefly below. More information on these transporter families is available in review articles on the topic (125–128).

8.2.2.1. SLC Transporters

The SLC transporters are part of a large superfamily of membrane proteins comprising approximately 300 *SLC* genes classified into 43 families (129). The SLC transporters have broad substrate specificities and mediate the transport of endogenous substances (e.g., amino acids, oligopeptides, sugars, organic anions and cations) as well as a wide variety of drugs.

Within the SLC superfamily genetic variants are best-characterized for the OATPs, which are encoded by the *SLCO* genes. The OATPs are expressed in multiple tissues including liver, gut, and blood-brain barrier. To avoid confusion, as new members of the OATP family are identified, a systematic nomenclature based on evolutionary conservation and amino acid sequence identities has been accepted by the HUGO Gene Nomenclature Committee and is commonly used (18). The human OATP family consists of 11 members (130).

OATP1B1 is one of the major uptake transporters expressed at the basolateral membrane of hepatocytes and has a critical role in the transport of several drugs, including rifampin, the statins (atorvastatin, rosuvastatin, pravastatin, and others), and repaglinide. As such, it is probably the best-characterized member of the OATP family. The OATP1B1 protein is encoded by *SLCO1B1* (previously known as *OATP-C*). Several allelic variants of *SLCO1B1* have been detected and are associated with decreased transporter activity (e.g., *SLCO1B1*1b, *2, *3, *4, *5, *6, *9, *13, *15*, and **17*) (131–133). Furthermore, polymorphisms within the *SLCO1B1* gene are clinically relevant since they affect the disposition (and, in some cases, the effect) of several drugs including pravastatin, fexofenadine, atrasentan, and repaglinide (134–140).

8.2.2.2. ABC Transporters

The human ABC superfamily of transporters is encoded by 49 genes classified into seven subfamilies based on sequence homology (nutrigene.4t.com/humanabc.htm). Members of this gene family encode proteins that confer multidrug resistance to cancer chemotherapy.

The most prominent member of the ABC-transporter family is P-glycoprotein (P-gp), which is the product of the multidrug resistance 1 gene (*MDR1/ABCB1*). The gene encoding P-gp was originally named *MDR1* for its ability to transport various anticancer agents such as docetaxel, etoposide, paclitaxel, topotecan, and vinblastine out of the cell, thereby conferring the phenotype of multidrug resistance. However, P-gp has since been shown to have a broad substrate specificity and is involved in the transport of a wide range of therapeutic agents including antiarrhythmics (e.g., digoxin, verapamil), antibiotics (e.g., erythromycin), immunosuppressants (cyclosporine, tacrolimus), and many others (128). P-gp is expressed in multiple tissues including the brain, small and large intestines, kidney, liver, and testes. P-gp expressed in the blood-brain barrier acts to protect the central nervous system by limiting the entry of drugs and other toxins. P-gp expressed in the intestine pumps drugs out of the cells into the lumen, thus limiting bioavailability.

Over 30 SNPs have been detected within the *ABCB1* gene (141–145). Much research has focused on three SNPs that are in linkage disequilibrium—two synonymous SNPs (1236$C \rightarrow T$ in exon 12, 3435$C \rightarrow T$ in exon 26) and a nonsynonymous SNP (2677$G \rightarrow T$, resulting in an amino acid change at position 893, A893S). Pharmacogenetic studies investigating the relationship between these *ABCB1* polymorphisms and the PK of digoxin, fexofenadine, and cyclosporine A, all well-characterized P-gp substrates, have yielded inconsistent results—in some cases the polymorphisms are suggested to increase activity, in other cases, to decrease activity, and in some cases have no effect at all [(146) and references therein]. Therefore, the effect of currently known *ABCB1* polymorphisms on drug PK remains unclear. Inconsistent results could arise if the polymorphisms tested are not themselves functional, but are simply markers for other functional sites. In such a case, varying patterns of linkage disequilibrium between populations could generate discrepant results. Likewise, the presence of multiple transport pathways available for the disposition of a particular drug can decrease the impact of disruption in any one pathway (146).

8.3. GENOMIC BIOMARKERS FOR PK STUDIES

In general, the pharmacogenetics of drug-metabolizing enzymes have been well studied at the mechanistic level, and PG-PK studies are becoming commonplace in the drug development process. In 2005 the FDA released a guidance for the submission of pharmacogenomic data (Guidance for Industry—Pharmacogenomic Data Submission; http://www.fda.gov/cder/guidance/6400fnl.pdf) to facilitate research in the area and encourage the submission of data to the regulatory agency (the guidance is further discussed in Chapter 9). In this guidance, the FDA defined a "valid" genomic biomarker as a "biomarker that is measured in an analytical test system with well established performance characteristics and for which there is an established scientific framework or body of evidence that elucidates the physiologic, toxicologic, pharmacologic, or clinical significance of the test results."

Certain drug-metabolizing enzymes, including CYP2D6, CYP2C9, CYP2C19, UGT1A1, and TPMT, are "known valid" enzyme biomarkers. Genetic information on these known biomarkers can be found in the product labeling for several drugs (See "Table of Valid Genomic Biomarkers in the Context of Approved Drug Labels"; www.fda.gov/cder/genomics/genomic-biomarkers-table.htm). "Known valid" biomarkers have a high level of clinical validation attached to them, while other CYPs and drug transport proteins that have not attained this burden of proof yet are referred to as "exploratory" biomarkers. Only by including these "exploratory" biomarkers in PG-PK analysis and generating data illustrating genotype-phenotype correlations can they ultimately transition to "known valid" biomarker status. A few examples of "valid biomarkers" found in current drug labels are described below.

8.3.1. Warfarin, CYP2C9, and VKORC1

Warfarin is a coumarin anticoagulant that is widely prescribed for prevention and treatment of myocardial infarction, thromboembolism, and strokes (147, 148). Warfarin dosing is typically done by trial and error, where an initial dose of warfarin ranging from 2 to 10 mg is selected based on therapeutic indication and other extrinsic factors. The maintenance dose is then obtained by adjusting the initial dose while monitoring the **International Normalized Ratio** (INR) until it reaches and stabilizes in a target range. A wide inter-individual variability in dose response is observed in warfarin treatment, and maintenance doses range from 4 to 90 mg (149, 150). The narrow therapeutic index of warfarin, combined with a wide inter-individual variability in dose response results in increased risk of bleeding due to excessive anticoagulation, especially early on in therapy.

Coumarin anticoagulants act by inhibiting the synthesis of vitamin K-dependent clotting factors. Specifically, warfarin is thought to interfere with the synthesis of clotting factors by inhibition of the C1 subunit of the vitamin K epoxide reductase complex (VKORC1), which is involved in recycling reduced vitamin K (151, 152). Warfarin is manufactured as a racemic mixture of S- and R-warfarin, with S-warfarin being approximately three to five times more effective at inhibiting VKORC1 than R-warfarin (153). Polymorphisms within *VKORC1* are associated with lower dose requirements for warfarin (154). S-warfarin is metabolized by CYP2C9 to 6- and 7-hydroxywarfarin, which is subsequently excreted in the bile. Individuals with variant *CYP2C9* alleles, *CYP2C9*2* and *CYP2C9*3*, have decreased S-warfarin clearance, are at increased risk of bleeding, and require a lower daily dose of warfarin (155–157). While *CYP2C9*2* and *CYP2C9*3* have been extensively studied for warfarin metabolism, less information is available for other, rarer, variants (e.g., *CYP2C9*5, CYP2C9*6, CYP2C9*8, CYP2C9*11*), but these are also expected to confer impaired warfarin metabolism (158, 159). Overall, approximately 30% of the variability in warfarin maintenance dose can be explained by polymorphisms within *VKORC1*, and approximately 40% of the variability can be explained by polymorphisms within *VKORC1* and *CYP2C9* (154). The FDA has recently updated the warfarin label to include this pharmacogenetic information (http://packageinserts.bms.com/pi/pi-coumadin.pdf), but genetic testing is not required for patients initiating warfarin therapy. Several companies already have or are in the process of developing diagnostic tests for *CYP2C9* and *VKORC1* genotyping, and these tests are also available through various genotyping service providers.

8.3.2. Irinotecan and UGT1A1

Irinotecan is a camptothecin derivative that shows antitumor activity via topoisomerase inhibition (160). Irinotecan is converted to its active metabolite SN-38 by carboxylesterase enzymes in the liver, and SN-38 is subsequently conjugated, primarily by UGT1A1, for elimination in bile and urine (161). The dose-limiting toxicities of irinotecan include severe diarrhea and neutropenia. These toxicities are attributed to increased levels of the active metabolite SN-38. Reduced expression of UGT1A1 (due to polymorphisms within the gene) is associated with lower SN-38 glucuronidation and increased toxicity (111, 162–164).

As already mentioned, the *UGT1A1*1* (wild type) allele has six "TA" dinucleotide repeats within its promoter while the *UGT1A1*28* allele has seven "TA" dinucleotide repeats resulting in reduced expression of the enzyme. Approximately 10% of the North American population is homozygous for the *UGT1A1*28* allele. Patients who are homozygous for the *UGT1A1*28* allele have lower UGT1A1 activity resulting in higher levels of SN-38, and consequently are at increased risk for neutropenia. The current drug label suggests a reduced initial dose of irinotecan for these patients. Results for patients who are heterozygous for the *UGT1A1*28* allele are mixed, but, in general, these individuals are thought to tolerate normal doses of irinotecan. A diagnostic test for *UGT1A1*28* genotype was approved by the FDA in 2005, but genetic testing is not required before irinotecan therapy.

8.4. UTILITY OF PG-PK STUDIES IN EARLY CLINICAL TRIALS

Pharmacogenetic analyses in early clinical trials (such as first-in-human and phase I studies) provide drug developers with knowledge of the pathways that affect disposition of the compound and that contribute to variability in PK and, potentially, response.

Some information regarding the disposition pathways of an NCE may be available from in vitro metabolism work conducted during preclinical studies. However, these studies are usually performed on a limited subset of the metabolizing pathways available to the NCE and do not inform on overall in vivo disposition in humans. PG-PK analyses in early clinical trials can identify the multiple factors that affect the disposition of the NCE. Unlike in vitro drug metabolism studies, where an assay for each enzyme needs to be set up independently, usually with different cofactors, conditions, or technologies, genotyping of clinical trial participants for multiple genes can be performed simultaneously, with a single technology. In this sense, the scope of a PG-PK study can encompass a larger number of drug-metabolizing enzymes and drug transport proteins than typical in vitro drug metabolism characterization. Information generated from these early clinical studies aids the planning of future drug interaction and bridging studies and with the interpretation of clinical PK data. An association between a functional genetic variant and the PK parameters of a compound implicates the gene product in disposition of the compound. Therefore, a pharmacogenetic study can potentially inform on the role and clinical importance of many different drug disposition pathways. The utility of information obtained from these early PG-PK studies for drug developers is discussed below.

Pharmacogenetic studies in early clinical trials provide information to improve pharmacokinetic-pharmacodynamic modeling for the selection of doses for phase II and III trials. For example, if dose-exposure relationships differ between groups based on genotype, this information is valuable in dose selection for further studies (19). Furthermore, if the difference in exposure between genetically identified groups is very large, one dose would not be feasible for the entire population. In such a case a genetic test could be required before dosing, which in turn may support the decision to discontinue development of the particular NCE in favor of another (see Box 8-4 for illustration of a PG-PK example). Once dose-response relationships are established, incorporation of pharmacogenetic information can allow for rational dose adjustments in groups of individuals to optimize therapy.

BOX 8-4 *Illustration of the Utility of Pharmacogenetic Information in Drug Development*

In a phase I single-dose study of approximately 70 subjects, considerable variability (coefficient of variation >40%) in PK parameters (**AUC** and $\mathbf{C_{max}}$) for drug X is observed. To determine whether there is a genetic basis for the variability, genotyping is performed on a panel of drug-metabolizing enzymes and drug transport proteins (~300 SNPs in 100 genes). An association is observed between PK parameters (**AUC** and $\mathbf{C_{max}}$) and *CYP2C19* genotype; a regression model with age, sex, weight, and race as covariates shows a significant effect of genotype ($P < 0.001$). Individuals classified as CYP2C19 poor metabolizers (PMs) based on genotype (homozygous for the *2 or *3 alleles) have approximately threefold higher **AUC** (as shown on left) and $\mathbf{C_{max}}$ (not shown) compared with subjects classified as extensive metabolizers (in this case classified as *CYP2C19*1* homozygotes or heterozygotes). *CYP2C19* is extensively characterized at the molecular genetic level, and genotype-metabolizer phenotype correlations are well established. It is also known that the frequency of *CYP2C19* PMs differs between ethnic groups; specifically, the frequency of PMs among Asians is approximately 15%, while the frequency is only approximately 3% among white and black subjects (20).

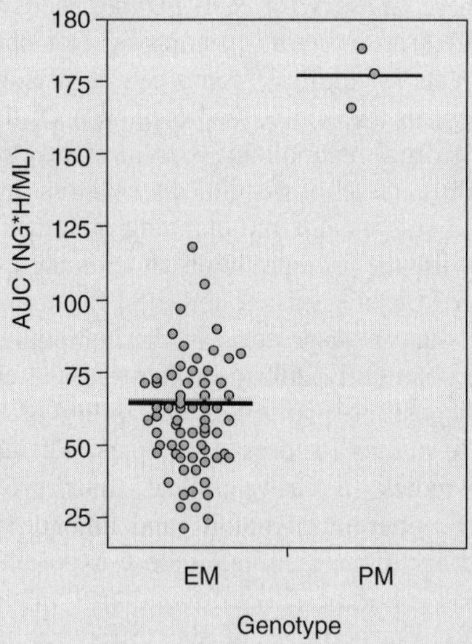

Assuming that the PG-PK correlation shown above is replicated in an independent study, this pharmacogenetic information can inform the development path of drug X as follows.

From the phase I study data it is apparent that CYP2C19 PMs have higher exposures to drug X. In this case, a bridging study, which is required to show pharmacokinetic equivalence between different geographic populations, would fail if performed in Japan, unless genotype information was incorporated in the study design.
Since an association is observed between *CYP2C19* genotype and the PK of the drug X, drug interaction studies with CYP2C19 substrates and/or inhibitors may be warranted. Higher exposures in CYP2C19 PMs may not be a concern if drug X has a wide therapeutic index. Otherwise, *CYP2C19* genotype frequency should be used as a factor in population PK-PD modeling for dose selection. If one dose cannot be used to achieve targeted exposures in both EMs and PMs, and there is a safety concern associated with higher exposures, then the pharmaceutical company may chose to continue development only in CYP2C19 EMs (and require genotyping prior to therapy—see Chapter 9 for codevelopment of a companion test) or else discontinue development of compound X in favor of a backup compound that is known to be metabolized by several pathways.

Information on PG-PK association may help to direct drug-drug interaction studies, since, in effect, the exposures observed in individuals with null or reduced-function alleles of a drug-metabolizing enzyme can be predictive of exposures observed with drug-drug interactions affecting the same enzyme. For example, drug exposure in a group of CYP2D6 PMs would be expected to be similar to exposure in a group of CYP2D6 EMs administered a CYP2D6 inhibitor. Therefore, in the PG-PK analysis if no difference is observed in the PK parameters between CYP2D6 PMs and CYP2D6 EMs, then the need to conduct drug-drug interaction studies between CYP2D6 substrates/inhibitors may be obviated. The absence of an association between any known, functional genotype and drug PK suggests that multiple pathways are involved in the disposition of the NCE and that drug-drug interactions are unlikely to be a concern further along in the development process. Conversely, if an association is observed between the PK of a drug and a particular genetic variant for a gene that was not tested in the in vitro metabolism studies, then additional studies to determine the effects of an inhibitor of that gene's product may be required. Furthermore, if a particular genotype is associated with PK, then pharmacogenetic data may inform dose ranging studies, studies in special populations, and drug-drug interaction studies.

PG-PK analyses can inform and direct the design of bridging studies. The allele frequencies of most functional polymorphisms in drug-metabolizing enzymes differ among geographic regions and ethnic populations (20, 165). If a PG-PK analysis identifies an association between one of these genetic factors and the PK parameters of a compound, then it is conceivable that in a different ethnic population, with different allele frequencies, the same dose will result in different mean exposures. Currently the requirement for bridging studies is to show pharmacokinetic equivalence. Therefore, understanding the underlying genetic contribution to variability in drug disposition will be valuable in designing successful regional bridging studies.

The ultimate impact of a PG-PK study will depend on the magnitude of difference in exposure between genotype groups, and on whether an association exists between exposure and effect (therapeutic or adverse).

8.5. LIMITATIONS OF PG-PK STUDIES

Early clinical trials are relatively small, and typical first-in-human and phase I studies usually enroll under a hundred individuals. For example, in a phase I trial of 60 individuals, if a polymorphism has a minor allele frequency of 5%, then based on the Hardy–Weinberg equation, approximately six individuals will be heterozygotes and the remaining will be homozygous for the wild-type allele. In other words, no individual with a homozygous variant genotype is likely to be observed. In such cases study power is adequate only for rather large differences in PK between the genotype. However, these studies are still useful since in the early stages of drug development large differences between populations will affect the decision to pursue the compound.

Further information on the functionality and clinical relevance of polymorphisms within the drug-metabolizing enzymes and drug transport proteins will serve to facilitate the interpretation and enhance the utility of PG-PK studies. While much is already known about the pharmacogenetics of drug-metabolizing enzymes, more information about drug transport proteins is required.

8.6. GENOTYPING TECHNOLOGIES

The advancements in genotyping technologies in the past several years have allowed for the broad utilization of pharmacogenetics in early clinical trials. When contemplating pharmacogenetic studies for association with PK, several questions arise: Is more better? How many, and which genes should be tested? And, more importantly, which sequence variations should be tested within these genes?

Ultra-high-throughput technologies are generally not warranted for PG-PK studies, since typically these studies are done as part of early clinical trials, which enroll fewer than one hundred individuals. Moreover, since the goal of these studies is to identify pathways that affect the disposition of the compound, the list for genotyping should be limited to PK-related genes. A reasonable place to start might be with all genes involved in drug metabolism and transport and, within these genes, the genotyping of all known, functional sequence polymorphisms.

There are myriad genotyping platforms available that combine high throughput and multiplexing capabilities with low cost per genotype. In addition, service providers supporting various platforms are available if in-house genotyping is unfeasible.

Several genotyping platforms specifically offer assays for drug-metabolizing enzymes and drug transport proteins [for, e.g., the Affymetrix® Drug

Metabolizing Enzymes and Transporters (DMET) panel, the AmpliChip® CYP450 Test (Roche Diagnostics), and TaqMan® SNP Genotyping Assays (Applied Biosystems)], but such assays can also be custom designed for other genotyping platforms [for, e.g., Illumina GoldenGate Genotyping and Pyrosequencing (Biotage)]. A wide range exists in the multiplexing and throughput capabilities of genotyping platforms, in the genotyping costs per sample, and in cost of the equipment required to perform the assay. For example, the AmpliChip® CYP450 Test (Roche Diagnostics) identifies only *CYP2D6* and *CYP2C19* genotypes (approximately 30 SNPs), whereas the Affymetrix® Drug Metabolizing Enzymes and Transporters (DMET) panel interrogates over 1000 SNPs in over 100 genes. The choice of platform/assays will depend on the study, since one platform may be too narrow in scope while another may be too broad. If the panel includes **tag SNPs** it may be too broad for an early PG-PK study. If other than causative SNPs are tested and shown to associate with a PK phenotype, then considerable amount of work will be needed to determine the molecular basis for the effect.

High sequence homology among genes (and pseudogenes) within drug-metabolizing enzyme and drug transporter families represents a considerable technological challenge for the development genotyping assays. Most current assay technologies employ primer extension, hybridization, or enzymatic cleavage methods for allele discrimination, followed by allele detection based on fluorescence, chemiluminescence, or mass (166). The interested reader is referred to review articles for more detailed discussions of genotyping technologies (166–168).

8.7. CONCLUSION

Individuals differ in their response to drugs. Pharmacogenetic studies can facilitate understanding of the sources of interindividual variability in drug response, allowing for rational dose adjustments for optimization of therapy. The knowledge accumulated through sequencing of the human genome, combined with advances in genomic technologies, has led to increased understanding of the genetic factors affecting drug disposition. Implementation of this knowledge and technology at the appropriate time during drug development can provide useful information for understanding and managing variability in response. Regulatory agencies, including the FDA and EMEA, have been proactive advocates in encouraging the use of pharmacogenetics during drug development. To this end, the release of guidances for pharmacogenomic data submission and active dialog between regulatory agencies and pharmaceutical companies are facilitating the integration of pharmacogenetic studies into the drug development continuum.

BOX 8-5 *Glossary of Terms*

Area under the curve (AUC) is a measure of the overall amount of drug in the body. It is the integral of the drug level over time (for example, from 0 to infinity, or any other specified time). An example of the AUC for a drug administered extravascularly (e.g., orally) is shown.

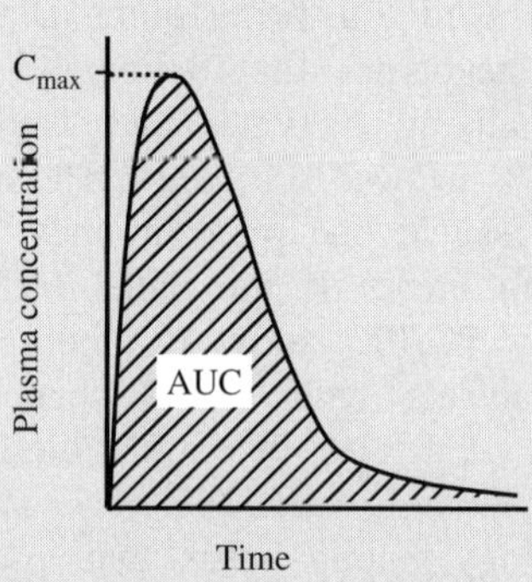

C_{max} is the maximum concentration of drug in the blood. The C_{max} of a drug taken orally is shown.

International normalized ratio (INR) is a measure of the time it takes for blood to clot, compared to an average.

Linkage disequilibrium is the nonrandom association between two or more alleles such that certain combinations of alleles are more likely to occur together on a chromosome than other combinations of alleles

SNP is an inherited single base pair alteration in DNA sequence that is found in at least 1% of the population.

Tag SNP is a SNP that is representative for a region with high linkage disequilibrium.

BOX 8-6 *List of Abbreviations and Definitions of Terms*

ANOVA: Analysis of variance

AUC: Area under the plasma concentration-time curve

BCRP: Breast cancer resistance protein

CAR: Constitutive androstane receptor

C_{max}: Maximum observed plasma concentration

CL/F: Clearance

CNT: Concentrative nucleoside transporter

CYP: Cytochrome P450

DNA: Deoxyribonucleic acid

EM: Extensive metabolizer

FMO: Flavin monooxygenase

IM: Intermediate metabolizer

MRP: Multidrug resistance protein

NAT: Arylamine *N*-acetyltransferase

NCE: New chemical entity

OAT: Organic anion transporter

OATP: Organic anion transporting polypeptide

OCT: Organic cation transporter

OCTN: Novel organic cation transporter

PEPT: Oligopeptide transporter

PG: Pharmacogenetics

P-gp: P-glycoprotein

PK: Pharmacokinetics

PM: Poor metabolizer

PXR: Pregnane X receptor

SNP: Single nucleotide polymorphism

SULT: Sulfotransferase

UGT: UDP-glucuronosyltransferase

UM: Ultrarapid metabolizer

REFERENCES

1. Vogel, F. Moderne problem der humangenetik. *Ergeb Inneren Med Kinderheilkunde* 1959;12:52–125.
2. Gonzalez, FJ, Skoda, RC, Kimura, S, Umeno, M, Zanger, UM, Nebert, DW, Gelboin, HV, Hardwick, JP, and Meyer, UA. Characterization of the common genetic defect in humans deficient in debrisoquine metabolism. *Nature* 1988;331:442–6.

3. Vatsis, KP, Martell, KJ, and Weber, WW. Diverse point mutations in the human gene for polymorphic *N*-acetyltransferase. *Proc Natl Acad Sci USA* 1991;88:6333–7.
4. Vatsis, KP and Weber, WW. *Human Acetyltransferases*. Berlin: Springer-Verlag; 1994. pp. 109–30.
5. Eichelbaum, M, Reetz, KP, Schmidt, EK, and Zekorn, C. The genetic polymorphism of sparteine metabolism. *Xenobiotica* 1986;16:465–81.
6. Eichelbaum, M, Spannbrucker, N, Steincke, B, and Dengler, HJ. Defective N-oxidation of sparteine in man: a new pharmacogenetic defect. *Eur J Clin Pharmacol* 1979;16:183–7.
7. Lesko, LJ and Woodcock, J. Translation of pharmacogenomics and pharmacogenetics: a regulatory perspective. *Nat Rev Drug Discov* 2004;3:763–9.
8. Beyth, RJ, Milligan, PE, and Gage, BF. Risk factors for bleeding in patients taking coumarins. *Curr Hematol Rep* 2002;1:41–9.
9. Bruford, EA, Lush, MJ, Wright, MW, Sneddon, TP, Povey, S, and Birney, E. The HGNC Database in 2008: a resource for the human genome. *Nucleic Acids Res* 2007.
10. Nelson, DR. Cytochrome P450 nomenclature. *Methods Mol Biol* 1998;107:15–24.
11. Nelson, DR. Cytochrome P450 nomenclature, 2004. *Methods Mol Biol* 2006;320:1–10.
12. Nelson, DR, Koymans, L, Kamataki, T, Stegeman, JJ, Feyereisen, R, Waxman, DJ, Waterman, MR, Gotoh, O, Coon, MJ, Estabrook, RW, Gunsalus, IC, and Nebert, DW. P450 superfamily: update on new sequences, gene mapping, accession numbers and nomenclature. *Pharmacogenetics* 1996;6:1–42.
13. Nelson, DR, Zeldin, DC, Hoffman, SM, Maltais, LJ, Wain, HM, and Nebert, DW. Comparison of cytochrome P450 (CYP) genes from the mouse and human genomes, including nomenclature recommendations for genes, pseudogenes and alternative-splice variants. *Pharmacogenetics* 2004;14:1–18.
14. Mackenzie, PI, Bock, KW, Burchell, B, Guillemette, C, Ikushiro, S, Iyanagi, T, Miners, JO, Owens, IS, and Nebert, DW. Nomenclature update for the mammalian UDP glycosyltransferase (UGT) gene superfamily. *Pharmacogenet Genomics* 2005;15:677–85.
15. Mackenzie, PI, Owens, IS, Burchell, B, Bock, KW, Bairoch, A, Belanger, A, Fournel- Gigleux, S, Green, M, Hum, DW, Iyanagi, T, Lancet, D, Louisot, P, Magdalou, J, Chowdhury, JR, Ritter, JK, Schachter, H, Tephly, TR, Tipton, KF, and Nebert, DW. The UDP glycosyltransferase gene superfamily: recommended nomenclature update based on evolutionary divergence. *Pharmacogenetics* 1997;7:255–69.
16. Blanchard, RL, Freimuth, RR, Buck, J, Weinshilboum, RM, and Coughtrie, MW. A proposed nomenclature system for the cytosolic sulfotransferase (SULT) superfamily. *Pharmacogenetics* 2004;14:199–211.
17. Vatsis, KP, Weber, WW, Bell, DA, Dupret, JM, Evans, DA, Grant, DM, Hein, DW, Lin, HJ, Meyer, UA, Relling, MV, et al. Nomenclature for N-acetyltransferases. *Pharmacogenetics* 1995;5:1–17.
18. Hagenbuch, B and Meier, PJ. Organic anion transporting polypeptides of the OATP/SLC21 family: phylogenetic classification as OATP/SLCO superfamily, new nomenclature and molecular/functional properties. *Pflugers Arch* 2004;447:653–65.
19. Katz, DA, Murray, B, Bhathena, A, and Sahelijo, L. Defining drug disposition determinants: a pharmacogenetic—pharmacokinetic strategy. *Nat Rev Drug Discov* 2008;7:293–305.
20. Andersson, T, Flockhart, DA, Goldstein, DB, Huang, SM, Kroetz, DL, Milos, PM, Ratain, MJ, and Thummel, K. Drug-metabolizing enzymes: evidence for clinical utility of pharmacogenomic tests. *Clin Pharmacol Ther* 2005;78:559–81.
21. Ma, Q and Lu, AY. CYP1A induction and human risk assessment: an evolving tale of in vitro and In vivo studies. *Drug Metab Dispos* 2007;35:1009–16.

22. Diaz, D, Fabre, I, Daujat, M, Saint Aubert, B, Bories, P, Michel, H, and Maurel, P. Omeprazole is an aryl hydrocarbon-like inducer of human hepatic cytochrome P450. *Gastroenterology* 1990;99:737–47.
23. Komatsu, T, Yamazaki, H, Shimada, N, Nakajima, M, and Yokoi, T. Roles of cytochromes P450 1A2, 2A6, and 2C8 in 5-fluorouracil formation from tegafur, an anticancer prodrug, in human liver microsomes. *Drug Metab Dispos* 2000;28:1457–63.
24. Nakajima, M, Yamamoto, T, Nunoya, K, Yokoi, T, Nagashima, K, Inoue, K, Funae, Y, Shimada, N, Kamataki, T, and Kuroiwa, Y. Role of human cytochrome P4502A6 in C-oxidation of nicotine. *Drug Metab Dispos* 1996;24:1212–7.
25. Torchin, CD, McNeilly, PJ, Kapetanovic, IM, Strong, JM, and Kupferberg, HJ. Stereoselective metabolism of a new anticonvulsant drug candidate, losigamone, by human liver microsomes. *Drug Metab Dispos* 1996;24:1002–8.
26. Yamano, S, Tatsuno, J, and Gonzalez, FJ. The CYP2A3 gene product catalyzes coumarin 7-hydroxylation in human liver microsomes. *Biochemistry* 1990;29:1322–9.
27. Benowitz, NL, Griffin, C, and Tyndale, R. Deficient C-oxidation of nicotine continued. *Clin Pharmacol Ther* 2001;70:567.
28. Hadidi, H, Zahlsen, K, Idle, JR, and Cholerton, S. A single amino acid substitution (Leu160His) in cytochrome P450 CYP2A6 causes switching from 7-hydroxylation to 3-hydroxylation of coumarin. *Food Chem Toxicol* 1997;35:903–7.
29. Inoue, K, Yamazaki, H, and Shimada, T. CYP2A6 genetic polymorphisms and liver microsomal coumarin and nicotine oxidation activities in Japanese and Caucasians. *Arch Toxicol* 2000;73:532–9.
30. Kitagawa, K, Kunugita, N, Katoh, T, Yang, M, and Kawamoto, T. The significance of the homozygous CYP2A6 deletion on nicotine metabolism: a new genotyping method of CYP2A6 using a single PCR-RFLP. *Biochem Biophys Res Commun* 1999;262:146–51.
31. Nakajima, M, Yamagishi, S, Yamamoto, H, Yamamoto, T, Kuroiwa, Y, and Yokoi, T. Deficient cotinine formation from nicotine is attributed to the whole deletion of the CYP2A6 gene in humans. *Clin Pharmacol Ther* 2000;67:57–69.
32. Oscarson, M, McLellan, RA, Gullsten, H, Agundez, JA, Benitez, J, Rautio, A, Raunio, H, Pelkonen, O, and Ingelman-Sundberg, M. Identification and characterisation of novel polymorphisms in the CYP2A locus: implications for nicotine metabolism. *FEBS Lett* 1999;460: 321–7.
33. Fukami, T, Nakajima, M, Higashi, E, Yamanaka, H, McLeod, HL, and Yokoi, T. A novel CYP2A6*20 allele found in African-American population produces a truncated protein lacking enzymatic activity. *Biochem Pharmacol* 2005;70:801–8.
34. Ariyoshi, N, Sawamura, Y, and Kamataki, T. A novel single nucleotide polymorphism altering stability and activity of CYP2a6. *Biochem Biophys Res Commun* 2001;281:810–4.
35. Daigo, S, Takahashi, Y, Fujieda, M, Ariyoshi, N, Yamazaki, H, Koizumi, W, Tanabe, S, Saigenji, K, Nagayama, S, Ikeda, K, Nishioka, Y, and Kamataki, T. A novel mutant allele of the CYP2A6 gene (CYP2A6*11) found in a cancer patient who showed poor metabolic phenotype towards tegafur. *Pharmacogenetics* 2002;12:299–306.
36. Fukami, T, Nakajima, M, Higashi, E, Yamanaka, H, Sakai, H, McLeod, HL, and Yokoi, T. Characterization of novel CYP2A6 polymorphic alleles (CYP2A6*18 and CYP2A6*19) that affect enzymatic activity. *Drug Metab Dispos* 2005;33:1202–10.
37. Fukami, T, Nakajima, M, Yoshida, R, Tsuchiya, Y, Fujiki, Y, Katoh, M, McLeod, HL, and Yokoi, T. A novel polymorphism of human CYP2A6 gene CYP2A6*17 has an amino acid substitution (V365M) that decreases enzymatic activity in vitro and In vivo. *Clin Pharmacol Ther* 2004;76:519–27.

38. Kitagawa, K, Kunugita, N, Kitagawa, M, and Kawamoto, T. CYP2A6*6, a novel polymorphism in cytochrome p450 2A6, has a single amino acid substitution (R128Q) that inactivates enzymatic activity. *J Biol Chem* 2001;276:17830–5.
39. Oscarson, M, McLellan, RA, Asp, V, Ledesma, M, Bernal Ruiz, ML, Sinues, B, Rautio, A, and Ingelman- Sundberg, M. Characterization of a novel CYP2A7/CYP2A6 hybrid allele (CYP2A6*12) that causes reduced CYP2A6 activity. *Hum Mutat* 2002;20:275–83.
40. Nakajima, M, Fukami, T, Yamanaka, H, Higashi, E, Sakai, H, Yoshida, R, Kwon, JT, McLeod, HL, and Yokoi, T. Comprehensive evaluation of variability in nicotine metabolism and CYP2A6 polymorphic alleles in four ethnic populations. *Clin Pharmacol Ther* 2006;80:282–97.
41. Kwon, JT, Nakajima, M, Chai, S, Yom, YK, Kim, HK, Yamazaki, H, Sohn, DR, Yamamoto, T, Kuroiwa, Y, and Yokoi, T. Nicotine metabolism and CYP2A6 allele frequencies in Koreans. *Pharmacogenetics* 2001;11:317–23.
42. Nakajima, M, Kwon, JT, Tanaka, N, Zenta, T, Yamamoto, Y, Yamamoto, H, Yamazaki, H, Yamamoto, T, Kuroiwa, Y, and Yokoi, T. Relationship between interindividual differences in nicotine metabolism and CYP2A6 genetic polymorphism in humans. *Clin Pharmacol Ther* 2001;69:72–8.
43. Shimada, T, Yamazaki, H, and Guengerich, FP. Ethnic-related differences in coumarin 7-hydroxylation activities catalyzed by cytochrome P4502A6 in liver microsomes of Japanese and Caucasian populations. *Xenobiotica* 1996;26:395–403.
44. Zanger, UM, Klein, K, Saussele, T, Blievernicht, J, M, HH, and Schwab, M. Polymorphic CYP2B6: molecular mechanisms and emerging clinical significance. *Pharmacogenomics* 2007;8:743–59.
45. Ribaudo, HJ, Haas, DW, Tierney, C, Kim, RB, Wilkinson, GR, Gulick, RM, Clifford, DB, Marzolini, C, Fletcher, CV, Tashima, KT, Kuritzkes, DR, and Acosta, EP. Pharmacogenetics of plasma efavirenz exposure after treatment discontinuation: an Adult AIDS Clinical Trials Group Study. *Clin Infect Dis* 2006;42:401–7.
46. Haas, DW, Ribaudo, HJ, Kim, RB, Tierney, C, Wilkinson, GR, Gulick, RM, Clifford, DB, Hulgan, T, Marzolini, C, and Acosta, EP. Pharmacogenetics of efavirenz and central nervous system side effects: an Adult AIDS Clinical Trials Group study. *AIDS* 2004;18:2391–400.
47. Rodriguez-Novoa, S, Barreiro, P, Rendon, A, Jimenez- Nacher, I, Gonzalez- Lahoz, J, and Soriano, V. Influence of 516G>T polymorphisms at the gene encoding the CYP450–2B6 isoenzyme on efavirenz plasma concentrations in HIV-infected subjects. *Clin Infect Dis* 2005;40:1358–61.
48. Rotger, M, Colombo, S, Furrer, H, Bleiber, G, Buclin, T, Lee, BL, Keiser, O, Biollaz, J, Decosterd, L, and Telenti, A. Influence of CYP2B6 polymorphism on plasma and intracellular concentrations and toxicity of efavirenz and nevirapine in HIV-infected patients. *Pharmacogenet Genomics* 2005;15:1–5.
49. Kirchheiner, J and Brockmoller, J. Clinical consequences of cytochrome P450 2C9 polymorphisms. *Clin Pharmacol Ther* 2005;77:1–16.
50. Andersson, T, Holmberg, J, Rohss, K, and Walan, A. Pharmacokinetics and effect on caffeine metabolism of the proton pump inhibitors, omeprazole, lansoprazole, and pantoprazole. *Br J Clin Pharmacol* 1998;45:369–75.
51. Desta, Z, Zhao, X, Shin, JG, and Flockhart, DA. Clinical significance of the cytochrome P450 2C19 genetic polymorphism. *Clin Pharmacokinet* 2002;41:913–58.
52. Zanger, UM, Raimundo, S, and Eichelbaum, M. Cytochrome P450 2D6: overview and update on pharmacology, genetics, biochemistry. *Naunyn Schmiedebergs Arch Pharmacol* 2004;369:23–37.
53. Johansson, I, Lundqvist, E, Bertilsson, L, Dahl, ML, Sjoqvist, F, and Ingelman- Sundberg, M. Inherited amplification of an active gene in the cytochrome P450 CYP2D locus as a cause of ultrarapid metabolism of debrisoquine. *Proc Natl Acad Sci USA* 1993;90:11825–9.

54. Ingelman-Sundberg, M. Pharmacogenetics of cytochrome P450 and its applications in drug therapy: the past, present and future. *Trends Pharmacol Sci* 2004;25:193–200.
55. Gaedigk, A, Ndjountche, L, Divakaran, K, Dianne Bradford, L, Zineh, I, Oberlander, TF, Brousseau, DC, McCarver, DG, Johnson, JA, Alander, SW, Wayne Riggs, K, and Steven Leeder, J. Cytochrome P4502D6 (CYP2D6) gene locus heterogeneity: characterization of gene duplication events. *Clin Pharmacol Ther* 2007;81:242–51.
56. Ma, MK, Woo, MH, and McLeod, HL. Genetic basis of drug metabolism. *Am J Health Syst Pharm* 2002;59:2061–9.
57. Marsh, S and McLeod, HL. Pharmacogenetics and oncology treatment for breast cancer. *Expert Opin Pharmacother* 2007;8:119–27.
58. Kawanishi, C, Lundgren, S, Agren, H, and Bertilsson, L. Increased incidence of CYP2D6 gene duplication in patients with persistent mood disorders: ultrarapid metabolism of antidepressants as a cause of nonresponse. A pilot study. *Eur J Clin Pharmacol* 2004;59:803–7.
59. Kirchheiner, J and Seeringer, A. Clinical implications of pharmacogenetics of cytochrome P450 drug metabolizing enzymes. *Biochim Biophys Acta* 2007;1770:489–94.
60. Kirchheiner, J, Nickchen, K, Bauer, M, Wong, ML, Licinio, J, Roots, I, and Brockmoller, J. Pharmacogenetics of antidepressants and antipsychotics: the contribution of allelic variations to the phenotype of drug response. *Mol Psychiatry* 2004;9:442–73.
61. Bernard, S, Neville, KA, Nguyen, AT, and Flockhart, DA. Interethnic differences in genetic polymorphisms of CYP2D6 in the U.S. population: clinical implications. *Oncologist* 2006;11:126–35.
62. Bertz, RJ and Granneman, GR. Use of in vitro and In vivo data to estimate the likelihood of metabolic pharmacokinetic interactions. *Clin Pharmacokinet* 1997;32:210–58.
63. Ingelman-Sundberg, M. Human drug metabolising cytochrome P450 enzymes: properties and polymorphisms. *Naunyn Schmiedebergs Arch Pharmacol* 2004;369:89–104.
64. Floyd, MD, Gervasini, G, Masica, AL, Mayo, G, George, AL, Jr., Bhat, K, Kim, RB, and Wilkinson, GR. Genotype-phenotype associations for common CYP3A4 and CYP3A5 variants in the basal and induced metabolism of midazolam in European- and African-American men and women. *Pharmacogenetics* 2003;13:595–606.
65. Thummel, KE and Wilkinson, GR. In Vitro and In vivo drug interactions involving human CYP3A. *Annu Rev Pharmacol Toxicol* 1998;38:389–430.
66. Westlind, A, Lofberg, L, Tindberg, N, Andersson, TB, and Ingelman- Sundberg, M. Interindividual differences in hepatic expression of CYP3A4: relationship to genetic polymorphism in the 5′-upstream regulatory region. *Biochem Biophys Res Commun* 1999;259:201–5.
67. Wilkinson, GR. Cytochrome P4503A (CYP3A) metabolism: prediction of in vivo activity in humans. *J Pharmacokinet Biopharm* 1996;24:475–90.
68. Wilkinson, GR. Drug metabolism and variability among patients in drug response. *N Engl J Med* 2005;352:2211–21.
69. Ozdemir, V, Kalow, W, Tang, BK, Paterson, AD, Walker, SE, Endrenyi, L, and Kashuba, AD. Evaluation of the genetic component of variability in CYP3A4 activity: a repeated drug administration method. *Pharmacogenetics* 2000;10:373–88.
70. He, P, Court, MH, Greenblatt, DJ, and Von Moltke, LL. Genotype-phenotype associations of cytochrome P450 3A4 and 3A5 polymorphism with midazolam clearance In vivo. *Clin Pharmacol Ther* 2005;77:373–87.
71. Henderson, L, Yue, QY, Bergquist, C, Gerden, B, and Arlett, P. St John's wort (*Hypericum perforatum*): drug interactions and clinical outcomes. *Br J Clin Pharmacol* 2002;54:349–56.
72. Piscitelli, SC, Burstein, AH, Chaitt, D, Alfaro, RM, and Falloon, J. Indinavir concentrations and St John's wort. *Lancet* 2000;355:547–8.
73. Ruschitzka, F, Meier, PJ, Turina, M, Luscher, TF, and Noll, G. Acute heart transplant rejection due to Saint John's wort. *Lancet* 2000;355:548–9.

74. Kane, GC and Lipsky, JJ. Drug-grapefruit juice interactions. *Mayo Clin Proc* 2000;75:933–42.
75. Ray, WA, Murray, KT, Meredith, S, Narasimhulu, SS, Hall, K, and Stein, CM. Oral erythromycin and the risk of sudden death from cardiac causes. *N Engl J Med* 2004;351:1089–96.
76. Wrighton, SA, Ring, BJ, Watkins, PB, and VandenBranden, M. Identification of a polymorphically expressed member of the human cytochrome P-450III family. *Mol Pharmacol* 1989;36:97–105.
77. Kuehl, P, Zhang, J, Lin, Y, Lamba, J, Assem, M, Schuetz, J, Watkins, PB, Daly, A, Wrighton, SA, Hall, SD, Maurel, P, Relling, M, Brimer, C, Yasuda, K, Venkataramanan, R, Strom, S, Thummel, K, Boguski, MS, and Schuetz, E. Sequence diversity in CYP3A promoters and characterization of the genetic basis of polymorphic CYP3A5 expression. *Nat Genet* 2001;27:383–91.
78. Hustert, E, Haberl, M, Burk, O, Wolbold, R, He, YQ, Klein, K, Nuessler, AC, Neuhaus, P, Klattig, J, Eiselt, R, Koch, I, Zibat, A, Brockmoller, J, Halpert, JR, Zanger, UM, and Wojnowski, L. The genetic determinants of the CYP3A5 polymorphism. *Pharmacogenetics* 2001;11:773–9.
79. Katz, DA, Grimm, DR, Cassar, SC, Gentile, MC, Ye, X, Rieser, MJ, Gordon, EF, Polzin, JE, Gustavson, LE, Driscoll, RM, O'Dea R, F, Williams, LA, and Bukofzer, S. CYP3A5 genotype has a dose-dependent effect on ABT-773 plasma levels. *Clin Pharmacol Ther* 2004;75:516–28.
80. Macphee, IA, Fredericks, S, Tai, T, Syrris, P, Carter, ND, Johnston, A, Goldberg, L, and Holt, DW. Tacrolimus pharmacogenetics: polymorphisms associated with expression of cytochrome p4503A5 and P-glycoprotein correlate with dose requirement. *Transplantation* 2002;74:1486–9.
81. Zheng, H, Webber, S, Zeevi, A, Schuetz, E, Zhang, J, Bowman, P, Boyle, G, Law, Y, Miller, S, Lamba, J, and Burckart, GJ. Tacrolimus dosing in pediatric heart transplant patients is related to CYP3A5 and MDR1 gene polymorphisms. *Am J Transplant* 2003;3:477–83.
82. Zheng, H, Zeevi, A, Schuetz, E, Lamba, J, McCurry, K, Griffith, BP, Webber, S, Ristich, J, Dauber, J, Iacono, A, Grgurich, W, Zaldonis, D, McDade, K, Zhang, J, and Burckart, GJ. Tacrolimus dosing in adult lung transplant patients is related to cytochrome P4503A5 gene polymorphism. *J Clin Pharmacol* 2004;44:135–40.
83. Koukouritaki, SB, Simpson, P, Yeung, CK, Rettie, AE, and Hines, RN. Human hepatic flavin-containing monooxygenases 1 (FMO1) and 3 (FMO3) developmental expression. *Pediatr Res* 2002;51:236–43.
84. Chung, WG, Park, CS, Roh, HK, Lee, WK, and Cha, YN. Oxidation of ranitidine by isozymes of flavin-containing monooxygenase and cytochrome P450. *Jpn J Pharmacol* 2000;84:213–20.
85. Hamman, MA, Haehner- Daniels, BD, Wrighton, SA, Rettie, AE, and Hall, SD. Stereoselective sulfoxidation of sulindac sulfide by flavin-containing monooxygenases. Comparison of human liver and kidney microsomes and mammalian enzymes. *Biochem Pharmacol* 2000;60:7–17.
86. Mushiroda, T, Douya, R, Takahara, E, and Nagata, O. The involvement of flavin containing monooxygenase but not CYP3A4 in metabolism of itopride hydrochloride, a gastroprokinetic agent: comparison with cisapride and mosapride citrate. *Drug Metab Dispos* 2000;28:1231–7.
87. Sachse, C, Ruschen, S, Dettling, M, Schley, J, Bauer, S, Muller- Oerlinghausen, B, Roots, I, and Brockmoller, J. Flavin monooxygenase 3 (FMO3) polymorphism in a white population: allele frequencies, mutation linkage, and functional effects on clozapine and caffeine metabolism. *Clin Pharmacol Ther* 1999;66:431–8.
88. Parte, P and Kupfer, D. Oxidation of tamoxifen by human flavin-containing monooxygenase (FMO) 1 and FMO3 to tamoxifen-N-oxide and its novel reduction back to tamoxifen by human cytochromes P450 and hemoglobin. *Drug Metab Dispos* 2005;33:1446–52.
89. Hisamuddin, IM, Wehbi, MA, Chao, A, Wyre, HW, Hylind, LM, Giardiello, FM, and Yang, VW. Genetic polymorphisms of human flavin monooxygenase 3 in sulindac-mediated primary chemoprevention of familial adenomatous polyposis. *Clin Cancer Res* 2004;10:8357–62.

90. Hisamuddin, IM, Wehbi, MA, Schmotzer, B, Easley, KA, Hylind, LM, Giardiello, FM, and Yang, VW. Genetic polymorphisms of flavin monooxygenase 3 in sulindac-induced regression of colorectal adenomas in familial adenomatous polyposis. *Cancer Epidemiol Biomarkers Prev* 2005;14:2366–9.
91. Kang, JH, Chung, WG, Lee, KH, Park, CS, Kang, JS, Shin, IC, Roh, HK, Dong, MS, Baek, HM, and Cha, YN. Phenotypes of flavin-containing monooxygenase activity determined by ranitidine N-oxidation are positively correlated with genotypes of linked FM03 gene mutations in a Korean population. *Pharmacogenetics* 2000;10:67–78.
92. Mayatepek, E, Flock, B, and Zschocke, J. Benzydamine metabolism in vivo is impaired in patients with deficiency of flavin-containing monooxygenase 3. *Pharmacogenetics* 2004;14:775–7.
93. Evans, DA. *N*-acetyltransferase. *Pharmacol Ther* 1989;42:157–234.
94. Evans, DA, Manley, KA, and Mc, KV. Genetic control of isoniazid metabolism in man. *Br Med J* 1960;2:485–91.
95. Weber, WW and Cohen, SN. The mechanism of isoniazid acetylation by human *N*-acetyltransferase. *Biochim Biophys Acta* 1968;151:276–8.
96. Vatsis, KP and Weber, WW *Acetyltransferases*. Cambridge: Pergamon; 1997. p. 385–99.
97. Hein, DW, McQueen, CA, Grant, DM, Goodfellow, GH, Kadlubar, FF, and Weber, WW. Pharmacogenetics of the arylamine N-acetyltransferases: a symposium in honor of Wendell W. Weber. *Drug Metab Dispos* 2000;28:1425–32.
98. Perry, HM, Jr. Late toxicity to hydralazine resembling systemic lupus erythematosus or rheumatoid arthritis. *Am J Med* 1973;54:58–72.
99. Westphal, GA, Reich, K, Schulz, TG, Neumann, C, Hallier, E, and Schnuch, A. *N*-acetyltransferase 1 and 2 polymorphisms in para-substituted arylamine-induced contact allergy. *Br J Dermatol* 2000;142:1121–7.
100. Wolkenstein, P, Carriere, V, Charue, D, Bastuji- Garin, S, Revuz, J, Roujeau, JC, Beaune, P, and Bagot, M. A slow acetylator genotype is a risk factor for sulphonamide-induced toxic epidermal necrolysis and Stevens-Johnson syndrome. *Pharmacogenetics* 1995;5:255–8.
101. Hein, DW, Grant, DM, and Sim, E. Update on consensus arylamine *N*-acetyltransferase gene nomenclature. *Pharmacogenetics* 2000;10:291–2.
102. Guillemette, C. Pharmacogenomics of human UDP-glucuronosyltransferase enzymes. *Pharmacogenomics J* 2003;3:136–58.
103. Kiang, TK, Ensom, MH, and Chang, TK. UDP-glucuronosyltransferases and clinical drug-drug interactions. *Pharmacol Ther* 2005;106:97–132.
104. Levesque, E, Girard, H, Journault, K, Lepine, J, and Guillemette, C. Regulation of the UGT1A1 bilirubin-conjugating pathway: role of a new splicing event at the UGT1A locus. *Hepatology* 2007;45:128–38.
105. Girard, H, Levesque, E, Bellemare, J, Journault, K, Caillier, B, and Guillemette, C. Genetic diversity at the UGT1 locus is amplified by a novel 3′ alternative splicing mechanism leading to nine additional UGT1A proteins that act as regulators of glucuronidation activity. *Pharmacogenet Genomics* 2007;17:1077–89.
106. Maruo, Y, Iwai, M, Mori, A, Sato, H, and Takeuchi, Y. Polymorphism of UDP-glucuronosyltransferase and drug metabolism. *Curr Drug Metab* 2005;6:91–9.
107. Desai, AA, Innocenti, F, and Ratain, MJ. UGT pharmacogenomics: implications for cancer risk and cancer therapeutics. *Pharmacogenetics* 2003;13:517–23.
108. Aono, S, Yamada, Y, Keino, H, Hanada, N, Nakagawa, T, Sasaoka, Y, Yazawa, T, Sato, H, and Koiwai, O. Identification of defect in the genes for bilirubin UDP-glucuronosyl-transferase in a patient with Crigler-Najjar syndrome type II. *Biochem Biophys Res Commun* 1993;197:1239–44.

109. Aono, S, Adachi, Y, Uyama, E, Yamada, Y, Keino, H, Nanno, T, Koiwai, O, and Sato, H. Analysis of genes for bilirubin UDP-glucuronosyltransferase in Gilbert's syndrome. *Lancet* 1995;345:958–9.

110. Aono, S, Yamada, Y, Keino, H, Sasaoka, Y, Nakagawa, T, Onishi, S, Mimura, S, Koiwai, O, and Sato, H. A new type of defect in the gene for bilirubin uridine 5′-diphosphate-glucuronosyltransferase in a patient with Crigler-Najjar syndrome type I. *Pediatr Res* 1994;35:629–32.

111. Bosma, PJ, Chowdhury, JR, Bakker, C, Gantla, S, de Boer, A, Oostra, BA, Lindhout, D, Tytgat, GN, Jansen, PL, Oude Elferink, RP, et al. The genetic basis of the reduced expression of bilirubin UDP-glucuronosyltransferase 1 in Gilbert's syndrome. *N Engl J Med* 1995;333:1171–5.

112. Monaghan, G, Ryan, M, Seddon, R, Hume, R, and Burchell, B. Genetic variation in bilirubin UPD-glucuronosyltransferase gene promoter and Gilbert's syndrome. *Lancet* 1996;347:578–81.

113. Sato, H, Adachi, Y, and Koiwai, O. The genetic basis of Gilbert's syndrome. *Lancet* 1996;347:557–8.

114. Beutler, E, Gelbart, T, and Demina, A. Racial variability in the UDP–glucuronosyltransferase 1 (UGT1A1) promoter: a balanced polymorphism for regulation of bilirubin metabolism? *Proc Natl Acad Sci USA* 1998;95:8170–4.

115. Guillemette, C, De Vivo, I, Hankinson, SE, Haiman, CA, Spiegelman, D, Housman, DE, and Hunter, DJ. Association of genetic polymorphisms in UGT1A1 with breast cancer and plasma hormone levels. *Cancer Epidemiol Biomarkers Prev* 2001;10:711–4.

116. Akaba, K, Kimura, T, Sasaki, A, Tanabe, S, Ikegami, T, Hashimoto, M, Umeda, H, Yoshida, H, Umetsu, K, Chiba, H, Yuasa, I, and Hayasaka, K. Neonatal hyperbilirubinemia and mutation of the bilirubin uridine diphosphate-glucuronosyltransferase gene: a common missense mutation among Japanese, Koreans and Chinese. *Biochem Mol Biol Int* 1998;46:21–6.

117. Huang, CS, Luo, GA, Huang, ML, Yu, SC, and Yang, SS. Variations of the bilirubin uridine-diphosphoglucuronosyl transferase 1A1 gene in healthy Taiwanese. *Pharmacogenetics* 2000;10:539–44.

118. Ciotti, M, Marrone, A, Potter, C, and Owens, IS. Genetic polymorphism in the human UGT1A6 (planar phenol) UDP-glucuronosyltransferase: pharmacological implications. *Pharmacogenetics* 1997;7:485–95.

119. Glatt, H and Meinl, W. Pharmacogenetics of soluble sulfotransferases (SULTs). *Naunyn Schmiedebergs Arch Pharmacol* 2004;369:55–68.

120. Van Loon, J and Weinshilboum, RM. Human platelet phenol sulfotransferase: familial variation in thermal stability of the TS form. *Biochem Genet* 1984;22:997–1014.

121. Raftogianis, RB, Wood, TC, Otterness, DM, Van Loon, JA, and Weinshilboum, RM. Phenol sulfotransferase pharmacogenetics in humans: association of common SULT1A1 alleles with TS PST phenotype. *Biochem Biophys Res Commun* 1997;239:298–304.

122. Raftogianis, RB, Wood, TC, and Weinshilboum, RM. Human phenol sulfotransferases SULT1A2 and SULT1A1: genetic polymorphisms, allozyme properties, and human liver genotype-phenotype correlations. *Biochem Pharmacol* 1999;58:605–16.

123. Ning, B, Nowell, S, Sweeney, C, Ambrosone, CB, Williams, S, Miao, X, Liang, G, Lin, D, Stone, A, Ratnasinghe, DL, Manjanatha, M, Lang, NP, and Kadlubar, FF. Common genetic polymorphisms in the 5′-flanking region of the SULT1A1 gene: haplotypes and their association with platelet enzymatic activity. *Pharmacogenet Genomics* 2005;15:465–73.

124. Hebbring, SJ, Adjei, AA, Baer, JL, Jenkins, GD, Zhang, J, Cunningham, JM, Schaid, DJ, Weinshilboum, RM, and Thibodeau, SN. Human SULT1A1 gene: copy number differences and functional implications. *Hum Mol Genet* 2007;16:463–70.

125. Ho, RH and Kim, RB. Transporters and drug therapy: implications for drug disposition and disease. *Clin Pharmacol Ther* 2005;78:260–77.

126. Kerb, R. Implications of genetic polymorphisms in drug transporters for pharmacotherapy. *Cancer Lett* 2006;234:4–33.
127. Cascorbi, I. Role of pharmacogenetics of ATP-binding cassette transporters in the pharmacokinetics of drugs. *Pharmacol Ther* 2006;112:457–73.
128. Ieiri, I, Takane, H, Hirota, T, Otsubo, K, and Higuchi, S. Genetic polymorphisms of drug transporters: pharmacokinetic and pharmacodynamic consequences in pharmacotherapy. *Expert Opin Drug Metab Toxicol* 2006;2:651–74.
129. Hediger, MA, Romero, MF, Peng, JB, Rolfs, A, Takanaga, H, and Bruford, EA. The ABCs of solute carriers: physiological, pathological and therapeutic implications of human membrane transport proteinsIntroduction. *Pflugers Arch* 2004;447:465–8.
130. Marzolini, C, Tirona, RG, and Kim, RB. Pharmacogenomics of the OATP and OAT families. *Pharmacogenomics* 2004;5:273–82.
131. Tirona, RG, Leake, BF, Merino, G, and Kim, RB. Polymorphisms in OATP-C: identification of multiple allelic variants associated with altered transport activity among European- and African-Americans. *J Biol Chem* 2001;276:35669–75.
132. Nozawa, T, Nakajima, M, Tamai, I, Noda, K, Nezu, J, Sai, Y, Tsuji, A, and Yokoi, T. Genetic polymorphisms of human organic anion transporters OATP-C (SLC21A6) and OATP-B (SLC21A9): allele frequencies in the Japanese population and functional analysis. *J Pharmacol Exp Ther* 2002;302:804–13.
133. Nozawa, T, Minami, H, Sugiura, S, Tsuji, A, and Tamai, I. Role of organic anion transporter OATP1B1 (OATP-C) in hepatic uptake of irinotecan and its active metabolite, 7-ethyl-10-hydroxycamptothecin: in vitro evidence and effect of single nucleotide polymorphisms. *Drug Metab Dispos* 2005;33:434–9.
134. Niemi, M, Schaeffeler, E, Lang, T, Fromm, MF, Neuvonen, M, Kyrklund, C, Backman, JT, Kerb, R, Schwab, M, Neuvonen, PJ, Eichelbaum, M, and Kivisto, KT. High plasma pravastatin concentrations are associated with single nucleotide polymorphisms and haplotypes of organic anion transporting polypeptide-C (OATP- C, SLCO1B1). *Pharmacogenetics* 2004;14:429–40.
135. Niemi, M, Kivisto, KT, Hofmann, U, Schwab, M, Eichelbaum, M, and Fromm, MF. Fexofenadine pharmacokinetics are associated with a polymorphism of the SLCO1B1 gene (encoding OATP1B1). *Br J Clin Pharmacol* 2005;59:602–4.
136. Niemi, M. Role of OATP transporters in the disposition of drugs. *Pharmacogenomics* 2007;8:787–802.
137. Mwinyi, J, Johne, A, Bauer, S, Roots, I, and Gerloff, T. Evidence for inverse effects of OATP-C (SLC21A6) 5 and 1b haplotypes on pravastatin kinetics. *Clin Pharmacol Ther* 2004;75:415–21.
138. Morimoto, K, Oishi, T, Ueda, S, Ueda, M, Hosokawa, M, and Chiba, K. A novel variant allele of OATP-C (SLCO1B1) found in a Japanese patient with pravastatin-induced myopathy. *Drug Metab Pharmacokinet* 2004;19:453–5.
139. Niemi, M, Backman, JT, Kajosaari, LI, Leathart, JB, Neuvonen, M, Daly, AK, Eichelbaum, M, Kivisto, KT, and Neuvonen, PJ. Polymorphic organic anion transporting polypeptide 1B1 is a major determinant of repaglinide pharmacokinetics. *Clin Pharmacol Ther* 2005;77:468–78.
140. Katz, DA, Carr, R, Grimm, DR, Xiong, H, Holley- Shanks, R, Mueller, T, Leake, B, Wang, Q, Han, L, Wang, PG, Edeki, T, Sahelijo, L, Doan, T, Allen, A, Spear, BB, and Kim, RB. Organic anion transporting polypeptide 1B1 activity classified by SLCO1B1 genotype influences atrasentan pharmacokinetics. *Clin Pharmacol Ther* 2006;79:186–96.
141. Cascorbi, I, Gerloff, T, Johne, A, Meisel, C, Hoffmeyer, S, Schwab, M, Schaeffeler, E, Eichelbaum, M, Brinkmann, U, and Roots, I. Frequency of single nucleotide polymorphisms in the P-glycoprotein drug transporter MDR1 gene in white subjects. *Clin Pharmacol Ther* 2001;69:169–74.

142. Kim, RB, Leake, BF, Choo, EF, Dresser, GK, Kubba, SV, Schwarz, UI, Taylor, A, Xie, HG, McKinsey, J, Zhou, S, Lan, LB, Schuetz, JD, Schuetz, EG, and Wilkinson, GR. Identification of functionally variant MDR1 alleles among European Americans and African Americans. *Clin Pharmacol Ther* 2001;70:189–99.
143. Kroetz, DL, Pauli- Magnus, C, Hodges, LM, Huang, CC, Kawamoto, M, Johns, SJ, Stryke, D, Ferrin, TE, DeYoung, J, Taylor, T, Carlson, EJ, Herskowitz, I, Giacomini, KM, and Clark, AG. Sequence diversity and haplotype structure in the human ABCB1 (MDR1, multidrug resistance transporter) gene. *Pharmacogenetics* 2003;13:481–94.
144. Allabi, AC, Horsmans, Y, Issaoui, B, and Gala, JL. Single nucleotide polymorphisms of ABCB1 (MDR1) gene and distinct haplotype profile in a West Black African population. *Eur J Clin Pharmacol* 2005;61:97–102.
145. Marzolini, C, Paus, E, Buclin, T, and Kim, RB. Polymorphisms in human MDR1 (P-glycoprotein): recent advances and clinical relevance. *Clin Pharmacol Ther* 2004;75:13–33.
146. Chinn, LW and Kroetz, DL. ABCB1 pharmacogenetics: progress, pitfalls, and promise. *Clin Pharmacol Ther* 2007;81:265–9.
147. Hirsh, J, Dalen, J, Anderson, DR, Poller, L, Bussey, H, Ansell, J, and Deykin, D. Oral anticoagulants: mechanism of action, clinical effectiveness, and optimal therapeutic range. *Chest* 2001;119:8S–21S.
148. Hylek, EM, Go, AS, Chang, Y, Jensvold, NG, Henault, LE, Selby, JV, and Singer, DE. Effect of intensity of oral anticoagulation on stroke severity and mortality in atrial fibrillation. *N Engl J Med* 2003;349:1019–26.
149. Gage, BF, Eby, C, Milligan, PE, Banet, GA, Duncan, JR, and McLeod, HL. Use of pharmacogenetics and clinical factors to predict the maintenance dose of warfarin. *Thromb Haemost* 2004;91:87–94.
150. Voora, D, McLeod, HL, Eby, C, and Gage, BF. The pharmacogenetics of coumarin therapy. *Pharmacogenomics* 2005;6:503–13.
151. Li, T, Chang, CY, Jin, DY, Lin, PJ, Khvorova, A, and Stafford, DW. Identification of the gene for vitamin K epoxide reductase. *Nature* 2004;427:541–4.
152. Rieder, MJ, Reiner, AP, Gage, BF, Nickerson, DA, Eby, CS, McLeod, HL, Blough, DK, Thummel, KE, Veenstra, DL, and Rettie, AE. Effect of VKORC1 haplotypes on transcriptional regulation and warfarin dose. *N Engl J Med* 2005;352:2285–93.
153. Takahashi, H and Echizen, H. Pharmacogenetics of warfarin elimination and its clinical implications. *Clin Pharmacokinet* 2001;40:587–603.
154. Wadelius, M, Chen, LY, Downes, K, Ghori, J, Hunt, S, Eriksson, N, Wallerman, O, Melhus, H, Wadelius, C, Bentley, D, and Deloukas, P. Common VKORC1 and GGCX polymorphisms associated with warfarin dose. *Pharmacogenomics J* 2005;5:262–70.
155. Herman, D, Locatelli, I, Grabnar, I, Peternel, P, Stegnar, M, Mrhar, A, Breskvar, K, and Dolzan, V. Influence of CYP2C9 polymorphisms, demographic factors and concomitant drug therapy on warfarin metabolism and maintenance dose. *Pharmacogenomics J* 2005;5:193–202.
156. Sanderson, S, Emery, J, and Higgins, J. CYP2C9 gene variants, drug dose, and bleeding risk in warfarin-treated patients: a HuGEnet systematic review and meta-analysis. *Genet Med* 2005;7:97–104.
157. Higashi, MK, Veenstra, DL, Kondo, LM, Wittkowsky, AK, Srinouanprachanh, SL, Farin, FM, and Rettie, AE. Association between CYP2C9 genetic variants and anticoagulation-related outcomes during warfarin therapy. *JAMA* 2002;287:1690–8.
158. Allabi, AC, Gala, JL, Horsmans, Y, Babaoglu, MO, Bozkurt, A, Heusterspreute, M, and Yasar, U. Functional impact of CYP2C95, CYP2C96, CYP2C98, and CYP2C911 In vivo among black Africans. *Clin Pharmacol Ther* 2004;76:113–8.

159. Kidd, RS, Curry, TB, Gallagher, S, Edeki, T, Blaisdell, J, and Goldstein, JA. Identification of a null allele of CYP2C9 in an African-American exhibiting toxicity to phenytoin. *Pharmacogenetics* 2001;11:803–8.
160. Kawato, Y, Aonuma, M, Hirota, Y, Kuga, H, and Sato, K. Intracellular roles of SN-38, a metabolite of the camptothecin derivative CPT-11, in the antitumor effect of CPT-11. *Cancer Res* 1991;51:4187–91.
161. Gupta, E, Lestingi, TM, Mick, R, Ramirez, J, Vokes, EE, and Ratain, MJ. Metabolic fate of irinotecan in humans: correlation of glucuronidation with diarrhea. *Cancer Res* 1994;54:3723–5.
162. Ando, Y, Saka, H, Asai, G, Sugiura, S, Shimokata, K, and Kamataki, T. UGT1A1 genotypes and glucuronidation of SN-38, the active metabolite of irinotecan. *Ann Oncol* 1998;9:845–7.
163. Iyer, L, Das, S, Janisch, L, Wen, M, Ramirez, J, Karrison, T, Fleming, GF, Vokes, EE, Schilsky, RL, and Ratain, MJ. UGT1A1*28 polymorphism as a determinant of irinotecan disposition and toxicity. *Pharmacogenomics J* 2002;2:43–7.
164. Innocenti, F, Undevia, SD, Iyer, L, Chen, PX, Das, S, Kocherginsky, M, Karrison, T, Janisch, L, Ramirez, J, Rudin, CM, Vokes, EE, and Ratain, MJ. Genetic variants in the UDP-glucuronosyltransferase 1A1 gene predict the risk of severe neutropenia of irinotecan. *J Clin Oncol* 2004;22:1382–8.
165. Engen, RM, Marsh, S, Van Booven, DJ, and McLeod, HL. Ethnic differences in pharmacogenetically relevant genes. *Curr Drug Targets* 2006;7:1641–8.
166. Kim, S and Misra, A. SNP genotyping: technologies and biomedical applications. *Annu Rev Biomed Eng* 2007;9:289–320.
167. Ragoussis, J. Genotyping technologies for all. *Drug Discov Today Technol* 2006;3:115–22.
168. Walker, EJ and Siminovitch, KA. Primer: genomic and proteomic tools for the molecular dissection of disease. *Nat Clin Pract Rheumatol* 2007;3:580–9.
169. Nebert, DW. Suggestions for the nomenclature of human alleles: relevance to ecogenetics, pharmacogenetics and molecular epidemiology. *Pharmacogenetics* 2000;10:279–90.
170. King, BP, Leathart, JB, Mutch, E, Williams, FM, and Daly, AK. CYP3A5 phenotype-genotype correlations in a British population. *Br J Clin Pharmacol* 2003;55:625–9.
171. Roy, JN, Lajoie, J, Zijenah, LS, Barama, A, Poirier, C, Ward, BJ, and Roger, M. CYP3A5 genetic polymorphisms in different ethnic populations. *Drug Metab Dispos* 2005;33:884–7.
172. Xie, HG, Wood, AJ, Kim, RB, Stein, CM, and Wilkinson, GR. Genetic variability in CYP3A5 and its possible consequences. *Pharmacogenomics* 2004;5:243–72.
173. Daly, AK. Significance of the minor cytochrome P450 3A isoforms. *Clin Pharmacokinet* 2006;45:13–31.
174. Robarge, JD, Li, L, Desta, Z, Nguyen, A, and Flockhart, DA. The star-allele nomenclature: retooling for translational genomics. *Clin Pharmacol Ther* 2007;82:244–8.
175. Kimchi-Sarfaty, C, Oh, JM, Kim, IW, Sauna, ZE, Calcagno, AM, Ambudkar, SV, and Gottesman, MM. A "silent" polymorphism in the MDR1 gene changes substrate specificity. *Science* 2007;315:525–8.
176. Sauna, ZE, Kimchi- Sarfaty, C, Ambudkar, SV, and Gottesman, MM. Silent polymorphisms speak: how they affect pharmacogenomics and the treatment of cancer. *Cancer Res* 2007;67:9609–12.
177. Sauna, ZE, Kimchi- Sarfaty, C, Ambudkar, SV, and Gottesman, MM. The sounds of silence: synonymous mutations affect function. *Pharmacogenomics* 2007;8:527–32.

Chapter 9

Overview of Regulatory Developments and Initiatives Related to the Use of Genomic Technologies in Drug Discovery and Development

9.1. INTRODUCTION TO RECENT REGULATORY DEVELOPMENTS IN THE GENOMIC AREA

There is a general consensus that the appropriate use of genomic technologies in drug discovery and development can help address two current major problems in health care, namely, the low productivity of pharmaceutical research and development (R&D) and the high proportion of patients who do not benefit from drugs or experience adverse events upon treatment. Throughout this book, we provide multiple examples that demonstrate or support this position. Ultimately, drug companies will have to find ways to execute on these promises and successfully integrate genomic technologies to improve their R&D processes and thus benefit the patients. Because of their function in developing and enforcing standards for drugs and diagnostic devices, regulatory agencies are in a unique position to foster the use of novel technologies that can positively impact the discovery and use of drugs with improved efficacy and safety. Not surprisingly, both the

Genomics in Drug Discovery and Development, by Dimitri Semizarov and Eric Blomme

U.S. and European regulatory authorities (FDA and EMEA, respectively) issued in 2004 and 2005 their own strategic position papers, explicitly expressing their concerns about the current drug pipeline and listing opportunities to better the success of drug development and to improve safety and efficacy of future and current medicines. These documents, known as the Critical Path White Paper and the Road Map, cover topics that go well beyond genomic technologies. Therefore, we will not discuss these position papers in their entirety. The readers are, nonetheless, highly encouraged to consult these two important milestone documents.

Both the FDA's Critical Path White Paper and the EMEA's Road Map have positioned genomics at the heart of a broader strategy to ameliorate drug discovery and development. Hence, in its Critical Path report issued in 2004 (http://www.fda.gov/oc/initiatives/criticalpath/whitepaper.html), the FDA officially acknowledged that "the emerging techniques of pharmacogenomics and proteomics show great promise for contributing biomarkers to target responders, monitoring clinical response, and biomarker targets of drug effectiveness." Furthermore, in its Opportunity List, the FDA has placed an emphasis on the role of genomic biomarkers in the development of personalized medicine. Similarly, the EMEA Road Map (http://www.emea.europa.eu/htms/general/direct/roadmap/roadmapintro.htm) addresses its concern with the current drug pipeline problem and indicates that "initiatives should focus on addressing the encountered difficulties in the development stage by exploring innovative approaches in drug development." However, while the Road Map does identify pharmacogenomics as an important group of novel technologies, it does not explicitly suggest that it necessarily represents a solution to the current decline in pharmaceutical R&D productivity.

As should be clear after reading the previous chapters of this book, the bulk of genomic data currently generated during discovery and development is mostly exploratory in nature and is usually not sufficiently mature to affect the approval or clinical use of novel medicines. However, the situation is rapidly evolving, and in an effort to promote the use of genomic technologies in drug development and to familiarize regulatory authorities with these emerging technologies, several draft or final guidance or position documents have been issued in the last few years. In this chapter, we provide a brief overview of some of these documents issued by regulatory authorities, as well as comment on opinions officially expressed by members of regulatory bodies. For simplicity, this chapter is organized around the major regulatory documents released so far and the various initiatives largely initiated by the Critical Path White Paper. Table 9.1 provides a list of all documents that are discussed in this chapter as well as additional guidance documents that are relevant to the subject. The URL for each document is provided in the table, as well as in the text to facilitate access for the reader. Since a good comprehension of these documents requires understanding of the terminology, a short definition of some key terms and frequently used acronyms is also presented in Boxes 9.1 and 9.2. In general, these documents are quite specific, but also leave some room for interpretation. It is not our intention

Table 9.1 URL of Useful Regulatory Guidance Documents

Title	URL
FDA Guidance Document Home Page	http://www.fda.gov/cder/guidance
FDA CDER list of Guidance Documents	http://www.fda.gov/cder/guidance/CompList122007.pdf
FDA Pharmacogenomic Data Submissions	http://www.fda.gov/cder/guidance/6400fnl.pdf
Attachment to FDA Pharmacogenomic Data Submissions	http://www.fda.gov/cder/guidance/6400fnlAttch.pdf
FDA Pharmacogenomic Data Submissions—Companion Guidance	http://www.fda.gov/cder/guidance/7735dft.htm
FDA Drug-Diagnostic Co-Development Draft Concept Paper	http://www.fda.gov/cder/genomics/pharmacoconceptfn.pdf
EMEA position paper on terminology in pharmacogenetics	http://www.emea.europa.eu/pdfs/human/press/pp/307001en.pdf
EMEA leaflet, Understanding the Terminology Used in Pharmacogenetics	http://www.emea.europa.eu/pdfs/human/pharmacogenetics/384204en.pdf
EMEA Guideline on Pharmacogenetics Briefing Meetings	http://www.emea.europa.eu/pdfs/human/pharmacogenetics/2022704en.pdf
Guiding Principles of Joint FDA EMEA Voluntary Genomic Data Submissions (VGDSs)	http://www.fda.gov/cder/genomics/FDAEMEA.pdf
ICH E15 Terminology in Pharmacogenomics	http://www.fda.gov/cder/guidance/7619dft.pdf
FDA CDRH Device Class Definition	http://www.fda.gov/cdrh/devadvice/3132.html
FDA Draft Guidance on In Vitro Diagnostic Multivariate Index Assays	http://www.fda.gov/cdrh/oivd/guidance/1610.pdf

to provide a personal interpretation of these documents. Rather, we here restrict our overview to contents and conclude with an outlook on the future impact of genomic technologies on the regulatory approval and clinical use of novel medicines.

BOX 9-1 *Glossary of Terms*

Biomarker: A parameter that is objectively measured and evaluated as an indication of normal or pathogenic process, or of a pharmacological response to a therapeutic intervention

Briefing Meeting: An optional informal meeting between a sponsor and regulators at the EMEA for the purpose of sharing scientific and technical information

Clinical utility: The elements that need to be considered when evaluating the risks and benefits in diagnosing or predicting risk for an event (drug response, presence or risk of a health condition)

Genotyping: Test to detect a specific, known genetic variant

Interdisciplinary Pharmacogenomics Review Group (IPRG): FDA group that reviews all VGDSs submitted to the FDA and consult, upon request, on the required GDSs

In Vitro Diagnostic Device (IVD): An "in vitro" reagent and any component part or accessory which is intended for use in the diagnosis of disease or other conditions, in man or other animals. [Section 201(h) of the Federal, Food, Drug, and Cosmetic Act] or "those reagents, instruments, and systems intended for use in the diagnosis of disease or other conditions, including a determination of the state of health, in order to cure, mitigate, treat, or prevent disease or its sequelae.... Such products are intended for use in the collection, preparation and examination of specimens taken form the human body." (21 CFR 809.3)

Pharmaceutical and Medical Devices Agency (PDMA): Japanese regulatory authority

Pharmacogenetics Working Party (PGWP): A permanent working party of the EMEA that provides recommendations on all matters relating directly or indirectly to Pharmacogenetics. This party plays a role similar to that of FDA's IPRG.

Pharmacogenetics: The study of interindividual genetic variation to drug response

Pharmacogenomics: The application of genomics technologies (i.e., technologies to measure genetic variation and gene expression alterations) to the study of human variability in drug response

Pharmacogenetic test: An assay used to determine interindividual variations in DNA sequence associated with differences in drug absorption and disposition (pharmacokinetics), drug action (pharmacodynamics) or with adverse reactions

Pharmacogenomic test: An assay used to determine interindividual variations in the mRNA expression levels of a gene set (gene expression signature) that is correlated with pharmacological function, therapeutic response or an adverse event

Probable valid biomarker: A biomarker that is measured in an analytical test system with well-established performance characteristics and for which there is a scientific framework or body of evidence that appears to elucidate the physiological, toxicological, pharmacological, or clinical significance of the test results. A probable valid biomarker may not have reached the status of a known valid marker because of any one of the following reasons: The data elucidating its significance may have been generated within a single company and may not be available for public scientific scrutiny.

Valid biomarker: A biomarker that is measured in an analytical test system with well-established performance characteristics and for which there is widespread agreement in the medical or scientific community about the physiological, toxicological, pharmacological, or clinical significance of the results

Toxicogenomics: The application of gene expression technologies to predict and characterize toxicity

BOX 9-2 *Acronyms*

BLA: Biologic License Application

CBER: FDA Center for Biological Evaluation and Research

CDER: FDA Center for Drug Evaluation and Research

CDRH: FDA Center for Devices and Radiological Health

CLIA: Clinical Laboratory Improvements Amendments Committee

EMEA: European Medicines Evaluation Agency

FDA: Food and Drug Administration, United States of America

GDS: A Genomic Data Submission

ICH: International Conference on Harmonization

IND: Investigational New Drug

IPRG: FDA Interdisciplinary Pharmacogenomics Review Group.

IVD: In Vitro Diagnostic Device

NDA: New Drug Application

OCP: FDA Office of Combination Products

PDMA: Pharmaceutical and Medical Devices Agency, Japan

PGWP: EMEA Pharmacogenetics Working Party

VGDS: Voluntary Genomic Data Submission

9.2. FDA GUIDANCE ON PHARMACOGENOMIC DATA SUBMISSION

Recognizing the potential of pharmacogenomics to maximize the effectiveness and minimize the risk of novel medicines, the FDA issued a guidance in March 2005 for the regulatory submission of pharmacogenomic data (http://www.fda.gov/cder/guidance/6400fnl.pdf) and created a "Genomics at FDA" webportal (http://www.fda.gov/cder/genomics/). This web site provides up-to-date regulatory and background information on genomics, and the reader is highly encouraged to visit this resource (1). The pharmacogenomic guidance reflects the effort of the FDA to promote the use of genomic technologies in drug development and is also designed to enhance the agency's knowledge of these emerging technologies. This guidance also represents a partial answer to the previously perceived lack of clarity on the regulatory response to pharmacogenomic data. Indeed, uncertainties related to how these complex and novel data would be used by the FDA in the drug application review process have been considered a significant factor in the delay of new pharmacogenomic products and in the reluctance of drug developers to apply pharmacogenomics during the FDA-regulated phases of drug development.

In finalizing this guidance document, FDA has openly cooperated with the various stakeholders, using a broad process of public consultation and organized appropriate forums to focus on the major issues and principles that the document should cover. Summaries of the outcomes of these various public forums have been published elsewhere (2–6). The pharmaceutical industry has welcomed this guidance, as it represents an important stepping-stone in the development of genomics-based biomarkers and the use of genomics-based safety data. In addition, this guidance provided reassurance to companies that exploratory genomic data would not bring negative regulatory consequences, an important aspect for the wider acceptance of this new technology. Since the guidance covers the requirements for both voluntary submissions and regulatory decision-making, we discuss these two aspects separately.

9.2.1. Voluntary Genomic Data Submission (VGDS)

A key objective of this guidance was to encourage sponsors to share genomic data in a voluntary process outside the formal regulatory mechanism. Indeed, the agency recognized that while pharmaceutical companies were generating pharmacogenomic data, most of these data were exploratory in nature and would therefore not be part of formal regulatory submissions. Anticipating a growing role of genomic data in drug development, the agency was cognizant of the importance to prepare its staff to appropriately evaluate future submissions, and this preparation entailed a certain level of familiarity with a variety of relevant scientific issues. The idea of voluntary submissions was first brought up in a workshop on pharmacogenetics/pharmacogenomics in drug development and regulatory decision-making under what was then referred as the **"safe harbor"** concept (7, 8). The Voluntary Genomic Data Submission (VGDS) process provides a mechanism by which genomic data can be submitted outside the formal regulatory approval process. The expected outcomes of a VGDS include an opportunity for the agency to learn how the industry is using genomic data in discovery and development, to prepare its scientific staff for later inclusion of these complex and novel data in the regulatory system, and to encourage the application of genomic technologies by fostering their use in discovery and development without prejudice to future drug submissions. It was also expected that this process would allow the industry to influence drug regulators regarding the use of genomic data in regulatory reviews. It should be mentioned that this process is quite innovative, since it represents the first time that the industry has been encouraged to share exploratory data on a voluntary basis outside the formal approval system.

VGDSs can be used for genomic data generated and applied at any stage of the drug discovery and development process. Submissions are free of charge and reviewed without prejudice to subsequent formal submissions. In other words, these submissions do not impact the subsequent independent review of a related formal submission, but they also cannot be used as a formal presubmission review process. Hence, companies cannot request agreements on items such as clinical protocols or other studies for approval of an application.

To ensure a clear separation with the formal review process, VGDSs are reviewed by a cross-center **Interdisciplinary Pharmacogenomic Review Group (IPRG)** that does not include staff involved in an associated formal submission. This group consists of staff from the device (CDRH) and drug (CDER and CBER) sections of the FDA and has the goal of creating the scientific and regulatory framework necessary for reviewing genomic data. From its inception, it has included among others, several senior FDA officials, such as Deputy Commissioner Dr. Janet Woodcock, Dr. Larry Lesko (Director of the Office of Clinical Pharmacology and Biopharmaceuticals, or OCPB), and Dr. Steven Gutman (Director of the Office of In Vitro Diagnostics, or OIVD). As of January 2008, Dr. Felix Frueh, Associate Director for Genomics in OCPB, chairs the IPRG. For more information, the reader is referred to the FDA Web site that reviews the Frequently Asked Questions regarding VGDS (http://www.fda.gov/cder/genomics/FAQ.htm). All VGDS data are protected from disclosure either outside the FDA or to review divisions, are routed directly to the IPRG, and are stored on a secured, separate server. These data are not distributed outside the IPRG without prior agreement of the sponsor and are not to be used for regulatory decision-making.

The IPRG reviews all VGDSs and is the primary point of contact for the sponsor during and after the review process on matters relating to the submission. To facilitate the review process, the IPRG can meet with the sponsor. Sponsor meetings can be requested by the sponsor or the FDA. After review, the IPRG writes a VGDS report, and a summary report of the VGDS review is sent to the sponsor. A meeting with the IPRG can be requested by the sponsor to discuss the findings in the report.

In general, the VGDS experience appears to be a success for both sponsors and the FDA. It has been well received by the pharmaceutical industry. The first VGDS submission was received in March 2004 and has been followed by over two dozen submissions and many sponsor meetings (1). According to figures disclosed, the submissions covered a wide range of therapeutic areas (cancer, obesity, depression, hypertension, and Alzheimer disease), and their scientific scope encompassed biomarker qualification, genotyping devices, microarrays, study design, biostatistics, and enrichment design (1). Although the complexity of the submissions was initially low, more recent submissions contained large and highly complex data sets (1). The program has recently been expanded to VXDS, where "X" stands simply for "exploratory." The objective is to also capture all the other "omics"-type exploratory technologies and to take an integrated (systems biology) approach to the use of these "omics" data in drug development. The FDA recently received the first of such data sets in which the bridging of these technologies has been the focus (1).

It is clear that VGDS submissions have been an effective approach for the FDA to gain access to genomic data and other information that otherwise would not have been available. These data encompass multiple therapeutic, scientific, and technical areas. Sponsor meetings have created a unique opportunity to share approaches on various topics, such as the evaluation of complex data sets, and have allowed sponsors to familiarize FDA scientists with genomic data and their interpretation. In fact, sponsors can request that staff from review divisions take part in the VGDS. At the same time, sponsors have gained by learning about the regulatory decision-making process and the expectations involving genomic data. Ultimately, this could prevent delays in future submissions containing required genomic data. Hence, several sponsors have used the VGDS process as a stepping-stone to present the same or related data in a regulatory context to the FDA later, such as in a phase II meeting or in protocol assessments for phase III studies (1). This approach provides an opportunity for sponsors to understand how the FDA may react to the use of this information in the context of regulatory decision-making. Such VXDS may center around questions such as "Has the marker been developed appropriately?", "Were the most critical experimental considerations taken?", or "Is the approach to use the marker in a prospective study appropriate?" (1).

A summary of the experience with VGDS at the FDA has recently been published (1). The non-clinical VGDSs consisted mostly of toxicogenomic data that were used as screening tools (genomic biomarkers) for compound prioritization (what we refer to as predictive toxicogenomics in Chapter 5) or as a tool to derive more detailed insights into the molecular mechanisms of toxicity (what we define as mechanistic toxicogenomics in Chapter 5). Clinical VGDSs covered the use of specific gene variants or gene expression signatures as markers for use in patient stratification in clinical trials.

9.2.2. Pharmacogenomic Data Submission

While the VGDS process was clearly an important part of the issued guidance, an increasing number of regulatory submissions include genomic data, and clear guidelines for genomic data submissions were needed. Similar to VGDS data, non-voluntary submissions are becoming more complex (shifting from candidate gene approaches to whole genome studies), and it is anticipated that in the near future a significant number of investigational new drugs (INDs) and new drug applications (NDAs) will contain pharmacogenomic information critical to clinical development (1). The FDA Guidance on Pharmacogenomic Data Submission clarifies the policy of the agency on the use of pharmacogenomic data in the drug application review process and covers the application of genomic concepts and technologies to non-clinical, clinical pharmacology, and clinical studies. More specifically, it provides guidelines to sponsors on pharmacogenomic data submission requirements, the format and procedure for data submission, and how the data will be used in regulatory decision-making. This guidance does not define a new regulatory process for the use of genomic data in drug development, but outlines how genomic data will be used within the existing framework. In other words, the agency recognizes that the scientific questions and issues addressed by genomic technologies are not novel and can be integrated in current regulatory practices. The reader is referred to the most recent guidance documents for more details. Here, we restrict our discussion to major concepts and use several selected examples as illustrations.

Examples of materials that would be appropriate for voluntary submissions or required for submissions can be found on the genomics site of the FDA at http://www.fda.gov/cder/guidance/6400fnlAttch.pdf. In general terms, genomic data for which formal submission is required include the following:

1. Data used for decision-making within a specific trial
2. Data to support scientific arguments about the mechanism of action, dose selection, safety, or efficacy
3. Data that will support registration or labeling language
4. Data generated on previously validated biomarkers

Useful algorithms are provided as appendices in the guidance to help evaluate whether or not submissions are required for INDs, NDAs, or BLAs (Biologic License Applications). The guidance also defines **pharmacogenomic tests** as follows: "An assay intended to study interindividual variations in whole-genome or candidate gene, single-nucleotide polymorphism (SNP) maps, haplotype markers, or alterations in gene expression or inactivation that may be correlated with pharmacological and therapeutic response. In some cases, the pattern or profile of change is the relevant biomarker, rather than changes in individual markers." This implies that gene expression signatures may represent validated biomarkers.

Central to this guidance is the definition of what constitutes a validated biomarker. Hence, the guidance makes a distinction between what the agency refers to as a "known valid biomarker" and what it regards as a "probable valid biomarker." According to the guidance, known valid biomarkers are measured in an analytical test system with well-established performance characteristics, and for which an established scientific framework or body of evidence exists to understand the significance of the test results. In addition, for a known valid biomarker, a widespread agreement exists in the medical or scientific community about the significance of the physiological, toxicological, pharmacological, or clinical significance of the results. The example provided by the guidance concerns genetic tests that distinguish allelic variants of drug-metabolizing enzymes, such as cytochrome P450 2D6 (CYP2D6) and thiopurine methyltransferase. Since the consequences for drug metabolism of genetic variation in these enzymes are well understood in the scientific community and are reflected in certain approved drug labels, those would be considered to be well established and, therefore, valid biomarkers. In contrast, a probable valid biomarker, albeit also measured in a robust analytical test, has only apparent clinical validity. In other words, available data are supportive, but not conclusive, of clinical validity or there is no current consensus or independent verification to confirm the significance of the test results. An example would be a genomic biomarker developed by a sponsor and not available for public scientific scrutiny or for independent verification.

This distinction illustrates the enormous amount of work and improvement that will be needed to make novel genomic biomarkers suitable for regulatory decision-making. In fact, it suggests that most pharmacogenomic biomarkers are probably insufficiently validated to be used in regulatory decision-making. Indeed, most published candidate biomarkers seldom become qualified for various reasons. On the one hand, there is probably an insufficient understanding in the scientific community of the appropriate regulatory pathway to qualification for use. On the other hand, there is, as we discuss below, a lack of clear incentives for sponsors to spearhead qualification work given the required financial resources and unclear returns on investment. On the positive side, however, various consortia are pooling resources in an effort to qualify novel broad biomarkers, and we cover this development in more detail below (Section 9.7). Similarly, the agency is establishing a pilot structure to start a qualification process for biomarkers in drug development (Section 9.6). Furthermore, technical issues associated with analytical genomic platforms are being comprehensively evaluated. Data released so far suggest that issues such as quality control of laboratory procedures and interplatform variability should not limit the future qualification of novel genomic biomarkers (9–12).

9.2.3. International Harmonization

While the FDA has led the way on issuing guidance and opinions on genomic data submission, regulatory authorities of other geographic areas have also been active in this area. In particular, in 2001, the EMEA published a position paper on terminology in pharmacogenetics with regard to classification of samples used in pharmacogenetic research (http://www.emea.europa.eu/pdfs/human/press/pp/

307001en.pdf). This was followed by the publication of a leaflet entitled "Understanding the Terminology Used in Pharmacogenetics" (http://www.emea.europa.eu/pdfs/human/pharmacogenetics/384204en.pdf). In addition, the EMEA has created a voluntary submission process referred to as **pharmacogenomics briefing meetings**. This process is similar, yet not identical, to the VGDS and was initially described in a concept paper covering the purpose and scope of the initiative and further defined in an EMEA Guideline on Pharmacogenetics Briefing Meetings (http://www.emea.europa.eu/pdfs/human/pharmacogenetics/2022704en.pdf). This guidance specifies the format of the meetings and of the submissions and the type of genomic data that the European regulators expect to receive. In these briefing meetings, the **Pharmacogenomics Working Party (PGWP)** plays a role quite similar to that performed by the IPRG. The PGWP brings together expertise across the EMEA and consists of a membership divided evenly between academic and regulatory scientists with expertise in diverse scientific, ethical, and regulatory matters relevant to the new genomic technologies in medicinal product development and assessment. For these briefing meetings, however, a clear demarcation between voluntary submission and ensuing formal submission is not present, since PGWP members can potentially be involved in formal submissions. EMEA held 14 pharmacogenetic data submission meetings between July 2003 and December 2005. There were two meetings with pharmacogenetic test manufacturers and 12 with the pharmaceutical industry.

As part of a wider process of collaboration between the FDA and the EMEA, joint meetings can now also be held for voluntary submissions. The first such bilateral VGDS project occurred in 2005. Information on how the FDA and the EMEA process requests for joint FDA-EMEA voluntary genomic data submission (VGDS) briefing meetings can be found in "The Guiding Principles of Processing Joint FDA EMEA Voluntary Genomic Data Submissions (VGDSs) within the Framework of the Confidentiality Arrangement" (http://www.fda.gov/cder/genomics/FDAEMEA.pdf). The guiding principles describe how the agencies process voluntary submissions and the associated briefing meetings. Briefly, VGDS briefing packages are sent directly to and reviewed by the IPRG and PGWP, and all submissions are held confidential to the extent permitted by law. After the joint meeting, a draft summary of the meeting is created by the sponsor and sent to both agencies 1 week after the joint VGDS meeting. A final summary of the meeting is ultimately issued jointly by the FDA and the EMEA and sent to the sponsor 3 weeks after the meeting date. These final summaries are again kept confidential to the extent permitted by law. Given the overall purpose and scope of these submissions, these joint meetings can only be beneficial for both regulators and sponsors.

In Japan, the PMDA, the Japanese regulatory authority, also issued a guidance document in 2005 on the use of pharmacogenomic data in regulatory submissions. This guidance asked pharmaceutical companies to voluntarily provide details of their past, current, and future use of pharmacogenomics in clinical trials (13). What is sought is not detailed scientific data, but rather general information about the type and scope of studies being conducted. These data are considered voluntary submissions and, as such, are not intended to be used in regulatory decision-making. Like the EMEA and FDA processes, the objective is for the PMDA to develop a sufficient understanding of how pharmacogenomics is used in drug development to allow for the generation of an appropriate guidance regarding the inclusion of genomic data in formal submissions.

Finally, the FDA, EMEA, and PDMA are working on harmonizing the submission of pharmacogenomic data through the ICH process. We briefly discussed the ICH in Chapter 5 to describe the types of preclinical toxicology studies required for regulatory submissions. As you may recall, the ICH (International Conference on Harmonization of Technical Requirements for Registration of Pharmaceuticals for Human Use) is a joint regulatory-industry undertaking created in 1990 for international harmonization of regulatory guidance documents across Europe, Japan and the US. The objective is to normalize the various activities and their design during the development of compounds that are registered in multiple regions in order to reduce the redundancy and cost of approval in different regions. The ICH is an ongoing project with major international conferences held on an annual basis. Pharmacogenomics was on the agenda of the 2007 ICH conference and had previously been addressed informally in other progress sessions. It is likely that this work will result in further harmonized guidance related to format and contents of pharmacogenomic data submissions or design of pharmacogenomic clinical trials. In November 2007, ICH Topic E15 (Definitions for Genomic Biomarkers, Pharmacogenomics, Pharmacogenetics, Genomic Data, and Sample Coding Categories; http://www.fda.gov/cder/guidance/7619dft.pdf) was issued. This guideline contains definitions of key terms in the disciplines of pharmacogenomics and pharmacogenetics, namely, genomic biomarkers, pharmacogenomics, pharmacogenetics, and genomic data and sample coding categories. This guidance was drafted to ensure that consistent definitions of terminology are being applied across all constituents of the ICH.

9.3. PHARMACOGENOMIC DATA SUBMISSIONS: DRAFT COMPANION GUIDANCE

In November 2006, FDA released a preliminary concept paper entitled "Recommendations for the Generation and Submission of Genomic Data" that was relevant to pharmacogenomic data submissions (http://www.fda.gov/cder/genomics/conceptpaper_20061107.pdf.). This concept paper was followed in August 2007 by a draft guidance that is intended to be utilized as a companion guidance to the "Guidance for Industry on Pharmacogenomic Data Submissions" (http://www.fda.gov/cder/guidance/7735dft.pdf). This document is based on the agency's experience with voluntary genomic data submissions as well as its review of numerous protocols and data submitted under IND applications, NDAs, and BLAs. The recommendations provided by the guidance are intended to foster scientific progress in the field of pharmacogenomics and to facilitate the use of pharmacogenomic data in drug development. The FDA believes that this guidance will be beneficial to sponsors considering the submission of either VGDSs or marketing submissions containing genomic data.

This document is indeed a useful addition for sponsors working with genomic data, as it provides some clarity on a variety of technical and analytical aspects. The guidance covers microarray-generated gene expression data and genotyping data and addresses proficiency testing, genomic data in clinical studies and non-clinical toxicology studies, and format for data submission. This guidance is strongly influenced by the work and experience of FDA scientists at the National Center for Toxicological Research (NCTR) (http://www.fda.gov/nctr/). Some of this work has been published elsewhere (9, 14–17).

Regarding gene expression data derived from microarrays, the document reviews several critical methodological issues that need to be considered when submitting microarray data. A range of useful recommendations is provided to ensure the use of high-quality RNA, and these cover topics like sample handling, RNA isolation, RNA storage, and RNA quality control. Other technical considerations, such as labeling reactions, hybridizations, or scanner settings, are also discussed extensively. Finally, the document lists required information for detection of differentially expressed genes, analysis methods, or analysis tools.

In the section on genotyping, methods are briefly mentioned without specifics, since there are a wide variety of methods currently available and likely to be developed in the future. Rather the document focuses on technical aspects, such as DNA isolation, handling, and characterization, and recommends information to include in the genotyping report.

The section on proficiency testing is intended to address low-quality microarray data related to a lack of technical proficiency of laboratories. Hence, the agency expects to receive evidence demonstrating the competency of the laboratory that generated the data included in a genomic submission. These data could consist of quality control metrics or standards that enable within-laboratory testing, but also of various indicators of overall laboratory proficiency. In particular, the guidance mentioned two FDA-led initiatives that have developed and characterized reference RNA samples for proficiency testing. In the first initiative, mixed tissue pools of rat RNA samples were evaluated to generate known differences in tissue-selective genes (18). The tissue pools consisted of two mixtures containing different proportions of RNA from each of four rat tissues (brain, liver, kidney, and testes). The data generated from testing three biological replicates of these two pools at eight laboratories on three array formats provide a benchmark set for both laboratory and data processing performance assessments. The second initiative, called the MicroArray Quality Control (MAQC) Project (http://www.fda.gov/nctr/science/centers/toxicoinformatics/maqc/index.htm), developed two human reference materials that were extensively evaluated on multiple microarray platforms (11, 14). These human reference materials are commercially available for use by laboratories to assess ability to reproduce MAQC data. The document also provides recommendations on experimental design for proficiency testing and suggests that testing be repeated throughout the year, such that consistency of the laboratory over time can be evaluated.

9.4. DRUG-DIAGNOSTIC CO-DEVELOPMENT CONCEPT PAPER

In April 2005, the FDA released a draft concept paper on Drug-Diagnostic Co-Development (http://www.fda.gov/cder/genomics/pharmacoconceptfn.pdf). This concept paper reflects preliminary thoughts from the agency on how to prospectively co-develop a drug and a diagnostic test in a scientifically robust and efficient way. The intent of this concept paper was to stimulate discussion and to solicit input from the public in an effort to develop a guidance. The reasoning was that a guidance document would assist in advancing the field of pharmacogenomics in a manner that will benefit both drug development programs and the public health.

Throughout this book, we have provided countless examples of genomics-based biomarkers that could be used with benefits for differential diagnosis of a disorder, for identification of patients likely to respond to a specific drug or at risk for specific adverse events, or for monitoring responses to drugs. These genomic biomarkers are at the heart of what is commonly referred to as personalized medicine. Pathways already exist for the approval of drug-test combinations, as illustrated by the well-publicized case of Herceptin® (trastuzumab, a drug used for the treatment of metastatic breast cancer in patients with overexpression of the HER2 protein) or Gleevec® (imatinib, a drug for the treatment of chronic myeloid leukemia). However, it is clear that for a new drug to come to market with an associated diagnostic test, a key challenge and potential barrier to success is simultaneous review, approval, and launch of the drug and test (13). This is further complicated by the fact that different regions have processes with varying degrees of alignment for combined drug/test review and approval. For example, in Europe, the registration of drugs and diagnostics is under the auspices of disparate agencies with separate procedures and legislation. In the U.S., both components are FDA-governed, with in vitro diagnostic devices being regulated through the Office of In Vitro Diagnostic Devices (OIVD). Thus there exists a need for better clarity in the approval pathway for a drug combined with a pharmacogenomic test, and this concept paper sets out what the agency may consider an appropriate process for co-developing a drug and a test together. It is beyond the scope of this chapter to go into the details of this document and we only briefly review its main contents here.

With the rapidly evolving technology, co-development of a drug and a diagnostic test is an area that may involve regulation of products across several FDA centers, such as CDER, CDRH, CBER, or OCP. Ideally, the diagnostic test could become an integral part of the drug development process, whereby the clinical phase could be used to demonstrate the clinical validity of the diagnostic test such that the drug would be cross-labeled for use with the test. However, this ideal model is associated with non-negligible process and scientific issues, especially for multiplex technologies. The concept paper addresses issues related to the development of **in vitro diagnostics (IVDs)** for mandatory use in decision-making about drug selection for patients in clinical practice. Both the test and the drug would be used in the clinical management of the patient. The diagnostic test may be used to identify patients most likely to respond to a drug, patients most likely to fail to respond to a drug, and/or patients most likely to exhibit adverse events that might contraindicate drug administration. The test may also be used to assist in understanding mechanisms of a disease or in determining how to enrich or select patient populations in order to conduct more rapid and predictable clinical trials for new therapies. However, this document does not cover the use of one test (e.g., CYP2D6 alleles) with multiple drugs or of several tests developed for serial or parallel use with a single drug. Furthermore, this paper does not address optional or exploratory tests that are not intended for further development [for instance, a biomarker used to study the pharmacokinetic/pharmacodynamic (PK/PD) relationship in early development] or those that do not affect the results of clinical trials (e.g., those that are used to understand mechanisms of disease). The paper addresses processes and procedures for submitting and reviewing a co-developed drug-test product and outlines the scientific issues relating to analytical and clinical validation and to evaluation of clinical utility.

The document recommends determination of the co-development pathway for the in vitro diagnostic early in development and discussions between the FDA reviewing centers, the OCP, and the manufacturers of both the diagnostic and the therapeutic drug, as appropriate. These pre-IND/IDE processes are outlined in existing FDA guidance documents. Ideally, a diagnostic test for subsequent pivotal efficacy and/or safety studies should be developed and analytically validated early in the drug development process to allow clinical validation and determination of the clinical utility of the test during the late-stage clinical trials. Consequently, the design of clinical studies should take into account statistical considerations for both the drug and the test. Alternatively, clinical trial specimens can be banked in optimal storage conditions to enable subsequent test development and/or retrospective hypothesis generation or confirmation of test performance.

Analytical validation studies should be conducted to evaluate the performance characteristics of the assay, where applicable, for each analyte claimed in the clinical use statement. In some cases, the device configuration used during certain drug trials for efficacy and safety may not be ideal for commercial use in clinical practice. Major changes to a device platform can, nonetheless, be validated with an independent prospective clinical data set or, alternatively, by testing retrospectively banked specimens from the original studies. If a multianalyte diagnostic test, such as a multigene marker, is used, the degree of analytical validation may depend on the number of features or readouts represented on the test. Hence, if the feature number is relatively low, each feature can be validated. In contrast, for large numbers of features (for instance, a large-scale microarray platform), it is obviously not feasible to verify each feature. In that situation, typical measures (e.g., accuracy, precision, analytical specificity, and analytical sensitivity) of the assay may be studied by using the system as a whole to prove the validity of the diagnostic test.

If the diagnostic test can be studied in early drug development (phase I or II clinical trials), the utility of the biomarker can be better assessed and prespecification of all key analytical and clinical validation aspects can be completed before the subsequent (late phase II and phase III trials) clinical studies. These aspects include the intended population and selection of diagnostic cutoff points intended to delineate what constitutes a positive result and, when appropriate, what the equivocal zones of decision-making are. The cutoff point that defines a positive and a negative result should be selected before performing the pivotal clinical drug/diagnostic study or the studies designed to provide evidence of adequate clinical test validation and clinical utility.

The guidance underscores that when a drug is linked to a diagnostic test their performance is interdependent. When a new diagnostic is being considered for use in selecting patients to receive or to avoid a particular drug therapy or to stratify patients in some other way, two distinct, but related, issues should be addressed. The first is clinical test validation, which evaluates the ability of the test to select patients with the biomarkers of interest. The second is clinical test utility, which interrogates the ability of the test to result in patient selection that will improve the benefit-risk ratio of the drug in the selected group versus the non-selected group. In other words, clinical utility addresses the risks and benefits to the patient associated with use of the test. Studies should be conducted to ensure that there is evidence to support both the use of the test analytically in patients and the use of the drug in test-positive and test-negative subgroups. For example, HER2 testing is not used for the purpose of detecting the presence of HER2 per se in biological samples (analytical validity), but to identify patients likely to respond to treatment with Herceptin® (clinical validity) and to ensure that patients receive optimum treatment choices (clinical utility). A definitive clinical study for a drug used in conjunction with a predictive biomarker would be one that allows for assessment of a drug's safety and efficacy (i.e., risk/benefit) as well as verification of the clinical utility of the biomarker in guiding the drug's use, including appropriate patient selection. The results obtained from well-controlled trials provide information on the predictive results of the diagnostic test as they relate to drug response (safety and/or efficacy), as well as on any differential in the drug effect in patients who tested positive and negative with the diagnostic test and between drug and placebo. The document extensively discusses alternative clinical trial designs to meet these objectives. It is noteworthy that this document does not address the likely situation in which the biomarker is not identified until late during clinical development, such that characteristics of the diagnostic tests cannot be specified until the phase III clinical studies. That situation represents a non-negligible challenge for which more guidance is warranted.

It is relevant here to indicate that the EMEA has also approved Herceptin® and, as such, can approve drugs co-developed with a diagnostic test. However, in contrast to the FDA, the EMEA does not have a diagnostics division and does not regulate diagnostic tests. This regulation resides in constituent member states. Consequently, the EMEA can only recommend the use of testing as part of a drug label, but cannot mandate the use of a particular diagnostic test. Therefore, the EMEA is unlikely to issue guidance on drug-diagnostic co-development, and this regulatory gap may impact harmonization through the IHC process.

9.5. REGULATIONS FOR IN VITRO DIAGNOSTIC ASSAYS

9.5.1. General Overview of Regulatory Pathways for Devices in the U.S.

Pharmacogenomic tests are considered in vitro diagnostics (IVDs), and IVDs are regarded as medical devices by the FDA. Since such diagnostic tests are quite relevant to genomics, we here briefly review some regulations related to diagnostic tests. Most of this overview is centered around U.S. regulations, with only occasional references to other regions. This section is not intended to be an in-depth review of regulations pertinent to all diagnostic tests, and the reader is encouraged to consult current guidance documents and more appropriate references for additional information.

In the U.S., IVDs are regulated by the FDA's Center for Devices and Radiological Health (CDRH). Diagnostic tests are assigned to one of three regulatory classes based on the level of control necessary to assure the safety and effectiveness of the device. The three classes are designated as Class I (low risk), Class II (moderate risk), and Class III (high risk). Additional information can be found at the following URL: http://www.fda.gov/cdrh/devadvice/3132.html.

Class I medical devices are subject to the least regulatory control. They present minimal potential for harm to the user and are often simpler in design than Class II or Class III devices. Class I devices are subject to "General Controls," as are Class II and Class III devices. These controls include, for instance, registration, listing, and good manufacturing practices to assure the safety and effectiveness of low-risk devices. Devices in this category include things like tongue depressors or examination gloves.

Class II medical devices typically require FDA clearance of a premarket notification submission (referred to as a *510(k)*). A **510(k)** is a premarket submission made to FDA to demonstrate that the device to be marketed is "substantially equivalent" (SE) to a legally marketed device with the same intended use (a predicate device). Class II devices are those for which general controls alone are insufficient to assure safety and effectiveness, and existing methods are available to provide such assurances. In addition to complying with general controls, Class II devices are also subject to special controls, such as special labeling requirements, mandatory performance standards, and postmarket surveillance. Examples of Class II devices include some equipment used for medical imaging (such as X-ray machines) or tests for detection of specific gene DNA mutations.

Class III is the most stringent regulatory category for devices, and Class III devices typically require the submission of an application for **Premarket Approval (PMA)**. PMA is the FDA process of scientific and regulatory review to evaluate the safety and effectiveness of Class III medical devices. Class III devices are those for which insufficient information exists to assure safety and effectiveness solely through general or special controls. These devices are usually those that support or sustain human life, are of substantial importance in preventing impairment of human health, or present a potential, unreasonable risk of illness or injury. For instance, the FDA classified the Roche AmpliChip® CYP450 Test as a Class III device requiring PMA. This microarray-based test system can be used to identify a patient's genotype for cytochromes P450 2D6 and 2C19 (CYP2D6 and CYP2C19) and, based on this analysis, provides their predicted phenotype (poor, intermediate, extensive, or ultrarapid metabolizer). The AmpliChip® CYP450 Test represented the first FDA-cleared test for analysis of CYP2D6 and CYP2C19.

It is noteworthy that FDA premarket review does not represent the only regulatory path for a laboratory test to be offered in the U.S. Hence, laboratories developing tests at a single site for use at their site alone can develop and offer so-called **"in-house"** or **"home-brew"** assays, or to use a term more recently coined, **laboratory-developed tests (LDT)**. These tests fall under the regulations of the **Clinical Laboratories Improvement Amendments (CLIA)** of 1988, which parallel, yet are different from, those of the FDA. In 1997, the FDA increased its oversight of in-house tests by implementing a program for application to the materials used as the active ingredients for these tests, the so-called **analyte-specific reagents (ASRs)**. ASRs must be registered and listed, be made under good manufacturing practices, and be subject to reporting of adverse events. Some genomic assays may be appropriately offered through this path; however, as discussed below, these assays would have to be relatively straightforward in design and interpretation (e.g., genotyping of a single gene variant).

9.5.2. Draft Guidance for Industry, Clinical Laboratories, and FDA Staff on In Vitro Diagnostic Multivariate Index Assays

In July 2007, the FDA released a Draft Guidance for Industry, Clinical Laboratories, and FDA Staff on In Vitro Diagnostic Multivariate Index Assays (http://www.fda.gov/cdrh/oivd/guidance/1610.pdf). This draft guidance addresses the definition and regulatory status of a class of in vitro Diagnostic Devices referred to as In Vitro Diagnostic Multivariate Index Assays (IVDMIAs). The draft guidance also addresses premarket pathways and postmarket requirements with respect to IVDMIAs.

In the eyes of the regulators, a **"device"** is "an instrument, apparatus, implement, machine, contrivance, implant, in vitro reagent, or other similar or related article, including any component, part, or accessory, which is intended for use in the diagnosis of disease or other conditions, or in the cure, mitigation, treatment, or prevention of disease, in man or other animals." Furthermore,"an IVDMIA is a device that combines the values of multiple variables using an interpretation function to yield a single, patient-specific result (e.g., a 'classification,' 'score,' 'index,' etc.), that is intended for use in the diagnosis of disease or other conditions, or in the cure, mitigation, treatment or prevention of disease, and provides a result whose derivation is non-transparent and cannot be independently derived or verified by the end user." In other words, an IVDMIA combines data from a number of input variables to generate a classification or a score (usually a univariate result) that typically cannot be independently analyzed by the clinician. This guidance is relevant here, since gene expression signatures fall under the definition of IVDMIAs. In contrast, most genotype determinations on a single gene or a limited number of genes would not fall under the category of IVDMIAs, since in that situation genotype identification typically has an established association with the phenotype of interest and the device does not incorporate a unique interpretation function, but rather provides standard interpretation of the individual variables that clinicians could do themselves. Likewise, chromosomal copy number determination would not be considered an IVDMIA for the same reasons.

The agency considers that IVDMIAs raise significant issues of safety and effectiveness, because the clinical validity of the claims is not transparent to patients, laboratorians, and clinicians who order these tests and because IVDMIAs frequently have a high-risk intended use. Consequently, the FDA considers that there is a need to regulate these devices to ensure that the IVDMIA is safe and effective for its intended use. In its draft guidance document, the FDA seeks to identify IVDMIAs as a discrete category of devices, and to clarify that, even when offered as in-house tests, IVDMIAs must meet pre- and postmarket device requirements under the Food, Drug and Cosmetic Act and FDA regulations, including premarket review requirements in the case of most Class II and III devices.

As for all devices, the draft guidance indicates that the FDA will take a risk-based approach to the regulation of IVDMIAs. To assist IVDMIA manufacturers in complying with device regulatory requirements, the FDA intends to exercise enforcement discretion with respect to certain requirements during an initial transition period following publication of the final version of this guidance. Further details can be found in the guidance.

As with all regulated devices, IVDMIA manufacturers are required to register with the FDA and to list, or identify to the FDA, the IVDMIAs they are marketing. The registration and listing requirement is a means of keeping the FDA advised of who is manufacturing devices, and of the types of devices an establishment is manufacturing or marketing. Each manufacturer who intends to market a Class I or Class II device that is not otherwise exempt must submit a 510(k) to the FDA.

9.6. BIOMARKER QUALIFICATION

As we discussed in Section 9.1, the FDA guidance on Pharmacogenomic Data Submission defines what type of pharmacogenomic data should be submitted to the FDA depending on the classification of the biomarker included in the submission. Data on known valid biomarkers are required in a submission, in contrast to data on probable valid biomarkers. As a reminder, **known valid biomarkers** are measured in an analytical test system with well-established performance characteristics, and there is widespread agreement in the medical or scientific community about the physiological, toxicological, pharmacological, or clinical significance of the results associated with these markers. In contrast, **probable valid biomarkers** are defined as biomarkers that are measured in an analytical test system with well-established performance characteristics, and for which there is a scientific framework or body of evidence that appears to elucidate the physiological, toxicological, pharmacological, or clinical significance of the test results. Furthermore, an additional category includes **exploratory biomarkers**, which represent potential precursors for probable or known valid biomarkers. As hypothesis-generating agents, exploratory biomarkers can be used to fill in gaps of knowledge about disease targets and variability in drug response, to bridge the results of animal model studies to what may be expected in the clinic, or to select new compounds based on anticipated lower cost–benefit ratios in drug development (19). However, the guidance does not provide a regulatory path to convert exploratory biomarkers into valid biomarkers and probable valid into known valid biomarkers. In two recent publications, FDA scientists have somewhat addressed this regulatory gap by proposing a process for accurate and comprehensive qualification of biomarkers (19, 20).

Despite the definition provided in the guidance, it may not always be clear whether or not a biomarker is valid. To address this lack of clarity, the Genomics Group in the OCPB has assembled a Web-Based Table of Valid Genomic Biomarkers in the Context of Approved Drug Labels (http://www.fda.gov/cder/genomics/genomic_biomarkers_table.htm.). This table sets precedents in the definition of clinical biomarkers in specific contexts and in the "valid" classification justified within the label context. This table is useful because it contains examples of text that accurately defines biomarker context and provides an updated list of valid biomarkers (20).

The two publications acknowledge that the traditional process, through which biomarkers have been customarily introduced and accepted in drug development, is quite inefficient and that improvements are needed. The FDA has, therefore, established a pilot structure to start a qualification process for biomarkers in drug development. This structure is designed around the IPRG to allow contributions of expertise from different FDA Centers, such as the CDER, CBER, CDRH, and NCTR. In particular, the IPRG has created a specific review function for the assessment of biomarker qualification data sets: **the IPRG Biomarker Qualification Review Team**.

The IPRG Biomarker Qualification Review Team is responsible for the evaluation of study protocols and for the review of study results related to the qualification of novel biomarkers of drug safety. This panel develops recommendations and guidance for the submission of biomarker data, assesses the original biomarker context proposal through a VXDS, and then evaluates the qualification study protocol together with the sponsor to reach a consensus protocol. Finally, this panel reviews qualification study results and drafts a recommendation for the clinical divisions. Useful algorithms illustrating the overall biomarker qualification pilot process are available in these two publications.

9.7. CURRENT INITIATIVES RELEVANT TO PHARMACOGENOMICS

One of the most visible successes emerging from the FDA's Critical Path White Paper has been the creation of various consortia or initiatives that should foster and accelerate the use of genomic technologies in drug development. These various groups are not necessarily focused on genomic technologies, but because of their likely impact on the drug pipeline, genomic solutions are frequently one of their top priorities. These efforts have brought together drug regulators, pharmaceutical companies, and academic institutions, and it is hoped that these concerted efforts will trigger rapid advances in the use of genomic technologies in drug discovery and development.

We review some of these consortia in Chapters 2 and 4, when discussing the reproducibility of the microarray technology. For instance, the **External RNA Controls Consortium (ERCC)** was created to develop external controls for gene expression experiments. This consortium brings together numerous industry, academic, and governmental organizations and has produced a specification document for external RNA controls that can be found at the following URL: www.nist.gov. Likewise, the **Microarray Quality Control (MAQC) Project** (http://www.fda.gov/nctr/science/centers/toxicoinformatics/maqc/index.htm) is an FDA-led consortium that has demonstrated the reliability and precision of current commercial microarray platforms in the context of biomarkers. In addition, this collective effort has developed two human reference materials that were extensively evaluated on multiple microarray platforms and are recommended as a tool for proficiency testing of laboratories (11, 14).

Most of these recent collaborations have been promoted through the **Critical Path Institute**. Based in Tucson, Arizona, the Critical Path Institute (C-Path; http://www.c-path.org.) was established in 2005 as an independent non-profit research and education institute to facilitate collaboration among its founding partners: the FDA, the University of Arizona, and SRI International. The Institute has received financial support from various sources (the State of Arizona, the City of Tucson, Pima County, regional municipalities, foundations, organizations, and private individuals). For instance, the **Predictive Safety Testing Consortium** (PSTC; http://www.fda.gov/oc/

initiatives/criticalpath/projectsummary/consortium.html) brings together participants from industry, academia, and government to identify and clinically qualify safety biomarkers. The objectives of this consortium are to improve the understanding of the safety profile of compounds before clinical trials, to develop novel tools for improving preclinical drug safety assessment, to identify early signals for predicting and preventing postmarketing safety issues, and to speed and/or reduce the cost of preclinical drug safety evaluation. The FDA, while not a member of the partnership, assists it in an advisory capacity. The work of the PSTC is coordinated by the Critical Path Institute and is expected to lead in the next few years to multiple preclinical qualification packages that will be useful for both drug developers and drug regulators to fine-tune the biomarker qualification process. This consortium enables pharmaceutical companies to share knowledge and resources, and should result in a more expedient qualification of broad biomarkers of toxicity for use in the scientific community.

Several additional projects relevant to genomic technologies are currently coordinated under the auspices of the Critical Path Institute. For instance, the Cancer Biomarkers project is one of the major projects for a national collaboration called the Oncology Biomarker Qualification Initiative formed by the FDA, the National Cancer Institute, and the Centers for Medicaid and Medicare Services to advance the development of personalized treatment and diagnostic testing. For example, the Critical Path Institute is coordinating a project to bring pharmaceutical and diagnostic companies together with government agencies in a project that will help guide lung cancer therapy. Likewise, in partnership with the University of Utah, the Critical Path Institute is evaluating genetic tests for their ability to predict safer and more effective doses of the anticoagulant warfarin (Coumadin®).

Additional consortia are also being created outside the Critical Path Institute. For instance, the international **Serious Adverse Events Consortium (SAEC)** has recently been launched to identify genetic markers related to severe, life-threatening adverse drug reactions. The SAEC is a non-profit corporation comprised of leading pharmaceutical companies and academic institutions with scientific and strategic input from the FDA. The two initial research programs address drug-related liver toxicity and a rare but serious drug-related skin condition called Stevens–Johnson syndrome. The FDA is providing consultation on the design and conduct of SAEC studies. The SAEC will also consult the EMEA and other national regulatory bodies for guidance on its efforts.

9.8. FUTURE IMPACT OF GENOMIC DATA ON DRUG DEVELOPMENT

In general, drug producers and regulators agree that pharmacogenomics holds significant promise to address the current drug pipeline problem and to develop medicines with improved patient benefits. However, it is important to realize that the technology is still in its infancy and that significant challenges remain before this promise translates into novel medicines with improved efficacy and lowered safety liabilities. These challenges are due not only to scientific but also regulatory and logistical issues.

On the scientific aspect, one must acknowledge that identifying and validating genomics-based biomarkers is far from a trivial task. Most diseases and responses to therapeutic agents are complex in nature, with contributions from an individual's genetic background but also from the environment. Moreover, on the genetic aspect, most instances will involve the contribution of multiple, relatively small effects from multiple genes and gene products that are likely to vary over time. Hence, the effect of genetic variation on drug response or drug toxicity will most often be subtle and challenging to detect. The same will likely be true for gene expression signatures to predict drug response or adverse events in the clinics. In other words, identifying and validating genomics-based biomarkers may represent in most cases a lengthy and costly endeavor that may not necessarily translate into shorter, simpler, and faster drug development programs.

In this chapter, we have covered the concept of co-development of a drug and a diagnostic test. From this brief discussion, it should be clear that the industry and the regulatory agencies are still fleshing out this concept, and that more work is needed to optimize the development and regulatory reviews of these combination products. This will not be a trivial task, since it requires the integration of diagnostic development processes into drug development processes. Because of this additional complexity, it is likely that the co-development approach will be mostly limited to oncology applications, to drugs that require careful dosing because of severe adverse events (warfarin would represent a good example of that situation), or to drugs that may result in irreversible damaging effects (such as the tardive dyskinesia observed with antipsychotic agents) (21). There are already several pharmacogenomic tests that are recommended for use with certain drugs, and this trend is likely to continue. However, it is unknown whether these tests will become required before drug treatment, and in all likelihood their use will be required only in the case of potential adverse drug reactions that may not be manageable or may be too costly to manage.

Finally, logistical issues should certainly not be underestimated. In Chapter 2, we stressed the importance of being able to interrogate formalin-fixed, paraffin-embedded tissues, since this method of preparation is the standard practice used in pathology laboratories. In general, the same is true for all novel approaches: Their impact on clinical practice is often more dependent on one's ability to integrate them into the current health care structure than on their economic or clinical value. Furthermore, in drug and diagnostic co-development, the additional costs involved in developing these tests may be difficult to justify if appropriate economic incentives are not evident. This last point is non-negligible. At a time of productivity concerns, novel technologies that result in even more R&D expenses without compensatory increases in revenues are unlikely to be enticing enough to justify their use. This may require changes in payment and reimbursement systems or in extended commercial exclusivity. On the positive side, however, genomic technologies may be used to rescue some drugs that do not meet standard requirements for efficacy and/or safety in the general population by identifying a subpopulation of patients who would benefit greatly from treatment or by detecting a minority group that may experience undesirable side effects after treatment.

Genomic technologies clearly represent the most prominent change in drug discovery and development that has occurred in the last decade, and applying these technologies should result in a spectrum of significant positive outcomes on the drug discovery and development process. Specific applications range, among others, from discovery and validation of novel therapeutic targets to risk assessment, design of improved clinical trials, and development of diagnostic tests for improved dosing recommendations. It is, however, difficult at this point to objectively predict the role that such a rapidly evolving technology will have in regulatory decision-making. A research environment amenable to the use of pharmacogenomics can only be beneficial, and the regulatory agencies have clearly been proactive in promoting this environment through public consultation, the VGDS process, and development of harmonized regulatory guidance. Over the next few years, it is anticipated that the use of pharmacogenomics will consistently increase, with a corresponding increased number of NDAs or other marketing authorization applications.

Nevertheless, it is also important to acknowledge the uncertainties associated with these still immature technologies. As we have seen throughout this book, there is no single model, but multiple possible alternatives for the integration and application of pharmacogenomics in drug discovery and development. This is particularly evident with toxicogenomics. When the concept of toxicogenomics was initially introduced, expectations were high and to some extent unrealistic: The technology was perceived as a better way to predict human toxicology and as a tool to completely revolutionize the current toxicology practices (22). The technology and analytical tools have since considerably improved, but they still have not reached a stage of maturity such that their use has become a mainstream approach in toxicology. Furthermore, most stakeholders would now agree that genomic technologies will not replace or revolutionize most current approaches, but will supplement them. Rather than revolutionizing the drug discovery and development process, these technologies, if successfully integrated, will improve our decision-making and ultimately contribute to better medicines for more pathological conditions at more sustainable costs. The extent of this improvement is difficult to fathom at this point, and the incremental value will likely be closely related to the ability of R&D organizations to execute on the promises of genomics.

What is known, however, is that there will be a growing interest in pharmacogenomic data from the regulators, especially in instances where there is significant variability in drug response, or evidence of serious adverse events. Therefore, the pharmaceutical and diagnostic industries can only benefit from increased interactions with the regulatory authorities and policy groups. These continued interactions will facilitate the acceptance of these technologies in the industry and the subsequent development of medicines and tests beneficial to the patient.

REFERENCES

1. Orr MS, Goodsaid F, Amur S, Rudman A, Frueh FW. The experience with voluntary genomic data submissions at the FDA and a vision for the future of the voluntary data submission program. *Clin Pharmacol Ther* 2007;81:294–297.
2. Lesko LJ, Salerno RA, Spear BB, Anderson DC, Anderson T, Brazell C, Collins J, Dorner A, Essayan D, Gomez-Mancilla B, Hackett J, Huang SM, Ide S, Killinger J, Leighton J, Mansfield E, Meyer R, Ryan SG, Schmith V, Shaw P, et al. Pharmacogenetics and pharmacogenomics in drug development and regulatory decision making: report of the first FDA-PWG-PhRMA-DruSafe Workshop. *J Clin Pharmacol* 2003;43:342–358.
3. Salerno RA, Lesko LJ. Pharmacogenomic data: FDA voluntary and required submission guidance. *Pharmacogenomics* 2004;5:503–505.
4. Salerno RA, Lesko LJ. Pharmacogenomics in drug development and regulatory decision-making: the Genomic Data Submission (GDS) proposal. *Pharmacogenomics* 2004;5:25–30.
5. Leighton JK, DeGeorge J, Jacobson-Kram D, MacGregor J, Mendrick D, Worobec A. Pharmacogenomic data submissions to the FDA: non-clinical case studies. *Pharmacogenomics* 2004;5:507–511.
6. Trepicchio WL, Williams GA, Essayan D, Hall ST, Harty LC, Shaw PM, Spear BB, Wang SJ, Watson ML. Pharmacogenomic data submissions to the FDA: clinical case studies. *Pharmacogenomics* 2004;5:519–524.
7. Lesko LJ, Woodcock J. Pharmacogenomic-guided drug development: regulatory perspective. *Pharmacogenomics J* 2002;2:20–24.
8. Hackett JL, Lesko LJ. Microarray data—the US FDA, industry and academia. *Nat Biotechnol* 2003;21:742–743.
9. Fuscoe JC, Tong W, Shi L. QA/QC issues to aid regulatory acceptance of microarray gene expression data. *Environ Mol Mutagen* 2007;48:349–353.
10. Guo L, Lobenhofer EK, Wang C, Shippy R, Harris SC, Zhang L, Mei N, Chen T, Herman D, Goodsaid FM, Hurban P, Phillips KL, Xu J, Deng X, Sun YA, Tong W, Dragan YP, Shi L. Rat toxicogenomic study reveals analytical consistency across microarray platforms. *Nat Biotechnol* 2006;24:1162–1169.
11. Shi L, Reid LH, Jones WD, Shippy R, Warrington JA, Baker SC, Collins PJ, de Longueville F, Kawasaki ES, Lee KY, Luo Y, Sun YA, Willey JC, Setterquist RA, Fischer GM, Tong W, Dragan YP, Dix DJ, Frueh FW, Goodsaid FM et al. The MicroArray Quality Control (MAQC) project shows inter- and intraplatform reproducibility of gene expression measurements. *Nat Biotechnol* 2006;24:1151–1161.
12. Shi L, Tong W, Fang H, Scherf U, Han J, Puri RK, Frueh FW, Goodsaid FM, Guo L, Su Z, Han T, Fuscoe JC, Xu ZA, Patterson TA, Hong H, Xie Q, Perkins RG, Chen JJ, Casciano DA. Cross-platform comparability of microarray technology: intra-platform consistency and appropriate data analysis procedures are essential. *BMC Bioinformatics* 2005;6 Suppl 2:S12.
13. Hall ST, Abbott N, Schmith G, Brazell C. Pharmacogenetics in drug development: regulatory and clinical considerations. *Drug Dev Res* 2004;62:102–111.
14. Tong W, Lucas AB, Shippy R, Fan X, Fang H, Hong H, Orr MS, Chu TM, Guo X, Collins PJ, Sun YA, Wang SJ, Bao W, Wolfinger RD, Shchegrova S, Guo L, Warrington JA, Shi L. Evaluation of external RNA controls for the assessment of microarray performance. *Nat Biotechnol* 2006;24:1132–1139.

15. Han T, Melvin CD, Shi L, Branham WS, Moland CL, Pine PS, Thompson KL, Fuscoe JC. Improvement in the reproducibility and accuracy of DNA microarray quantification by optimizing hybridization conditions. *BMC Bioinformatics* 2006;7 Suppl 2:S17.
16. Shi L, Tong W, Su Z, Han T, Han J, Puri RK, Fang H, Frueh FW, Goodsaid FM, Guo L, Branham WS, Chen JJ, Xu ZA, Harris SC, Hong H, Xie Q, Perkins RG, Fuscoe JC. Microarray scanner calibration curves: characteristics and implications. *BMC Bioinformatics* 2005;6 Suppl 2:S11.
17. Shi L, Tong W, Goodsaid F, Frueh FW, Fang H, Han T, Fuscoe JC, Casciano DA. QA/QC: challenges and pitfalls facing the microarray community and regulatory agencies. *Expert Rev Mol Diagn* 2004;4:761–777.
18. Thompson KL, Rosenzweig BA, Pine PS, Retief J, Turpaz Y, Afshari CA, Hamadeh HK, Damore MA, Boedigheimer M, Blomme E, Ciurlionis R, Waring JF, Fuscoe JC, Paules R, Tucker CJ, Fare T, Coffey EM, He Y, Collins PJ, Jarnagin K et al. Use of a mixed tissue RNA design for performance assessments on multiple microarray formats. *Nucleic Acids Res* 2005;33:e187.
19. Goodsaid F, Frueh F. Process map proposal for the validation of genomic biomarkers. *Pharmacogenomics* 2006;7:773–782.
20. Goodsaid F, Frueh F. Biomarker qualification pilot process at the US Food and Drug Administration. *AAPS J* 2007;9:E105–E108.
21. Arranz MJ, de Leon J. Pharmacogenetics and pharmacogenomics of schizophrenia: a review of last decade of research. *Mol Psychiatry* 2007;12:707–747.
22. Lovett RA. Toxicogenomics. Toxicologists brace for genomics revolution. *Science* 2000; 289:536–537.

Index

Genomics in Drug Discovery and Development, by Dimitri Semizarov and Eric Blomme